LATTICE GAS METHODS
FOR PARTIAL
DIFFERENTIAL EQUATIONS

LATTICE GAS METHODS FOR PARTIAL DIFFERENTIAL EQUATIONS

A VOLUME OF LATTICE GAS
REPRINTS AND ARTICLES, INCLUDING
SELECTED PAPERS FROM THE WORKSHOP
ON LARGE NONLINEAR SYSTEMS,
HELD AUGUST, 1987
IN LOS ALAMOS, NEW MEXICO

Edited by

Gary D. Doolen
Los Alamos National Laboratory

Uriel Frisch
Observatoire de Nice, France

Brosl Hasslacher
Los Alamos National Laboratory

Steven Orszag
Princeton University

Stephen Wolfram
University of Illinois

Proceedings Volume IV

SANTA FE INSTITUTE
STUDIES IN THE SCIENCES OF COMPLEXITY

Routledge
Taylor & Francis Group

LONDON AND NEW YORK

First published 1990 by Westview Press

Published 2018 by Routledge
52 Vanderbilt Avenue, New York, NY 10017
2 Park Square, Milton Park, Abingdon, Oxon OX14 4RN

Routledge is an imprint of the Taylor & Francis Group, an informa business

Library of Congress Cataloging-in-Publication Data

Lattice Gas Methods for Partial Differential Equations
edited by Gary D. Doolen ... [et al.]
    p.   cm.—(Santa Fe Institute studies in the sciences of complexity; v. 4)
Proceedings of the Workshop on Large Nonlinear Systems, Los Alamos, New Mexico
Includes index.
    ISBM 0-201-15679-2.—IBSN 0-201-13232-X (pbk.)
    1. Differential equations, Partial.   2. Lattice gas. I. Doolen, Gary D. II. Series.
QA374.L34 1989   515.3'53–dc19   89-30691

ISBN 13: 978-0-367-00287-9 (hbk)
ISBN 13: 978-0-367-15274-1 (pbk)

About the Santa Fe Institute

The *Santa Fe Institute* (SFI) is a multidisciplinary graduate research and teaching institution formed to nurture research on complex systems and their simpler elements. A private, independent institution, SFI was founded in 1984. Its primary concern is to focus the tools of traditional scientific disciplines and emerging new computer resources on the problems and opportunities that are involved in the multidisciplinary study of complex systems—those fundamental processes that shape almost every aspect of human life. Understanding complex systems is critical to realizing the full potential of science, and may be expected to yield enormous intellectual and practical benefits.

All titles from the *Santa Fe Institute Studies in the Sciences of Complexity* series will carry this imprint which is based on a Mimbres pottery design (circa A.D. 950–1150), drawn by Betsy Jones.

Santa Fe Institute Studies in the Sciences of Complexity

PROCEEDINGS VOLUMES

Volume	Editor	Title
I	David Pines	Emerging Syntheses in Science, 1987
II	Alan S. Perelson	Theoretical Immunology, Part One, 1988
III	Alan S. Perelson	Theoretical Immunology, Part Two, 1988
IV	Gary D. Doolen et al.	Lattice Gas Methods of Partial Differential Equations
V	Philip W. Anderson et al.	The Economy as an Evolving Complex System
VI	Christopher G. Langton	Artificial Life: Proceedings of an Interdisciplinary Workshop on the Synthesis and Simulation of Living Systems
VII	George I. Bell & Tom Marr	Computers and DNA

LECTURES VOLUMES

Volume	Editor	Title
I	Daniel L. Stein	Lectures in the Sciences of Complexity

Contributors to This Volume

Bruce J. Bayly
Applied and Computational Mathematics, Princeton University

P.-M. Binder
Center for Nonlinear Studies, Los Alamos National Laboratory

B. M. Boghosian
Thinking Machines Corporation

Hudong Chen
Center for Nonlinear Studies, Los Alamos National Laboratory

Shiyi Chen
Center for Nonlinear Studies, Los Alamos National Laboratory

P. Clavin
Combustion Research Laboratory, University of Provence

Andre Clouqueur
Laboratoire de Physique, École Normale Supérieure

A. Despain
JASON, Mitre Corporation

Dominique d'Humières
Physics Laboratory, École Normale Supérieure

K. Diemer
Center for Nonlinear Studies, Los Alamos National Laboratory

Gary D. Doolen
Center for Nonlinear Studies, Los Alamos National Laboratory

Uriel Frisch
Centre National de la Recherche Scientifique, Observatoire de Nice

Castor Fu
Center for Nonlinear Studies, Los Alamos National Laboratory

Brosl Hasslacher
Los Alamos National Laboratory

Tudatsugu Hatori
Department of Physics and Astronomy, Dartmouth College

Victor Hayot
Ohio State University

Michel Hénon
Centre National de la Recherche Scientifique, Observatoire de Nice

K. Hunt
Center for Nonlinear Studies, Los Alamos National Laboratory

Leo P. Kadanoff
The Research Institutes, The University of Chicago

Jeffrey M. Keller
Massachusetts Institute of Technology

Pierre Lallemand
Physics Laboratory, École Normale Supérieure

Y. C. Lee
Center for Nonlinear Studies, Los Alamos National Laboratory

C. D. Levermore
Lawrence Livermore Laboratory

Norman Margolus
MIT Laboratory for Computer Science

William H. Matthaeus
Bartol Research Institute, University of Delaware

C. E. Max
JASON, Mitre Corporation

Guy R. McNamara
The Research Institutes, The University of Chicago

David Montgomery
Dartmouth College

Steven A. Orszag
Princeton University

Yves Pomeau
Physics Laboratory, École Normale Supérieure

Jean-Pierre Rivet
École Normale Supérieure

Daniel H. Rothman
Massachusetts Institute of Technology

Geoffrey Searby
Laboratoire de recherche sur la combustion

Tsutomu Shimomura
Los Alamos National Laboratory

Tommaso Toffoli
MIT Laboratory for Computer Science

Stephen Wolfram
Center for Complex Systems Research and
University of Illinois

Victor Yakhot
Princeton University

Gianluigi Zanetti
The Research Institutes, The University of Chicago

Gary Doolen
Los Alamos National Laboratory, Los Alamos, NM 87545

Preface

MOTIVATION

Although the idea of using discrete methods for modeling partial differential equations occurred very early, the actual statement that cellular automata techniques can approximate the solutions of hydrodynamic partial differential equations was first discovered by Frisch, Hasslacher, and Pomeau. Their description of the derivation, which assumes the validity of the Boltzmann equation, appeared in the *Physical Review Letters* in April 1986. It is the intent of this book to provide some overview of the directions that lattice gas research has taken from 1986 to early 1989.

There are several reasons for the recent rapid growth in lattice gas research. The method provides very high resolution. Problems with 5,000,000,000 cells can now be run on a CRAY Y/MP. Also, the algorithm is totally parallel. This parallel feature is easily exploited on existing computers. In addition, an enormous gain can be made by constructing dedicated hardware. Already, inexpensive dedicated boards are available which allow small PCs to run lattice gas problems near CRAY speeds. Dedicated boards are now planned for delivery in 1990 which are expected to be a thousand times more powerful. It is possible to build with existing technology a dedicated machine which has the complexity of existing CRAYs but which would execute lattice gas algorithms many millions of times faster. One should interpret this impressive gain in computer speed cautiously. For periodic problems on existing machines, lattice gas methods are slower than spectral methods at least by an order of magnitude. But for complicated boundary conditions, lattice gas methods can solve problems which are not solvable by other methods. An example is flow through porous media.

At least four separate communities are closely following lattice gas developments with quite different expectations. The molecular dynamics community considers the lattice gas method to be a minimal-bit strategy for solving Newton's equations of motion using orders of magnitude more particles than are usually simulated. Finite difference theorists consider the method to be an over-restricted set of finite difference equations, rightly expecting many finite difference diseases to be amplified. Statistical mechanicians hope to use the method to gain new insight into the relation between microdynamics and macrodynamics. Parallel computer hardware scientists see the method as the simplest and fastest totally parallel algorithm with broad applications.

BOOK OVERVIEW

In the section labeled "Basic Papers," a *Scientific-American*-level reprint of an article by Shimomura, Doolen, Hasslacher, and Fu highlights the principal advantages and some of the limitations of lattice gas methods. Derivations of the hydrodynamic equations are outlined in articles by Wolfram and by Frisch, d'Humieres, Hasslacher, Lallemand, Pomeau, and Rivet. The article by Diemer et al. gives some details of the viscosity and mean free path for several two-dimensional and three-dimensional models. The article by Hénon derives the analytic formula for viscosity from the definition of viscosity. The section on computer hardware begins with a paper which outlines the impressive speed which could be obtained by building a dedicated lattice gas computer with existing hardware. The following article by Margolus and Toffoli describes the CAM-6, an existing computer board which enables a personal computer to achieve supercomputer speeds for small lattice gas systems. This article also describes the CAM-8, a dedicated board now in the design phase. The article by Clouqueur and d'Humières describes the capabilities of the RAP-1, a board with capabilities similar to those of the CAM-6.

The section on hydrodynamic studies and application papers contains a selection of some of the many papers on hydrodynamic applications of the lattice gas method. The power of the method to use minimal computer memory and to conserve exactly the required constants of the motion have been studied in several of these reprints. Many of the articles describe detailed computer simulations, but to date there have not been any articles which describe the codes themselves. A few translations of original French articles are included in this section.

The section on more partial differential equations gives a small sampling of a few of the lattice gas algorithms developed to solve partial differential equations which differ considerably from the Navier-Stokes equations.

Finally, a bibliography including most of the papers which reference the original Frisch, Hasslacher and Pomeau article is provided which includes some abstracts. An attempt was made to include articles through April 1989.

FUTURE CHALLENGES

One can think of the lattice gas method as filling a niche between molecular dynamic methods and continuum methods. A present challenge is to determine the boundaries of parameter space where lattice gas methods are most appropriate. It is possible to add complexity to these methods until the results become indistinguishable from both molecular dynamics and continuum methods. However, the method becomes slower as the complexity increases. At the present time, the lattice gas method appears to be ideal for describing flow through porous media.

The lattice gas algorithm can be shown to approximately solve the Navier-Stokes equations in the long wavelength limit. But the algorithm can go far below this limit and possibly give considerable insight into the correct macroscopic treatment in situations where gradients are important and also in situations where the mean-free path is not negligible. The understanding of how to go from the microscopic rules to the correct macroscopic equations remains a challenge, and the lattice gas has much to contribute here.

Another challenge is to determine the fastest lattice algorithm. At present, table look-up methods are very fast. For example, 300,000,000 sites can now be updated each second on a CRAY X/MP 416. The table size grows, however, as 2 to the Nth power, where N is the number of bits required at each site. For 24-bit models, 16-million-word tables are required. It appears that reduced-size tables are possible, but the restrictions which they place on the collision rules may significantly limit the range of allowed viscosities. Several studies are in progress to determine the class of algorithms which ought to be implemented in the type of dedicated lattice-gas computer described in the first article in this book in the section on computer hardware.

The most significant challenge is the implementation of the lattice gas algorithm on large-scale dedicated hardware. The gain in speed over existing computers is a factor of the order of many millions. This opportunity does not belong exclusively to lattice gas techniques but applies to all algorithms which have a parallel implementation.

INTENDED READERSHIP

The book is assembled to show potential users and lattice gas scientists what research has been completed and to give some indication of the utility and limitations of lattice gas models. Most articles should be readable by students who are considering entering the field or who are contemplating the application or extension of these methods to their favorite problems.

ACKNOWLEDGMENTS

Many people have contributed to the creation of this volume. Any attempt to list them all would necessarily fall short. At various stages, advice and guidance were received gratefully from Uriel Frisch, Brosl Hasslacher, Stephen Orszag, and Stephen Wolfram. Much of the bibliographical work was provided by Dominique d'Humieres and Tsutomu Shimomura. Finally, the essential coordination and typesetting by Ronda Butler-Villa of the Santa Fe Institute is very much appreciated.

Gary Doolen
Center for Nonlinear Studies
Los Alamos National Laboratory
May 1989

Contents

More Partial Differential Equations   451

Basic Papers

Tsutomu Shimomura, Gary D. Doolen, Brosl Hasslacher, and Castor Fu
Los Alamos National Laboratory, Los Alamos, NM 87545

Calculations Using Lattice Gas Techniques

This paper originally appeared in *Los Alamos Science*, Special Issue, 1987, pages 201–210.

Over the last few years the tantalizing prospect of being able to perform hydrodynamic calculations orders-of-magnitude faster than present methods allow has prompted considerable interest in lattice gas techniques. A few dozen published papers have presented both advantages and disadvantages, and several groups have studied the possibilities of building computers specially designed for lattice gas calculations. Yet the hydrodynamics community remains generally skeptical toward this new approach. The question is often asked, "What calculations can be done with lattice gas techniques?" Enthusiasts respond that in principle the techniques are applicable to any calculation, adding cautiously that increased accuracy requires increased computational effort. Indeed, by adding more particle directions, more particles per site, more particle speeds, and more variety in the interparticle scattering rules, lattice gas methods can be tailored to achieve better and better accuracy. So the real problem is one of tradeoff: How much accuracy is gained by making lattice gas methods more complex, and what is the computational price of those complications? That problem has not yet been well studied. This paper and most of the research to date focus on the

Lattice Gas Methods for Partial Differential Equations, SFI SISOC,
Eds. Doolen et al., Addison-Wesley Publishing Co., 1990

simplest lattice gas models in the hope that knowledge of them will give some insight into the essential issues.

We begin by examining a few of the features of the simple models. We then display results of some calculations. Finally, we conclude with a discussion of limitations of the simple models.

FEATURES OF SIMPLE LATTICE GAS METHODS

We will discuss in some depth the memory efficiency and the parallelism of lattice gas methods, but first we will touch on their simplicity, stability, and ability to model complicated boundaries.

Computer codes for lattice gas methods are enormously simpler than those for other methods. Usually the essential parts of the code are contained in only a few dozen lines of FORTRAN. And those few lines of code are much less complicated than the several hundred lines of code normally required for two- and three-dimensional hydrodynamic calculations.

There are many hydrodynamic problems that cause most standard codes (such as finite-difference codes, spectral codes, and particle-in-cell codes) to crash. That is, the code simply stops running because the algorithm becomes unstable. Stability is not a problem with the codes for lattice gas methods. In addition, such methods conserve energy and momentum exactly, with no roundoff errors.

Boundary conditions are quite easy to implement for lattice gas methods, and they do not require much computer time. One simply chooses the cells to which boundary conditions apply and updates those cells in a slightly different way. One of three boundary conditions is commonly chosen: bounce-back, in which the directions of the reflected particles are simply reversed; specular, in which mirror-like reflection is simulated; or diffusive, in which the directions of the reflected particles are chosen randomly.

We consider next the memory efficiency of the lattice gas method. When the 2-dimensional hydrodynamic lattice gas algorithm is programmed on a computer with a word length of, say, 64 bits (such as the Cray X-MP), two impressive efficiencies occur. The first arises because every single bit of memory is used equally effectively. Coined "bit democracy" by von Neumann, such efficient use of memory should be contrasted with that attainable in standard calculations, where each number requires a whole 64-bit word. The lattice gas is "bit democratic" because all that one needs to know is whether or not a particle with a given velocity direction exists in a given cell. Since the number of possible velocity directions is six and no two particles in the same cell can have the same direction, only six bits of information are needed to completely specify the state of a cell. Each of those six bits corresponds to one of the six directions and is set to 1 if the cell contains a particle with that direction and to 0 otherwise. Suppose we designate the six directions by A, B, C, D, E, and F as shown in Figure 1. We associate each bit in the 64-bit word A with a different cell, say, the first 64 cells in the first row. If the first cell contains (does

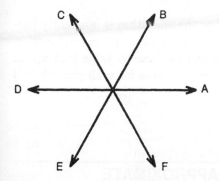

FIGURE 1 Six directions designated as A through F corresponding to six bits used to specify the state of a cell.

not contain) a particle with direction A, we set the first bit in A to 1 (0). Similarly, we pack information about particles in the remaining 63 cells with direction A into the remaining 63 bits of A. The same scheme is used for the other five directions. Consequently, all the information for the first 64 cells in the first row is contained in the six words, A, B, C, D, E, and F. Note that all bits are equally important and all are fully utilized.

To appreciate the significance of such efficient use of memory, consider how many cells can be specified in the solid-state storage device presently used with the Cray X-MP/416 at Los Alamos. That device stores 512,000,000 64-bit words. Since the necessary information for $10\frac{2}{3}$ cells can be stored in each word, the device can store information for about 5,000,000,000 cells, which corresponds to a two-dimensional lattice with 100,000 cells along one axis and 50,000 cells along the other. That number of cells is a few orders-of-magnitude greater than the number normally treated when other methods are used. (Although such high resolution may appear to be a significant advantage of the lattice gas method, some averaging over cells is required to obtain smooth results for physical quantities such as velocity and density.)

The second efficiency is related to the fact that lattice gas operations are bit oriented rather than floating-point-number oriented and therefore execute more naturally on a computer. Most computers can carry out logic operations bit by bit. For example, the result of the logic operation AND on the 64-bit words A and B is a new 64-bit word in which the ith bit has a value of 1 only if the ith bits of both A and B have values of 1. Hence in one clock cycle a logic operation can be performed on information for 64 cells. Since a Cray X-MP/416 includes eight logical function units, information for 8 times 64, or 512, cells can be processed during each clock cycle, which lasts about 10 nanoseconds. Thus information for 51,200,000,000 cells can be processed each second. The two-dimensional lattice gas models used so far require from about thirty to one hundred logic operations to implement the scattering rules and about another dozen to move the particles to the next cells. So the number of cells that can be updated each second by logic operations is near 500,000,000. Cells can also be updated by table-look-up methods. The authors have a table-look-up code for three-dimensional hydrodynamics that processes about 30,000,000 cells per second.

A final feature of the lattice gas method is that the algorithm is inherently parallel. The rules for scattering particles within a cell depend only on the combination of particle directions in that cell. The scattering can be done by table look-up, in which one creates and uses a table of scattering results—one for each possible cell configuration. Or it can be done by logic operation.

USING LATTICE GAS METHODS TO APPROXIMATE HYDRODYNAMICS

In August 1985 Frisch, Hasslacher, and Pomeau demonstrated that one can approximate solutions to the Navier-Stokes equations by using lattice gas methods, but their demonstration applied only to low-velocity incompressible flows near equilibrium. No one knew whether more interesting flows could be approximated. Consequently, computer codes were written to determine the region of validity of the lattice gas method. Results of some of the first simulations done at Los Alamos and of some later simulations are shown in Figures 1 through 6. (Most of the early calculations were done on a Celerity computer, and the displays were done on a Sun workstation.) All the results indicate qualitatively correct fluid behavior.

Plate 1a (see color plates) demonstrates that a stable trailing vortex pattern develops in a 2-dimensional lattice gas flowing past a plate. Figure 1b shows that without a three-particle scattering rule, which removes the spurious conservation of momentum along each line of particles, no vortex develops.

Plate 2 shows that stable vortices develop in a lattice gas at the interface between fluids moving in opposite directions. The Kelvin-Helmholtz instability is known to initiate such vortices. The fact that lattice gas methods should stimulate vortex evolution was reassuring and caused several scientists to begin to study the new method.

Plate 3 shows the complicated wake that develops behind the V-shaped wedge in a uniform-velocity flow.

Plate 4 shows the periodic oscillation of a low-velocity wake behind a cylinder. With a Reynolds number of 76, the flow has a stable period of oscillation that slowly grows to its asymptotic limit.

Plate 5 shows a flow with a higher Reynolds number past an ellipse. The wake here becomes chaotic and quite sensitive to details of the flow.

Plate 6 shows views of a 3-dimensional flow around a square plate, which was one of the first results from Los Alamos in 3-dimensional lattice gas hydrodynamic simulations.

Rivet and Frisch and other French scientists have developed a similar code that measures the kinematic shear viscosity numerically; the results compare well with theoretical predictions.

The lattice gas calculations of a group at the University of Chicago (Kadanoff, McNamara, and Zanetti) for two-dimensional flow through a channel agree with the known parabolic velocity profile for low-velocity channel flows.

The above calculations, and many others, have established some confidence that qualitative features of hydrodynamic flows are simulated by lattice-gas methods. Problems encountered in detailed comparisons with other types of calculations are discussed in the next section.

LIMITATIONS OF SIMPLE LATTICE GAS MODELS

As we discussed earlier, lattice gas methods can be made more accurate by making them more complicated—by, for example, adding more velocity directions and magnitudes. But the added complications degrade the efficiency. We mention in this section some of the difficulties (associated with limited range of speed, velocity dependence of the equation of state, and noisy results) encountered in the simplest lattice-gas models.

The limited range of flow velocities is inherent in a model that assumes a single speed for all particles. The sound speed in such models can be shown to be about two-thirds of the particle speed. Hence flows in which the Mach number (flow speed divided by sound speed) is greater than 1.5 cannot be simulated. This difficulty is avoided by adding particles with a variety of speeds.

The limited range of velocities also restricts the allowed range of Reynolds numbers. For small Reynolds numbers (0 to 1000) the flow is smooth, for moderate Reynolds numbers (2000 to 6000) some turbulence is observed, and for high Reynolds numbers (10,000 to 10,000,000) extreme turbulence occurs. Since the effective viscosity, ν, is typically about 0.2 in two-dimensional problems, the Reynolds number scales with the characteristic length, l, allowed by computer memory. Currently the upper bound on l is of the order of 100,000.

The velocity dependence of the equation of state is unusual and is a consequence of the inherent Fermi-Dirac distribution of the lattice gas. The low-velocity equation of state for a lattice gas can be written as $p = \frac{1}{2}\rho \left(1 - \frac{1}{2}v^2\right)$, where p is the pressure, ρ is the density, and v is the flow speed. Thus, for constant-pressure flows, regions of higher velocity flows have higher densities.

The velocity dependence of the equation of state is related to the fact that lattice gas models lack Galilean invariance. The standard Navier-Stokes equation for incompressible fluids is

$$\frac{\partial v}{\partial t} + v \cdot \nabla v = -\nabla p + \nu \nabla^2 v.$$

But in the incompressible, low-velocity limit the single-speed hexagonal lattice gas follows the equation

$$\frac{\partial v}{\partial t} + g(\rho)v \cdot \nabla v = -\nabla p + \nu\nabla^2 v,$$

where

$$g(\rho) = \frac{3 - \rho}{6 - \rho}$$

and ρ is the average number of particles per cell. The extra factor $g(\rho)$ requires special treatment. The conventional way to adjust for the fact that $g(\rho)$ does not equal unity (as it does in the Navier-Stokes equation) is to simply scale the time, t, and the viscosity, ν, by the factor $g(\rho)$ as follows: $t' = g(\rho)t$ and $\nu' = \nu/g(\rho)$. (The pressure must also be scaled.) Hence a density-dependent scaling of the time, the viscosity, and the pressure is required to bring the lattice gas model into a form that closely approximates the hydrodynamics of incompressible fluids in the low-velocity limit.

Finally, the discreteness of the lattice gas approximation introduces noise into the results. One method of smoothing the results for comparison with other methods is to average in space and time. In practice, spatial averages are taken over 64, 256, 512, or 1024 neighboring cells for time-dependent flows in two dimensions. For steady-state flows, time averaging is done. The details of noise reduction are complicated, but they must be addressed in each comparison calculation. The presence of noise is both a virtue and a defect. Noise ensures that only robust (that is, physical) singularities survive, whereas in standard codes, which are subject to less noise, mathematical artifacts can produce singularities. On the other hand, the noise in the model can trigger instabilities.

CONCLUSION

In the last few years lattice gas methods have been shown to simulate the qualitative features of hydrodynamic flows in two and three dimensions. Precise comparisons with other methods of calculation remain to be done, but it is believed that the accuracy of the lattice gas method can be increased by making the models more complicated. But how complicated they have to be to obtain the desired accuracy is an unanswered question.

Calculations based on the simple models are extremely fast and can be made several orders-of-magnitude faster by using special-purpose computers, but the models must be extended to get quantitative results with an accuracy greater than 1 percent. Significant research remains to be done to determine the accuracy of a given lattice gas method for a given flow problem.

NOTE ADDED IN PROOF: Recently Kadanoff, McNamara, and Zanetti reported precise comparisons between theoretical predictions and lattice gas simulations. They used a seven-bit hexagonal model on a small automaton universe to simulate forced two-dimensional channel flow for long times. Three tests were used to probe the hydrodynamic and statistical-mechanical behavior of the model. The tests determined (1) the profile of momentum density in the channel, (2) the equation of state given by the statistical mechanics of the system, and (3) the logarithmic divergence in the viscosity (a famous effect in two-dimensional hydrodynamics and a deep test of the accuracy of the model in the strong nonlinear regime).

The results were impressive. First, to within the accuracy of the simulation, there is no discrepancy between the parabolic velocity profile predicted by macroscopic theory and the lattice gas simulation data. Second, the equation of state derived from theory fits the simulation data to better than 1 percent. Finally, the measured logarithmic divergence in the viscosity as a function of channel width agrees with prediction. These results are at least one order-of-magnitude more accurate than any previously reported calculations.

REFERENCES

1. University of Chicago, preprint, October 1987

U. Frisch†, B. Hasslacher‡, and Y. Pomeau*

†Centre National de la Recherche Scientifique, Observatoire de Nice, 06003 Nice Cedex, France; ‡Theoretical Division and Center for Nonlinear Studies, Los Alamos National Laboratory, Los Alamos, New Mexico 87545; and *Centre National de la Recherche Scientifique, Ecole Normale Supérieure, 75231 Paris Cedex, France, and Service de Physique Théorique, Centre d'Etudes Nucléaires de Saclay, 91191 Gif-sur-Yvette, France

Lattice-Gas Automata for the Navier-Stokes Equation

This paper originally appeared in *Physical Review Letters*, Volume 56, Number 14, April 7, 1986, pp. 1505–1508.

We show that a class of deterministic lattice gases with discrete Boolean elements simulates the Navier-Stokes equation, and can be used to design simple, massively parallel computing machines.

The relatively recent availability of sophisticated interactive digital simulation has led to considerable progress in the unraveling of universal features of complexity generated by nonlinear dynamical systems with few degrees of freedom. In contrast, nonlinear systems with many degrees of freedom, e.g., high-Reynolds-number flow, are understood only on a quite superficial level,[1] and are likely to remain so, unless they can be explored in depth, e.g., by interactive simulation. This is many orders of magnitude beyond the capacity of existing computational resources. There are similar limitations on our ability to simulate many other multidimensional field theories.

Massively parallel architectures and algorithms are needed to avoid the ultimate computation limits of the speed of light and various solid-state constraints. Also, when parameter space must be explored quickly and extreme accuracy is unnecessary, a floating-point representation may not be efficient. For example, to compute the drag due to turbulent flow past an obstacle with a modest accuracy of 5 bits, common experience in computational fluid dynamics shows that intermediate computations require from 32 to 64 bits. Floating-point representations hierarchically favor bits in the most significant places,[2] which is a major cause of numerical instability. In principle, schemes which give bits equal weight would be preferable. Because of roundoff noise, a floating-point calculation can run away to unphysical regimes, in an attempt to treat each bit equally.

A simulation strategy can be devised which both is naturally parallel and treats all bits on an equal footing, for systems which evolve by discrete cellular automaton rules, with only local interactions.[3] This avoids the complex switching networks which limit the computational power of conventional parallel arrays.

There has been speculation that various physically interesting field equations can be approximated by the large-scale behavior of suitably chosen cellular automata.[4] We shall here construct lattice-gas automata which asymptotically go over to the incompressible 2-D and 3-D Navier-Stokes equations.

To understand the physics behind lattice gases, we first point out that a fluid can be described on three levels: the molecular level at which motion, usually Hamiltonian, is reversible; the kinetic level, in the irreversible low-density Boltzmann approximation; and the macroscopic level, in the continuum approximation. At the first two levels of description, the fluid is near thermodynamic equilibrium. In the last there are free thermodynamic variables: local density, momentum, temperature, etc. A macroscopic description of the fluid comes about by a patching together of equilibria which are varying slowly in space and time, implying continuum equations for thermodynamic variables as consistency conditions. This was first realized by Maxwell,[5] and put in final form by Chapman and Enskog.[6]

There are many ways of building microscopic models that lead to a given set of continuum equations. It is known that one can build two- and three-dimensional Boltzmann models, with a small number of velocity vectors, which, in the continuum limit, reproduce quite accurately major fluid dynamical features (e.g., shock waves in a dilute gas, etc.[7]). Such Boltzmann models are fundamentally probabilistic, discrete only in velocity, but continuous in space and time. In contrast, we will use lattice-gas models, which have a completely discrete phase space and time and therefore may be viewed as made of "Boolean molecules."

The simplest case is the Hardy, de Pazzis, and Pomeau model[8] (hereafter called HPP) which has an underlying regular, square, two-dimensional lattice with unit link lengths. At each vertex, there are up to four molecules of equal mass, with unit speed, whose velocities point in one of the four link directions. The simultaneous occupation of a vertex by identical molecules is forbidden. Time is also discrete. The update is as follows. First, each molecule moves one link, to the nearest vertex to which its velocity was pointing. Then, any configuration of exactly two molecules moving in opposite directions at a vertex (head-on collisions) is replaced by another one at right angles to the original. All other configurations are left unchanged. The HPP model has a number of important properties.[8] The crucial one is the existence of thermodynamic equilibria. No ergodic theorem is known, but relaxation to equilibrium has been demonstrated numerically.[8] These equilibria have free *continuous* parameters, namely, the average density and momentum. The equilibrium distribution functions are completely factorized over vertices and directions, being independent of vertex position, but dependent on direction, unless the mean momentum vanishes. When the density and momentum are varied slowly in space and time, "macrodynamical" equations emerge which differ from the nonlinear Navier-Stokes equations in three respects.

The discrepancies may be classified as (1) lack of Galilean invariance, (2) lack of isotropy, and (3) a crossover dimension problem. Galilean invariance is by definition broken by the lattice; consequently, thermodynamic equilibria with different velocities cannot be related by a simple transformation. This is reflected by the nonlinear term in the momentum equation, containing a momentum flux tensor, which not only has quadratic terms in the hydrodynamic velocity **u**, as it should be in the Navier-Stokes equation, but also has nonlinear corrections to arbitrarily high order in the velocity. However, these terms are negligible at low Mach number, a condition which also guarantees incompressibility. The HPP automaton is invariant under $\pi/2$ rotations. Such a lattice symmetry is insufficient to insure the isotropy of the fourth-degree tensor relating momentum flux to quadratic terms in the velocity. Finally, crossover dimension is a general property of two-dimensional hydrodynamics, when thermal noise is added to the Navier-Stokes equation or to the HPP version of it. Simply put, the viscosity develops a logarithmic scale dependence, which is a dimensional crossover phenomenon, common in phase transitions and field theory.[9] In three dimensions, this difficulty does not exist.

Focusing on the isotropy problem, we note that for the HPP model, the momentum flux tensor has the form

$$P_{\alpha\beta} = p\delta_{\alpha\beta} + T_{\alpha\beta\gamma\epsilon}u_\gamma u_\epsilon + O(u^4). \tag{1}$$

Here $p = \rho/2$ is the pressure; terms odd in **u** vanish by parity. The tensor T is, by construction, pairwise symmetric in both (α, β) and (γ, ϵ). Observe that when the underlying microworld is *two-dimensional* and invariant under the hexagonal rotation group (multiples of $\pi/3$), the tensor T is isotropic and Eq. (1) takes the form

$$P_{\alpha\beta} = (p + \mu u^2)\delta_{\alpha\beta} + \lambda u_\alpha u_\beta + O(u^4), \tag{2}$$

with suitable scalar factors λ and μ. At low Mach number, this is the correct form for the Navier-Stokes equation. This observation appears to be new. So, in two dimensions, we will use a triangular instead of a square lattice. Each vertex then has a hexagonal neighborhood (Figure 1). We will call this model the hexagonal lattice gas (HLG). The setup is the same as in the HPP lattice gas, except for modified collision rules. A suitable set is one given by Harris,[10] in connection with a discrete Boltzmann model, supplemented by a Fermi exclusion condition, of single occupation of each Boolean state. The Fermi-modified Harris rules are as follows: Number the six links out of any vertex counterclockwise, with an index i, defined on the integers (mod6). There are both two- and three-body collisions. For two-body collisions, we have $(i, i + 3)$ goes to (a) $(i + 1, i - 2)$ or (b) $(i - 1, i + 2)$. Type a and b outcomes have equal *a priori* weights. For three-body collisions we have $(i, i + 2, i - 2)$ goes to $(i + 3, i + 1, i - 1)$. In these rules, it is assumed that no incident link to a vertex is populated, other than the ones given as initial states. All other configurations remain unaffected by collisions. These rules are designed to conserve particle number and momentum at each vertex, i.e., a total of three scalar conservation relations. Without three-body collisions, there would be four scalar conservation relations, namely mass and momentum along each of the three lattice directions.

Note that the HPP rules are invariant under duality (interchange of particles and holes), whereas the present rules are not. Duality can be restored by addition of suitable four-particle collision rules, but we will not use them here.

We display a variant of this model where at most one particle is allowed to remain at rest at each vertex. The rest particles are labeled by an asterisk and the previous rules are supplemented with $(i, i + 2)$ goes to $(i - 2, *)$ and $(i, *)$ goes to $(i + 2, i - 2)$. Additional variations on the model allow one to define a nontrivial temperature. The remainder of this discussion is concerned only with the basic (HLG) model.

We briefly outline how the hexagonal lattice gas leads to the two-dimensional Navier-Stokes equations. A detailed derivation will be presented elsewhere.[11] Let N_i be the average population at a vertex with velocity in the direction i. The average

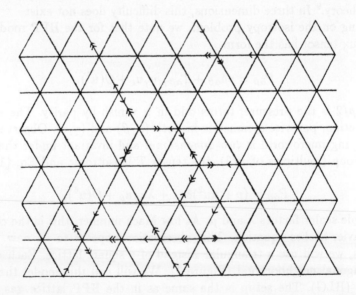

FIGURE 1 Triangular lattice with hexagonal symmetry and hexagonal lattice-gas rules. Particles at time t and $t + 1$ are marked by single and double arrows, respectively.

is over a macroscopic space-time region so that N_i depends slowly on space and time variables. We define a slowly varying density ρ and momentum ρu by

$$\rho = \sum_i N_i, \qquad \rho u = \sum_i N_i c_i, \tag{3}$$

where c_i is a unit vector in the direction i. Locally, for a given ρ and u, the N_i's can be computed from both these definitions and the detailed-balance equations at thermodynamic equilibrium, which are too involved to present here. This gives a Fermi-Dirac distribution:

$$N_i = \left\{1 + \exp\left[\alpha(\rho, u) + \beta(\rho, u) c_i \cdot u\right]\right\}^{-1}. \tag{4}$$

In general, α and β satisfy equations with no simple solutions. However, when $u = 0$, it is obvious by symmetry that $N_i = \rho/6$. Therefore, α and β can be expanded in a Taylor series around $u = 0$. The result can be used to compute mass and momentum flux to first order in the macroscopic gradients. Second-order terms in the gradients (viscous terms) are obtained by Green-Kubo relations or by a Chapman-Enskog expansion.[12] The following set of hydrodynamic equations is thus obtained:

$$\frac{\partial \rho}{\partial t} + \nabla \cdot (\rho u) = 0, \tag{5}$$

$$\frac{\partial}{\partial t}(\rho u_\alpha) + \sum_\beta \frac{\partial}{\partial x_\beta}\left[g(\rho)\rho u_\alpha u_\beta + O(u^4)\right]$$

$$= -\frac{\partial}{\partial x_\alpha}p + \eta_1(\rho)\nabla^2 u_\alpha + \eta_2(\rho)\frac{\partial}{\partial x_\alpha}\nabla \cdot u, \tag{6}$$

with $g(\rho) = (\rho - 3)/(\rho - 6)$ and $p = \rho/2$. $\eta_1(\rho)$ and $\eta_2(\rho)$ are the shear and bulk viscosities.[12] Deletion of the nonlinear and viscous terms gives the wave equation for sound waves propagating isotropically with a speed equal to the "velocity of light" (here set equal to 1) over $\sqrt{2}$, just as for a two-dimensional photon gas. These sound waves have been observed in simulations on the MIT cellular automaton machine by Margolus, Toffoli, and Vichniac.[13] They used lattice-gas models that yield the same wave equation as above.

The nonlinear system (5) and (6) goes over to the incompressible Navier-Stokes equation by the following limiting procedure: Let the Mach number $M = u\sqrt{2}$ tend to zero, and the hydrodynamic scale L tend to infinity, while keeping their product fixed. As in the usual derivation of the incompressible limit, density fluctuations become irrelevant, except in the pressure term; also, the continuity equation (5) reduces to $\nabla \cdot u = 0$. Thus, the factor $g(\rho)$ is to leading order a constant and may, for $0 < \rho < 3$, be absorbed in a rescaled time. The resulting Reynolds number is

$$N_R = \frac{ML\rho g(\rho)}{\sqrt{2}\eta_1(\rho)}. \tag{7}$$

Note that Galilean invariance, which does not hold at the lattice level, is restored macroscopically.

A straightforward lift of the hexagonal lattice-gas model from two into three dimensions does not work. The reason is that the regular space-filling simplex with the greatest symmetry in three dimensions is the face centered cubic, with twelve equal-speed velocity directions out of each vertex. Unfortunately, the relevant tensors such as $T_{\alpha\beta\gamma\epsilon}$ in Eq. (1) depend now on three constants. This induces a spurious, isotropy-breaking term in the Navier-Stokes equation, proportional to $(\partial/\partial x_\alpha)u_\alpha^2$ (no summation on α).

This obstacle may be removed by a splitting method. The nonlinear term in the three-dimensional Navier-Stokes equation is recast as the sum of two terms, each containing spurious elements and each realizable on a different lattice (for example, a face-centered-cubic lattice and a regular cubic lattice).

In lattice-gas models, as in general cellular automata (CA's), boundary conditions are very easy to implement. Specular reflection of molecules gives so-called "free slip" boundary conditions for the hydrodynamic velocity **u**. "Rigid" boundary conditions are obtained either by random scattering of particles back into the incoming half plane from a locally planar boundary, or by specular reflection from a microscale roughened version of the macroscopic boundary.

We mention some practical limitations on lattice-gas models. For the hydrodynamic description to hold, there must be a scale separation between the smallest hydrodynamic scale and the lattice link length; as we shall see, this requirement is automatically satisfied. Lattice-gas models must be run at moderate Mach numbers M (say, 0.3 to 0.5), to remain incompressible, and to avoid spurious high-order nonlinear terms. For fixed Mach numbers, the largest Reynolds number associated with a D-dimensional lattice with $O(N)$ sites in each direction is $O(N)$. This is because in our units, the kinematic viscosity of the hexagonal lattice gas is $O(1)$. From standard turbulence theory,[14] it follows that the dissipation scale is $O(N^{1/2})$ in 2D and $O(N^{1/4})$ in 3D. This insures the required scale separation at large Reynolds numbers. It would, however, be desirable to reduce the scale separation, especially in 2D, to avoid excessive storage requirements compared to conventional incompressible floating-point simulations (in the latter, the mesh can be taken comparable to the dissipation scale).

For this, we observe that the viscosity in the lattice gas is decreased by a factor P if we subdivide each cell into a sublattice with links P times smaller. We note also that the sublattice need not be similar to the original lattice. It must have the same collision rules, to preserve local thermodynamic equilibria, but the geometry does not matter since macroscopic quantities may be considered uniform over the cell. Thus, all the sublattice vertices in a given cell may be regarded as indistinguishable and can be coded in $O(\ln P)$ rather than $O(P^D)$ bits; interactions occur between randomly chosen vertex pairs within cells and between neighboring cells, the latter being less frequent by a factor $O(1/P)$.

Simulations of the models discussed here, done on general-purpose computers and exhibiting a variety of known two-dimensional hydrodynamic phenomena, have been made by d'Humières, Lallemand, and Shimomura.[15]

We have given a concrete hydrodynamical example of how CA's can be used to simulate classical nonlinear fields. We expect that further CA implementations will be found for the Navier-Stokes equation and other problems, not necessarily based on thermalized lattice gases and possibly less constrained than ours.

S. Wolfram stimulated our interest in cellular automata as a possible new approach to turbulence phenomena. Acknowledgments are also due to T. Bloch, R. Caflish, D. d'Humières, R. Gatignol, R. Kraichnan, P. Lallemand, N. Margolus, D. Nelson, J. L. Oneto, S. A. Orszag, J. P. Rivet, T. Shimomura, Z. S. She, B. Shraiman, T. Toffoli, and G. Vichniac, as well as the following: Woods Hole Geophysical Fluid Dynamics Summer Program (U.F., Y.P.); Aspen Center for Physics, 1985 Chaos Workshop (B.H.); and Service de Physique Théorique, Centre d'Etudes Nucleaires de Saclay (B.H.). This work was supported in part by National Science Foundation Grant No. 8442384.

REFERENCES

1. Frisch, U. *Phys. Scr.* **T9** (1985):131.
2. Knuth, D. E. *The Art of Computer Programming: Semi-Numerical Algorithms*, vol. 2. Reading, MA: Addison-Wesley, 1981, 238.
3. Toffoli, T. *Physica* (Amsterdam) **10D** (1984):117; Wolfram, S. *Nature* **311** (1984):419.
4. Pomeau, Y. *J. Phys. A* **17** (1984):L415; Vichniac, G. *Physica* (Amsterdam) **10D** (1984):96, and references therein; Margolus, N. *Physica* (Amsterdam) **10D** (1984):81.
5. Maxwell, J. C. *The Scientific Papers*, vol. 2. Cambridge, England: Cambridge Univ. Press, 1890, 681.
6. Uhlenbeck, G. E., and G. W. Ford. "Lectures in Statistical Mechanics." *Lectures in Applied Math.*, vol. 1. Providence, RI: American Mathematical Society, 1963.
7. Broadwell, J. E. *Phys. Fluids* **7** (1964):1243; Gatignol, R. "Théorie Cinétique des Gaz à Répartition Discrète des Vitesses." *Lecture Notes in Physics*, vol. 36. Berlin: Springer, 1964.
8. Hardy, J., and Y. Pomeau. *J. Math. Phys.* **13** (1972):1042; Hardy, J., Y. Pomeau, and O. de Pazzis. *J. Math. Phys.* **14** (1973):1746; Hardy, J., O. de Pazzis, and Y. Pomeau. *Phys. Rev. A* **13** (1976):1949.
9. Forster, D., D. R. Nelson, and M. J. Stephen. *Phys. Rev. A.* **16** (1977):732.
10. Harris, S. *Phys. Fluids* **9** (1966):1328.
11. Frisch, U., D. d'Humières, B. Hasslacher, P. Lallemand, Y. Pomeau, and J.-P. Rivet. "Lattice Gas Hydrodynaamics in Two and Three Dimensions." Reprinted in this volume.
12. Rivet, J., and U. Frisch. *C.R. Seances Acad. Sci., Ser. 2* **302** (1986):267.
13. Margolus, N., T. Toffoli, and G. Vichniac. Private communication; *Massachusetts Institute of Technology Technical Memo No. LCS-TM-296*, 1984.
14. Kolmogorov, A. N. *C.R. (Dokl.) Acad. Sci. USSR* **30** (1941):301, 538; Kraichnan, R. *Phys. Fluids* **10** (1967):1417; Batchelor, G. K. *Phys. Fluids* **12**, Suppl. 2 (1969):233.
15. d'Humières, D., P. Lallemand, and T. Shimomura. "Lattice Gas Cellular Automata, A New Experimental Tool for Hydrodynamics," to be published.

Stephen Wolfram

The Institute for Advanced Study, Princeton, New Jersey and Thinking Machines Corporation, 245 First Street, Cambridge, Massachusetts 02142; present address Center for Complex Systems Research, and Departments of Physics, Mathematics and Computer Science, University of Illinois, 508 South Sixth Street, Champaign, Illinois 61820

Cellular Automaton Fluids 1: Basic Theory

This paper originally appeared in the *Journal of Statistical Physics*, volume 45, nos. 3/4 (1986), pp. 471–526.

Continuum equations are derived for the large-scale behavior of a class of cellular automaton models for fluids. The cellular automata are discrete analogues of molecular dynamics, in which particles with discrete velocities populate the links of a fixed array of sites. Kinetic equations for microscopic particle distributions are constructed. Hydrodynamic equations are then derived using the Chapman-Enskog expansion. Slightly modified Navier-Stokes equations are obtained in two and three dimensions with certain lattices. Viscosities and other transport coefficients are calculated using the Boltzmann transport equation approximation. Some corrections to the equations of motion for cellular automaton fluids beyond the Navier-Stokes order are given.

Key words: Cellular automata; derivation of hydrodynamics; molecular dynamics; kinetic theory; Navier-Stokes equations.

1. INTRODUCTION

Cellular automata (e.g., Refs. 1 and 2) are arrays of discrete cells with discrete values. Yet sufficiently large cellular automata often show seemingly continuous macroscopic behavior (e.g., Refs. 1 and 3). They can thus potentially serve as models for continuum systems, such as fluids. Their underlying discreteness, however, makes them particularly suitable for digital computer simulation and for certain forms of mathematical analysis.

On a microscopic level, physical fluids consist of discrete particles. But on a large scale, they, too, seem continuous, and can be described by the partial differential equations of hydrodynamics (e.g., Ref. 4). The form of these equations is in fact quite insensitive to microscopic details. Changes in molecular interaction laws can affect parameters such as viscosity, but do not alter the basic form of the macroscopic equations. As a result, the overall behavior of fluids can be found without accurately reproducing the details of microscopic molecular dynamics.

This paper is the first in a series which considers models of fluids based on cellular automata whose microscopic rules give discrete approximations to molecular dynamics.[1] The paper uses methods from kinetic theory to show that the macroscopic behavior of certain cellular automata corresponds to the standard Navier-Stokes equations for fluid flow. The next paper in the series[16] describes computer experiments on such cellular automata, including simulations of hydrodynamic phenomena.

Figure 1 shows an example of the structure of a cellular automaton fluid model. Cells in an array are connected by links carrying a bounded number of discrete "particles." The particles move in steps and "scatter" according to a fixed set of deterministic rules. In most cases, the rules are chosen so that quantities such as particle number and momentum are conserved in each collision. Macroscopic variations of such conserved quantities can then be described by continuum equations.

Particle configurations on a microscopic scale are rapidly randomized by collisions, so that a local equilibrium is attained, described by a few statistical average quantities. (The details of this process will be discussed in a later paper.) A master equation can then be constructed to describe the evolution of average particle densities as a result of motion and collisions. Assuming slow variations with position and time, one can then write these particle densities as an expansion in terms of macroscopic quantities such as momentum density. The evolution of these quantities is determined by the original master equation. To the appropriate order in the

---

[1]This work has many precursors. A discrete model of exactly the kind considered here was discussed in Ref. 6. A version on a hexagonal lattice was introduced in Ref. 7, and further studied in Refs. 8,9. Related models in which particles have a discrete set of possible velocities, but can have continuously variable positions and densities, were considered much earlier.[10,14] Detailed derivations of hydrodynamic behavior do not, however, appear to have been given even in these cases (see, however, e.g., Ref. 15).

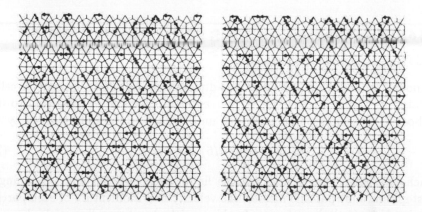

FIGURE 1 Two successive microscopic configurations in the typical cellular automaton fluid model discussed in Section 2. Each arrow represents a discrete "particle" on a link of the hexagonal grid. Continuum behavior is obtained from averages over large numbers of particles.

expansion, certain cellular automaton models yield exactly the usual Navier-Stokes equations for hydrodynamics.

The form of such macroscopic equations is in fact largely determined simply by symmetry properties of the underlying cellular automaton. Thus, for example, the structure of the nonlinear and viscous terms in the Navier-Stokes equations depends on the possible rank three and four tensors allowed by the symmetry of the cellular automaton array. In two dimensions, a square lattice of particle velocities gives anisotropic forms for these terms.[6] A hexagonal lattice, however, has sufficient symmetry to ensure isotropy.[7] In three dimensions, icosahedral symmetry would guarantee isotropy, but no crystallographic lattice with such a high degree of symmetry exists. Various structures involving links beyond nearest neighbors on the lattice can instead be used.

Although the overall form of the macroscopic equations can be established by quite general arguments, the specific coefficients which appear in them depend on details of the underlying model. In most cases, such transport coefficients are found from explicit simulations. But, by using a Boltzmann approximation to the master equation, it is possible to obtain some exact results for such coefficients, potentially valid in the low-density limit.

This paper is organized as follows. Section 2 describes the derivation of kinetic and hydrodynamic equations for a particular sample cellular automaton fluid model. Section 3 generalizes these results and discusses the basic symmetry conditions necessary to obtain standard hydrodynamic behavior. Section 4 then uses the Boltzmann equation approximation to investigate microscopic behavior and obtain results for transport coefficients. Section 5 discusses a few extensions of the model. The Appendix gives an SMP program[17] used to find macroscopic equations for cellular automaton fluids.

2. MACROSCOPIC EQUATIONS FOR A SAMPLE MODEL

2.1. STRUCTURE OF THE MODEL

The model[7] is based on a regular two-dimensional lattice of hexagonal cells, as illustrated in Figure 1. The site at the center of each cell is connected to its six neighbors by links corresponding to the unit vectors e_1 through e_6 given by

$$e_a = \left(\cos(2\pi a/6), \sin(w\pi a/6) \right). \tag{2.1.1}$$

At each time step, zero or one particle lie on each directed link. Assuming unit time steps and unit particle masses, the velocity and momentum of each particle is given simply by its link vector e_a. In this model, therefore, all particles have equal kinetic energy, and have zero potential energy.

The configuration of particles evolves in a sequence of discrete time steps. At each step, every particle first moves by a displacement equal to its velocity e_a. Then the particles on the six links at each site are rearranged according to a definite set of rules. The rules are chosen to conserve the number and total momentum of the particles. In a typical case, pairs of particles meeting head on might scatter through 60°, as would triples of particles 120° apart. The rules may also rearrange other configurations, such as triples of particles meeting asymmetrically. Such features are important in determining parameters such as viscosity, but do not affect the form of the macroscopic equations derived in this section.

To imitate standard physical processes, the collision rules are usually chosen to be microscopically reversible. There is therefore a unique predecessor, as well as a unique successor, for each microscopic particle configuration. The rules for collisions in each cell thus correspond to a simple permutation of the possible particle arrangements. Often the rules are self-inverse. But, in any case, the evolution of a complete particle configuration can be reversed by applying inverse collision rules at each site.

The discrete nature of the cellular automaton model makes such precise reversal in principle possible. But the rapid randomization of microscopic particle configurations implies that very complete knowledge of the current configuration is needed. With only partial information, the evolution may be effectively irreversible.[8,19]

2.2. BASIS FOR KINETIC THEORY

Cellular automaton rules specify the precise deterministic evolution of microscopic configurations. But if continuum behavior is seen, an approximate macroscopic description must also be possible. Such a description will typically be a statistical one, specifying not, for example, the exact configuration of particles, but merely the probabilities with which different configurations appear.

A common approach is to consider ensembles in which each possible microscopic configuration occurs with a particular probability (e.g., Ref. 18). The reversibility of the microscopic dynamics ensures that the total probability for all configurations in the ensemble must remain constant with time. The probabilities for individual configurations may, however, change, as described formally by the Liouville equation.

An ensemble is in "equilibrium" if the probabilities for configurations in it do not change with time. This is the case for an ensemble in which all possible configurations occur with equal probability. For cellular automata with collision rules that conserve momentum and particle number, the subsets of this ensemble that contain only those configurations with particular total values of the conserved quantities also correspond to equilibrium ensembles.

If the collision rules effectively conserved absolutely no other quantities, then momentum and particle number would uniquely specify an equilibrium ensemble. This would be the case if the system were ergodic, so that starting from any initial configuration, the system would eventually visit all other microscopic configurations with the same values of the conserved quantities. The time required would, however, inevitably be exponentially long, making this largely irrelevant for practical purposes.

A more useful criterion is that starting from a wide range of initial ensembles, the system evolves rapidly to ensembles whose statistical properties are determined solely from the values of conserved quantities. In this case, one could assume for statistical purposes that the ensemble reached contains all configurations with these values of the conserved quantities, and that the configurations occur with equal probabilities. This assumption then allows for the immediate construction of kinetic equations that give the average rates for processes in the cellular automaton.

The actual evolution of a cellular automaton does not involve an ensemble of configurations, but rather a single, specific configuration. Statistical results may nevertheless be applicable if the behavior of this single configuration is in some sense "typical" of the ensemble.

This phenomenon is in fact the basis for statistical mechanics in many different systems. One assumes that appropriate space or time averages of an individual configuration agree with averages obtained from an ensemble of different configurations. This assumption has never been firmly established in most practical cases; cellular automata may in fact be some of the clearest systems in which to investigate it.

The assumption relies on the rapid randomization of microscopic configurations, and is closely related to the second law of thermodynamics. At least when statistical or coarse-grained measurements are made, configurations must seem progressively more random, and must, for example, show increasing entropies. Initially ordered configurations must evolve to apparent disorder.

The reversibility of the microscopic dynamics nevertheless implies that ordered initial configurations can always in principle be reconstructed from a complete knowledge of these apparently disordered states. But just as in pseudorandom sequence generators or cryptographic systems, the evolution may correspond to a sufficiently complex transformation that any regularities in the initial conditions cannot readily be discerned. One suspects in fact that no feasibility simple computation can discover such regularities from typical coarse-grained measurements.[19,20]

As a result, the configurations of the system seem random, at least with respect to standard statistical procedures.

While most configurations may show progressive randomization, some special configurations may evolve quite differently. Configurations obtained by computing time-reversed evolution from ordered states will, for example, evolve back to ordered states. Nevertheless, one suspects that the systematic construction of such "antithermodynamic" states must again require detailed computations of a complexity beyond that corresponding to standard macroscopic experimental arrangements.

Randomization requires that no additional conserved quantities are present. For some simple choices of collision rules, spurious conservation laws can nevertheless be present, as discussed in Section 4.5. For most of the collision rules considered in this paper, however, rapid microscopic randomization does seem to occur.

As a result, one may use a statistical ensemble description. Equilibrium ensembles in which no statistical correlations are present should provide adequate approximations for many macroscopic properties. At a microscopic level, however, the deterministic dynamics does lead to correlations in the detailed configurations of particles.[2] Such correlations are crucial in determining local properties of the system. Different levels of approximation to macroscopic behavior are obtained by ignoring correlations of different orders.

Transport and hydrodynamic phenomena involve systems whose properties are not uniform in space and time. The uniform equilibrium ensembles discussed above cannot provide exact descriptions of such systems. Nevertheless, so long as macroscopic properties vary slowly enough, collisions should maintain approximate local equilibrium, and should make approximations based on such ensembles accurate.

2.3. KINETIC EQUATIONS

An ensemble of microscopic particle configurations can be described by a phase space distribution function which gives the probability for each complete configuration. In studying macroscopic phenomena, it is, however, convenient to consider reduced distribution functions, in which an average has been taken over most degrees of freedom in the system. Thus, for example, the one-particle distribution function $f_a(\mathbf{x}, t)$ gives the probability of finding a particle with velocity $\mathbf{e}_a$ at position $\mathbf{x}$ and time t, averaged over all other features of the configuration (e.g., Ref. 23).

Two processes lead to changes in f_a with time: motion of particles from one cell to another, and interactions between particles in a given cell. A master equation can be constructed to describe these processes.

---

[2]The kinetic theory approach used in this paper concentrates on average particle distribution functions. An alternative but essentially equivalent approach concentrates on microscopic correlation functions (e.g., Refs. 21, 22).

In the absence of collisions, the cellular automaton rules imply that all particles in a cell at position $\mathbf{X}$ with velocity $\mathbf{e}_a$ move at the next time step to the adjacent cell at position $\mathbf{X} + \mathbf{e}_a$. As a result, the distribution function evolves according to

$$f_a(\mathbf{X} + \mathbf{e}_a, T + 1) = f_a(\mathbf{X}, T). \tag{2.3.1}$$

For large lattices and long time intervals, position and time may be approximated by continuous variables. One may define, for example, scaled variables $\mathbf{x} = \delta_x \mathbf{X}$ and $t = \delta_t T$, where δ_x, $\delta_t \ll 1$. In terms of these scales variables, the difference equation (2.3.1) becomes

$$f_a(\mathbf{x} + \mathbf{e}_a \delta_x, t + \delta_t) - f_a(\mathbf{x}, t) = 0. \tag{2.3.2}$$

In deriving macroscopic transport equations, this must be converted to a differential equation. Carrying out a Taylor expansion, one obtains[24]

$$\delta_t \partial_t f_a + \delta_x \mathbf{e}_a \cdot \nabla f_a + \frac{1}{2} \delta_t^2 \partial_{tt} f_a + \delta_x \delta_t (\mathbf{e}_a \cdot \nabla) \partial_t f_a + \frac{1}{2} \delta_x^2 (\mathbf{e}_a \cdot \nabla)^2 f_a + O(\delta^3) = 0. \tag{2.3.3}$$

If all variations in the f_a are assumed small, and certainly less than $O(1/\delta_x, 1/\delta_t)$, it suffices to keep only first-order terms in δ_x, δ_t. In this way one obtains the basic transport equation

$$\partial_t f_a(\mathbf{x}, t) + \mathbf{e}_a \cdot \nabla f_a(\mathbf{x}, t) = 0. \tag{2.3.4}$$

This has the form of a collisionless Boltzmann transport equation for f_a (e.g., Ref. 25). It implies, as expected, that f_a is unaffected by particle motion in a spatially uniform system.

Collisions can, however, change f_a even in a uniform system, and their effect can be complicated. Consider, for example, collisions that cause particles in directions $\mathbf{e}_1$ and $\mathbf{e}_4$ to scatter in directions $\mathbf{e}_2$ and $\mathbf{e}_5$. The rate for such collisions is determined by the probability that particles in directions $\mathbf{e}_1$ and $\mathbf{e}_4$ occur together in a particular cell. This probability is defined as the joint two-particle distribution function $\tilde{F}^{(2)}_{14}$. The collisions deplete the population of particles in direction $\mathbf{e}_1$ at a rate $\tilde{F}^{(2)}_{14}$. Microscopic reversibility guarantees the existence of an inverse process, which increases the population of particles in direction $\mathbf{e}_1$ at a rate given in this case by $\tilde{F}^{(2)}_{25}$. Notice that in a model where there can be at most one particle on each link, the scattering to directions $\mathbf{e}_2$ and $\mathbf{e}_5$ in a particular cell can occur only if no particles are already present on these links. The distribution function $\tilde{F}$ is constructed to include this effect, which is mathematically analogous to the Pauli exclusion principle for fermions.

The details of collisions are, however, irrelevant to the derivation of macroscopic equations given in this section. As a result, the complete change due to collisions in a one-particle distribution function f_a will for now be summarized by a simple "collision term" Ω_a, which in general depends on two-particle and higher-order distribution functions. (In the models considered here, Ω_a is always entirely local, and cannot depend directly on, for example, derivatives of distribution functions.) In terms of Ω_a, the kinetic equation (2.3.3) extended to include collisions becomes

$$\partial_t f_a + \mathbf{e}_a \cdot \nabla f_a = \Omega_a . \qquad (2.3.5)$$

With the appropriate form for Ω_a, this is an exact statistical equations for f_a (at least to first order in δ).

But the equation is not in general sufficient to determine f_a. It gives the time evolution of f_a in terms of the two-particle and higher-order distribution functions that appear in Ω_a. The two-particle distribution function then in turn satisfies an equation involving three-particle and higher-order distribution functions, and so on. The result is the exact BBGKY hierarchy of equations,[23] of which Eq. (2.3.5) is the first level.

The Boltzmann transport equation approximates (2.3.5) by assuming that Ω_a depends only on one-particle distribution functions. In particular, one may make a "molecular chaos" assumption that all sets of particles are statistically uncorrelated before each collision, so that multiple-particle distribution functions can be written as products of one-particle ones. The distribution function $\tilde{F}_{14}^{(2)}$ is thus approximated as $f_1 f_4 (1 - f_2)(1 - f_3)(1 - f_5)(1 - f_6)$. The resulting Boltzmann equations will be used in Section 4. In this section, only the general form (2.3.5) is needed.

The derivation of Eq. (2.3.5) has been discussed here in the context of a cellular automaton model in which particles are constrained to lie on the links of a fixed array. In this case, the maintenance of terms in (2.3.3) only to first order in δ_x, δ_t is an approximation, and corrections can arise, as discussed in Section 2.5.[24] Equation (2.3.5) is, however, exact for a slightly different class of models, in which particles have a discrete set of possible velocities, but follow continuous trajectories with arbitrary spatial positions. Such "discrete velocity gases" have often been considered, particularly in studies of highly rarefied fluids, in which the mean distance between collisions is comparable to the overall system size.[11,14]

2.4. CONSERVATION LAWS

The one-particle distribution functions typically determine macroscopic average quantities. In particular, the total particle density n is given by

$$\sum_a f_a = n . \qquad (2.4.1)$$

while the momentum density $n\mathbf{u}$, where $\mathbf{u}$ is the average fluid velocity, is given by

$$\sum_a \mathbf{e}_a f_a = n\mathbf{u} . \qquad (2.4.2)$$

The conservation of these quantities places important constraints on the behavior of the f_a.

In a uniform system $\nabla f_a = 0$, so that Eq. (2.3.5) becomes

$$\partial_t f_a = \Omega_a \qquad (2.4.3)$$

and Eqs. (2.4.1) and (2.4.2) imply

$$\sum_a \Omega_a = 0, \qquad (2.4.4)$$

$$\sum_a \mathbf{e}_a \Omega_a = 0. \qquad (2.4.5)$$

Using the kinetic equation (2.3.5), Eq. (2.4.4) implies

$$\partial_t \sum_a f_a + \sum_a \mathbf{e}_a \cdot \nabla f_a = 0. \qquad (2.4.6)$$

With the second term in the form $\nabla \cdot (\sum \mathbf{e}_a f_a)$, Eq. (2.4.6) can be written exactly in terms of macroscopic quantities as

$$\partial_t n + \nabla \cdot (n\mathbf{u}) = 0, \qquad (2.4.7)$$

this is the usual continuity equation, which expresses the conservation of fluid. It is a first example of a macroscopic equation for the average behavior of a cellular automaton fluid.

Momentum conservation yields the slightly more complicated equation

$$\partial_t \sum_a \mathbf{e}_a f_a + \sum_a \mathbf{e}_a (\mathbf{e}_a \cdot \nabla f_a) = 0. \qquad (2.4.8)$$

Defining the momentum flux density tensor

$$\Pi_{ij} = \sum_a (\mathbf{e}_a)_i (\mathbf{e}_a)_j f_a. \qquad (2.4.9)$$

Eq. (2.4.8) becomes

$$\partial_t (nu_i) + \partial_j \Pi_{ij} = 0. \qquad (2.4.10)$$

No simple macroscopic result for Π_{ij} can, however, be obtained directly from the definitions (2.4.1) and (2.4.2).

Equations (2.4.7) and (2.4.10) have been derived here from the basic transport equation (2.3.5). However, as discussed in Section 2.3, this transport equation is only an approximation, valid to first order in the lattice scale parameters δ_x, δ_t.[24] Higher-order versions of (2.4.7) and (2.4.10) may be derived from the original Taylor expansion (2.3.3), and in some cases, correction terms are obtained.[24]

Assuming $\delta_x = \delta_t = \delta$, Eq. (2.4.6) to second order becomes

$$\sum_a \left[(\partial_t + \mathbf{e}_a \cdot \nabla) + \frac{1}{2}\delta(\partial_t + \mathbf{e}_a \cdot \nabla)^2 \right] = 0. \qquad (2.4.11)$$

Writing the $O(\delta)$ term in the form

$$\partial_t \sum_a (\partial_t + \mathbf{e}_a \cdot \nabla) f_a + \sum_a (\partial_t + \mathbf{e}_a \cdot \nabla) \mathbf{e}_a f_a, \qquad (2.4.12)$$

this term is seen to vanish for any f_a that satisfies the first-order equations (2.4.7) and (2.4.10). Lattice discretization effects thus do not affect the continuity equation (2.4.7), at least to second order.

Corrections do, however, appear at this order in the momentum equation (2.4.10). To second order, Eq. (2.4.8) can be written as

$$\sum_a (\partial_t + \mathbf{e}_a \cdot \nabla) \mathbf{e}_a f_a + \frac{1}{2} \delta \partial_t \sum_a (\partial_t + \mathbf{e}_a \cdot \nabla) \mathbf{e}_a f_a$$
$$+ \frac{1}{2} \delta \sum_a [\mathbf{e}_a \cdot \nabla \partial_t + (\mathbf{e}_a \cdot \nabla)^2] \mathbf{e}_a f_a = 0. \qquad (2.4.13)$$

The second term vanishes if f_a satisfies the first-order equation (2.4.8). The third term, however, contains piece trilinear in the $\mathbf{e}_a$, which gives a correction to the momentum equation (2.4.10).[24]

2.5. CHAPMAN-ENSKOG EXPANSION

If there is local equilibrium, as discussed in Section 2.2., then the microscopic distribution functions $f_a(\mathbf{x}, t)$ should depend, on average, only on the macroscopic parameters $\mathbf{u}(\mathbf{x}, t)$ and $n(\mathbf{x}, t)$ and their derivatives. In general, this dependence may be very complicated. But in hydrodynamic processes, $\mathbf{u}$ and n vary only slowly with position and time. In addition, in the subsonic limit, $|\mathbf{u}| \ll 1$.

With these assumptions, one may approximate the f_a by a series or Chapman-Enskog expansion in the macroscopic variables. To the order required for standard hydrodynamic phenomena, the possible terms are

$$f_a = f \left\{ 1 + c^{(1)} \mathbf{e}_a \cdot \mathbf{u} + c^{(2)} \left[(\mathbf{e}_a \cdot \mathbf{u})^2 - \frac{1}{2} |\mathbf{u}|^2 \right] \right.$$
$$\left. + c_\nabla^{(2)} \left[(\mathbf{e}_a \cdot \nabla)(\mathbf{e}_a \cdot \mathbf{u}) - \frac{1}{2} \nabla \cdot \mathbf{u} \right] + \cdots \right\} \qquad (2.5.1)$$

where the $c^{(i)}$ are undetermined coefficients. The first three terms here represent the change in microscopic particle densities as a consequence of changes in macroscopic fluid velocity; the fourth term accounts for first-order dependence of the particle densities on macroscopic spatial variations in the fluid velocity. The structures of these terms can be deduced merely from the need to form scalar quantities f_a from the vectors $\mathbf{e}_a$, $\mathbf{u}$ and ∇.

The relation

$$\sum_a (\mathbf{e}_a)_i (\mathbf{e}_a)_j = \frac{M}{d} \delta_{ij} \qquad (2.5.2)$$

where here $M = 6$ and $d = 2$, and i and j are space indices, has been used in Eq. (2.5.1) to choose the forms of the $|u|^2$ and ∇u terms so as to satisfy the constraints (2.4.1) and (2.4.2), independent of the values of the coefficients $c^{(2)}$ and $c_\nabla^{(2)}$. In terms of (2.5.1), Eq. (2.4.1) yield immediately

$$f = n/6 \tag{2.5.3}$$

while (2.4.2) gives

$$c^{(1)} = 2 . \tag{2.5.4}$$

The specific values of $c^{(2)}$ and $c_\nabla^{(2)}$ can be determined only by explicit solution of the kinetic equation (2.3.5) including collision terms. (Some approximate results for these coefficients based on the Boltzmann transport equation will be given in Section 4.) Nevertheless, the structure of macroscopic equations can be derived from (2.5.1) without knowledge of the exact values of these parameters.

For a uniform equilibrium system with $\mathbf{u} = 0$, all the f_a are given by

$$f_a = f = n/6 . \tag{2.5.5}$$

In the case, the momentum flux tensor (2.4.9) is equal to the pressure tensor, given, as in the standard kinetic theory of gases, by

$$P_{ij} = \sum_a (e_a)_i (e_a)_j f = \frac{1}{2} n \delta_{ij} \tag{2.5.6}$$

where the second equality follows from Eq. (2.5.2). Note that this form is spatially isotropic, despite the underlying anisotropy of the cellular automaton lattice. This result can be deduced from general symmetry considerations, as discussed in Section 3. Equation (2.5.6) gives the equation of state relating the scalar pressure to the number density of the cellular automaton fluid:

$$p = n/2 . \tag{2.5.7}$$

When $\mathbf{u} \neq 0$, Π_{ij} can be evaluated in the approximation (2.5.1) using the relations

$$\sum_a (e_a)_i (e_a)_j (e_a)_k = 0 \tag{2.5.8}$$

and

$$\sum_a (e_a)_i (e_a)_j (e_a)_k (e_a)_l = \frac{M}{d(d+2)} (\delta_{ij}\delta_{kl} + \delta_{ik}\delta_{jl} + \delta_{il}\delta_{jk}) . \tag{2.5.9}$$

The result is

$$\Pi_{ij} = \frac{n}{2}\delta_{ij} + \frac{n}{4}c^{(2)}\left[u_i u_j - \frac{1}{2}|\mathbf{u}|^2\delta_{ij}\right] + \frac{n}{4}c_\nabla^{(2)}\left[\partial_i u_j - \frac{1}{2}\nabla \cdot \mathbf{u}\right] . \tag{2.5.10}$$

Substituting the result into Eq. (2.4.10), one obtains the final macroscopic equation

$$\partial_t(n\mathbf{u}) + \frac{1}{4}nc^{(2)}\left\{(\mathbf{u}\cdot\nabla)\mathbf{u} + \left[\mathbf{u}(\nabla\cdot\mathbf{u}) - \frac{1}{2}|\mathbf{u}|^2\right]\right\}$$
$$= -\frac{1}{2}\nabla n - \frac{1}{8}nc_\nabla^{(2)}\nabla^2\mathbf{u} - \frac{1}{4}\Xi \qquad (2.5.11)$$

where

$$\Xi = \mathbf{u}(\mathbf{u}\cdot\nabla)\left(nc^{(2)}\right) - \frac{1}{2}|\mathbf{u}|^2\nabla\left(nc^{(2)}\right) + (\mathbf{u}\cdot\nabla)\left(nc_\nabla^{(2)}\right) - \frac{1}{2}(\nabla\cdot\mathbf{u})\nabla\left(nc_\nabla^{(2)}\right). \quad (2.5.12)$$

The form (2.5.10) for Π_{ij} follows exactly from the Chapman-Enskog expansion (2.5.1). But to obtain Eq. (2.5.11), one must use the momentum equation (2.4.10). Equation (2.4.13) gives corrections to this equations that arise at second order in the lattice size parameter δ. These corrections must be compared with other effects included in Eq. (2.5.11). The rescaling $\mathbf{x} = \delta_x\mathbf{X}$ implies that spatial gradient terms in the Chapman-Enskog expansion can be of the same order as the $O(\delta_x)$ correction terms in Eq. (2.4.13). When the $\mathbf{e}_a \cdot \mathbf{u}$ term in the Chapman-Enskog expansion (2.5.1) for the f_a is substitute into the last term of Eq. (2.4.13), it gives a contribution[24]

$$\Psi = -\frac{1}{16}nc^{(1)}\nabla^2\mathbf{u} = -\frac{1}{8}n\nabla^2\mathbf{u} \qquad (2.5.13)$$

to the right-hand side of Eq. (2.5.11). Note that Ψ depends solely on the choice of $\mathbf{e}_a$, and must, for example, vary purely linearly with the particle density f.

2.6. NAVIER-STOKES EQUATION

The standard Navier-Stokes equation for a continuum fluid in d dimensions can be written in the form

$$\partial_t(n\mathbf{u}) + \mu n(\mathbf{u}\cdot\nabla)\mathbf{u} = -\nabla p + \eta\nabla^2\mathbf{u} + \left(\zeta + \frac{1}{d}\eta\right)\nabla(\nabla\cdot\mathbf{u}) \qquad (2.6.1)$$

where p is pressure, and η and ζ are, respectively, shear and bulk viscosities (e.g., Ref. 27). The coefficient μ of the convective term is usually constrained to have value 1 by Galilean invariance. Note that the coefficient of the last term in Eq. (2.6.1) is determined by the requirement that the term in Π_{ij} proportional to η be traceless.[27,57]

The macroscopic equation (2.5.11) for the cellular automaton fluid is close to the Navier-Stokes form (2.6.1). The convective and viscous terms are present, and have the usual structure. The pressure term appears according to the equation of state (2.5.7). There are, however, a few additional terms.

Terms proportional to $\mathbf{u}\nabla n$ must be discounted, since they depend on features of the microscopic distribution functions beyond those included in the Chapman-Enskog expansion (2.5.1). The continuity equation (2.4.7) shows that terms proportional to $\mathbf{u}(\nabla\cdot\mathbf{u})$ must also be neglected.

The term proportional to $\nabla|\mathbf{u}|^2$ remains, but can be combined with the ∇n term to yield an effective pressure term which includes fluid kinetic energy contributions.

The form of the viscous terms in (2.5.11) implies that for the cellular automaton fluid, considered here, the bulk viscosity is given by

$$\zeta = 0. \tag{2.6.2}$$

The value of η is determined by the coefficient $c_\nabla^{(2)}$ that appears in the microscopic distribution function (2.5.1), according to

$$\eta = n\nu = -\frac{1}{8}nc_\nabla^{(2)} \tag{2.6.3}$$

where ν is the kinematic viscosity. An approximate method of evaluating $c_\nabla^{(2)}$ is discussed in Section 4.6.

The convective term in Eq. (2.5.11) has the same structure as in the Navier-Stokes equation (2.6.1), but includes a coefficient

$$\mu = \frac{1}{4}c^{(2)} \tag{2.6.4}$$

which is not in general equal to 1. In continuum fluids, the covariant derivative usually has the form $D_t = \partial_t + \mathbf{u} \cdot \nabla$ implied by Galilean invariance. The cellular automaton fluid acts, however, as a mixture of components, each with velocities $\mathbf{e}_a$, and these components can contribute with different weights to the covariant derivatives of different quantities, leading to convective terms with different coefficients.

The usual coefficient of the convective term can be recovered in Eq. (2.6.1) and thus Eq. (2.5.11) by a simple rescaling in velocity: setting

$$\tilde{\mathbf{u}} = \mu\mathbf{u} \tag{2.6.5}$$

the equation for $\tilde{\mathbf{u}}$ has coefficient 1 for the $(\tilde{\mathbf{u}} \cdot \nabla)\tilde{\mathbf{u}}$ term.

Small perturbations from a uniform state may be represented by a linearized approximation to Eqs. (2.4.7) and (2.5.11), which has the standard sound wave equation form, with a sound speed obtained from the equation of state (2.5.7) as

$$c = 1/\sqrt{2}. \tag{2.6.6}$$

The form of the Navier-Stokes equation (2.6.1) is usually obtained by simple physical arguments. Detailed derivations suggest, however, that more elaborate equations may be necessary, particularly in two dimensions (e.g., Ref. 28). The Boltzmann approximation used in Section 4 yields definite values for $c^{(2)}$ and $c_\nabla^{(2)}$. Correlation function methods indicate, however, that additional effects yield logarithmically divergent contributions to $c_\nabla^{(2)}$ in two dimensions (e.g., Ref. 29). The full viscous term in this case may in fact be of the rough form $\nabla^2 \log(\nabla^2)\mathbf{u}$.

2.7. HIGHER-ORDER CORRECTIONS

The derivation of the Navier-Stokes form (2.5.11) neglects all terms in the Chapman-Enskog expansion beyond those given explicitly in Eq. (2.5.1). This approximation is expected to be adequate only when $|\mathbf{u}| \ll c$. Higher-order corrections may be particularly significant for supersonic flows involving shocks (e.g., Ref. 30).

Since the dynamics of shocks are largely determined just by conservation laws (e.g., Ref. 27), they are expected to be closely analogous in cellular automaton fluids and in standard continuum fluids. For $|\mathbf{u}|/c \gtrsim 2$, however, shocks become so strong and thin that continuum descriptions of physical fluids can no longer be applied in detail (e.g., Ref. 14). The structure of shocks in such cases can apparently be found only through consideration of explicit particle dynamics.[11,14]

In the transonic flow regime $|\mathbf{u}| \approx c$, however, continuum equations may be used, but corrections to the Navier-Stokes form may be significant. A class of such corrections can potentially be found by maintaining terms $O(u^3)$ and higher in the Chapman-Enskog expansion (2.5.1).

In the homogeneous fluid approximation $\nabla \mathbf{u} = 0$, one may take

$$
\begin{aligned}
f_a = f\Big\{ & 1 + c^{(1)}\mathbf{e}_a \cdot \mathbf{u} + c^{(2)}\big[(\mathbf{e}_a \cdot \mathbf{u})^2 + \sigma_2|\mathbf{u}|^2\big] \\
& + c^{(3)}\big[(\mathbf{e}_a \cdot \mathbf{u})^3 + \sigma_3|\mathbf{u}|^2(\mathbf{e}_a \cdot \mathbf{u})\big] \\
& + c^{(4)}\big[(\mathbf{e}_a \cdot \mathbf{u})^4 + \sigma_{4,1}|\mathbf{u}|^2(\mathbf{e}_a \cdot \mathbf{u})^2 + \sigma_{4,2}|\mathbf{u}|^4\big] + \cdots \Big\}
\end{aligned}
\tag{2.7.1}
$$

The constraints (2.4.1) and (2.4.2) imply

$$
c^{(1)} = d \tag{2.7.2}
$$

$$
\sigma_2 = -\frac{1}{d} \tag{2.7.3}
$$

$$
\sigma_3 = -\frac{3}{d+2} \tag{2.7.4}
$$

$$
\frac{3}{d(d+2)} + \frac{1}{d}\sigma_{4,1} + \sigma_{4,2} = 0 \tag{2.7.5}
$$

where d is the space dimension, equal to two for the model of this section.

Corrections to (2.5.11) can be found by substituting (2.7.1) in the kinetic equation (2.4.8). For the hexagonal lattice model, one obtains, for example,

$$
\begin{aligned}
& \partial_t(nu_x) + \frac{1}{4}nc^{(2)}(u_x\partial_x u_x + u_x\partial_y u_y + u_y\partial_y u_x - u_y\partial_x u_y) \\
& + \frac{1}{8}nc^{(4)}\Big\{ \big[(5 + 4\sigma_{4,1})u_x^3 - 3u_x u_y^2\big]\partial_x u_x \\
& \quad + \big[(3 + 2\sigma_{4,1})u_y^3 + (3 + 6\sigma_{4,1})u_x^2 u_y\big]\partial_y u_x \\
& \quad - \big[(3 + 4\sigma_{4,1})u_y^3 + 3u_x^2 u_y\big]\partial_x u_y \\
& \quad + \big[(1 + 2\sigma_{4,1})u_x^3 + (9 + 6\sigma_{4,1})u_x u_y^2\big]\partial_y u_y \Big\} = 0.
\end{aligned}
\tag{2.7.6}
$$

The $O(u^2)$ term in Eq. (2.7.6) has the isotropic form given in Eq. (2.5.11). The $O(u^4)$ term is, however, anisotropic.

To obtain an isotropic $O(u^4)$ term, one must generalize the model, as discussed in Section 3. One possibility is to allow vectors e_a corresponding to corners of an M-sided polygon with $M > 6$. In this case, the continuum equation deduced from the Chapman-Enskog expansion (2.7.1) becomes

$$\partial_t(n\mathbf{u}) + \frac{1}{4}nc^{(2)}\left[(\mathbf{u}\cdot\nabla)\mathbf{u} + \mathbf{u}(\nabla\cdot\mathbf{u}) - \frac{1}{2}\nabla|\mathbf{u}|^2\right]$$
$$+ \frac{1}{4}nc^{(4)}(1+\sigma_{4,1})\left\{|\mathbf{u}|^2[(\mathbf{u}\cdot\nabla)\mathbf{u} + \mathbf{u}(\nabla\cdot\mathbf{u}) - \nabla|\mathbf{u}|^2] + \mathbf{u}(\mathbf{u}\cdot\nabla)|\mathbf{u}|^2\right\} = 0.$$

$$(2.7.7)$$

This gives a definite form for the next-order corrections to the convective part of the Navier-Stokes equation.

Corrections to the viscous part can be found by including terms proportional to $\nabla\mathbf{u}$ in the Chapman-Enskog expansion (2.7.1). The possible fourth-order terms are given by contractions of $u_i u_j \delta_k u_l$ with products of $(e_a)_m$ or δ_{mn}. They yield a piece in the Chapman-Enskog expansion of the form

$$c_\nabla^{(4)}\big[\tau_1(\mathbf{e}_a\cdot\mathbf{u})^2(\mathbf{e}_a\cdot\nabla)(\mathbf{e}_a\cdot\mathbf{u}) + \tau_2|\mathbf{u}|^2(\mathbf{e}_a\cdot\nabla)(\mathbf{e}_a\cdot\mathbf{u})$$
$$+ \tau_3(\mathbf{e}_a\cdot\mathbf{u})(\mathbf{u}\cdot\nabla)(\mathbf{e}_a\cdot\mathbf{u}) + \tau_4(\mathbf{e}_a\cdot\mathbf{u})^2(\nabla\cdot\mathbf{u}) + \tau_5|\mathbf{u}|^2(\nabla\cdot\mathbf{u})\big]$$

$$(2.7.8)$$

where Eq. (2.4.1) implies the constraints (for $d = 2$)

$$\tau_1 + 2\tau_3 = 0, \qquad (2.7.9)$$

$$\tau_1 + 4\tau_2 + 4\tau_4 + 8\tau_5 = 0. \qquad (2.7.10)$$

The resulting continuum equations may be written in terms of vectors formed by contractions of $u_i u_j \partial_k \partial_l u_m$ and $u_i \partial_j u_k \partial_l u_m$. The complete result is

$$\partial_t(n\mathbf{u}) + \frac{1}{4}nc^{(2)}\left[(\mathbf{u}\cdot\nabla)\mathbf{u} + \mathbf{u}(\nabla\cdot\mathbf{u}) - \frac{1}{2}\nabla|\mathbf{u}|^2\right]$$

$$+ \frac{1}{4}nc^{(4)}(1+\sigma_{4,1})\left\{|\mathbf{u}|^2\left[(\mathbf{u}\cdot\nabla)\mathbf{u} + \mathbf{u}(\nabla\cdot\mathbf{u}) - \nabla|\mathbf{u}|^2\right] + \mathbf{u}(\mathbf{u}\cdot\nabla)|\mathbf{u}|^2\right\}$$

$$= \frac{1}{8}nc_\nabla^{(2)}\nabla^2\mathbf{u}$$

$$- \frac{1}{32}nc_\nabla^{(4)}\Bigg[\left((\tau_1 - 4\tau_2 + 12\tau_4)\mathbf{u}(\nabla\cdot\mathbf{u})^2 - (\tau_1 - 4\tau_2 + 4\tau_4)\right.$$

$$\times \mathbf{u}\Big\{\nabla[(\mathbf{u}\cdot\nabla)\mathbf{u}] - (\mathbf{u}\cdot\nabla)(\nabla\cdot\mathbf{u})\Big\} + 8\tau_4\Big\{[(\mathbf{u}\cdot\nabla)\mathbf{u}]\cdot\nabla\Big\}\mathbf{u}$$

$$+ \frac{1}{2}(\tau_1 + 4\tau_2)[(\nabla|\mathbf{u}|^2)\cdot\nabla]\mathbf{u}$$

$$+ 2\tau_1\mathbf{u}\left[\frac{1}{2}\nabla\cdot(\nabla|\mathbf{u}|^2) - \mathbf{u}\cdot(\nabla^2\mathbf{u})\right] - 4\tau_4(\nabla\cdot\mathbf{u})\nabla|\mathbf{u}|^2$$

$$+ \left\{8\tau_4\left[\mathbf{u}(\mathbf{u}\cdot\nabla)(\nabla\cdot\mathbf{u}) - \frac{1}{2}|\mathbf{u}|^2\nabla(\nabla\cdot\mathbf{u})\right]\right.$$

$$+ 2\tau_1\mathbf{u}[\mathbf{u}\cdot(\nabla^2\mathbf{u})] + 4\tau_2|\mathbf{u}|^2\nabla^2\mathbf{u}\Bigg\}\Bigg]$$

$$(2.7.11)$$

where, on the right-hand side, the first group of terms are all $O((\nabla\mathbf{u})^2)$, while the second group are $O(\nabla\nabla\mathbf{u})$. Further corrections involve higher derivative terms, such as $u_i\partial_j\partial_k\partial_l u_m$. For a channel flow with $u_x = ax^2$, $u_y = 0$, the time time-independent terms in Eq. (2.7.11) have an x component

$$\frac{1}{4}ac_\nabla^{(2)} + \frac{5}{8}a^3x^4c_\nabla^{(4)}(\tau_1 + 2\tau_2 + 2\tau_4) + \frac{1}{2}a_2x^3c^{(2)} + a^4x^7x^{(4)}(1+\sigma_{4,1}) \quad (2.7.12)$$

and zero y component.

3. SYMMETRY CONSIDERATIONS

3.1. TENSOR STRUCTURE

The form of the macroscopic equations (2.4.7) and (2.5.11) depends on few specific properties of the hexagonal lattice cellular automaton model. The most important properties relate to the symmetries of the tensors

$$\mathbf{E}^{(n)}_{i_1 i_2 \cdots i_n} = \sum_a (e_a)_{i_1} \cdots (e_a)_{i_n} . \tag{3.1.1}$$

These tensors are determined in any cellular automaton fluid model simply from the choice of the basic particle directions e_a. The momentum flux tensor (2.4.9) is given in terms of them by

$$\Pi_{ij} = f\Bigg(\mathbf{E}^{(2)}_{ij} + c^{(1)} \mathbf{E}^{(3)}_{ijk} u_k + c^{(2)} \Big[\mathbf{E}^{(4)}_{ijkl} u_k u_l + \sigma \mathbf{E}^{(2)}_{ij} u_k u_k \Big]$$

$$c^{(2)}_\nabla \Big[\mathbf{E}^{(4)}_{ijkl} \partial_k u_l + \sigma \mathbf{E}^{(2)}_{ij} \partial_k u_k \Big] \Bigg) \tag{3.1.2}$$

where repeated indices are summed, and to satisfy the conditions (2.4.1) and (2.4.2)

$$\sigma = -\mathbf{E}^{(4)}_{ijkk} / \mathbf{E}^{(2)}_{ij} . \tag{3.1.3}$$

The basic condition for standard hydrodynamic behavior is that the tensors $\mathbf{E}^{(n)}$ for $n \leq 4$ which appear in (3.1.2) should be isotropic. From the definition (3.1.1), the tensors must always be invariant under the discrete symmetry group of the underlying cellular automaton array. What is needed is that they should in addition be invariant under the full continuous rotation group.

The definition (3.1.1) implies that the $\mathbf{E}^{(n)}$ must be totally symmetric in their space indices. With no further conditions, the $\mathbf{E}^{(n)}$ could have $\binom{n+d-1}{n}$ independent components in d space dimensions. Symmetries in the underlying cellular automaton array provide constraints which can reduce the number of independent components.

Tensors that are invariant under all rotations and reflections (or inversions) can have only one independent component. Such invariance is obtained with a continuous set of vectors e_a uniformly distributed on the unit sphere. Invariance up to finite n can also be obtained with certain finite sets of vectors e_a.

Isotropic tensors $\mathbf{E}^{(n)}$ obtained with sets of M vectors e_a in d space dimensions must take the form

$$\mathbf{E}^{(2n+1)} = 0 \tag{3.1.4}$$

$$\mathbf{E}^{(2n)} = \frac{M}{d(d+2) \cdots (d+2n-2)} \Delta^{(2n)} \tag{3.1.5}$$

where

$$\Delta_{ij}^{(2)} = \delta_{ij} \tag{3.1.6}$$

$$\Delta_{ijkl}^{(4)} = \delta_{ij}\delta_{kl} + \delta_{ik}\delta_{jl} + \delta_{il}\delta_{jk} \tag{3.1.7}$$

and in general $\Delta^{(2n)}$ consists of a sum of all the $(2n-1)!!$ possible products of Kronecker delta symbols of pairs on indices, given by the recursion relation

$$\Delta_{i_1 i_2 \cdots i_{2n}}^{(2n)} = \sum_{j=2}^{2n} \delta_{i_1 i_j} \Delta_{i_2 \cdots i_{j-1} i_{j+1} \cdots i_{2n}}^{(2n-2)}. \tag{3.1.8}$$

The form of the $\Delta^{(2n)}$ can also be specified by giving their upper simplicial components (whose indices form a nonincreasing sequence). Thus, in two dimensions,

$$\Delta^{(4)} = [3, 0, 1, 0, 3] \tag{3.1.9}$$

where the 1111, 2111, 2211, 2221, and 2222 components are given. In three dimensions,

$$\Delta^{(4)} = [3, 0, 1, 0, 3, 0, 0, 0, 0, 1, 0, 1, 0, 0, 3]. \tag{3.1.10}$$

Similarly,

$$\Delta^{(6)} = [5, 0, 1, 0, 1, 0, 5] \tag{3.1.11}$$

and

$$\Delta^{(6)} = [15, 0, 3, 0, 3, 0, 15, 0, 0, 0, 0, 0, 0, 3, 0, 1, 0, 3, 0, 0, 0, 0, 3, 0, 3, 0, 0, 15] \tag{3.1.12}$$

in two and three dimensions, respectively.

For isotropic sets of vectors e_a, one finds from (3.1.5)

$$\frac{1}{M}\sum_a (e_a \cdot v)^{(2n)} = Q_{2n}|v|^{2n} = \frac{(2n-1)!!}{d(d+2)\cdots(d+2n-2)}|v|^{2n} \tag{3.1.13}$$

so that for $d = 2$

$$Q_2 = \frac{1}{2}, \qquad Q_4 = \frac{3}{8}, \qquad Q_6 = \frac{5}{16}, \qquad Q_8 = \frac{35}{128} \tag{3.1.14}$$

while for $d = 3$

$$Q_{2n} = \frac{1}{2n+1}. \tag{3.1.15}$$

Similarly,

$$\frac{1}{M}\sum_a (e_a \cdot v)^{2n} e_a \cdot v = Q_{2n}|v|^{2n} v. \tag{3.1.16}$$

In the model of Section 2, all the particle velocities e_a are fundamentally equivalent, and so are added with equal weight in the tensor (3.1.1). In some cellular automaton fluid models, however, one may, for example, allow particle velocities e_a with unequal magnitudes (e.g., Ref. 31). The relevant tensors in such cases are

$$E_{i_1 i_2 \cdots i_n}^{(n)} = \sum_a w(|e_a|^2)(e_a)_{i_1}\cdots(e_a)_{i_n} \tag{3.1.17}$$

where the weights $w(|e_a|^2)$ are typically determined from coefficients in the Chapman-Enskog expansion.

3.2. POLYGONS

As a first example, consider a set of unit vectors e_a corresponding to the vertices of a regular M-sided polygon:

$$e_a = \left(\cos \frac{2\pi a}{M}, \sin \frac{2\pi a}{M} \right) \tag{3.2.1}$$

For sufficiently large M, any tensor $\mathbf{E}^{(n)}$ constructed from these e_a must be isotropic. Table 1 gives the conditions on M necessary to obtain isotropic $\mathbf{E}^{(n)}$. In general, it can be shown that $\mathbf{E}^{(n)}$ is isotropic if and only if M does not divide any integers $n, n-2, n-4, \ldots$.[32] Thus, for example, $\mathbf{E}^{(n)}$ must be isotropic whenever $n > M$.

In the case $M = 6$, corresponding to the hexagonal lattice considered in Section 2, the $\mathbf{E}^{(n)}$ are isotropic up to $n = 5$. The macroscopic equations obtained in this case thus have the usual hydrodynamic form. However, a square lattice, with $M = 4$, yields an anisotropic $\mathbf{E}^{(4)}$, given by

$$\mathbf{E}^{(4)}|_{M=4} \; 2\delta^{(4)} \tag{3.2.2}$$

where $\delta^{(n)}$ is the Kronecker delta symbol with n indices. The macroscopic equation obtained in this case is

$$\delta_t(nu_x) + \frac{1}{2}nc^{(2)}(u_x\partial_x u_x - u_y\partial_x u_y)$$

$$= -\frac{1}{2}\partial_x n - \frac{1}{8}nc_\triangledown^{(2)}(\partial_{xx}u_x - \partial_{xy}u_y) - \frac{1}{4}(u_x^2 - u_y^2)\partial_x\left(nc^{(2)}\right)$$

$$- \frac{1}{8}(\partial_x u_x - \partial_y u_y)\partial_x\left(nc_\triangledown^{(2)}\right)$$

TABLE 1 Conditions for the Tensors $\mathbf{E}^{(n)}$ of Eq. (3.1.1) to be Isotropic with the Lattice Vectors e_a chosen to Correspond to the Vertices of Regular M-Sided Polygons

$\mathbf{E}^{(2)}$	$M > 2$
$\mathbf{E}^{(3)}$	$M \geq 2, M \neq 3$
$\mathbf{E}^{(4)}$	$M > 2, M \neq 4$
$\mathbf{E}^{(5)}$	$M \geq 2, M \neq 3, 5$
$\mathbf{E}^{(6)}$	$M > 4, M \neq 6$
$\mathbf{E}^{(7)}$	$M \geq 2, M \neq 3, 5, 7$

TABLE 2 Isotropy of the Tensors $\mathbf{E}^{(n)}$ with e_a Chosen as the M Vertices of Regular Polyhedra[1]

	e_a	M	$\mathbf{E}^{(2)}$	$\mathbf{E}^{(3)}$	$\mathbf{E}^{(4)}$	$\mathbf{E}^{(5)}$	$\mathbf{E}^{(6)}$
Tetrahedron	$(1,1,1)$, cyc: $(1,-1,-1)$	4	Y	N	N	N	N
Cube	$(\pm 1, \pm 1, \pm 1)$	8	Y	Y	N	Y	N
Octahedron	cyc: $(\pm 1, 0, 0)$	6	Y	Y	N	Y	N
Dodecahedron	$(\pm 1, \pm 1, \pm 1)$, cyc: $(0, \pm \phi^{-1}, \pm \phi)$	20	Y	Y	Y	Y	N
Icosahedron	cyc: $(0, \pm \phi, \pm 1)$	12	Y	Y	Y	Y	N

[1] In the forms for e_a (which are given without normalization), the notation "cyc:" indicates all cyclic permutations. (All possible combinations of signs are chosen in all cases.) ϕ is the golden ratio $(1 + \sqrt{5})/2 \approx 1.618$.

which does not have the standard Navier-Stokes forms.[6][3]

On a hexagonal lattice, $\mathbf{E}^{(4)}$ is isotropic, but $\mathbf{E}^{(6)}$ has the component form

$$\mathbf{E}^{(6)} \,|\, _{M=6} = \frac{1}{16}[33, 0, 3, 0, 9, 0, 27] \tag{3.2.4}$$

which differs from the isotropic result (3.1.11). The corrections (2.7.6) to the Navier-Stokes equation are therefore anisotropic in this case.

3.3. POLYHEDRA

As three-dimensional examples, one can consider vectors e_a corresponding to the vertices of regular polyhedra. Only for the five Platonic solids are all the $|e_a|^2$ equal. Table 2 gives results for the isotropy of the $\mathbf{E}^{(n)}$ in these cases. Only for the icosahedron and dodecahedron is $\mathbf{E}^{(4)}$ found to be isotropic, so that the usual hydrodynamic equations are obtained. As in two dimensions, the $\mathbf{E}^{(2n)}$ for the cube are all proportional to a single Kronecker delta symbol over all indices.

In five and higher dimensions, the only regular polytopes are the simplex, and the hypercube and its dual.[34] These give isotropic $\mathbf{E}^{(n)}$ only for $n < 3$, and for $n < 4$ and $n < 4$, respectively.

In four dimensions, there are three additional regular polytopes,[34] specified by Schläfi symbols $\{3, 4, 3\}$, $\{3, 3, 5\}$, and $\{5, 3, 3\}$. (The elements of these lists

[3]Note that even the linearized equation for sound waves is anisotropic on a square lattice. The waves propagate isotropically, but are damped with an effective viscosity that varies with direction, and can be negative.[33]

give the number of edges around each vertex, face, and 3-cell, respectively.) The $\{3,4,3\}$ polytope has 24 vertices with coordinates corresponding to permutations of $(\pm 1, \pm 1, 0, 0)$. It yields $\mathbf{E}^{(n)}$ that are isotropic up to $n = 4$. The $\{3,3,5\}$ polytope has 120 vertices corresponding to $(\pm 1, \pm 1, \pm 1, \pm 1)$, all permutations of $(\pm 2, 0, 0, 0)$, and even-signature permutations of $(\pm\phi, \pm 1, \phi^{-1}, 0)$, where $\phi = (1 + \sqrt{5})/2$. The $\{5,3,3\}$ polytope is the dual of $\{3,3,5\}$. Both yield $\mathbf{E}^{(n)}$ that are isotropic up to $n = 8$.

3.4. GROUP THEORY

The structure of the $\mathbf{E}^{(n)}$ was found above by explicit calculations based on particular choices for the e_a. The general form of the results is, however, determined solely by the symmetries of the set of e_a. A finite group $\mathbf{G}$ of transformations leaves the e_a invariant. (For the hexagonal lattice model of Section 2, it is the hexagonal group S_6.) In general $\mathbf{G}$ is a finite subgroup of the d-dimensional rotation group $O(d)$.

The e_a form the basis for a representation of $\mathbf{G}$, as do their products $\mathbf{E}^{(n)}$. If the representation $\mathbf{R}^{(n)}$ carried by the $\mathbf{E}^{(n)}$ is irreducible, then the $\mathbf{E}^{(n)}$ can have only one independent component, and must be rotationally invariant. But $\mathbf{R}^{(n)}$ is in general reducible. The number of irreducible representations that it contains gives the number of independent components of $\mathbf{E}^{(n)}$ allowed by invariance under $\mathbf{G}$.

This number can be found using the method of characters (e.g., Refs. 35 and 36). Each class of elements of $\mathbf{G}$ in a particular representation $\mathbf{R}$ has a character that receives a fixed contribution from each irreducible component of $\mathbf{R}$. Characters for the representation $\mathbf{R}^{(n)}$ of $\mathbf{G}$ can be found by first evaluating them for arbitrary rotations, and then specializing to the particular sets of rotations (typically through angles of the form $2\pi/k$) that appear in $\mathbf{G}$. To find characters for arbitrary rotations, one writes the $\mathbf{E}^{(n)}$ as sums of completely traceless tensors $\mathbf{U}^{(n)}$ which form irreducible representations of $O(d)$ (e.g., Ref. 37):

$$\mathbf{E}^{(n)} = \mathbf{U}^{(n)} + \mathbf{U}^{(n-2)} + \cdots \mathbf{U}^{(0)} . \tag{3.4.1}$$

The characters of the $\mathbf{E}^{(n)}$ are then sums of the characters $\chi^{(m)}$ for the irreducible tensors $\mathbf{U}^{(m)}$. For proper rotations through an angle ϕ, the $\chi^{(m)}$ are given by (e.g., Ref. 37)

$$\chi^{(m)}(\phi) = e^{2\pi i m\phi} \qquad (d = 2)$$

$$\chi^{(m)}(\phi) = \frac{\sin\left[(2m+1)\phi/2\right]}{\sin\phi/2} \qquad (d = 3) . \tag{3.4.2}$$

The resulting characters for the representations $\mathbf{R}^{(n)}$ formed by the $\mathbf{E}^{(n)}$ are given in Table 3.

The number of irreducible representations in $\mathbf{R}^{(n)}$ can be found as usual by evaluating the characters for each class in $\mathbf{R}^{(n)}$ (e.g., Ref. 35). Consider as an

TABLE 3 Characters of Transformations of Totally Symmetric Rank n Tensors $\mathbf{E}^{(n)}$ in d Dimensions[1]

Dimension	Rank	Character
2	2	$4c^2 - 1$
2	4	$(4c^2 + 2c - 1)(4c^2 - 2c - 1)$
3	2	$4c^2 + 2c$
3	4	$(2c + 1)(2c - 1)(4c^2 + 2c - 1)$

[1] $c = \cos(\phi)$, where ϕ is the rotation angle. For improper rotations in three dimensions $\pi - \phi$ must be used.

example the case of $\mathbf{R}^{(4)}$ with $\mathbf{G}$ the octahedral group $\mathbf{O}$. This group has classes E, $8C_3$, $9C_2$, $6C_4$, where E represents the identity, and C_k represents a proper rotation by $\phi = 2\pi/k$ about a k-fold symmetry axis. The characters for these classes in the representation $\mathbf{R}^{(4)}$ can be found from Table 3. Adding the results, and dividing by the total number of classes in $\mathbf{G}$, one finds that $\mathbf{R}^{(4)}$ contains exactly two irreducible representations of $\mathbf{O}$. Rank 4 symmetric tensors can thus have up to two independent components while still being invariant under the octahedral group.[38]

In general, one may consider sets of vectors $\mathbf{e}_a$ that are invariant under any point symmetry group. Typically, the larger the group is, the smaller the number of independent components in the $\mathbf{E}^{(n)}$ can be. In two dimensions, there are an infinite number of point groups, corresponding to transformations of regular polygons. There are only a finite of nontrivial additional point groups in three dimensions. The largest is the group $\mathbf{Y}$ of symmetries of the icosahedron (or dodecahedron). Second largest is the cubic group $\mathbf{E}$. As seen in Table 2, only $\mathbf{Y}$ guarantees isotropy of all tensors $\mathbf{E}^{(n)}$ up to $n = 4$ (compare Ref. 39).

It should be noted, however, that such group-theoretic considerations can only give upper bounds on the number of independent components in the $\mathbf{E}^{(n)}$. The actual number of independent components depends on the particular choice of the $\mathbf{e}_a$, and potentially on the values of weight such as those in Eq. (3.1.16).

3.5. REGULAR LATTICES

If the vectors $\mathbf{e}_a$ correspond to particle velocities, the possible displacements of particles at each time step must be of the form $\sum_a k_a \mathbf{e}_a$. In discrete velocity gases, particle positions are not constrained. But in a cellular automaton model, they are usually taken to correspond to the sites of a regular lattice.

Only a finite number of such "crystallographic" lattices can be constructed in any space dimension (e.g., Refs. 40 and 41). As a result, the point symmetry groups that can occur are highly constrained. In two dimensions, the most symmetrical lattices are square and hexagonal ones. In three dimensions, the most symmetrical are hexagonal and cubic. The group-theoretic arguments of Section 3.4 suffice to show that in two dimensions, hexagonal lattices must give tensors $E^{(n)}$ that are isotropic up to $n = 4$, and so yield standard hydrodynamic equations (2.5.11). In three dimensions, group-theoretic arguments alone fail to establish the isotropy of $E^{(4)}$ for hexagonal and cubic lattices. A system with icosahedral point symmetry would be guaranteed to yield an isotropic $E^{(4)}$, but since it is not possible to tesselate three-dimensional space with regular icosahedra, no regular lattice with such a large point symmetry group can exist.

Crystallographic lattices are classified not only by point symmetries, but also by the spatial arrangement of their sites. The lattices consist of "unit cells" containing a definite arrangement of sites, which can be repeated to form a regular tesselation. In two dimensions, five distinct such Bravais lattice structures exists; in three dimensions, there are 14 (e.g., Refs. 40 and 41).

Sites in these lattices can correspond directly to the sites in a cellular automaton. The links which carry particles in cellular automaton fluid models are obtained by joining pairs of sites, usually in a regular arrangement. The link vectors give the velocities e_a of the particles.

In the simplest cases, the links join each site to its nearest neighbors. The regularity of the lattice implies that in such cases, all the e_a are of equal length, so that all particles have the same speed.

TABLE 4 Forms of the Tensors $E^{(n)}$ for the Most Symmetrical Three-Dimensional Bravais Lattices[1]

Cubic	e_a	M	$E^{(2)}$	$E^{(4)}$	$E^{(6)}$
Primitive	cyc: $(\pm1,0,0)$	6	$2\delta^{(2)}$	$2\delta^{(4)}$	$2\delta^{(6)}$
Body-centered	$(\pm1,\pm1,\pm1)$	8	$8\delta^{(2)}$	$8(\Delta^{(4)} - 2\delta^{(4)})$	$8(\Delta^{(6)} - 2\Delta^{(4,2)} + 15\delta^{(6)})$
Face-centered	cyc: $(\pm1,\pm1,0)$	12	$8\delta^{(2)}$	$4(\Delta^{(4)} - \delta^{(4)})$	$4(\Delta^{(4,2)} - 13\delta^{(6)})$

[1] The basic vectors e_a (used here without normalization) are taken to join each site with its M nearest neighbors. $\delta^{(n)}$ represents the Kronecker delta symbol of n indices; $\Delta^{(n)}$ represents the rotationally tensor defined in Eqs. (3.1.6)–(3.1.8). $\Delta^{(n,m)}$ is the sum of all possible products of pairs of Kronecker delta symbols with n and m indices, respectively.

TABLE 5 Sequence of Simple Lie Groups Whose Sets of Root Vectors Yield Optimal Lattices for Sphere Packing in d Dimensions[1]

d	Group		M	n_{max}
1	A_1	$SU(2)$	2	
2	A_2	$SU(3)$	6	4
3	A_3	$SU(4)$	12	2
4	D_4	$SO(8)$	24	4
5	D_5	$SO(10)$	40	2
6	E_6		72	0

[1] These lattices may also yield maximal isotropy for the tensors $\mathbf{E}^{(n)}$. Results are given for the maximum even n at which the $\mathbf{E}^{(n)}$ are found to be isotropic. The root vectors are given in Ref. 45.

For two-dimensional square and hexagonal lattices, the e_a with this nearest neighbor arrangement have the form (3.2.1). The results of Section 3.2 then show that with hexagonal lattices, such e_a give $\mathbf{E}^{(n)}$ that are isotropic up to $n = 4$, and so yield the standard hydrodynamic continuum equations (2.6.1).

Table 4 gives the forms of $\mathbf{E}^{(n)}$ for the most symmetrical three-dimensional lattices with nearest neighbor choices for the e_a. None yield isotropic $\mathbf{E}^{(4)}$ (compare Ref. 38).

The hexagonal and face-centered cubic lattices, which have the largest point symmetry groups in two and three dimensions, respectively, are also the lattices that give the densest packings of circles and spheres (e.g., Ref. 42). One suspects that in more than three dimensions (compare Ref. 43) the lattices with the largest point symmetry continue to be those with the densest sphere packing. The spheres are placed on lattice sites; the positions of their nearest neighbors are defined by a Voronoi polyhedron or Wigner-Seitz cell. The densest sphere packing is obtained when this cell, and thus the nearest neighbor vectors e_a, are closest to forming a sphere. In dimensions $d \leq 8$, it has been found that the optimal lattices for sphere packings are those based on the sets of root vectors for a sequence of simple Lie groups (e.g., Ref. 44). Results on the isotropy of the tensors $\mathbf{E}^{(n)}$ for these lattices are given in Table 5.

More isotropic sets of $\mathbf{e}_a$ can be obtained by allowing links to join sites on the lattice beyond nearest neighbors.[31] On a square lattice, one may, for example, include diagonal links, yielding a set of vectors

$$\mathbf{e}_a = (0, \pm 1), (\pm 1, 0), (\pm 1, \pm 1). \tag{3.5.1}$$

Including weights $w(|\mathbf{e}_a|^2)$ as in Eq. (3.1.16), this choice of $\mathbf{e}_a$ yields

$$\mathbf{E}^{(2)} = 2\left[w(1) + 2w(2)\right]\delta^{(2)}, \tag{3.5.2}$$

$$\mathbf{E}^{(4)} = 4w(2)\Delta^{(4)} + 2\left[w(1) - 4w(2)\right]\delta^{(4)}. \tag{3.5.3}$$

If the ratio of particles on diagonal and orthogonal links can be maintained so that

$$w(1) = 4w(2), \tag{3.5.4}$$

then Eq. (3.5.3) shows that $\mathbf{E}^{(4)}$ will be isotropic. This choice effectively weights the individual vectors $(0, \pm 1)$ and $(\pm 1, 0)$ with a factor $\sqrt{2}$. As a result, the vectors (3.5.1) are effectively those for a regular octagon, given by Eq. (3.2.1) with $M = 8$.

Including all 24 $\mathbf{e}_a$ with components $|(\mathbf{e}_a)_i| \leq 2$ on a square lattice, one obtains

$$\mathbf{E}^{(2)} = 2\left[w(1) + 2w(2) + 4w(4) + 10w(5) = 8w(8)\right]\delta^{(2)}, \tag{3.5.5}$$

$$\mathbf{E}^{(4)} = 4\left[w(2) + 8w(5) + 16w(8)\right]\Delta^{(4)}$$
$$+ 2\left[w(1) - 4w(2) + 16w(4) - 14w(5) - 64w(8)\right]\delta^{(4)}, \tag{3.5.6}$$

$$\mathbf{E}^{(6)} = \frac{4}{3}\left[w(2) + 20w(5) + 64w(8)\right]\Delta^{(6)}$$
$$+ 2\left[w(1) - 8w(2) + 64w(4) - 70w(5) - 512w(8)\right]\delta^{(6)}. \tag{3.5.7}$$

With $w(5) = w(8) = 0$, $\mathbf{E}^{(4)}$ and $\mathbf{E}^{(6)}$ are isotropic if

$$\frac{w(2)}{w(1)} = \frac{3}{8}, \qquad \frac{w(4)}{w(1)} = \frac{1}{32}. \tag{3.5.8}$$

They cannot both be isotropic if $w(4)$ also vanishes.

In three dimensions, one may consider a cubic lattice with sites at distances 1, $\sqrt{2}$, and $\sqrt{3}$ joined. The $\mathbf{e}_a$ in this case contain all those for primitive, face-centered, and body-centered cubic lattices, as given in Table 4. The $\mathbf{E}^{(n)}$ can then be deduced from the results of Table 4 and are given by

$$\mathbf{E}^{(2)} = 2\left[w(1) + 4w(2) + 4w(3)\right]\delta^{(2)}, \tag{3.5.9}$$

$$\mathbf{E}^{(4)} = 4\left[w(2) + 2w(3)\right]\Delta^{(4)} + 2\left[w(1) - 2w(2) - 8w(3)\right]\delta^{(4)}, \tag{3.5.10}$$

$$\mathbf{E}^{(6)} = 8w(2)\Delta^{(6)} + 4\left[w(2) - 4w(3)\right]\Delta^{(4,2)} + 2\left[w(1) - 26w(2) + 64w(3)\right]\delta^{(6)}. \tag{3.5.11}$$

Isotropy of $\mathbf{E}^{(4)}$ is obtained when

$$w(1) = 2w(2) + 8w(3) \tag{3.5.12}$$

and of $\mathbf{E}^{(6)}$ when

$$w(1) = 10w(2) = 40w(3). \tag{3.5.13}$$

Notice that (3.5.12) and (3.5.13) cannot simultaneously be satisfied by any nonzero choice of weights. Nevertheless, so long as (3.5.12) holds, isotropic hydrodynamic behavior is obtained in this three-dimensional cellular automaton fluid. Isotropic $\mathbf{E}^{(6)}$ can be obtained by including in addition vectors e_a of the form $(\pm 2, 0, 0)$ (and permutations), and choosing

$$w(2) = \frac{1}{2}w(1), \qquad w(3) = \frac{1}{8}w(1), \qquad w(4) = \frac{1}{16}w(1). \tag{3.5.14}$$

The weights in Eq. (3.1.17) give the probabilities for particles with different speeds to occur. These probabilities are determined by microscopic equilibrium conditions. They can potentially be controlled by using different collision rules on different time steps (as discussed in Section 4.9). Each set of collision rules can, for example, be arranged to yield each particle speed with a certain probability. Then the frequency with which different collision rules are used can determine the densities of particles with different speeds.

3.6. IRREGULAR LATTICES

The general structure of cellular automaton fluid models considered here requires that particles can occur only at definite positions and with definite discrete velocities. But the possible particle positions need not necessarily correspond with the sites of a regular lattice. The directions of particle velocities should be taken from the directions of links. But the particle speeds may consistently be taken independent of the lengths of links.

As a result, one may consider constructing cellular automaton fluids on quasi-lattices (e.g., Ref. 46), such as that illustrated in Figure 2. Particle velocities are taken to follow the directions of the links, but to have unit magnitude, independent of the spatial lengths of the links. Almost all intersections involve just two links, and so can support only two-particle interactions. These intersections occur at a seemingly irregular set of points, perhaps providing a more realistic model of collisions in continuum fluids.

Color Plates

FIGURE 1a Flow past a plate with periodic boundary conditions.

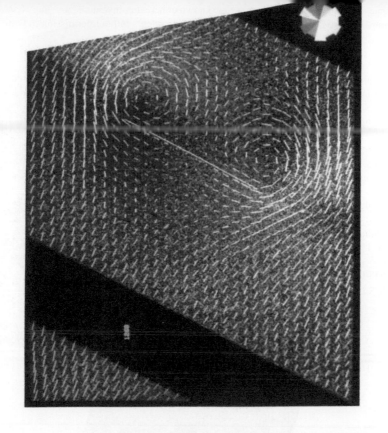

Figure 1b. The same simulation as that
in Figure 1a but with no three-
... result, spurious

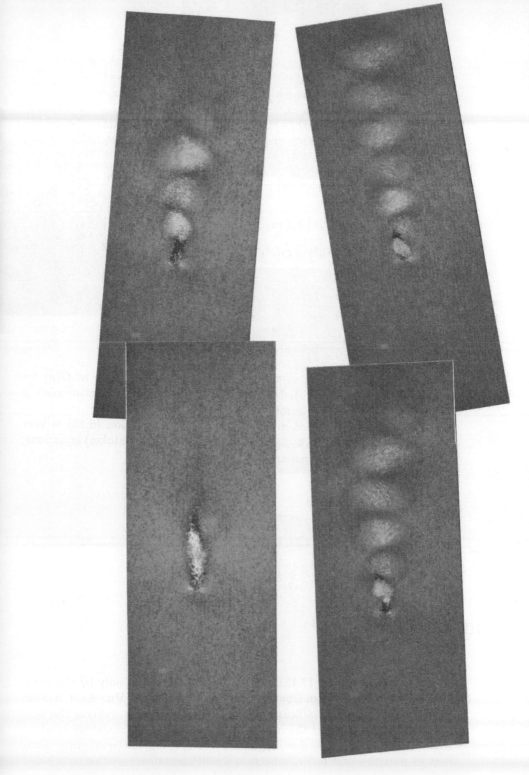

Figure 3. A wake grows behind a wedge. The flow is from left to right with periodic boundary conditions. [See the top row...]

Figure 5. A turbulent wake grows behind an ellipse being dragged through a fluid consisting of 11 million particles and 8 million cells. The ellipse is composed of about 2400 cells in which the velocity directions of the entering particles are reversed. The flow has periodic right and left boundaries. (An infinite sequence of equivalent ellipses exists to the left and

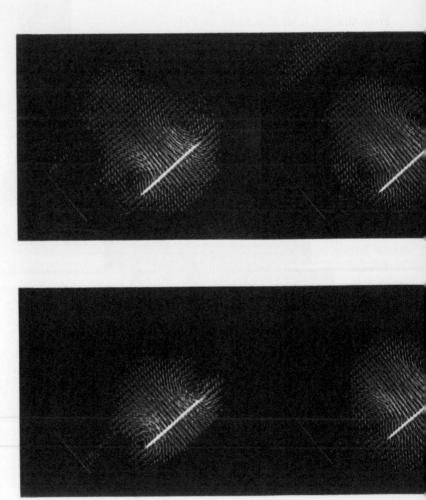

Figure 6. Cross sections (printed here) and three-dimensional views (not included in this reprint) show the development of vortices behind a square plate.

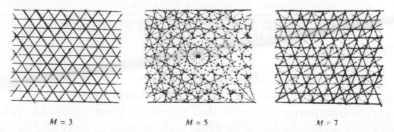

$M = 3$ $M = 5$ $M = 7$

FIGURE 2 Lattices and quasilattices constructed from grids oriented in the directions of the vertices of regular M-sided polygons. An appropriate dual of the $M = 5$ pattern is the Penrose aperiodic tiling.

The possible e_a on regular lattices are highly constrained, as discussed in Section 3.5. But it is possible to construct quasilattices which yield any set of e_a. Given a set of generator vectors g_a, one constructs a grid of equally spaced lines orthogonal to each of them.[47] The directions of these lines correspond to the e_a.

If the tangents of the angles between the g_a are rational, then these lines must eventually form a periodic pattern, corresponding to a regular lattice. But if, for example, the g_a correspond to the vertices of a pentagon, then the pattern never becomes exactly periodic, and only a quasilattice is obtained. A suitable dual of the quasilattice gives in fact the standard Penrose aperiodic tiling.[48]

In three dimensions, one may form grids of planes orthogonal to generator vectors g_a. Possible particle positions and velocities are obtained from the lines in which these planes intersect.

Continuum equations may be derived for cellular automaton fluids on quasilattices by the same methods as were used for regular lattices above. But by appropriate choices of generator vectors, three-dimensional quasilattices with effective icosahedral point symmetry may be obtained, so that isotropic fluid behavior can be obtained even with a single particle speed.

Quasilattices yield an irregular array of particle positions, but allow only a limited number of possible particle velocities. An entirely random lattice would also allow arbitrary particle velocities. Momentum conservation cannot be obtained exactly with discrete collision rules on such a lattice, but may be arranged to hold on average.

4. EVALUATION OF TRANSPORT COEFFICIENTS

4.1. INTRODUCTION

Section 2 gave a derivation of the general form of the hydrodynamic equations for a sample cellular automaton fluid model. This section considers the evaluation of the specific transport coefficients that appear in these equations. While these coefficients may readily be found by explicit simulation, as discussed in the second paper in this series, no exact mathematical procedure is known for calculating them. This section considers primarily an approximation method based on the Boltzmann transport equation. The results obtained are expected to be accurate for certain transport coefficients at low particle densities.[4]

4.2. BASIS FOR BOLTZMANN TRANSPORT EQUATION

The kinetic equation (2.3.5) gives an exact result for the evolution of the one-particle distribution function f_a. But the collision terms Ω_a in this equation depends on two-particle distribution functions, which in turn depend on higher-order distribution functions, forming the BBGKY hierarchy of kinetic equations. To obtain explicit results for the f_a one must close or truncate this hierarchy.

The simplest assumption is that there are no statistical correlations between the particles participating in any collision. In this case, the multiparticle distribution functions that appear in Ω_a can be replaced by products of one-particle distribution functions f_a, yielding an equation of the standard Boltzmann transport form, which can in principle be solved explicitly for the f_a.

Even if particles were uncorrelated before a collision, they must necessarily show correlations after the collision. As a result, the factorization of multiparticle distribution functions used to obtain the Boltzmann transport equation cannot formally remain consistent. At low densities, it may nevertheless in some cases provide an adequate approximation.

Correlations produced by a particular collision are typically important only if the particles involved collide again before losing their correlations. At low densities, particles usually travel large distances between collisions, so that most collisions involve different sets of particles. The particles involved in one collision will typically suffer many other collisions before meeting again, so that they are unlikely to maintain correlations. At high densities, however, the same particles often undergo many successive collisions, so that correlations can instead be amplified.

In the Boltzmann transport equation approximation, correlations and deviations from equilibrium decay exponentially with time. Microscopic perturbations may, however, lead to collective, hydrodynamic, effects, which decay only as a power of time.[29] Such effects may lead to transport coefficients that are nonanalytic functions of density and other parameters, as mentioned in Section 2.6.

---

[4]Some similar results have been obtained by a slightly different method in Ref. 49.

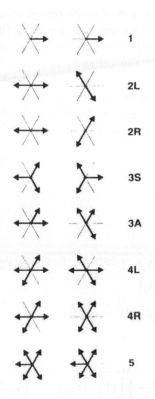

1

2L

2R

3S

3A

4L

4R

5

FIGURE 3   Possible types of initial and final states for collisions in the cellular automaton fluid model of Section 2.

4.3. CONSTRUCTION OF BOLTZMANN TRANSPORT EQUATION

This subsection describes the formulation of the Boltzmann transport equation for the sample cellular automaton fluid model discussed in Section 2.

The possible classes of particle collisions in this model are illustrated in Figure 3. The rules for different collisions within each class are related by lattice symmetries. But, as illustrated in Figure 3, several choices of overall rules for each class are often allowed by conservation laws.

In the simplest case, the same rule is chosen for a particular class of collisions at every site. But it is often convenient to allow different choices of rules at different sites. Thus, for example, there could be a checkerboard arrangement of sites on which two-body collisions lead alternately to scattering to the left and to the right. In general, on may apply a set of rules denoted by k at some fraction γ_k of the sites in a cellular automaton. (A similar procedure was mentioned in Section 3.5 as a means for obtaining isotropic behavior on three-dimensional cubic lattices.) The randomness of microscopic particle configurations suggests that the γ_k should serve merely to change the overall probabilities for different types of collisions.

The term Ω_a in the kinetic equation (2.3.5) for f_a is a sum of terms representing possible collisions involving particles of type a. Each term gives the change in the number of type a particles due to a particular type of collisions, multiplied by the probability for the arrangement of particles involved in the collision to occur. In the Boltzmann equation approximation, the probability for a particular particle arrangement is taken to be a simple product of the densities f_b for particles that should be present, multiplied by factors $(1 - f_c)$ for particles that should be absent.

The complete Boltzmann transport equation for the model of Section 2 thus becomes

$$\partial_t f_a + \mathbf{e}_a \cdot \nabla f_a = \Omega_a \tag{4.3.1}$$

where

$$
\begin{aligned}
\Omega = & \left[\gamma_{2L} \Lambda(1,4) + (\gamma_2 - \gamma_{2L}) \Lambda(2,5) \right] - \gamma_2 \Lambda(0,3) \\
& + \gamma_{3S} \left[\Lambda(1,3,5) - \Lambda(0,2,4) \right] \\
& + \gamma_{3A} \left[\Lambda(2,4,5) + \Lambda(1,2,5) - \Lambda(0,3,5) - \Lambda(0,2,3) \right. \\
& \left. + \Lambda(1,4,5) + \Lambda(1,2,4) - \Lambda(0,3,4) - \Lambda(0,1,3) \right] \\
& + \left[\gamma_4 \Lambda(1,2,4,5) - \gamma_{4L} \Lambda(0,2,3,5) - (\gamma_4 - \gamma_{4L}) \Lambda(0,1,3,4) \right].
\end{aligned}
\tag{4.3.2}
$$

Here

$$
\Lambda_a(i_1, i_2, \dots, i_k) = \frac{f_{a+i_1}}{1 - f_{a+i_1}} \frac{f_{a+i_2}}{1 - f_{a+i_2}} \cdots \frac{f_{a+i_k}}{1 - f_{a+i_k}} \prod_{j=1}^{M} (1 - f_{a+j}) \tag{4.3.3}
$$

where all indices on the f_b are evaluated modulo M, and in this case $M = 6$. Note that in Eq. (4.3.2), the index a has been dropped on both Ω and Λ.

The Boltzmann transport equations for any cellular automaton fluid model have the overall form of Eqs. (4.3.1) and (4.3.2). In a more general case, the simple addition of constant i_j to the indices a in the definition of Λ can be replaced by transformations with appropriate lattice symmetry group operations.

Independent of the values of the γ_k, Ω_a is seen to satisfy the momentum and particle number constraints (2.4.4) and (2.4.5).

In the following calculations it is often convenient to maintain arbitrary values for the γ_k so as to trace the contributions of different classes of collisions. But to obtain a form for Ω_a that is invariant under the complete lattice symmetry group, one must take

$$\gamma_{2L} = \gamma_{2R} = \frac{1}{2} \gamma_2, \tag{4.3.4}$$

$$\gamma_{4L} = \gamma_{4R} = \frac{1}{2} \gamma_4. \tag{4.3.5}$$

4.4. LINEAR APPROXIMATION TO BOLTZMANN TRANSPORT EQUATION

In studying macroscopic behavior, one assumes that the distribution functions f_a differ only slightly from their equilibrium values, as in the Chapman-Enskog expansion (2.5.1). The f_a may thus be approximated as

$$f_a = f(1 - \phi_a) \qquad (|\phi_a| \ll 1). \tag{4.4.1}$$

With this approximation, the collision term Ω_a in the Boltzmann transport equation may be approximated by a power series expansion in the ϕ_a:

$$\Omega_a = \sum_b \omega_{ab}^{(1)} \phi_b + \sum_{b,c} \omega_{abc}^{(2)} \phi_b \phi_c + \cdots \tag{4.4.2}$$

The matrix $\omega^{(1)}$ here is analogous to the usual linearized collision operator (e.g., Ref. 26). Notice that for a cellular automaton fluid model with collisions involving at most K particles, the expansion (4.4.2) terminates at $O(\phi^K)$.

Microscopic reversibility immediately implies that the tensors $\omega^{(n)}$ are all completely symmetric in their indices. The conservation laws (2.4.4) and (2.4.5) yield conditions on all the $\omega^{(n)}$ of the form

$$\sum_{abc\ldots} \omega_{abc\ldots}^{(n)} = 0, \tag{4.4.3}$$

$$\sum_{abc\ldots} e_a \omega_{abc\ldots}^{(n)} = 0. \tag{4.4.4}$$

In the particular case of $\omega^{(1)}$, the more stringent conditions

$$\sum_b \omega_{ab}^{(1)} = 0 \tag{4.4.5}$$

and

$$\sum_b e_a \omega_{ab}^{(1)} = 0 \tag{4.4.6}$$

also apply.

In the model of Section 2, all particle types a are equivalent up to lattice symmetry transformations. As a result, $\omega_{(a+1)bc\ldots}^{(n)}$ is always given simply by a cyclic shift of $\omega_{abc\ldots}^{(n)}$, so that the complete form of the $\omega^{(n)}$ can be determined from the first row $\omega_{1bc\ldots}^{(n)}$. The $\omega^{(n)}$ are thus circulant tensors (e.g., Ref. 50), and the values of their components depend only on numerical differences between their indices, evaluated modulo M.

Expansion of (4.3.2) now yields

$$
\begin{aligned}
\omega_{ab}^{(1)} = f^2(1-f)\text{circ}\Big\{ &- \big[\gamma_2\bar{f}^2 + (\gamma_{3S} + 4\gamma_{3A})\bar{f}f + \gamma_4 f^2\big], \\
&\gamma_{2L}\bar{f}^2 + (\gamma_{3S} + 2\gamma_{3A})\bar{f}f + \gamma_{4L}f^2, \\
&(1-\gamma_{2L})\bar{f}^2 + (-\gamma_{3S} + 2\gamma_{3A})\bar{f}f + (\gamma_4 - \gamma_{aL})f^2, \\
&- \big[\gamma_2\bar{f}^2 + (-\gamma_{3S} + 4\gamma_{3A})\bar{f}f + \gamma_4 f^2\big] \\
&\gamma_{2L}\bar{f}^2 + (-\gamma_{3S} + 2\gamma_{3A})\bar{f}f + \gamma_{4L}f^2, \\
&(1-\gamma_{2L})\bar{f}^2 + (\gamma_{3S} + 2\gamma_{3A})\bar{f}f + (\gamma_4 - \gamma_{4L})f^2\Big\}
\end{aligned}
\tag{4.4.7}
$$

where $\bar{f} = (1-f)$. Taking for simplicity $\gamma_2 = 1$, $\gamma_{2L} = 1/2$, $\gamma_{3S} = 1$, $\gamma_{3A} = \gamma_{4i} = 0$, one finds

$$
\omega_{ab}^{(1)} = f^2(1-f)^2\text{circ}\left[-1, \frac{1}{2}(1+f), \frac{1}{2}(1-3f), 2f-1, \frac{1}{2}(1-3f), \frac{1}{2}(1+f)\right]
\tag{4.4.8}
$$

$$
\omega_{abc}^{(2)} = \frac{1}{2}f^2(1-f)\text{circ}
\begin{bmatrix}
0 & -f(f-1) & f(3f-1) \\
-f(f-1) & 0 & 2f(f-1) \\
f(3f-1) & 2f(f-1) & 0 \\
-2(f-1)(2f-1) & -f(5f-3) & -f(f-1) \\
f(3f-1) & (f-1)(2f-1) & 2f(3f-2) \\
-f(f-1) & -2f^2 & (f-1)(2f-1)
\end{bmatrix}
$$

$$
\begin{bmatrix}
-2(f-1)(2f-1) & f(3f-1) & -f(f-1) \\
-f(5f-3) & (f-1)(2f-1) & -2f^2 \\
-f(f-1) & 2f(3f-2) & (f-1)(2f-1) \\
0 & -f(f-1) & -f(5f-3) \\
-f(f-1) & 0 & 2f(f-1) \\
-f(5f-3) & 2f(f-1) & 0
\end{bmatrix}
\tag{4.4.9}
$$

4.5. APPROACH TO EQUILIBRIUM

In a spatially uniform system close to equilibrium, one may use a linear approximation to the Boltzmann equation (4.3.1):

$$
\partial_t(f\phi_a) = \sum_b \omega_{ab}^{(1)}\phi_b .
\tag{4.5.1}
$$

This equation can be solved in terms of the eigenvalues and eigenvectors of the matrix $\omega_{ab}^{(1)}$. The circulant property of $\omega_{ab}^{(1)}$ considerably simplified the computations required.

An $M \times M$ circulant matrix U_{ab} can in general be written in the form[50]

$$U_{ab} = \mathbf{u}\big[(a - b + 1) \bmod M\big] = U_{11}I + U_{12}\Pi + \cdots + U_{1M}\Pi^{(M-1)} \qquad (4.5.2)$$

where

$$\Pi = \mathrm{circ}[0, 1, 0, 0, \ldots, 0] \qquad (4.5.3)$$

is an $M \times M$ cyclic permutation matrix, and I is the $M \times M$ identity matrix. From this representation, it follows that all $M \times M$ circulants have the same set of right eigenvectors $\mathbf{v}_c$, with components given by

$$(\mathbf{v}_c)_a = \frac{1}{\sqrt{M}} \exp \frac{2\pi i(c - 1)(a - 1)}{M}. \qquad (4.5.4)$$

Writing

$$\Gamma(z) = \sum_{a=1}^{M} U_{1a} z^{a-1} \qquad (4.5.5)$$

the corresponding eigenvalues are found to be

$$\lambda_c = \Gamma\left(\exp\left[\frac{2\pi i(c-1)}{M}\right]\right). \qquad (4.5.6)$$

Using these results, the eigenvectors of $\omega_{ab}^{(1)}$ for the model of Section 2 are found to be

$$\mathbf{v}_1 = \frac{1}{\sqrt{6}}(1, 1, 1, 1, 1, 1)$$

$$\mathbf{v}_2 = \frac{1}{\sqrt{6}}(1, \sigma, -\sigma^*, -1, -\sigma, \sigma^*) = (\mathbf{v}_6)^*$$

$$\mathbf{v}_3 = \frac{1}{\sqrt{6}}(1, -\sigma^*, -\sigma, 1, -\sigma^*, -\sigma) = (\mathbf{v}_5)^*$$

$$\mathbf{v}_4 = \frac{1}{\sqrt{6}}(1, -1, 1, -1, 1, -1) \qquad (4.5.7)$$

$$\frac{1}{2}(\mathbf{v}_2 + \mathbf{v}_6) = \frac{1}{2\sqrt{6}}(2, 1, -1, -2, -1, 1)$$

$$\frac{1}{2i}(\mathbf{v}_2 + \mathbf{v}_6) = \frac{\sqrt{3}}{2\sqrt{6}}(0, 1, 1, 0, -1, -1)$$

where

$$\sigma = \exp(i\pi/3) = \frac{1}{2}(1 + i\sqrt{3})$$

and the corresponding eigenvalues are

$$\lambda_1 = 0$$
$$\lambda_2 = 0$$
$$\lambda_3 = -3f^2(1-f)\left\{\left[\gamma_2(1-f)^2 + 4\gamma_{3A}f(1-f) + \gamma_4 f^2\right]\right.$$
$$\left. -\frac{4i}{\sqrt{3}}\left[(1-f)^2\left(\frac{\gamma_2}{2} - \gamma_{2L}\right) + f^2\left(\frac{\gamma_4}{2} - \gamma_{4L}\right)\right]\right\} \tag{4.5.8}$$
$$\lambda_4 = -6\gamma_{3S}f^3(1-f)^2$$
$$\lambda_5 = (\lambda_3)^*$$
$$\lambda_6 = 0$$

Combinations of the ϕ_a corresponding to eigenvectors with zero eigenvalue are conserved with time according to Eq. (4.5.1). Three such combinations are associated with the conservation laws (2.4.1) and (2.4.2). $\mathbf{v}_1$ corresponds to $\sum_a \phi_a$, which is the total particle number density $(\mathbf{v}_2 - \mathbf{v}_6)/2$ and $(\mathbf{v}_2 - \mathbf{v}_6)/2i$ correspond, respectively, to the x and y components of the momentum density $\sum_a e_a \phi_a$.

The ϕ_a may always be written as sums of pieces proportional to each of the orthogonal eigenvectors $\mathbf{v}_c$ of Eq. (4.5.7):

$$\phi_a = \sum_c \psi_c(\mathbf{v}_c)_a \tag{4.5.9}$$

The coefficients ψ_1, $(\psi_2 + \psi_6)/2$, and $(\psi_2 - \psi_6)/2i$ give the values of the conserved particle and momentum densities in this representation, and remain fixed with time.

The general solution of Eq. (4.5.1) is given in terms of Eq. (4.5.9) by

$$\psi_c(t) = \psi_c(0)e^{\lambda_c t} \tag{4.5.10}$$

Equation (4.5.8) shows that for any positive choices of the γ_k, all nonzero λ_c have negative real parts. As a result, the associated ψ_c must decay exponentially with time. Only the combinations of ϕ_a associated with conserved quantities survive at large times.

This result supports the local equilibrium assumption used for the derivation of hydrodynamic equations in Section 2. It implies that regardless of the initial average densities ϕ_a, collisions bring the system to an equilibrium that depends only on the values of the macroscopic conserved quantities (2.4.1) and (2.4.2). One may thus expect to be able to describe the local state of the cellular automaton fluid on time scales large compared to $|\lambda_c|^{-1}(\lambda_c \neq 0)$ solely in terms of these macroscopic conserved quantities. [Section 4.2 nevertheless mentioned some effects not accounted for by the Boltzmann equation (4.3.1) that can slow the approach to equilibrium.]

One notable feature of the results (4.5.8) is that they imply that the final equilibrium values of the ϕ_a are not affected by the choice of the parameters γ_{2L} and

γ_{4L}, which determine the mixtures of two- and four-particle collisions with different chiralities. When the rate for collisions with different chiralities are unequal, however, λ_3 and λ_5 acquire imaginary parts, which lead to damped oscillations in the ϕ_a as a function of time.

When all the types of collisions illustrated in Figure 3 can occur, Eq. (4.5.8) implies that momentum and particle number are indeed the only conserved quantities. If, however, only two-particle collisions are allowed, then there are additional conserved quantities. In fact, whenever symmetric three-particle collisions are absent, so that $\gamma_{3S} = 0$, Eq. (4.5.8) implies that the quantities

$$Q_i = \sum_{a=i}^{M/2+i} f_a \qquad (4.5.11)$$

where the index a is evaluated modulo $M = 6$, is conserved. Thus, independent of the value of γ_{2L}, the total momenta on the two sides of any line (not along a lattice direction) through the cellular automaton must independently be conserved.

If three-particle symmetric collisions are absent, the cellular automaton thus exhibits a spurious additional conservation law, which prevents the attainment of standard local equilibrium, and modifies the hydrodynamic behavior discussed in Section 2. Section 4.8 considers some general conditions which avoid such spurious conservation laws.

4.6. EQUILIBRIUM CONDITIONS AND TRANSPORT COEFFICIENTS

Section 4.5 discussed the solution of the Boltzmann transport equation for uniform cellular automaton fluids. This section considers nonuniform fluids, and gives some approximate results for transport coefficients.

The Chapman-Enskog expansion (2.5.1) gives the general form for approximations to the microscopic distribution functions f_a. The coefficients $c^{(2)}$ and $c_\nabla^{(2)}$ that appear in this expansion can be estimated using the Boltzmann transport equation (4.3.1) from the microscopic equilibrium condition

$$\partial_t f_a = 0. \qquad (4.6.1)$$

In estimating $c^{(2)}$, one must maintain terms in Ω_a to the second order in ϕ_b, but one can neglect spatial variation in the ϕ_a. As a result, the Boltzmann equation (4.3.1) becomes

$$\sum_b \omega_{ab}^{(1)} \phi_b + \sum_{b,c} \omega_{abc}^{(2)} \phi_b \phi_c = 0. \qquad (4.6.2)$$

substituting forms for the ϕ_a from the Chapman-Enskog expansion (2.5.1), one obtains

$$c^{(2)} \sum_b \omega_{ab}^{(1)} (\mathbf{u} \cdot \mathbf{e}_b)^2 + [c^{(1)}]^2 \sum_{b,c} \omega_{abc}^{(2)} (\mathbf{u} \cdot \mathbf{e}_b)(\mathbf{u} \cdot \mathbf{e}_c) = 0 \qquad (4.6.3)$$

where $c^{(1)} = 2$ according to Eq. (2.5.4). Using the forms for $\omega^{(1)}$ and $\omega^{(2)}$ determined by the expansion of Eq. (4.3.2), one finds that the two terms in (4.6.3) show exactly the same dependence on the γ_k. The final result for $c^{(2)}$ is thus independent of the γ_k, and is given by

$$c^{(2)} = 2(1 - 2f)/(1 - f). \tag{4.6.4}$$

In the Boltzmann equation approximation, this implies that the coefficient μ of the $n(\mathbf{u} \cdot \nabla)\mathbf{u}$ term in the hydrodynamic equation (2.6.1) is $(1 - 2f)/[2(1 - f)]$. Notice that, as discussed in Section 2.6, this coefficient is not in general equal to 1.

The value of the coefficient $c_\nabla^{(2)}$ can be found by a slightly simpler calculation, which depends only on the linear part $\omega_{ab}^{(1)}$ of the expansion of the collision term Ω_a. Keeping now first-order spatial derivatives of the ϕ_a, one can determine $c_\nabla^{(2)}$ from the equilibrium condition

$$\sum_b \omega_{ab}^{(1)} \phi_b = f\mathbf{e}_a \cdot \nabla \phi_b \tag{4.6.5}$$

which yields

$$\sum_b c_\nabla^{(2)} \omega_{ab}^{(1)} (\mathbf{e}_b \cdot \nabla)(\mathbf{e}_b \cdot \mathbf{u}) = c^{(1)} f(\mathbf{e}_a \cdot \nabla)(\mathbf{e}_a \cdot \mathbf{u}). \tag{4.6.6}$$

With the approximations used, Eq. (2.4.7) implies that $\nabla \cdot \mathbf{u} = 0$. Then Eq. (4.6.6) gives the result

$$c_\nabla^{(2)} = -2\left\{ 12f(1 - f)[\gamma_2(1 - f)^2 + 4\gamma_{3A}f(1 - f) + \gamma_4 f^2] \right\}^{-1}. \tag{4.6.7}$$

Using Eq. (2.6.3), this gives the kinematic viscosity of the cellular automaton fluid in the Boltzmann equation approximation as

$$v = \left\{ 12f(1 - f)[\gamma_2(1 - f)^2 + 4\gamma_{3A}f(1 - f) + \gamma_4 f^2] \right\}^{-1}. \tag{4.6.8}$$

Some particular values are

$$
\begin{aligned}
v &= [12f(1 - f)^3]^{-1} & (\gamma_2 = 1, \gamma_{3A} = \gamma_4 = 0); \\
v &= [12f(1 - f)(1 + 2f - 2f^2)]^{-1} & (\gamma_2 = \gamma_{3A} = \gamma_4 = 1).
\end{aligned}
\tag{4.6.9}
$$

For $f = 1/6$ one obtains in these cases $v \approx 0.86$ and $v \approx 0.47$, respectively, while for $f = 1/3$, $v \approx 0.84$ and $n \approx 0.26$.

4.7. A GENERAL NONLINEAR APPROXIMATION

At least for homogeneous systems, Boltzmann's H theorem (e.g., Ref. 51) yields a general form for the equilibrium solution of the full nonlinear Boltzmann equation (4.3.1). The H function can be defined as

$$H = \sum_a \tilde{f}_a \log(\tilde{f}_a) \tag{4.7.1}$$

where

$$\tilde{f}_a = \frac{f_a}{1 - f_a}. \tag{4.7.2}$$

This definition is analogous to that used for Fermi-Dirac particles (e.g., Refs. 51, 52): the factors $(1 - f_a)$ account for the exclusion of more than one particle on each link, as in Eq. (4.3.1). The microscopic reversibility of (4.3.1) implies that when the equilibrium condition $\partial_t H = 0$ holds, all products $\tilde{f}_{a_1} \tilde{f}_{a_2} \ldots$ must be equal for all initial and final sets of particles $\{a_1, a_2, \ldots\}$ that can participate in collisions. As a result, the $\log(\tilde{f}_a)$ must be simple linear combinations of the quantities conserved in the collisions. If only particle number and momentum are conserved, and there are no spurious conserved quantities such as (4.5.11), the $\tilde{f}_a$ can always be written in the form[7,49,54]

$$\tilde{f}_a = \exp(-\alpha - \beta \mathbf{u} \cdot \mathbf{e}_a). \tag{4.7.3}$$

The one-particle distribution functions thus have the usual Fermi-Dirac form

$$f_a = [1 + \exp(\alpha + \beta \mathbf{u} \cdot \mathbf{e}_a)]^{-1} \tag{4.7.4}$$

where α and β are in general functions of the conserved quantities n and $|\mathbf{u}|^2$.
For small $|\mathbf{u}|^2$, one may write

$$\alpha = \alpha_0 + \alpha_1 |\mathbf{u}|^2 + \cdots, \qquad \beta = \beta_0 + \beta_1 |\mathbf{u}|^2 + \cdots \tag{4.7.5}$$

These expansions can be substituted into Eq. (4.7.4) and the results compared with the Chapman-Enskog expansion (2.7.1).
For $\mathbf{U} = 0$, one finds immediately the "fugacity relation"

$$\exp(-\alpha_0) = \frac{f}{1 - f}. \tag{4.7.6}$$

Then, from the expansion (related to that for generating Euler polynomials)

$$(1 + \xi e^x)^{-1} = \frac{1}{1 + \xi} - \frac{\zeta}{(1 + \xi)^2} x - \frac{\xi(1 - \xi)}{2(1 + \xi)^3} x^2$$
$$- \frac{\xi(1 - 4\xi + \xi^2)}{6(1 + \xi)^4} x^3 - \frac{\xi(1 - \xi)(1 - 10\xi + \xi^2)}{24(1 + \xi)^5} x^5 + \cdots \tag{4.7.7}$$

together with the constraints (2.7.3)–(2.7.5), one obtains (for $d = 2$)

$$\beta_0 = -\frac{2}{1-f}$$

$$\alpha_1 = \frac{1-2f}{(1-f)^2}$$

$$\beta_1 = \frac{1-2f+2f^2}{(1-f)^3} \tag{4.7.8}$$

$$\alpha_2 = -\frac{(1-2f)(3-4f+4f^2)}{16(1-f)^4}$$

where it has been assumed that the e_a form an isotropic set of unit vectors, satisfying Eq. (3.1.5). The complete Chapman-Enskog expansion (2.7.1) then becomes

$$
\begin{aligned}
f_a = f\Bigg\{ & 1 + d\mathbf{u}\cdot\mathbf{e}_a + \frac{d^2}{2}\frac{1-2f}{1-f}\left[(\mathbf{u}\cdot\mathbf{e}_a)^2 - \frac{1}{d}|\mathbf{u}|^2\right] \\
& + \frac{d^3}{6}\frac{1-6f+6f^2}{(1-f)^2}\left[(\mathbf{u}\cdot\mathbf{e}_a)^3 - \frac{3}{d+2}|\mathbf{u}|^2(\mathbf{u}\cdot\mathbf{e}_a)\right] \\
& + \frac{1}{48}\frac{1-2f}{(1-f)^3}\big[32(1-12f+12f^2)(\mathbf{u}\cdot\mathbf{e}_a)^4 \\
& + 384f(1-f)|\mathbf{u}|^2(\mathbf{u}\cdot\mathbf{e}_a)^2 + 3(11-36f+36f^2)|\mathbf{u}|^4\big] + \cdots \Bigg\}
\end{aligned}
\tag{4.7.9}
$$

where for the last term it has been assumed that $d = 2$.

The result (4.6.4) for $c^{(2)}$ follows immediately from this expansion. For cellular automaton fluid models with $\mathbf{E}^{(6)}$ isotropic, the continuum equation (2.7.7) holds. The results for the coefficients that appear in this equation can be obtained from the approximation (4.7.9), and have the simple forms

$$c^{(2)} = \frac{d^2(1-2f)}{2(1-f)} \tag{4.7.10}$$

$$c^{(4)}(1+\sigma_{4,1}) = \frac{2(1-2f)}{3(1-f)^3} \qquad (d=2) \tag{4.7.11}$$

These results allow an estimate of the importance of the next-order corrections to the Navier-Stokes equations included in Eq. (2.7.7). They suggest that the corrections may be important whenever $|\mathbf{u}/(1-f)^2|^2$ is not small compared to 1. The corrections can thus potentially be important both at high average velocities and high particle densities.

The hexagonal lattice model of Section 2 yields a continuum equation of the form (2.7.6), with an anisotropic $O(\mathbf{u}^2 \nabla \mathbf{u})$ term. Equation (4.7.9) gives in this case

$$
\begin{aligned}
\partial_t(6fu_x) &+ \frac{3f(1-2f)}{1-f}(u_x\partial_x u_x + u_x\partial_y u_y + u_y\partial_y u_x - u_y\partial_x u_y) \\
&+ \frac{f(1-2f)}{4(1-f)^3}\Big\{[(55 - 84f + 84f^2)u_x^2 + 3(13 + 4f - 4f^2)u_y^2]u_x\partial_x u_x \\
&+ 2[(1 + 12f - 12f^2)u_x^2 + 9(1-2f)^2 u_y^2]u_x\partial_y u_y \\
&+ 6[(1 + 12f - 12f^2)u_x^2 + (1-2f)^2]u_y\partial_y u_x \\
&+ 3[(13 + 4f - 4f^2)u_x^2 + (13 - 28f + 28f^2)u_y^2]u_y\partial_x u_y\Big\} = 0
\end{aligned}
\tag{4.7.12}
$$

$$
\begin{aligned}
\partial_t(6fu_y) &+ \frac{3f(1-2f)}{1-f}(-u_x\partial_x u_x + u_x\partial_x u_y + u_y\partial_x u_x - u_y\partial_y u_y) \\
&+ \frac{f(1-2f)}{4(1-f)^3}\Big\{[(35 - 36f + 36f^2)u_x^2 + 3(17 - 44f + 44f^2)u_y^2]u_x\partial_y u_x \\
&+ 2[(1 + 12f - 12f^2)u_x^2 + 9(1-2f)^2 u_y^2]u_x\partial_x u_y \\
&+ 6[(1 + 12f - 12f^2)u_x^2 + (1-2f)^2]u_y\partial_x u_x \\
&+ 3[(17 - 44f + 44f^2)u_x^2 + (17 - 12f + 12f^2)u_y^2]u_y\partial_y u_y\Big\} = 0
\end{aligned}
\tag{4.7.13}
$$

The $O(\mathbf{u}\nabla\mathbf{u})$ term is as given in Eq. (2.5.11). The $O(\mathbf{u}^3\nabla\mathbf{u})$ terms are anisotropic, and are not even invariant under exchange of x and y coordinates ($\pi/2$ rotation). For small densities f, Eqs. (4.7.12) and (4.7.13) become

$$
\begin{aligned}
\partial_t(u_x) &+ \frac{1}{2}(u_x\partial_x u_x + u_x\partial_y u_y + u_y\partial_y u_x - u_y\partial_x u_y) \\
&+ \frac{1}{24}[(55u_x^2 + 39u_y^2)u_x\partial_x u_x + 2(u_x^2 + 9u_y^2)u_x\partial_y u_y \\
&+ 6(u_x^2 + u_y^2)u_y\partial_y u_x + 39(u_x^2 + u_y^2)u_y\partial_x u_y] = 0
\end{aligned}
\tag{4.7.14}
$$

$$
\begin{aligned}
\partial_t(u_y) &+ \frac{1}{2}(-u_x\partial_x u_x + u_x\partial_x u_y + u_y\partial_x u_x - u_y\partial_y u_y) \\
&+ \frac{1}{24}[(35u_x^2 + 51u_y^2)u_x\partial_y u_x + 2(u_x^2 + 9u_y^2)u_x\partial_x u_y \\
&+ 6(u_x^2 + u_y^2)u_y\partial_x u_x + 51(u_x^2 + u_y^2)] = 0
\end{aligned}
\tag{4.7.15}
$$

The results (4.7.10) and (4.7.11) follow from the Fermi-Dirac particle distribution (4.7.4). If instead an arbitrary number of particles were allowed at each site, the equilibrium particle distribution (4.7.4) would take on the Maxwell-Boltzmann form

$$
f_a \exp(-\alpha - \beta\mathbf{u}\cdot\mathbf{e}_a).
\tag{4.7.16}
$$

With this simpler form, more complete results for f_a as a function of n and u can be found. Results which are isotropic to all orders in $\mathbf{u}$ can be obtained only for an infinite set of possible particle directions, parameterized, say, by a continuous angle θ. In this case, the number and momentum densities (2.4.1) and (2.4.2) may be obtained as integrals

$$\frac{1}{2\pi}\int_0^{2\pi} f(\theta)d\theta = f, \tag{4.7.17}$$

$$\frac{1}{2\pi}\int_0^{2\pi} e(\theta)f(\theta)d\theta = f\mathbf{u}. \tag{4.7.18}$$

With the distribution (4.7.16), these integrals become

$$\frac{1}{2\pi}\int_0^{2\pi} \exp(-\alpha - \beta u\cos\theta)d\theta = e^{-x}I_0(\beta u) = f \tag{4.7.19}$$

$$\frac{1}{2\pi}\int_0^{2\pi} \exp(-\alpha - \beta u\cos\theta)\cos\theta\mathbf{u}/u\,d\theta = -e^{-x}I_1(\beta u)\mathbf{u}/u = f\mathbf{u} \tag{4.7.20}$$

where $u = |\mathbf{u}|$, and the $I_\nu(z)$ are modified Bessel functions (e.g., Ref. 53)

$$I_\nu(z) = \sum_{n=0}^{\infty}\frac{(z/2)^{\nu+2n}}{n!(\nu+n)!}, \quad I_0(0) = 1, \quad I_\nu(0) = 0 \quad (\nu > 0) \tag{4.7.21}$$

$$\int_{-1}^{1}(1-x^2)^{\nu-1/2}e^{-zx}dx = \pi z^{-\nu}(2\nu-1)!!I_x(z) \tag{4.7.22}$$

$$I_{\nu-1}(z) - I_{\nu+1}(z) - I_{\nu+1}(z) = (2\nu/z)I_\nu(z) \tag{4.7.23}$$

The rapid convergence of the series (4.7.21) means that Eqs. (4.7.19) and (4.7.20) provide highly accurate approximations even for a small number of discrete directions e_a. [For example, with $M = 6$, $\alpha = 0$, $\beta = 1$, and $\mathbf{u} = (1,1)$, the error in Eq. (4.7.19) is less than 10^{-9}.]

For the simple distribution (4.7.16) the momentum flux density tensor (2.4.9) may be evaluated in direct analogy with Eqs. (4.7.19) and (4.7.20) as

$$\Pi_{ij} = e^{-x}\left[\frac{1}{2}I_0(\beta u)\delta_{ij} + I_2(\beta u)\left(\frac{u_iu_j}{u^2} - \frac{1}{2}\delta_{ij}\right)\right]. \tag{4.7.24}$$

Using the recurrence relation (4.7.23), and substituting the results (4.7.19) and (4.7.20), this may be rewritten in the form

$$\Pi_{ij} = f\left[\frac{1}{2}I_0(\beta u)\delta_{ij} + \left(1 + \frac{2}{\beta}\right)\left(\frac{u_iu_j}{u^2} - \frac{1}{2}\delta_{ij}\right)\right]. \tag{4.7.25}$$

Combining Eqs. (4.7.19) and (4.7.20), one finds that the function $\beta(f, u)$ is independent of f, and can be determined from the implicit equation

$$\frac{I_1(\beta u)}{I_0(\beta u)} = \frac{I_0'(\beta u)}{I_0(\beta u)} = -u. \tag{4.7.26}$$

Expanding in powers of u^2, as in Eq. (4.7.5), yields

$$\beta(u) = \sum_{n=0} \beta_n u^{2n} \tag{4.7.27}$$

$$\beta_0 = -2, \quad \beta_1 = -1, \quad \beta_2 = -\frac{5}{6}, \quad \beta_3 = -\frac{19}{24}, \quad \beta_4 = -\frac{143}{180}$$

Equation (4.7.19) then gives

$$\alpha(u, f) = \sum_{n=0} \alpha_n u^{2n} \tag{4.7.28}$$

$$\exp(-\alpha_0) = f, \quad \alpha_1 = 1, \quad \alpha_2 = \frac{3}{4}, \quad \alpha_3 = \frac{25}{36}, \quad \alpha_4 = \frac{133}{192}$$

In the limit $u \to 1$, $\beta \to -\infty$.

The above results immediately yield values for the transport coefficients $c^{(n)}$ in the Chapman-Enskog expansion:

$$c^{(2)} = 2 \tag{4.7.29}$$

$$c^{(4)}(1 + \sigma_{4,1}) = \frac{2}{3} \tag{4.7.30}$$

independent of density. Equation (4.7.29) implies that the coefficient μ of the convective term in the Navier-Stokes equation (2.6.1) is equal to 1/2. The deviation from the Galilean invariant result 1 is associated with the constraint of fixed speed particles.

Figure 4 shows the exact result for $\beta(u)$ obtained from Eq. (4.7.26), compared with series expansions to various orders. Significant deviations from the $O(u^2)$ "Navier-Stokes" approximation are seen for $u \gtrsim 0.4$.

For Fermi-Dirac distributions of the form (4.7.4), the integrals (4.7.19) and (4.7.20) can only be expressed as infinite sums of Bessel functions.

4.8. OTHER MODELS

The results obtained so far can be generalized directly to a large class of cellular automaton fluid models.

In the main case considered in Section 3, particles have velocities corresponding to a set of M unit vectors e_a. If this set is invariant under inversion, then both e_a and $-e_a$ always occur. As a result, two particles colliding head on with velocities e_a and $-e_a$ can always scatter in any directions e_b and $-e_b$ with $b \neq a$. One simple possibility is to choose the rules at different sites so that each scattering direction

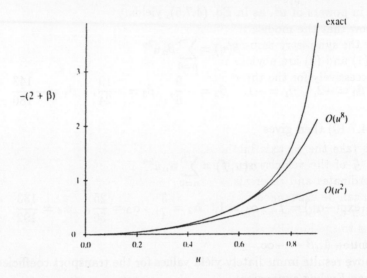

FIGURE 4 Dependence of $\beta(u)$ from Eq. (4.7.16) on the magnitude u of the macroscopic velocity. The results are for Maxwell-Boltzmann particles with unit speeds and arbitrary directions in two dimensions. The function $\beta(u)$ appears both in the microscopic distribution function (4.7.16) and in the macroscopic momentum flux tensor (4.7.25). The result for $\beta(u)$ from an exact solution of the implicit equation (4.7.26) is given, together with results from the series expansion (4.7.27). The $O(u^2)$ result corresponds to the Navier-Stokes approximation. Deviation from the exact results is seen for $u \gtrsim 0.4$.

occurs with equal probability. If only such two-particle collisions are possible (as in a low-density approximation), and only one particle is allowed on each link, then the Boltzmann transport equation becomes

$$\partial_t f_a + \mathbf{e}_a \cdot \nabla f_a = -f_a f_{\bar{a}} \prod_{c \neq a, \bar{a}} (1 - f_c) + \frac{1}{M - 2} \sum_{b \neq a, \bar{a}} f_b f_{\bar{b}} \prod_{c \neq b, \bar{b}} (1 - f_c) \quad (4.8.1)$$

where $f_{\bar{a}}$ is the distribution function for particles with direction $-\mathbf{e}_a$. To second order in the expansion (4.4.2) this gives

$$\partial_t (f\phi_a)_e - a \cdot \nabla (f\phi_a)$$

$$= f^2 (1-f)^{M-3} \left[-(\phi_a + \phi_{\bar{a}}) + \frac{1}{M-2} \sum_{b \neq a, \bar{a}} (\phi_b + \phi_{\bar{b}}) \right]$$

$$+ f^2 (1-f)^{M-4} (1-2f) \left[-\phi_a \phi_{\bar{a}} + \frac{1}{M-2} {\sum}'_{b, \bar{b} \neq a, \bar{a}} \phi_b \phi_{\bar{b}} \right]$$

$$+ f^3 (1-f)^{M-4} \left[-{\sum}'_{b, c \neq a, \bar{a}; c \neq \bar{b}} \phi_b \phi_c + \frac{1}{M-2} \sum_{b \neq a} (\phi_a \phi_b - \phi_{\bar{a}} \phi_{\bar{b}}) \right]$$

$$(4.8.2)$$

where $\sum'$ denotes summation over the triangular region in which the indices form a strictly increasing sequence.

The form of the ϕ_a for a homogeneous system can be obtained from the general equilibrium conditions of Section 4.7. The coefficients $c^{(n)}$ in the Chapman-Enskog expansion are then given by Eqs. (4.7.10) and (4.7.11). The convective transport coefficient μ in the Navier-Stokes equation (2.6.1) is thus given by

$$\mu = \frac{d^2 (1-2f)}{8(1-f)}. \qquad (4.8.3)$$

The $c_{\nabla}^{(n)}$ cannot be obtained by the methods of Section 4.7. But from Eq. (4.8.2) one may deduce immediately the linearized collision term

$$\omega_{ab}^{(1)} = f^2 (1-f)^{M-3} \text{circ}[\chi_a], \qquad (4.8.4)$$

$$\chi_1 = \chi_{1+M/2} = -1, \qquad \chi_a = \frac{2}{M-2}.$$

Then, in analogy with Eq. (4.6.8), the kinematic viscosity for the cellular automaton fluid is found to be

$$\nu = \frac{M-2}{2d(d+2)Mf^2) 1-f)^{M-3}}. \qquad (4.8.5)$$

For an icosahedral set of e_a, with $d = 3$ and $M = 12$, this yields

$$\nu = [36 f^2 (1-f)^9]^{-1}. \qquad (4.8.6)$$

Several generalizations may now be considered. First, one may allow not just one, but, say, up to κ particles on each link of the cellular automaton array. in the limit $\kappa \to \infty$ an arbitrary density of particles is thus allowed in each cell. The Boltzmann equation for this case is the same as (4.8.1), but with all $(1-f_c)$ factors omitted. The resulting transport coefficients are

$$\mu = \frac{1}{2}, \tag{4.8.7}$$

$$\nu = \frac{M-2}{2d(d+2)Mf^2}. \tag{4.8.8}$$

Another generalization is to allow collisions that involve more than two particles. The simplest such collisions are "composite" ones, formed by superposing collisions involving two or less particles. The presence of such collisions changes the values of transport coefficients, but cannot affect the basic properties of the model. The four-particle and asymmetric three-particle collisions in the hexagonal lattice model of Section 2 are examples of composite collisions. They increase the total collision rate, and thus, for example, decrease the viscosity, but do not change the overall macroscopic behavior of the model.

In general, collisions involving k particles can occur if the possible e_a are such that

$$\sum_{i=1}^{k} e_{a_i} = \sum_{i=1}^{k} e_{b_i} \tag{4.8.9}$$

for some sets of incoming and outgoing particles a_i and b_i. Cases in which all the a_i and b_i are distinct may be considered "elementary" collisions. In the hexagonal lattice model of Section 2, only two-particle and symmetric three-particle collisions are elementary.

No elementary three-particle collisions are possible on primitive and body-centered cubic three-dimensional lattices, or with e_a corresponding to the vertices of icosahedra or dodecahedra. For a face-centered cubic lattice, however, eight distinct triples of e_a sum to zero [an example is $(1,-1,0)+(0,1,-1)+(-1,0,1)$], so that elementary three-particle collisions are possible.

One feature of the hexagonal lattice model discussed in Section 2 is the existence of the conservation law (4.5.11) when elementary three-body symmetric collisions are absent. Such spurious conservation laws exist in any cellular automaton fluid model in which all particles have the same speed, and only two-particle collisions can occur. Elementary three-particle collisions provide one mechanism for avoiding these conservation laws and allowing the equilibrium of Section 4.7 to be attained.

4.9. MULTIPLE-SPEED MODELS

A further generalization is to allow particles with velocities e_a of different magnitudes. This generalization is significant not only in allowing two-particle collisions alone to avoid the spurious conservation laws of Section 4.5, but also in making it possible to obtain isotropic hydrodynamic behavior on cubic lattices, as discussed in Section 3.5.

One may define a kinetic energy $1/2|e_a|^2$ that differs for particles of different speeds. In studies of processes such as heat conduction, one must account for the conservation of total kinetic energy. In many cases, however, one considers systems in contact with a heat bath, so that energy need not be conserved in individual collisions.

In a typical case, one may then take pairs of particles with speed s_i colliding head on to give pairs of particles with some other speed s_j. In general, different collision rules may be used on different sites, typically following some regular pattern, as discussed in Section 4.3. Thus, for example, collisions between speed s_i particles may yield speed s_j particles at a fraction $\gamma_{i \to j}$ of the sites.

The number m_i of possible particles with speed $s_i = (|e_a|^2)^{1/2}$ that can occur at each site is determined by the structure of the lattice. The collision rules at different sites may be arranged, as in Section 4.8, to yield particles of a particular speed s_i with equal probabilities in each of the m_i possible directions.

In a homogeneous system, the probability f_i for a link with speed s_i to be populated should satisfy the master equation

$$\partial_t \tilde{f}_i = \sum_j \Gamma_{ij} \tilde{f}_j^2 = \sum_{j \neq i} (-\gamma_{i \to j} m_i \tilde{f}_i^2 + \gamma_{j \to i} m_j \tilde{f}_j^2) \tag{4.9.1}$$

where it is assumed that

$$\sum_j \gamma_{i \to j} = \sum_j \gamma_{j \to i} = 1 \tag{4.9.2}$$

and $\tilde{f}$ is the reduced particle density given by Eq. (4.7.2). With two speeds, Γ_{ij} becomes

$$\Gamma_{ij} = \begin{pmatrix} -\gamma_{1 \to 2} m_1 & \gamma_{2 \to 1} m_2 \\ \gamma_{1 \to 2} m_1 & -\gamma_{2 \to 1} m_2 \end{pmatrix}. \tag{4.9.3}$$

The solutions of Eq. (4.9.1) can be found in terms of the eigenvalues and corresponding eigenvectors of this matrix:

$$\lambda = 0: \qquad (\gamma_{2 \to 1} m_2, \gamma_{1 \to 2} m_1) \tag{4.9.4}$$
$$\lambda = -(\gamma_{1 \to 2} m_1 + \gamma_{2 \to 1} m_2): \qquad (-1, 1) \tag{4.9.5}$$

In the large-time limit, only the equilibrium eigenvector (4.9.4) should survive, giving a ratio of reduced particle densities

$$\frac{\tilde{f}_2^2}{\tilde{f}_1^2} = \frac{\gamma_{1 \to 2} m_1}{\gamma_{2 \to 1} m_2}. \tag{4.9.6}$$

For three-particle speeds, one finds the equilibrium conditions

$$\begin{aligned}
\frac{\tilde{f}_2^2}{\tilde{f}_1^2} &= \frac{m_1}{m_2} \frac{\gamma_{1 \to 2}\gamma_{3 \to 1} + \gamma_{1 \to 2}\gamma_{3 \to 2} + \gamma_{1 \to 3}\gamma_{3 \to 2}}{\gamma_{2 \to 1}\gamma_{3 \to 1} + \gamma_{2 \to 1}\gamma_{3 \to 2} + \gamma_{2 \to 3}\gamma_{3 \to 1}}, \\
\frac{\tilde{f}_3^2}{\tilde{f}_1^2} &= \frac{m_1}{m_3} \frac{\gamma_{1 \to 2}\gamma_{2 \to 3} + \gamma_{1 \to 3}\gamma_{2 \to 1} + \gamma_{1 \to 3}\gamma_{2 \to 3}}{\gamma_{2 \to 1}\gamma_{3 \to 1} + \gamma_{2 \to 1}\gamma_{3 \to 2} + \gamma_{2 \to 3}\gamma_{3 \to 1}}.
\end{aligned} \tag{4.9.7}$$

Different choices for the γ_k yield different equilibrium speed distributions. The probabilities f_i give the weights $w(s_i^2)$ that appear in Eq. (3.1.16). Equation (4.9.6) shows that by choosing

$$\gamma_{2\to1} \approx 2\gamma_{1\to2} \tag{4.9.8}$$

one obtains a ratio of weights for the model of Eq. (3.5.1) that satisfy the condition (3.5.5) for the isotropy of $\mathbf{E}^{(4)}$. [There is a small correction to equality in Eq. (4.9.8) associated with the difference between f and $\tilde{f}$.]

On a cubic lattice, one may similarly satisfy the condition (3.5.12) for the isotropy of $\mathbf{E}^{(4)}$ simply by taking $f_3 = 0$, and

$$\gamma_{2\to1} \approx 2\gamma_{1\to2} . \tag{4.9.9}$$

In this way, one may obtain approximate isotropic hydrodynamic behavior on a three-dimensional cubic lattice.

4.10. TAGGED PARTICLE DYNAMICS

In the discussion above, all the particles in the cellular automaton fluid were assumed indistinguishable. This section considers the behavior of a small concentration of special "tagged" particles.

The density g_a of tagged particles with direction $\mathbf{e}_a$ satisfies an equation of the Fokker-Planck type (e.g., Ref. 26):

$$\partial_t g_a + (\mathbf{e}_a \cdot \nabla) g_a = \Theta_a . \tag{4.10.1}$$

Assuming as in the Boltzmann equation approximation that there are no correlations between particles at different sites, the collision term of Eq. (4.10.1) may be written in the form

$$\Theta_a = \sum_b \theta_{ab} g_b \tag{4.10.2}$$

where θ_{ab} gives the probability that a particle that arrives at a particular site from direction $\mathbf{e}_b$ leaves in direction $\mathbf{e}_a$ with $a \neq b$. The probability is averaged over different arrangements of ordinary particles. Various deterministic rules may be chosen for collisions between ordinary and tagged particles. The simplest assumption is that on average the tagged particles take the place of any of the outgoing particles with equal probability.

Conservation of the total number of tagged particles implies

$$\sum_a g_a = gM . \tag{4.10.3}$$

The total momentum of tagged particles is not conserved; the background of ordinary particles acts like a "heat bath" which can exchange momentum with the tagged particles through the noise term θ_a. Assuming a uniform background fluid, one may make an expansion for the g_a of the form

$$g_a = (g + d^{(1)} \mathbf{e}_a \cdot \nabla g + \cdots).$$ (4.10.4)

The total number of tagged particles then satisfies the equation

$$\partial_t g + d^{(1)} \frac{1}{M} \mathbf{E}_{ij}^{(2)} \partial_i \partial_j g = 0$$ (4.10.5)

where the collision term disappears as a result of Eq. (4.10.3). With the $\mathbf{e}_a$ chosen so that $\mathbf{E}_{ij}^{(2)}$ is isotropic, Eq. (4.10.5) becomes the standard equation for self-diffusion,

$$\partial_t g = D \nabla^2 g$$ (4.10.6)

with the diffusion coefficient D given by

$$D = -\frac{1}{d} d^{(1)}.$$ (4.10.7)

The value of $D^{(1)}$ must be found by solving Eq. (4.10.1) for g_a using the approximation (4.10.4). The equilibrium condition for Eq. (4.10.1) in this case becomes

$$(\mathbf{e}_a \cdot \nabla) g = d^{(1)} \sum_b \theta_{ab} \mathbf{e}_b \cdot \nabla g.$$ (4.10.8)

Thus $-d^{(1)}$ is given in this approximation by the mean free path λ for particle scattering, so that the diffusion coefficient is given by the standard kinetic theory formula

$$D = \frac{1}{d} \lambda.$$ (4.10.9)

For the hexagonal lattice model of Section 2,

$$D = \left\{ 2f^2 (1-f)^2 \left[(1-f)^2 + (\gamma_{3S} + 4\gamma_{3A}) f(1-f) + \gamma_4 f^2 \right] \right\}^{-1}.$$ 4.10.10

5. SOME EXTENSIONS

The simple physical basis for cellular automaton fluid models makes it comparatively straightforward for them to include many of the physical effects that occur in actual fluid experiments.

Boundaries can be represented by special sites in the cellular automaton array. Collisions with boundaries conserve particle number, but not particle momentum. One possibility is to choose boundary collision rules that exactly reverse the velocities of all particles, so that particles in a layer close to the boundary have zero average momentum. This choice yields macroscopic "no slip" boundary conditions, appropriate for many solid surfaces (e.g., Ref. 27). For boundaries that consist of flat segments aligned along lattice directions, an alternative is to take particles to undergo "specular" reflection, yielding a zero average only for the transverse component of particle momentum, and giving "free slip" macroscopic boundary conditions. The roughness of surfaces may be modeled explicitly by including various combinations of these microscopic boundary conditions (corresponding, say, to different coefficients of accommodation).

Arbitrarily complex solid boundaries may be modeled by appropriate arrangements of boundary cells. To model, for example, a porous medium one can, for example, use a random array of "boundary" cells with appropriate statistical properties.

A net flux of fluid can be maintained by continually inserting particles on one edge with an appropriate average momentum and extracting particles on an opposite edge. The precise arrangement of the inserted particles should not affect the macroscopic properties of the system, since microscopic processes should rapidly establish a microscopically random state of local equilibrium. Large-scale inhomogeneities, perhaps representing "free stream turbulence" (e.g., Ref. 4), can be included explicitly.

External pressure and density constraints, whether static or time-dependent, can be modeled by randomly inserting or extracting particles so that local average particle densities correspond to the macroscopic distribution required.

External forces can be modeled by randomly changing velocities of individual particles so as to impart momentum to the fluid at the required average rate. Moving boundaries can then be modeled by explicit motion of the special boundary cells, together with the inclusion of an appropriate average momentum change for particles striking the boundary. Gravitational and other force fields can also be represented in a "quantized approximation" by explicit local changes in particle velocities.

Many other physical effects depend on the existence of surfaces that separate different phases of a fluid or distinct immiscible fluids. The existence of such surfaces requires collective ordering effects within the system. For some choices of parameters, no such ordering can typically occur. But as the parameters change, phase transitions may occur, allowing large correlated regions to form. Such phenomena will be studied elsewhere. (Surface tension effects have been observed in other two-dimensional cellular automata.[3])

---

6. DISCUSSION

Partial differential equations have conventionally formed the basis for mathematical models of continuum systems such as fluids. But only in rather simple circumstances can exact mathematical solutions to such equations be found. Most actual studies of fluid dynamics must thus be based on digital computer simulations, which use discrete approximations to the original partial differential equations (e.g., Ref. 55).

Cellular automata provide an alternative approach to modeling fluids and other continuum systems. Their basic constituent cells are discrete, and ideally suited to simulation by digital computers. Yet collections of large numbers of these cells can show overall continuum behavior. This paper has given theoretical arguments, that with appropriate rules for the individual cells, the overall behavior obtained should follow that described by partial differential equations for fluids.

The cellular automata considered give simple idealized models for the motion and collision of microscopic particles in a fluid as expected from the second law of thermodynamics, precise particle configurations are rapidly randomized, and may be considered to come to some form of equilibrium. In this equilibrium, it should be adequate to describe configurations merely in terms of probabilities that depend on a few macroscopic quantities, such as momentum and particle number, that are conserved in the microscopic particle interactions. Such averaged macroscopic quantities change only slowly relative to the rate of particle interactions. Partial differential equations for their behavior can be found from the transport equations for the average microscopic particle dynamics.

So long as the underlying lattice is sufficiently isotropic, many cellular automata yield in the appropriate approximation the standard Navier-Stokes equations for continuum fluids. The essential features necessary for the derivation of these equations are the conservation of a few macroscopic quantities, and the randomization of all other quantities, by microscopic particle interactions. The Navier-Stokes equations follow with approximations of low fluid velocities and velocity gradients. The simplicity of the cellular automaton model in fact makes it possible to derive in addition next-order corrections to these equations.

The derivation of hydrodynamic behavior from microscopic dynamics has never been entirely rigorous. Cellular automata can be considered as providing a simple example in which the necessary assumptions and approximations can be studied in detail. But strong support for the conclusions comes from explicit simulations of cellular automaton fluid models and the comparison of results with those from actual experiments. The next paper in this series will present many such simulations.

The cellular automaton method of this paper can potentially be applied to a wide variety of processes conventionally described by partial differential equations.

One example is diffusion. At a microscopic level, diffusion arises from random particle motions. The cellular automata used above can potentially reproduce diffusion phenomena, as discussed in Section 4.10. But much simpler cellular automaton rules should suffice. The derivation of the diffusion equation requires that the number of particles be conserved. But it is not necessary for total particle momentum to be conserved. Instead, particle directions should be randomized at each site. Such randomization can potentially be achieved by very simple cellular automaton rules, such as that of Ref. 20. Thus, one may devise cellular automaton methods for the solution of the diffusion equation,[56] which in turn gives a relaxation method for solving Laplace, Poisson, and related equations.

Whenever the physical basis for partial differential equations involves large numbers of particles or other components with local interactions, one can expect to derive an effective cellular automaton model. For systems such as electromagnetic or gravitational fields, such models can perhaps be obtained as analogues of lattice gauge theories.

APPENDIX: SMP PROGRAMS

This Appendix contains a sample SMP[17] computation of the macroscopic equations for the hexagonal lattice cellular automaton fluid model of Section 2.

The SMP definitions are as follows:

```
/* two-dimensional case */
d:2

/* define position and velocity vectors */
r:{x,y}
u:{ux,uy}

/* generate polygonal set of lattice vectors */
<ITrig
polygon[$n] :: (e:Ar[$n,{Cos[2Pi $/$n],Sin[2Pi $/$n]}])

/* calculate terms in number density, momentum vector and stress tensor */
suma[$x] :: Ex[Sum[$x,{a,1,Len[e]}]]
nterm[$f] :: suma[$f[a]]
uterm[$f] :: suma[e[a] $f[a]]
piterm[$f] :: suma[e[a]**e[a] $f[a]]

/* define vector analysis operators */
egrad[$x,$a] :: Sum[e[$a][i] Dt[$x,r[i]],{i,1,d}]
div[$x] :: Sum[Dt[$x[i],r[i]],{i,1,d}]

/* terms in Chapman-Ensokg expansion */
n : f Len[e]
ce0[$a] : f
ce1[$a] : f e[$a].u
ce2[$a] : f ((e[$a].u)^2 - u.u/2)
ce2d[$a] : f (egrad[e[$a].u,$a] - div[u]/2)
celist : {ce0,ce1,ce2,ce2d}

/* specify commutativity of second derivatives */
Dt[$f,$1,{$2_=(Ord[$2,$1]>0),1}] :: Dt[$f,$2,$1]

/* define printing of derivatives */
_Dt[Pr][[$1,$2]]::Fmt[{{0,0},{1,-1},{2,0}},D,$2,$1]
_Dt[Pr][[$1,$2,[$3,1]]]::Fmt[{{0,0},{1,-1},{2,-1},{3,0}},D,$2,$3,$1]
```

The following is a transcript of an interactive SMP session:

```
#I[1]::  <"cafluid.smp"  /* load definitions */

#I[2]::  polygon[6]  /* set up for hexagonal lattice */
```

$$\#0[2]:\quad \{\{1/2,\tfrac{\sqrt{3}}{2}\},\{-1/2,\tfrac{\sqrt{3}}{2}\},\{-1,0\},\{-1/2,-\tfrac{\sqrt{3}}{2}\},\{1/2,-\tfrac{\sqrt{3}}{2}\},\{1,0\}\}$$

```
#I[3]::  Map[nterm,celist]  /* find contributions to number density from
                               terms in Chapman-Enskog expansion */

#0[3]:   {6f,0,0,0}

#I[4]::  Map[uterm,celist]  /* find contributions to momentum vector */

#0[4]:   {{0,0},{3f ux,3f uy},{0,0},{0,0}}

#I[5]::  Map[piterm,celist]  /* stress tensor */
```

$$\#0[5]:* \quad \{\{\{3f,0\},\{0,3f\}\},\{\{0,0\},\{0,0\}\},$$

$$\left\{\left\{\frac{3f\,ux^2}{4}-\frac{3f\,uy^2}{4},\frac{3f\,ux\,uy}{2}\right\},\left\{\frac{3f\,ux\,uy}{2},\frac{-3f\,ux^2}{4}+\frac{3f\,uy^2}{4}\right\}\right\},$$

$$\left\{\left\{\frac{3f\,D_x ux}{4}-\frac{3f\,D_y uy}{4},\frac{3f\,D_y ux}{4}+\frac{3f\,D_x uy}{4}\right\},\right.$$

$$\left.\left\{\frac{3f\,D_y ux}{4}+\frac{3f\,D_x uy}{4},\frac{-3f\,D_x ux}{4}+\frac{3f\,D_y uy}{4}\right\}\right\}\}$$

```
#I[6]::  Dt[f,$$]:0 ;  /* make incompressibility approximation */

#I[7]::  Fac[Map[div,@5]]  /* contributions to momentum equation */
```

$$\#0[7]:* \quad \{\{0,0\},\{0,0\},\left\{\frac{3f\,(ux\,D_x ux + ux\,D_y uy + uy\,D_y ux - uy\,D_x uy)}{2},\right.$$

$$\frac{-3f\,(ux\,D_y ux - ux\,D_x uy - uy\,D_x ux - uy\,D_y uy)}{2}\Big\},$$

$$\left\{\frac{3f\,(D_{xx} ux + D_{yy} ux)}{4},\frac{3f\,(D_{xx} uy + D_{yy} uy)}{4}\right\}\}$$

ACKNOWLEDGMENTS

Many people have contributed in various ways to the material presented here. For general discussions I thank Uriel Frisch, Brosl Hasslacher, David Levermore, Steve Orszag, Yves Pomeau, and Victor Yakhot. For specific suggestions I thank Roger Dashen, Dominique d'Humières, Leo Kadanoff, Paul Martin, John Milnor, Steve Omohundro, Paul Steinhardt, and Larry Yaffe. Most of the calculations described here were made possible by using the SMP general-purpose computer mathematics system[17] I thank Thinking Machines Corporation for much encouragement and partial support of this work.

REFERENCES

1. Wolfram, S., ed. (1986), *Theory and Applications of Cellular Automata* (World Scientific).
2. Wolfram, S. (1984), "Cellular Automata as Models of Complexity," *Nature* **311**, 419.
3. Packard, N., and S. Wolfram (1985), "Two-Dimensional Cellular Automata," *J. Stat. Phys.* **38**, 901.
4. Tritton, D. J. (1977), *Physical Fluid Dynamics* (Van Nostrand).
5. Wood, W. W. (1975), "Computer Studies on Fluid Systems of Hard-Core Particles," *Fundamental Problems in Statistical Mechanics 3*, Ed. E. D. G. Cohen (North-Holland).
6. Hardy, J., Y. Pomeau, and O. de Pazzis (1973), "Time Evolution of a Two-Dimensional Model System. I. Invariant States and Time Correlation Functions," *J. Math. Phys.* **14**, 1746; J. Hardy, O. de Pazzis, and Y. Pomeau (1976), "Molecular Dynamics of a Classical Lattice Gas: Transport Properties and Time Correlation Functions," *Phys. Rev. A* **13**, 1949.
7. Frisch, U., B. Hasslacher, and Y. Pomeau (1986), "Lattice Gas Automata for the Navier-Stokes Equation," *Phys. Rev. Lett.* **56**, 1505.
8. Salem, J., and S. Wolfram (1986), "Thermodynamics and Hydrodynamics with Cellular Automata," *Theory and Applications of Cellular Automata*, Ed. S. Wolfram (World Scientific).
9. d'Humières, D., P. Lallemand, and T. Shimomura (1985), "An Experimental Study of Lattice Gas Hydrodynamics," *Los Alamos preprint LA-UR-85-4051;* D. d'Humières, Y. Pomeau, and P. Lallemand (1985), "Simulation d'Allées de Von karman Bidimensionnelles à l'Aide d'un Gaz sur Reseau," *C. R. Acad. Sci. Paris II* **301**, 1391.
10. Broadwell, J. (1964), "Shock Structure in a Simple Discrete Velocity Gas," *Phys. Fluids* **7**, 1243.

11. Cabannes, H. (1980), "The Discrete Boltzmann Equation," *Lecture Notes* (Berkeley).
12. Gatignol, R. (1975), *Theorie Cinetique des Gaz à Repartition Discrete de Vitesse* (Springer).
13. Hardy, J., and Y. Pomeau (1972), "Thermodynamics and Hydrodynamics for a Modeled Fluid," *J. Math. Phys.* **13**, 1042.
14. Harris, S. (1971), *The Boltzmann Equation* (Holt, Rinehart and Winston).
15. Caflisch, R., and G. Papanicolaou (1979), "The Fluid-Dynamical Limit of a Nonlinear Model Boltzmann Equation," *Commun. Pure Appl. Math.* **32**, 589.
16. Nemnich, B., and S. Wolfram, "Cellular Automaton Fluids 2: Basic Phenomenology," in preparation.
17. Wolfram, S. (1983), *SMP Reference Manual* (Inference Corporation, Los Angeles); S. Wolfram (1985), "Symbolic Mathematical Computation," *Commun. ACM* **28**, 390.
18. Sommerfeld, A. (1955), *Thermodynamics and Statistical Mechanics* (Academic Press).
19. Wolfram, S. (1985), "Origins of Randomness in Physical Systems," *Phys. Rev. Lett.* **55**, 449.
20. Wolfram, S. (1986), "Random Sequence Generation by Cellular Automata," *Adv. Appl. Math.* **7**, 123.
21. Boon, J. P., and S. Yip (1980), *Molecular Hydrodynamics* (McGraw-Hill).
22. Lifshitz, E. M., and L. P. Pitaevskii (1980), *Statistical Mechanics, Part 2* (Pergamon), chapter 9.
23. Liboff, R. (1969), *The Theory of Kinetic Equations* (Wiley).
24. Levermore, D., "Discretization Effects in the Macroscopic Properties of Cellular Automaton Fluids," in preparation.
25. Lifshitz, E. M., and L. P. Pitaevskii (1981), *Physical Kinetics* (Pergamon).
26. Resibois, P., and M. De Leener (1977), *Classical Kinetic Theory of Fluids* (Wiley).
27. Landau, L. D., and E. M. Lifshitz (1959), *Fluid Mechanics*.
28. Ernst, M. H., B. Cichocki, J. R. Dorfman, J. Sharma, and H. van Beijeren (1978), "Kinetic Theory of Nonlinear Viscous Flow in Two and Three Dimensions," *J. Stat. Phys.* **18**, 237.
29. Dorfman, J. R. (1975), "Kinetic and Hydrodynamic Theory of Time Correlation Functions," *Fundamental Problems in Statistical Mechanics 3*, Ed. E. D. G. Cohen (North Holland).
30. Courant, R., and K. O. Friedrichs (1948), *Supersonic Flows and Shock Waves* (Interscience).
31. Levermore, D., private communication.
32. Milnor, J., private communication.
33. Yakhot, V., B. Bayley, and S. Orszag (1986), "Analogy between Hyperscale Transport and Cellular Automaton Fluid Dynamics," *Princeton University preprint (February 1986)*.
34. Coxeter, H. S. M. (1963), *Regular Polytopes* (Macmillan).
35. Hammermesh, M. (1962), *Group Theory* (Addison-Wesley), chapter 9.

36. Landau, L. D., and E. M. Lifshitz (1977), *Quantum Mechanics* (Pergamon), chapter 12.
37. Boerner, H. (1970), *Representations of Groups* (North-Holland), chapter 7.
38. Landau, L. D., and E. M. Lifshitz (1975), *Theory of Elasticity* (Pergamon), section 10.
39. Levine, D., et al. (1985), "Elasticity and Dislocations in Pentagonal and Icosahedral Quasicrystals," *Phys. Rev. Lett.* **14**, 1520.
40. Landau, L. D., and E. M. Lifshitz (1978), *Statistical Physics* (Pergamon), chapter 13.
41. Vainshtein, B. K. (1981), *Modern Crystallography* (Springer), chapter 2.
42. Conway, J. H., and N. J. A. Sloane, to be published.
43. Schwarzenberger, R. L. E. (1980), *N-Dimensional Crystallography* (Pitman).
44. Milnor, J. (1976), "Hilbert's Problem 18: On Crystallographic Groups, Fundamental Domains, and on Sphere Packing," *Proc. Symp. Pure Math.* **28**, 491.
45. Wybourne, B. G. (1974), *Classical Groups for Physicists* (Wiley), p. 78; R. Slansky (1981), "Group Theory for Unified Model Building," *Phys. Rep.* **79**, 1.
46. Grunbaum, B., and G. C. Shepard, *Tilings and Patterns* (Freeman), in press; D. Levine and P. Steinhardt, "Quasicrystals I: Definition and Structure," *Univ. of Pennsylvania preprint.*
47. de Bruijn, N. G. (1981), "Algebraic Theory of Penrose's Non-Periodic Tilings of the Plane," *Nedl. Akad. Wetensch. Indag. Math.* **43**, 39; J. Socolar, P. Steinhardt, and D. Levine (1985), "Quasicrystals with Arbitrary Orientational Symmetry," *Phys. Rev. B.* **32**, 5547.
48. Penrose, R. (1979), "Pentaplexity: A Class of Nonperiodic Tilings of the Plane," *Math. Intelligencer* **2**, 32.
49. Rivet, J. P., and U. Frisch (1986), "Automates sur Gaz de Reseau dans l'Approximation de Boltzmann," *C. R. Acad. Sci. Paris II* **302**, 267.
50. Davis, P. J. (1979), *Circulant Matrices* (Wiley).
51. Landau, L. D., and E. M. Lifshitz (1978), *Statistical Physics* (Pergamon), chapter 5.
52. Kolb, E., and S. Wolfram (1980), "Baryon Number Generation in the Early Universe," *Nucl. Phys. B.* **172**, 224, Appendix A.
53. Gradshteyn, I. S., and I. M. Ryzhik (1965), *Table of Integrals, Series and Products* (Academic Press).
54. Frisch, U., private communication.
55. Roache, P. (1976), *Computational Fluid Mechanics* (Hermosa, Albuquerque).
56. Omohundro, S., and S. Wolfram (1985), unpublished.
57. d'Humières, D., private communication.

Uriel Frisch,* Dominique d'Humières,** Brosl Hasslacher[†] Pierre Lallemand** Yves Pomeau,[‡] and Jean-Pierre Rivet***

*CNRS, Observatoire de Nice, BP 139, 06003 Nice Cedex, France; **CNRS, Laboratoire de Physique de l'Ecole Normale Supérieure, 24 rue Lhomond, 75231 Paris Cedex 05, France; †Theoretical Division and Center for Nonlinear Studies, Los Alamos National Laboratory, Los Alamos, NM 87544, ‡CNRS, Laboratoire de Physique de l'Ecole Normale Supérieure, 24 rue Lhomond, 75231 Paris Cedex 05, France and Physique Théorique, Centre d'Études Nucléaires de Saclay, 91191 Gif-sur-Yvette, France; * * *Observatoire de Nice, BP 139, 06003 Nice Cedex, France and École Normale Supérieure, 45 rue d'Ulm, 75230 Paris Cedex 05, France

Lattice Gas Hydrodynamics in Two and Three Dimensions

This paper originally appeared in *Complex Systems* (1987), Volume 1, pages 649–707.

ERRATA

For equations (4.8) and (4.9), $f_F D$ should be f_{FD}. Equation (4.13) should read

$$N_i = f_{FD} + q_1 f'_{FD} \mathbf{u} \cdot \mathbf{c}_i + h_2 f'_{FD} \mathbf{u}^2 + \frac{1}{2} q_1^2 f''_{FD} (\mathbf{u} \cdot \mathbf{c}_i)^2 + O(u^3). \tag{4.13}$$

The line after Eq. (4.13) should read "Here, f_{FD}, f'_{FD}, and f''_{FD} are the values at...."

Equation (5.15) should read

$$\partial_{t_2}(\rho u_\alpha) + \partial_{1\beta} \left[\left(\psi(\rho) + \frac{D}{2c^2 b} \right) T_{\alpha\beta\gamma\delta} \partial_{1\gamma}(\rho u_\delta) \right] = O(u^2). \tag{5.15}$$

Equations (7.13) and (7.15) should read

$$\rho_0 \partial_t \mathbf{u} + \rho_0 g(\rho_0) \mathbf{u} \cdot \nabla \mathbf{u} = -c_s^2 \nabla \left(\rho' - \rho_0 g(\rho_0) \frac{u^2}{c^2} \right) + \rho_0 \nu(\rho_0) \nabla^2 \mathbf{u}$$

$$\nabla \cdot \mathbf{u} = 0 \tag{7.13}$$

$$\mathbf{r} = \epsilon^{-1} \mathbf{r}_1, \qquad t = \frac{1}{g(\rho_0)} \epsilon^{-2} T, \qquad \mathbf{u} = \epsilon \mathbf{U},$$

$$\rho' - \rho_0 g(\rho_0) \frac{u^2}{c^2} = \frac{\rho_0 g(\rho_0)}{c_s^2} \epsilon^2 P', \qquad \nu = g(\rho_0)\nu'. \tag{7.15}$$

Complex Systems **1** (1987) 649–707

Lattice Gas Hydrodynamics
in Two and Three Dimensions

Uriel Frisch
CNRS, Observatoire de Nice, BP 139, 06003 Nice Cedex, France

Dominique d'Humières
CNRS, Laboratoire de Physique de l'École Normale Supérieure,
24 rue Lhomond, 75231 Paris Cedex 05, France

Brosl Hasslacher
Theoretical Division and Center for Nonlinear Studies,
Los Alamos National Laboratories, Los Alamos, NM 87544, USA

Pierre Lallemand
CNRS, Laboratoire de Physique de l'École Normale Supérieure,
24 rue Lhomond, 75231 Paris Cedex 05, France

Yves Pomeau
CNRS, Laboratoire de Physique de l'École Normale Supérieure,
24 rue Lhomond, 75231 Paris Cedex 05, France
and
Physique Théorique, Centre d'Études Nucléaires de Saclay,
91191 Gif-sur-Yvette, France

Jean-Pierre Rivet
Observatoire de Nice, BP 139, 06003 Nice Cedex, France
and
École Normale Supérieure, 45 rue d'Ulm, 75230 Paris Cedex 05, France

Abstract. Hydrodynamical phenomena can be simulated by discrete lattice gas models obeying cellular automata rules [1,2]. It is here shown for a class of D-dimensional lattice gas models how the macro-dynamical (large-scale) equations for the densities of microscopically conserved quantities can be systematically derived from the underlying exact "microdynamical" Boolean equations. With suitable restrictions on the crystallographic symmetries of the lattice and after proper limits are taken, various standard fluid dynamical equations are obtained, including the incompressible Navier-Stokes equations in two and three dimensions. The transport coefficients appearing in the macrodynamical equations are obtained using variants of the fluctuation-dissipation theorem and Boltzmann formalisms adapted to fully discrete situations.

1. Introduction

It is known that wind or water tunnels can be indifferently used for testing low Mach number flows, provided the Reynolds numbers are identical. Indeed, two fluids with quite different microscopic structures can have the same macroscopic behavior because the *form* of the macroscopic equations is entirely governed by the microscopic conservation laws and symmetries. Although the values of the *transport coefficients* such as the viscosity may depend on the details of the microphysics, still, two flows with similar geometries and identical values for the relevant dimensionless transport coefficients are related by similarity.

Recently, such observations have led to a new simulation strategy for fluid dynamics: fictitious microworld models obeying discrete cellular automata rules have been found, such that two- and three-dimensional fluid dynamics are recovered in the macroscopic limit [1,2]. Cellular automata, introduced by von Neumann and Ulam [3], consist of a lattice, each site of which can have a finite number of states (usually coded by Boolean variables); the automaton evolves in discrete steps, the sites being simultaneously updated by a deterministic or nondeterministic rule. Typically, only a finite number of neighbors are involved in the updating of any site. A very popular example is Conway's Game of Life [4]. In recent years, there has been a renewal of interest in this subject (see, e.g., [5–7]), especially because cellular automata can be implemented in massively parallel hardware [8–10].

The class of cellular automata used for the simulation of fluid dynamics are here called "lattice gas models". Historically, they emerged from attempts to construct discrete models of fluids with varying motivations. The aim of *molecular dynamics* is to simulate the *real* microworld in order, for example, to calculate transport coefficients; one concentrates mass and momentum in discrete particles with continuous time, positions, and velocities and arbitrary interactions [11–14]. Discrete velocity models, introduced by Broadwell [15] (see also [16–20]), have been used mostly to understand rarefied gas dynamics. The velocity set is now finite, space and time are still continuous, and the evolution is probabilistic, being governed by Boltzmann scattering rules. To our knowledge, the first lattice gas model with fluid dynamical features (sound waves) was introduced by Kadanoff and Swift [21]. It uses a master-equation model with continuous time. The first fully deterministic lattice gas model (now known as HPP) with discrete time, positions, and velocities was introduced by Hardy, de Pazzis, and Pomeau [22,23] (see also related work in reference 24). The HPP model, a presentation of which will be postponed to section 2, was introduced to analyze, in as simple a framework as possible, fundamental questions in statistical mechanics such as ergodicity and the divergence of transport coefficients in two dimensions [23]. The HPP model leads to sound waves, which have been observed in simulations on the MIT cellular automaton machine [8]. The difficulties of the HPP model in coping with full fluid dynamics were overcome by Frisch, Hasslacher, and Pomeau [1] for the two-dimensional Navier-Stokes equations; models adapted to the three-dimensional case were introduced by d'Humières, Lallemand, and Frisch [2]. This has led to rapid development of the subject [25–47]. These papers are mostly concerned with lattice gas models leading to the Navier-Stokes equations. A number of other problems are known to be amenable to lattice gas models: buoyancy effects [48], seismic P-waves [49], magneto-hydrodynamics [50–52], reaction-diffusion models [53–55], interfaces and combustion phenomena [56,57], Burgers' model [58].

The aim of this paper is to present in detail and without unnecessary restrictions the theory leading from a simple class of D-dimensional "one-speed" lattice gas models to the continuum macroscopic equations of fluid dynamics in two and three dimensions. The extension of our approach to multi-speed models, including, for example, zero-velocity "rest particles", is quite straightforward; there will be occasional brief comments on such models. We now outline the paper in some detail while emphasizing some of the key steps. Some knowledge of nonequilibrium statistical mechanics is helpful for reading this paper, but we have tried to make the paper self-contained.

Section 2 is devoted to various lattice gas models and their symmetries. We begin with the simple fully deterministic HPP model (square lattice). We then go to the FHP model (triangular lattice) which may be formulated with deterministic or nondeterministic collision rules. Finally, we consider a general class of (usually) nondeterministic, one-speed models containing the pseudo-four-dimensional, face-centered-hypercubic (FCHC) model for use in three dimensions [2]. In this section, we also introduce various abstract symmetry assumptions, which hold for all three models (HPP, FHP, and FCHC) and which will be very useful in reducing the complexity of the subsequent algebra.

In section 3, we introduce the "microdynamical equations", the Boolean equivalent of Hamilton's equations in ordinary statistical mechanics. We then proceed with the probabilistic description of an ensemble of realizations of the lattice gas. At this level, the evolution is governed by a (discrete) Liouville equation for the probability distribution function.

In section 4, we show that there are equilibrium statistical solutions with no equal-time correlations between sites. Under some mildly restrictive assumptions, a Fermi-Dirac distribution is obtained for the mean populations which is universal, i.e., independent of collision rules. This distribution is parametrized by the mean values of the collision invariants (usually, mass and momentum).

Locally, mass and momentum are discrete, but the mean values of the density and mass current can be tuned continuously, just as in the "real world". Furthermore, space and time can be regarded as continuous by considering local equilibria, slowly varying in space and time (section 5). The matching of these equilibria leads to macroscopic PDEs for the conserved quantities.

The resulting "macrodynamical equations", for the density and mass current, are not invariant under arbitrary rotations. However, in section 6, we show that the essential terms in the macroscopic equations become isotropic as soon as the lattice gas has a sufficiently large crystallographic symmetry group (as is the case for the FHP and pseudo-four-dimensional models, but not for the HPP model).

When the necessary symmetries hold, fluid dynamical equations are derived in section 7. We consider various limits involving large scales and times and small velocities (compared to particle speed). In one limit, we obtain the equations of scalar sound waves; in another limit, we obtain the incompressible Navier-Stokes equations in two and three dimensions. It is noteworthy that Galilean invariance, which does not hold at the microscopic level, is restored in these limits.

In section 8, we show how to determine the viscosities of lattice gases. They can be expressed in terms of equilibrium space-time correlation functions via an adaptation to lattice gases of fluctuation-dissipation relations. This is done with a viewpoint of "noisy" hydrodynamics, which also brings out the crossover peculiarities of two dimensions, namely a residual weak scale-dependence of transport coefficients at large scales. Alternatively, fluctuation-dissipation relations can be obtained from the Liouville equation with a Green-Kubo formalism [43]. Fully explicit expressions for the viscosities can be derived via the "Lattice Boltzmann Approximation", not needed for any earlier steps. This is a finite-difference variant of the discrete-velocity Boltzmann approximation. The latter, which assumes continuous space and time variables, is valid only at low densities, while its lattice variant seems to capture most of the finite-density effects (with the exception of two-dimensional crossover effects). Further studies of the Lattice Boltzmann Approximation may be found in reference 42. Implications for the question of the Reynolds number are discussed at the end of the section.

Section 9 is the conclusion. Various questions are left for the appendices: detailed technical proofs, inclusion of body forces, catalog of results for various FHP models, proof of an H-theorem for the Lattice Boltzmann Approximation (due to M. Hénon).

2. Deterministic and nondeterministic lattice gas models

2.1 The HPP model

Let us begin with a heuristic construction of the HPP model [22–24]. Consider a two-dimensional square lattice with unit lattice constant as shown in figure 1. Particles of unit mass and unit speed are moving along the

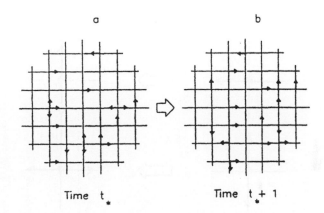

Figure 1: The HPP model. The black arrows are for cell-occupation. In (a) and (b) the lattice is shown at two successive times.

lattice links and are located at the nodes at integer times. Not more than one particle is to be found at a given time and node, moving in a given direction (exclusion principle). When two and exactly two particles arrive at a node from opposite directions (head-on collisions), they immediately leave the node in the two other, previously unoccupied, directions (see figure 2). These deterministic collision laws obviously conserve mass (particle number) and momentum and are the only nontrivial ones with these properties. Furthermore, they have the same discrete invariance group as the lattice.

The above definition can be formalized as follows. We take an L by L square lattice, periodically wrapped around (a nonessential assumption, made for convenience). Eventually, we will let $L \to \infty$. At each node, labeled by the discrete vector r_*, there are four cells labeled by an index i, defined modulo four. The cells are associated to the unit vectors c_i connecting the node to its four nearest neighbors (i increases counterclockwise). Each cell (r_*, i) has two states coded with a Boolean variable: $n_i(r_*) = 1$ for "occupied" and $n_i(r_*) = 0$ for "unoccupied". A *cellular automaton* updating rule is defined on the *Boolean field* $n_{\cdot} = \{n_i(r_*), i = 1, \ldots, 4, r_* \in \text{Lattice}\}$. It has two steps. Step one is *collision*: at each node, the four-bit states $(1, 0, 1, 0)$ and $(0, 1, 0, 1)$ are exchanged; all other states are left unchanged. Step two is *propagation*: $n_i(r_*) \to n_i(r_* - c_i)$. This two-step rule is applied at each integer time, t_*. An example of implementation of the rule, in which arrows stand for cell-occupation, is shown in figures 1a and b.

Collisions in the HPP model conserve mass and momentum locally,

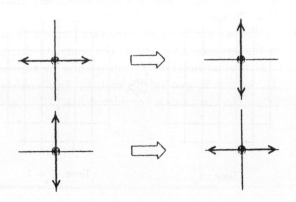

Figure 2: Collision rules for the HPP model.

whereas propagation conserves them globally. (Actually, momentum is conserved along each horizontal and vertical line, resulting in far too many conserved quantities for physical modeling.) If we attribute to each particle a kinetic energy $\frac{1}{2}$, the total kinetic energy is also conserved. Energy conservation is, however, indistinguishable from mass conservation and will not play any dynamical role. Models having an energy conservation law independent of mass conservation will not be considered in this paper (see [2,29]).

The dynamics of the HPP model is invariant under all discrete transformations that conserve the square lattice: discrete translations, rotations by $\frac{\pi}{2}$, and mirror symmetries. Furthermore, the dynamics is invariant under *duality*, that is exchange of 1's and 0's (particles and holes).

2.2   The FHP models

The FHP models I, II, and III (see below), introduced by Frisch, Hasslacher, and Pomeau [1] (see also [25–31,35,38–44,46]) are variants of the HPP model with a larger invariance group. The residing lattice is triangular with unit lattice constant (figure 3). Each node is now connected to its six neighbors by unit vectors c_i (with i defined modulo six) and is thus endowed with a six-bit state (or seven, see below). Updating involves again propagation (defined as for HPP) and collisions.

In constructing collision rules on the triangular lattice, we must consider

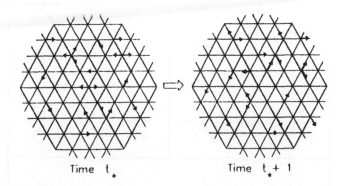

Figure 3: The FHP model with binary head-on and triple collisions at two successive times.

both *deterministic* and *nondeterministic* rules. For a head-on collision with occupied "input channels" $(i, i + 3)$, there are two possible pairs of occupied "output channels" such that mass and momentum are conserved, namely $(i + 1, i + 4)$ and $(i - 1, i - 4)$ (see figure 4a). We can decide always to choose one of these channels; we then have a deterministic model, which is *chiral*, i.e., not invariant under mirror-symmetry. Alternatively, we can make a nondeterministic (random) choice, with equal probabilities to restore mirror-symmetry. Finally, we can make a pseudo-random choice, dependent, for example, on the parity of a time or space index.

We must also consider *spurious conservation laws*. Head-on collisions conserve, in addition to total particle number, the difference of particle numbers in any pair of opposite directions $(i, i+3)$. Thus, head-on collisions on a triangular lattice conserve a total of four scalar quantities. This means that in addition to mass and momentum conservation, there is a spurious conservation law. The large-scale dynamics of such a model will differ drastically from ordinary hydrodynamics, unless the spurious conservation law is removed. One way to achieve this is to introduce triple collisions $(i, i + 2, i + 4) \rightarrow (i + 1, i + 3, i + 5)$ (see figure 4b).

Several models can be constructed on the triangular lattice. The simplest set of collision rules with no spurious conservation law, which will be called FHP-I, involves only (pseudo-random) binary head-on collisions and triple collisions. FHP-I is not invariant under duality (particle-hole exchange), but can be made so by inclusion of the duals of the head-on collisions (see figure 4c). Finally, the set of collision rules can be saturated (exhausted) by inclusion of head-on collisions with a "spectator" [59], that is, a particle which remains unaffected in a collision; figure 4d is an example of a head-on collision with a spectator present.

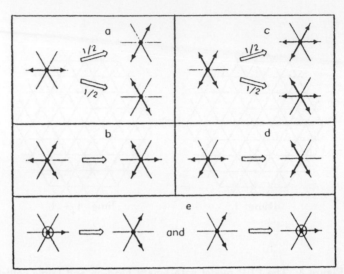

Figure 4: Collision rules for the FHP models: (a) head-on collision with two output channels given equal weights; (b) triple collision; (c) dual of head-on collision under particle-hole exchange; (d) head-on collision with spectator; (e) binary collisions involving one rest particle (represented by a circle).

The model, FHP-II, is a seven-bit variant of FHP-I including a zero-velocity "rest particle", the additional collision rules of figure 4e, and variants of the head-on and triple collisions of figures 4a and 4b with a spectator rest particle. Binary collisions involving rest particles remove spurious conservations, and do so more efficiently at low densities than triple collisions. Finally, model FHP-III is a collision-saturated version of FHP-II [31]. For simplicity, we have chosen not to cover the theory of models with rest particles in detail.

The dynamics of the FHP models are invariant under all discrete transformations that conserve the triangular lattice: discrete translations, rotations by $\pi/3$, and mirror symmetries with respect to a lattice line (we exclude here the chiral variants of the models).

2.3   The FCHC four-dimensional and the pseudo-four-dimensional models

Three dimensional regular lattices do not have enough symmetry to ensure macroscopic isotropy [1,2,39]. A suitable four-dimensional model has been introduced by d'Humières, Lallemand, and Frisch [2]. The residing lattice is face-centered-hypercubic (FCHC), defined as the set of signed integers (x_1, x_2, x_3, x_4) such that $x_1 + x_2 + x_3 + x_4$ is even. Each node is connected via links of length $c = \sqrt{2}$ to 24 nearest neighbors, having two coordinates differing by ± 1. Thus, the FCHC model has 24-bit states. The 24 possible velocity vectors are again denoted c_i; for the index i, there is no preferred ordering and we will leave the ordering unspecified. Propagation for the FCHC lattice gas is as usual. Collision rules should conserve mass and four-momentum while avoiding spurious conservations. This can be achieved with just binary collisions, but better strategies are known [32,33]. Non-deterministic rules involving transition probabilities are needed to ensure that the collisions and the lattice have the same invariance group (precise definitions are postponed to section 2.4).

The allowed transformations of the FCHC model are discrete transla-tions and those isometries generated by permutations of coordinates, rever-sal of one or several coordinates and symmetry with respect to the hyper-plane $x_1 + x_2 + x_3 + x_4 = 0$.

To model three-dimensional fluids and maintain the required isotropies, we define the pseudo-four-dimensional model [2] as the three-dimensional projection of an FCHC model with unit periodicity in the x_4-direction (see figure 5). It resides on an ordinary cubic lattice with unit lattice constant. The full four-dimensional discrete velocity structure is preserved as follows. There is one communication channel to the twelve next-nearest neighbors (corresponding to the twelve velocity vectors such that v_4, the fourth com-ponent of the velocity, vanishes) and there are two communication channels to the six nearest neighbors (corresponding respectively to velocities with $v_4 = \pm 1$). During the propagation phase, particles with $v_4 = \pm 1$ move to nearest neighbor nodes, while particles with $v_4 = 0$ move to next-nearest neighbors. The collision strategy is the same as for the FCHC model, so that four-momentum is conserved. The fourth component is not a spuri-ously conserved quantity because, in the incompressible limit, it does not effectively couple back to the other conserved quantities [2].

2.4   A general class of nondeterministic models

In most of this paper, we will work with a class of models (generally non-deterministic) encompassing all the above one-speed models. The relevant common aspects of all those models are now listed: there is a regular lat-tice, the nodes of which are connected to nearest neighbors through links of equal length; all velocity directions are in some sense equivalent and the velocity set is invariant under reversal; at each node there is a cell associ-ated with each possible velocity. This cell can be occupied by one particle at most; particles are indistinguishable; particles are marched forward in time by successively applying collision and propagation rules; collisions are purely local, having the same invariances as the velocity set; and collisions conserve only mass and momentum.

We now give a more formal definition of these one-speed models as cellular automata. Let us begin with the geometrical aspects. We take a D-dimensional Bravais lattice $\mathcal{L}$ in $\mathbf{R}^D$ of finite extension $O(L)$ in all directions (eventually, $L \to \infty$); the position vector r_* of any node of such a lattice is a linear combination with integer coefficients of D independent generating vectors [60]. We furthermore assume that there exists a set of

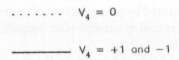

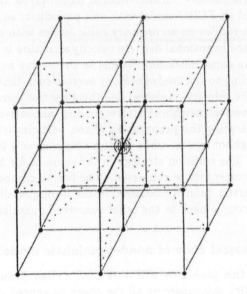

Figure 5: The pseudo-four-dimensional FCHC model. Only the neighborhood of one node is shown. Along the dotted links, connecting to next-nearest neighbors, at most one particle can propagate, with component $v_4 = 0$; along the thick black links, connecting to nearest neighbors, up to two particles can propagate, with components $v_4 = \pm 1$.

b "velocity vectors" c_i having equal modulus c, the *particle speed*. c_i has spatial components $c_{i\alpha}$ $(\alpha = 1, \ldots, D)$ [1] We require the following for c_i:

1. For any $r_* \in \mathcal{L}$, the set of the $r_* + c_i$'s is the set of nearest neighbors of r_*.

2. Any two nodes can be connected via a finite chain of nearest neighbors.

3. For any pair (c_i, c_j) there exists an element in the "crystallographic" group $\mathcal{G}$ of isometries globally preserving the set of velocity vectors, which maps c_i into c_j.

4. For any velocity vector c_i, we denote by $\mathcal{G}_i$ the subgroup of $\mathcal{G}$ which leaves c_i invariant and thus leaves its orthogonal hyperplane, Π_i, globally invariant; we assume that (a) there is no non-vanishing vector in Π_i invariant under all the elements of $\mathcal{G}_i$ and (b) the only linear transformations within the space Π_i commuting with all the elements of $\mathcal{G}_i$ are proportional to the identity.

Now, we construct the automaton. To each node r_* we attach a b-bit state $n(r_*) = \{n_i(r_*), i = 1, \ldots, b\}$, where the n_i's are Boolean variables. The updating of the "Boolean field", $n(.)$, involves two successive steps: collision followed by propagation. We choose this particular order for technical convenience; after a large number of iterations, it will become irrelevant which step was first.[2] Propagation is defined as

$$n_i(r_*) \to n_i(r_* - c_i).\tag{2.1}$$

The spatial shifting by c_i is performed on a periodically[3] wrapped around lattice with $O(L)$ sites in any direction; eventually, $L \to \infty$. Collision is the simultaneous application at each node of nondeterministic transition rules from an in-state $s = \{s_i, i = 1, \ldots, b\}$ to an out-state $s' = \{s'_i, i = 1, \ldots, b\}$. Each transition is assigned a probability $A(s \to s') \geq 0$, normalized to one ($\sum_{s'} A(s \to s') = 1 \; \forall s$), and depending only on s and s' and not on the node. The following additional assumptions are made.

5. Conservation laws: the *only* collections of b real numbers a_i such that

$$\sum_i (s'_i - s_i) A(s \to s') a_i = 0, \quad \forall s, s',\tag{2.2}$$

---

[1] In this paper, Greek and Roman indices refer respectively to components and velocity labels. Summation over repeated Greek indices, but not Roman ones, is implicit.

[2] For deterministic lattice gases, such as HPP, it is possible to bring out the reversibility of the updating rule by defining the state of the automaton at half-integer times, with particles located at the middle of links connecting nearest-neighbor nodes; updating then comprises half a propagation, followed by collision, followed by another half propagation [22].

[3] Other boundary conditions at the lattice edge can also be used—for example, "wind-tunnel" conditions [25,26,28].

are linear combinations of 1 (for all i) and of $c_{i1}, ..., c_{iD}$, i.e. a_i is related to mass and momentum conservation.

6. Invariance under all isometries preserving the velocity set:

$$A\big(g(s) \to g(s')\big) = A(s \to s'), \quad \forall g \in \mathcal{G}, \quad \forall s, s'. \tag{2.3}$$

7. Semi-detailed balance:

$$\sum_s A(s \to s') = 1, \quad \forall s'. \tag{2.4}$$

Various comments are now in order. Semi-detailed balance, also used in discrete velocity Boltzmann models [17], means that if before collision all states have equal probabilities, they stay so after collision. It is trivially satisfied when the collision rule is deterministic and one-to-one. There exists also a stronger assumption, detailed balance (that is $A(s \to s') = A(s' \to s)$), which will not be needed here. The HPP, FHP, and FCHP lattice gases satisfy the above assumptions (1) through (4). The proofs are given in Appendix A. The other assumptions (5) through (7) hold by construction with the exception of the chiral versions of FHP. The latter do not satisfy (6) because the collision rules are not invariant under the mirror-symmetries with respect to velocity vectors. Full $\mathcal{G}$-invariance holds for the *velocity set* of the pseudo-four-dimensional model, which is the same as for the FCHC model; however, the spatial structure is only invariant under the smaller group of the three-dimensional cubic lattice.

The invariance assumptions introduced above have important consequences for the transformation properties of vectors and tensors. The following definitions will be used. A tensor is said to be $\mathcal{G}$-invariant if it is invariant under any isometry in $\mathcal{G}$. A set of i-dependent tensors of order p $\{T_i = t_{i\alpha_1\alpha_2...\alpha_p}, i = 1, ..., b\}$ is said to be $\mathcal{G}$-invariant if any isometry in $\mathcal{G}$ changing c_i into c_j, changes T_i into T_j. Note that this is stronger than global invariance under the group $\mathcal{G}$. The velocity moment of order p is defined as $\sum_i c_{i\alpha_1} c_{i\alpha_2} \cdots c_{i\alpha_p}$.

We now list the transformation properties following from $\mathcal{G}$-invariance. The proofs are given in Appendix B.

P1 *Parity-invariance.* The set of velocity vectors is invariant under space-reversal.

P2 Any set of i-dependent vectors $v_{i\alpha}$, which is $\mathcal{G}$-invariant, is of the form $\lambda c_{i\alpha}$.

P3 Any set of i-dependent tensors $t_{i\alpha\beta}$, which is $\mathcal{G}$-invariant, is of the form $\lambda c_{i\alpha} c_{i\beta} + \mu \delta_{\alpha\beta}$.

P4 *Isotropy of second-order tensors.* Any $\mathcal{G}$-invariant tensor $t_{\alpha\beta}$ is of the form $\mu \delta_{\alpha\beta}$.

P5 Any $\mathcal{G}$-invariant third-order tensor vanishes.

P6 *Velocity moments.* Odd-order velocity moments vanish. The second-order velocity moment is given by

$$\sum_i c_{i\alpha}c_{i\beta} = \frac{bc^2}{D}\delta_{\alpha\beta}.\tag{2.5}$$

There is, in general, no closed form expression for even-order velocity moments beyond second order, with the assumptions made up to this point (see section 6).

3. Microdynamics and probabilistic description

3.1 Microdynamical equations

It is possible to give a compact representation of the "microdynamics", describing the application of the updating rules to the Boolean field. This is the cellular automaton analog of Hamilton's equations of motion in classical statististical mechanics. We begin with the HPP lattice gas (section 2.1). Let $n_i(t_*, r_*)$, as defined in section 2.1, denote the HPP Boolean field at the discrete time t_*. With i labeling the four cells of an HPP node, the collision rule can be formulated as follows: If the in-state has i and $i+2$ empty and $i+1$ and $i+3$ occupied, then the opposite holds in the out-state; similarly, if the in-state has $i+1$ and $i+3$ empty and i and $i+2$ occupied; otherwise, the content of cell i is left unchanged. Thus, the updating of the Boolean field may be written

$$n_i(t_* + 1, r_* + c_i) =$$
$$\left(n_i \wedge \neg(n_i \wedge n_{i+2} \wedge \neg n_{i+1} \wedge \neg n_{i+3})\right) \vee \left(n_{i+1} \wedge n_{i+3} \wedge \neg n_i \wedge \neg n_{i+2}\right)\tag{3.1}$$

where the whole r.h.s. is evaluated at t_* and r_*. The symbols $\wedge$, $\vee$, and $\neg$ stand for **AND, OR,** and **NOT** respectively. It is known that any Boolean relation can be recoded in arithmetic form ($\wedge$ becomes multiplication, $\neg$ becomes one minus the variable, etc.). In this way, we obtain

$$n_i(t_* + 1, r_* + c_i) = n_i(t_*, r_*) + \Delta_i(n).\tag{3.2}$$

The "collision function" $\Delta_i(n)$, which can take the values ± 1 and 0, describes the change in $n_i(t_*, r_*)$ due to collisions. For the HPP model, it depends only on i and on the set of n_j's at t_* and r_*, denoted n; it is given by

$$\Delta_i(n) =$$
$$n_{i+1}n_{i+3}(1 - n_i)(1 - n_{i+2}) - n_i n_{i+2}(1 - n_{i+1})(1 - n_{i+3}).\tag{3.3}$$

Equation (3.2) (with $\Delta_i(n)$ given by equation (3.3)) will be called the *microdynamical HPP equation*. It holds for arbitrary i (modulo four), for arbitrary integer t_*, and for arbitrary $r_* \in \mathcal{L}$ ($\mathcal{L}$ designates the lattice).

It is easy to extend the microdynamical formalism to other models. For FHP-I (section 2.2), we find that the collision function may be written (i is now defined modulo six)

$$
\begin{aligned}
\Delta_i(n) = \xi_{i,r_*} \quad & n_{i+1}n_{i+4}(1 - n_i)(1 - n_{i+2})(1 - n_{i+3})(1 - n_{i+5}) \\
+(1 - \xi_{i,r_*}) \quad & n_{i+2}n_{i+5}(1 - n_i)(1 - n_{i+1})(1 - n_{i+3})(1 - n_{i+4}) \\
- \quad & n_i n_{i+3}(1 - n_{i+1})(1 - n_{i+2})(1 - n_{i+4})(1 - n_{i+5}) \quad (3.4) \\
+ \quad & n_{i+1}n_{i+3}n_{i+5}(1 - n_i)(1 - n_{i+2})(1 - n_{i+4}) \\
- \quad & n_i n_{i+2}n_{i+4}(1 - n_{i+1})(1 - n_{i+3})(1 - n_{i+5})
\end{aligned}
$$

Here, ξ_{i,r_*} denotes a time- and site-dependent Boolean variable which takes the value one when head-on colliding particles are to be rotated counterclockwise and zero otherwise (remember, that there are two possible outcomes of such collisions). For the theory, the simplest choice is to assign equal probabilities to the two possibilities and to assume independence of all the ξ's. In practical implementations, other choices are often more convenient.

We now give the microdynamical equation for the general class of nondeterministic models defined in section 2.4. Propagation is as before. For the collision phase at a given node, it is convenient to sum over all 2^b in-states $s = \{s_i = 0 \text{ or } 1, \; i = 1, \ldots, b\}$ and 2^b out-states s'. The nondeterministic transitions are taken care of by the introduction at each time and node and for any pair of states (s, s') of a Boolean variable $\xi_{ss'}$ (time and space labels omitted for conciseness). We assume that

$$
\langle \xi_{ss'} \rangle = A(s \to s'), \quad \forall s, s', \tag{3.5}
$$

where $A(s \to s')$ is the transition probability introduced in section 2.4; the angular brackets denote averaging. We also assume that

$$
\sum_{s'} \xi_{ss'} = 1, \quad \forall s. \tag{3.6}
$$

Since the ξ's are Boolean, equation (3.6) means that, for a given in-state s and a given realization of $\xi_{ss'}$, one and only one out-state s' is obtained. It is now clear that the microdynamical equation can be written as

$$
n_i(t_* + 1, r_* + c_i) = \sum_{s,s'} s_i' \xi_{ss'} \prod_j n_j^{s_j}(1 - n_j)^{(1 - s_j)}. \tag{3.7}
$$

The factor s_i' ensures the presence of a particle in the cell i after the collision; the various factors in the product over the index j ensure that before the collision the pattern of n_j's matches that of s_j's. Using equation (3.7) and the identity

$$
\sum_s s_i \prod_j n_j^{s_j}(1 - n_j)^{(1 - s_j)} = n_i, \tag{3.8}
$$

we can rewrite the microdynamical equation in a form that brings out the collision function

$$
n_i(t_* + 1, r_* + c_i) = n_i + \Delta_i(n)
$$

$$
\Delta_i(n) = \sum_{s,s'} (s_i' - s_i) \xi_{ss'} \prod_j n_j^{s_j}(1 - n_j)^{(1 - s_j)}. \tag{3.9}
$$

In the sequel, it will often be useful to have a compact notation. We define the collision operator,

$$C : n_i(\mathbf{r}_*) \mapsto n_i(\mathbf{r}_*) + \Delta_i(n(\mathbf{r}_*)), \tag{3.10}$$

the streaming operator,

$$S : n_i(\mathbf{r}_*) \mapsto n_i(\mathbf{r}_* - \mathbf{c}_i), \tag{3.11}$$

and the evolution operator, the composition of the latter,

$$\mathcal{E} = S \circ C. \tag{3.12}$$

The entire updating can now be written as

$$n(t_* + 1, .) = \mathcal{E} n(t_*, .), \tag{3.13}$$

where the point in the second argument of the n's stands for all the space variables.

An interesting property of the microdynamical equation, not shared by the Hamilton equations of ordinary statistical mechanics, is that it remains meaningful for an *infinite* lattice, since the updating of any given node involves only a finite number of neighbors.

3.2   Conservation relations

Conservation of mass and momentum *at each node* in the collision process can be expressed by the following relations for the collision function:

$$\sum_i \Delta_i(n) = 0, \quad \forall n \in \{0, 1\}^b, \tag{3.14}$$

$$\sum_i \mathbf{c}_i \Delta_i(n) = 0, \quad \forall n \in \{0, 1\}^b, \tag{3.15}$$

where $\{0, 1\}^b$ denotes the set of all possible b-bit words. This implies important conservation relations for the Boolean field:

$$\sum_i n_i(t_* + 1, \mathbf{r}_* + \mathbf{c}_i) = \sum_i n_i(t_*, \mathbf{r}_*), \tag{3.16}$$

$$\sum_i \mathbf{c}_i n_i(t_* + 1, \mathbf{r}_* + \mathbf{c}_i) = \sum_i \mathbf{c}_i n_i(t_*, \mathbf{r}_*). \tag{3.17}$$

3.3   The Liouville equation

We now make the transition, traditional in statistical mechanics, from a deterministic to a probabilistic point of view. This can be obscured by the fact that some of our models are already probabilistic. So, let us assume for a while that the evolution operator is deterministic and invertible (as is the case for HPP).

Assuming that we have a finite lattice, we define the phase space, Γ, as the set of all possible *assignments* $s(.) = \{s_i(r_*), i = 1, \ldots, b, r_* \in \mathcal{L}\}$ of the Boolean field $n_i(r_*)$. A particular assignment of the Boolean field will be called a *configuration*. We now consider at time $t_* = 0$ an *ensemble* of initial conditions, each endowed with a probability $P(0, s(.)) \geq 0$, such that

$$\sum_{s(.) \in \Gamma} P(0, s(.)) = 1. \tag{3.18}$$

We let each configuration in the ensemble evolve according to the automaton updating rule, i.e., with the evolution operator $\mathcal{E}$ of equation (3.13). The latter being, here, invertible, conservation of probability is expressed as

$$P(t_* + 1, s(.)) = P(t_*, \mathcal{E}^{-1} s(.)). \tag{3.19}$$

This equation is clearly the analog of the Liouville equation of statistical mechanics, and will be given the same name. Alternatively, the Liouville equation can be written

$$P(t_* + 1, S s(.)) = P(t_*, C^{-1} s(.)). \tag{3.20}$$

To derive this, we have used equation (3.12) and put the streaming operator in the l.h.s., a form which will be more convenient subsequently.

In the *nondeterministic* case, we must enlarge the probability space to include not only the phase space of initial conditions, but the space of all possible choices of the Boolean variables $\xi(ss')$, which at each time and each node select the unique transition from a given in-state s (see section 3.1). Since the ξ's are independently chosen at each time, the entire Boolean field $n(t_*, .)$ is a Markov process (with deterministic rules, this process is degenerate). What we will continue to call the Liouville equation is actually the Chapman-Kolmogorov equation for this Markov process, namely

$$P(t_* + 1, S s'(.)) = \sum_{s(.) \in \Gamma} \prod_{r_* \in \mathcal{L}} A(s(r_*) \to s'(r_*)) P(t_*, s(.)). \tag{3.21}$$

This equation just expresses that the probability at $t_* + 1$ of a given (propagated) configuration $s'(.)$ is the sum of the probabilities at t_* of all possible original configurations $s(.)$ times the transition probability. The latter is a product, because we assumed that the ξ's are chosen independently at each node. In the deterministic case, $A(s(r_*) \to s'(r_*))$ selects the unique configuration $C^{-1} s'(.)$, so that equation (3.20) is recovered.

3.4 Mean quantities

Having introduced a probablistic description, we now turn to mean quantities. For an "observable" $q(n(t_*, .))$, which depends on the Boolean field at a single time, the mean is given by ensemble averaging over $P(t_*, s(.))$

$$\left\langle q\big(n(t_*, .)\big)\right\rangle = \sum_{s(.)\in\Gamma} q\big(s(.)\big) P\big(t_*, s(.)\big). \qquad (3.22)$$

An important role will be played in the sequel by the following mean quantities: the *mean population*

$$N_i(t_*, r_*) = \big\langle n_i(t_*, r_*)\big\rangle, \qquad (3.23)$$

the *density*, and the *mass current* (mean momentum)

$$\rho(t_*, r_*) = \sum_i N_i(t_*, r_*), \quad j(t_*, r_*) = \sum_i c_i N_i(t_*, r_*). \qquad (3.24)$$

Note that these are mean quantities per node, not per unit area or volume. The *density per cell* is defined as $d = \rho/b$. Finally, the *mean velocity* u is defined by

$$j(t_*, r_*) = \rho(t_*, r_*) u(t_*, r_*). \qquad (3.25)$$

Note that under duality (exchange of particles and holes), ρ changes into $b - \rho$, d into $1 - d$, , j into $-j$, and u into the "mean hole-velocity" $u_H = -u d/(1 - d)$.

Averaging of the microdynamical conservation relations 3.16 and 3.17 leads to conservation relations for the mean populations

$$\sum_i N_i(t_* + 1, r_* + c_i) = \sum_i N_i(t_*, r_*), \qquad (3.26)$$

$$\sum_i c_i N_i(t_* + 1, r_* + c_i) = \sum_i c_i N_i(t_*, r_*). \qquad (3.27)$$

4. Equilibrium solutions

It has been shown by Hardy, Pomeau, and de Pazzis [22] that the HPP model has very simple statistical equilibrium solutions (which they call invariant states) in which the Boolean variables at all the cells are independent. Such equilibrium solutions are the lattice gas equivalent of Maxwell states in classical statistical mechanics and are therefore crucial for deriving hydrodynamics. There are similar results for the general class of nondeterministic models introduced in section 2.4, which are now discussed.

4.1   Steady solutions of the Liouville equation

We are interested in equilibrium solutions, that is, steady-state solutions of the Liouville equation (3.21) for a finite, periodically wrapped around lattice. Collisions on the lattice are purely local (their impact parameter is zero). This suggests the existence of equilibrium solutions with no single-time spatial correlations. The lattice properties being translation-invariant, the distribution should be the same at each node. Thus, we are looking for equilibrium solutions of the form

$$P\big(s(.)\big) = \prod_{r_*} p\big(s(r_*)\big), \qquad (4.1)$$

where $p(s)$, the probability of a given state, is node-independent. Maximization of the entropy (see Appendix F) suggests that $p(s)$ should be completely factorized over all cells, that is, of the form

$$p(s) = \prod_i N_i^{s_i}(1 - N_i)^{(1-s_i)}. \qquad (4.2)$$

Note that $N_i^{s_i}(1 - N_i)^{(1-s_i)}$ is the probability of a Boolean variable with mean N_i.

Now, we must check that there are indeed solutions of the form that we have been guessing. Substitution of $P(s(.))$ given by equation (4.1) with $p(s)$ given by equation (4.2) into the Liouville equation (3.21) leads to

$$\prod_j N_j^{s'_j}(1 - N_j)^{(1-s'_j)} = \sum_s A(s \to s') \prod_j N_j^{s_j}(1 - N_j)^{(1-s_j)}, \quad \forall s', \quad (4.3)$$

where N_i is the mean population of cell i, independent of the node and of the time.

Equation (4.3) is a set 2^b (the number of different states) equations for b unknowns. The fact that it actually possesses solutions is nontrivial. Furthermore, these solutions can be completely described. Indeed, we have the following lemma.

Lemma 1. *The following statements are equivalent:*

1. *The N_i's are a solution of equation (4.3).*

2. *The N_i's are a solution of the set of b equations*

$$\sum_{ss'}(s'_i - s_i)A(s \to s') \prod_j N_j^{s_j}(1 - N_j)^{(1-s_j)} = 0, \quad \forall i. \qquad (4.4)$$

3. *The N_i's are given by the Fermi-Dirac distribution*

$$N_i = \frac{1}{1 + \exp(h + \mathbf{q} \cdot \mathbf{c}_i)}, \qquad (4.5)$$

where h is an arbitrary real number and $\mathbf{q}$ is an arbitrary D-dimensional vector.

The proof of the equivalence is given in Appendix C; it makes use of semi-detailed balance and the absence of spurious invariants. The most important consequence of the lemma is the *Universality Theorem*. Nondeterministic lattice gas models satisfying semi-detailed balance and having no spurious invariants admit universal equilibrium solutions, completely factorized over all nodes and all cells, with mean populations given by the Fermi-Dirac distribution (4.5), dependent only on the density ρ and the mass current $\mathbf{j} = \rho\mathbf{u}$, and independent of the transition probabilities $A(s \to s')$.

The proof follows from the observation that the Lagrange multipliers h and $\mathbf{q}$ of the Fermi-Dirac distribution can be calculated in terms of the density and the mass current through the relations

$$\rho = \sum_i N_i = \sum_i \frac{1}{1 + \exp(h + \mathbf{q} \cdot \mathbf{c}_i)}, \tag{4.6}$$

$$\rho\mathbf{u} = \sum_i N_i \mathbf{c}_i = \sum_i \mathbf{c}_i \frac{1}{1 + \exp(h + \mathbf{q} \cdot \mathbf{c}_i)}. \tag{4.7}$$

For the HPP model, this set of equations is reducible to a cubic polynomial equation, so that explicit solutions are known [22]. For the FHP model, explicit solutions are known only for special cases [61].

It is not particularly surprising for models that have a built-in exclusion principle (not more than one particle per cell) to obtain a Fermi-Dirac distribution at equilibrium. Note that the factorized equilibrium solutions remain meaningful on an infinite lattice. There is no proof at the moment that the only equilibrium solutions which are relevant in the limit of infinite lattices are of the above form, namely completely factorized (which then *implies* the Fermi-Dirac distribution). There is strong numerical evidence, for those models that have been simulated, that the Fermi-Dirac is the only relevant one [8,25,27].

4.2 Low-speed equilibria

In the "real world", equilibrium distributions with different mean velocities are simply related by a Galilean transformation. Galilean invariance does not hold at the microscopic level for a lattice gas; therefore, there is no simple relation between the equilibria with vanishing and nonvanishing mean velocity. For subsequent derivations of fluid dynamical equations, we will only need equilibria with low speeds, that is with $u = |\mathbf{u}| \ll c$, the particle speed. Such equilibria can be calculated perturbatively in powers of u.

We write the equilibrium distribution as

$$N_i = f_F D\big(h(\rho, \mathbf{u}) + \mathbf{q}(\rho, \mathbf{u}) \cdot \mathbf{c}_i\big), \tag{4.8}$$

where we have used the Fermi-Dirac function

$$f_F D(x) = \frac{1}{1 + e^x}. \tag{4.9}$$

We observe that

$$\mathbf{u} = 0 \Rightarrow N_i = \frac{\rho}{b} = d. \tag{4.10}$$

Indeed, by assumption (3) of section 2.4, there exists an isometry of the lattice exchanging any two velocity vectors c_i and c_j; the vector $\mathbf{u} = 0$ being also trivially invariant, the mean population N_i is independent of i. Thus, $f_{FD}(h(\rho, 0)) = d$ and $\mathbf{q}(\rho, 0) = 0$.

Furthermore, it follows from parity-invariance $(\mathbf{u} \to -\mathbf{u}, \; c_i \to -c_i)$ that

$$h(\rho, -\mathbf{u}) = h(\rho, \mathbf{u}), \quad \mathbf{q}(\rho, -\mathbf{u}) = -\mathbf{q}(\rho, \mathbf{u}). \tag{4.11}$$

We now expand h and $\mathbf{q}$ in powers of $\mathbf{u}$

$$h(\rho, \mathbf{u}) = h_0 + h_2 u^2 + O(u^4)$$

$$q_\alpha(\rho, \mathbf{u}) = q_1 u_\alpha + O(u^3), \tag{4.12}$$

where h_0, h_2, and q_1 depend on ρ. The fact that h_2 and q_1 are scalars rather than second-order tensors is a consequence of the isotropy of second-order tensors (property P4 of section 2.4). We substitute equation (4.12) into equation (4.8) and expand the mean populations in powers of $\mathbf{u}$

$$N_i = f_F D + q_1 f_F' D \mathbf{u} \cdot c_i + h_2 f_F' D u^2 + \frac{1}{2} q_1^2 f_F'' D (\mathbf{u} \cdot c_i)^2 + O(u^3). \tag{4.13}$$

Here, $f_F D$, $f_F' D$, and $f_F'' D$ are the values at h_0 of the Fermi-Dirac function and its first and second derivatives. From equation (4.13), we calculate the density $\rho = \sum_i N_i$ and the mass current $\rho \mathbf{u} = \sum_i c_i N_i$, using the velocity moment relations (P6 of section 2.4). Identification gives h_0, h_2, and q_1 in terms of ρ. This is then used to calculate the equilibrium mean population up to second order in $\mathbf{u}$; we obtain

$$N_i^{eq}(\rho, \mathbf{u}) = \frac{\rho}{b} + \frac{\rho D}{c^2 b} c_{i\alpha} u_\alpha + \rho G(\rho) Q_{i\alpha\beta} u_\alpha u_\beta + O(u^3) \tag{4.14}$$

where

$$G(\rho) = \frac{D^2}{2c^4 b} \frac{b - 2\rho}{b - \rho} \quad \text{and} \quad Q_{i\alpha\beta} = c_{i\alpha} c_{i\beta} - \frac{c^2}{D} \delta_{\alpha\beta}. \tag{4.15}$$

In equation (4.14), the superscript "eq" stresses that the mean population are evaluated at equilibrium.

Note that the coefficient $G(\rho)$ of the quadratic term vanishes for $\rho = b/2$, that is, when the density of particles and holes are the same. This result, which holds more generally for the coefficients of any even power of $\mathbf{u}$, follows by duality: N_i^{eq} goes into $1 - N_i^{eq}$ and $\mathbf{u}$ into $-\mathbf{u}$ at $\rho = b/2$. It does not matter whether or not the collision rules are duality-invariant, as long as they satisfy semi-detailed balance, since the equilibrium is then universal.

5. Macrodynamical equations

In the "real world", fluid dynamics may be viewed as the gluing of *local* thermodynamic equilibria with slowly varying parameters [62,63]. Lattice gases also admit equilibrium solutions.[4] These have continuously adjustable parameters, the mean values of the conserved quantities, namely mass and momentum. On a very large lattice, we can set up local equilibria with density and mass current slowly changing in space and time. From the conservation relations, we will derive by a multi-scale technique *macrody-namical* equations, that is, PDEs for the large scale and long-time behavior of density and mass current.

We consider a lattice gas satisfying all the assumptions of section 2.4. We denote by $\rho(r_*)$ and $u(r_*)$ the density and (mean) velocity[5] at lattice node r_*. We assume that these quantities are changing on a spatial scale ϵ^{-1} (in units of lattice constant). This requires that the lattice size L be itself at least $O(\epsilon^{-1})$. Eventually, we let $\epsilon \to 0$. The spatial change is assumed to be sufficiently regular to allow interpolations for the purpose of calculating derivatives.[6] When time and space are treated as continuous, they are denoted t and r. We further assume that the density is $O(1)$ and that the velocity is small compared to the particle speed c.[7] We expect the following phenomena:

1. relaxation to local equilibrium on time scale ϵ^0,

2. density perturbations propagating as sound waves on time scale ϵ^{-1},

3. diffusive (and possibly advective) effects on time scale ϵ^{-2}.

We thus use a three-time formalism: t_* (discrete), $t_1 = \epsilon t_*$, and $t_2 = \epsilon^2 t_*$, the latter two being treated as continuous variables. We use two space variables: r_* (discrete) and $r_1 = \epsilon r_*$ (continuous).

Let us denote by $N_i^{(0)}(r_*)$ the mean equilibrium populations based on the local value of ρ and u. They are given by equation (4.14). The actual mean populations $N_i(t,r)$ will be close to the equilibrium values and may be expanded in powers of ϵ:

$$N_i = N_i^{(0)}(t,r) + \epsilon N_i^{(1)}(t,r) + O(\epsilon^2). \tag{5.1}$$

The corrections should not contribute to the local values of density and mean momentum; thus,

$$\sum_i N_i^{(1)}(t,r) = 0 \quad \text{and} \quad \sum_i c_i N_i^{(1)}(t,r) = 0. \tag{5.2}$$

---

[4]The qualification "thermodynamic" is not so appropriate since there is no relevant energy variable.

[5]Henceforth, we will just write "velocity", since this *mean velocity* changes in space.

[6]The interpolations can be done via the Fourier representation if the lattice is periodic.

[7]Eventually, we will assume the velocity to be $O(\epsilon)$, but at this point it is more convenient to keep ϵ and u as independent expansion parameters.

We now start from the exact conservation relations (3.26) and (3.27) and expand both the N_i's and the finite differences in powers of ϵ. Note that all finite differences must be expanded to *second* order; otherwise, the viscous terms are not correctly captured. Time and space derivatives will be denoted ∂_t and $\partial_\mathbf{r} = \{\partial_\alpha, \alpha = 1, \ldots, D\}$. For the multi-scale formalism, we make the substitutions

$$\partial_t \to \epsilon \partial_{t_1} + \epsilon^2 \partial_{t_2} \quad \text{and} \quad \partial_\mathbf{r} \to \epsilon \partial_{\mathbf{r}_1}. \tag{5.3}$$

The components of $\partial_{\mathbf{r}_1}$ will be denoted $\partial_{1\alpha}$.

To leading order, $O(\epsilon)$, we obtain

$$\partial_{t_1} \sum_i N_i^{(0)} + \partial_{1\beta} \sum_i c_{i\beta} N_i^{(0)} = 0, \tag{5.4}$$

and

$$\partial_{t_1} \sum_i c_{i\alpha} N_i^{(0)} + \partial_{1\beta} \sum_i c_{i\alpha} c_{i\beta} N_i^{(0)} = 0. \tag{5.5}$$

We now substitute the equilibrium values (4.14) for the $N_i^{(0)}$'s and use the velocity moment relations **P6** of section 2.4. We obtain the "macrodynamical Euler equations"

$$\partial_{t_1} \rho + \partial_{1\beta}(\rho u_\beta) = 0, \tag{5.6}$$

and

$$\partial_{t_1}(\rho u_\alpha) + \partial_{1\beta} P_{\alpha\beta} = 0. \tag{5.7}$$

$P_{\alpha\beta}$ is the momentum-flux tensor,[8]

$$\begin{aligned} P_{\alpha\beta} &\equiv \sum_i c_{i\alpha} c_{i\beta} N_i^{:q} \\ &= \frac{c^2}{D} \rho \delta_{\alpha\beta} + \rho G(\rho) T_{\alpha\beta\gamma\delta} u_\gamma u_\delta + O(u^4), \end{aligned} \tag{5.8}$$

with

$$T_{\alpha\beta\gamma\delta} = \sum_i c_{i\alpha} c_{i\beta} Q_{i\gamma\delta}, \tag{5.9}$$

and $G(\rho)$ and $Q_{i\gamma\delta}$ given by equation (4.15) of section 4. Note that the correction term in the r.h.s. of equation (5.8) is $O(u^4)$ rather than $O(u^3)$; indeed, it follows from the parity-invariance of the lattice gas that first-order spatial derivative terms do not contain odd powers of u.

We now proceed to the next order, $O(\epsilon^2)$. We expand equations (3.26) and (3.27) to second order; collecting all $O(\epsilon^2)$ terms, we obtain

$$\partial_{t_2} \sum_i N_i^{(0)} + \frac{1}{2} \partial_{t_1} \partial_{t_1} \sum_i N_i^{(0)} + \partial_{t_1} \partial_{1\beta} \sum_i c_{i\beta} N_i^{(0)}$$

---

[8] Actually, this is only the *leading* order approximation to the momentum-flux.

$$+ \frac{1}{2}\partial_{1\beta}\partial_{1\gamma}\sum_i c_{i\beta}c_{i\gamma}N_i^{(0)} + \partial_{t_1}\sum_i N_i^{(1)} + \partial_{1\beta}\sum_i c_{i\beta}N_i^{(1)} = 0, \quad (5.10)$$

and

$$\partial_{t_2}\sum_i c_{i\alpha}N_i^{(0)} + \frac{1}{2}\partial_{t_1}\partial_{t_1}\sum_i c_{i\alpha}N_i^{(0)} + \partial_{t_1}\partial_{1\beta}\sum_i c_{i\alpha}c_{i\beta}N_i^{(0)}$$

$$+ \frac{1}{2}\partial_{1\beta}\partial_{1\gamma}\sum_i c_{i\alpha}c_{i\beta}c_{i\gamma}N_i^{(0)} + \partial_{t_1}\sum_i c_{i\alpha}N_i^{(1)}$$

$$+ \partial_{1\beta}\sum_i c_{i\alpha}c_{i\beta}N_i^{(1)} = 0. \quad (5.11)$$

By equation (5.2), $\sum_i N_i^{(1)} = 0$ and $\sum_i c_{i\alpha}N_i^{(1)} = 0$. For the $N_i^{(0)}$'s, we substitute their low-speed equilibrium form (4.14), leaving out $O(u^2)$ terms. Re-expressing derivatives of ρ and ρu with respect to t_1 in terms of space derivatives, using equations (5.6) and (5.7), we obtain

$$\partial_{t_2}\rho = 0 \quad (5.12)$$

and

$$\partial_{t_2}(\rho u_\alpha) + \partial_{1\beta}\left(\sum_i c_{i\alpha}c_{i\beta}N_i^{(1)} + \frac{D}{2c^2 b}T_{\alpha\beta\gamma\delta}\partial_{1\gamma}(\rho u_\delta)\right) = O(u^2). \quad (5.13)$$

Equation (5.12) tells us that there is no mass diffusion (there is a single species of particles). Equation (5.13) describes the momentum diffusion over long $(O(\epsilon^{-2}))$ time-scales. It has two contributions. The term involving $T_{\alpha\beta\gamma\delta}$ comes from particle propagation and we will comment on it later.

The other term in equation (5.13) involves the deviations $N_i^{(1)}$ from the equilibrium mean populations. $N_i^{(1)}$ vanishes when the equilibrium is uniform. It must therefore be a linear combination of gradients (with respect to r_1) of ρ and ρu. Linear response theory is needed to calculate the coefficients. At this point, we will only make use of symmetry arguments to reduce the number of coefficients. We assume that u is small, so that to leading order equilibria are invariant under the isometry group $\mathcal{G}$ of the lattice (see section 2.4). Since the gradient of ρ is a vector and the gradient of ρu is a second-order tensor, properties **P2** and **P3** of section 2.4 allow us to write

$$N_i^{(1)} = \sigma c_{i\alpha}\partial_{1\alpha}\rho + (\psi c_{i\alpha}c_{i\beta} + \chi\delta_{\alpha\beta})\partial_{1\alpha}(\rho u_\beta). \quad (5.14)$$

By equation (5.2), we have $\sigma = 0$ and $c^2\psi + D\chi = 0$. Note that ψ should depend on ρ, but not on u, since it is evaluated at u = 0. Substituting the expression for $N_i^{(1)}$ into equation (5.13), we obtain

$$\partial_{t_1}(\rho u_\alpha) + \partial_{1\beta}\left[\left(\psi(\rho) + \frac{D}{2c^2 b}\right)T_{\alpha\beta\gamma\delta}\partial_{1\gamma}(\rho u_\delta)\right] = O(u^2). \quad (5.15)$$

In the sequel, it will be more convenient to collapse the set of four equations, governing the evolution of ρ and ρu on $O(\epsilon^{-1})$ and $O(\epsilon^{-2})$ time-scales, into a pair of equations, written in terms of the original variables t and $\mathbf{r}$ (in their continuous version). We thus obtain the *macrodynamical equations*

$$
\begin{aligned}
\partial_t \rho \;&+\; \partial_\beta(\rho u_\beta) = 0, \\
\partial_t(\rho u_\alpha) \;&+\; \partial_\beta\left(\rho G(\rho)T_{\alpha\beta\gamma\delta}u_\gamma u_\delta + \tfrac{c^2}{D}\rho\delta_{\alpha\beta}\right) \\
&+\; \partial_\beta\left[\left(\psi(\rho) + \tfrac{D}{2c^2 b}\right)T_{\alpha\beta\gamma\delta}\partial_\gamma(\rho u_\delta)\right] \\
&=\; O(\epsilon u^3) + O(\epsilon^2 u^2) + O(\epsilon^3 u).
\end{aligned}
\tag{5.16}
$$

The equivalence of equations (5.16) and (5.17) to (5.6), (5.7), (5.12), and (5.15) follows by equation (5.3). Note that equation (16) is the standard density equation of fluid mechanics and that equation (5.17) already has a strong resemblance to the Navier-Stokes equations.

6. Recovering isotropy

The macrodynamical equations (5.16) and (5.17) are not fully isotropic. The presence of a lattice with discrete rotational symmetries is still felt through the tensor

$$
T_{\alpha\beta\gamma\delta} = \sum_i c_{i\alpha}c_{i\beta}Q_{i\gamma\delta} = \sum_i c_{i\alpha}c_{i\beta}\left(c_{i\gamma}c_{i\delta} - \frac{c^2}{D}\delta_{\gamma\delta}\right),
\tag{6.1}
$$

appearing in both the nonlinear and diffusive terms of (5.17). Furthermore, the higher-order terms in the r.h.s. of equation (5.17) have no reason to be isotropic. This should not worry us since they will eventually turn out to be irrelevant. Contrary to translational discreteness, rotational discreteness cannot go away under the macroscopic limit; the latter involves large scales but not in any way "large angles", since the group of rotations is compact.

We have seen in section 2.4 that tensors up to third order having the same invariance group $\mathcal{G}$ as the discrete velocity set are isotropic. Not so for tensors of fourth order such as $T_{\alpha\beta\gamma\delta}$. Indeed, for the HPP model (section 2.1), explicit calculation of the momentum-flux tensor, given by equation (5.8), is quite straightforward. The result is

$$
P_{11} =
$$

$$
\rho G(\rho)(u_1^2 - u_2^2) + \frac{\rho}{2} + O(u^4), \quad P_{22} = \rho G(\rho)(u_2^2 - u_1^2) + \frac{\rho}{2} + O(u^4), \tag{6.2}
$$

$$
P_{12} = P_{21} = 0,
\tag{6.3}
$$

with

$$
G(\rho) = \frac{2-\rho}{4-\rho}.
\tag{6.4}
$$

The only second-order tensors quadratic in the velocity being $u_\alpha u_\beta$ and $\mathbf{u} \cdot \mathbf{u}\, \delta_{\alpha\beta}$, the tensor $P_{\alpha\beta}$ is not isotropic.

In order to eventually obtain the Navier-Stokes equations, the tensor $T_{\alpha\beta\gamma\delta}$ given by equation (6.1) must be isotropic, that is, *invariant under the full orthogonal group*. This tensor is pairwise symmetrical in (α, β) and (γ, δ); from equation (6.1), it follows that it satisfies

$$\sum_\gamma T_{\alpha\beta\gamma\gamma} = 0, \qquad \sum_{\alpha\beta} T_{\alpha\beta\alpha\beta} = bc^4 \left(1 - \frac{1}{D}\right). \tag{6.5}$$

When the tensor $T_{\alpha\beta\gamma\delta}$ is isotropic, these properties uniquely constrain it to be of the following form:

$$T_{\alpha\beta\gamma\delta} = \frac{bc^4}{D(D+2)} \left(\delta_{\alpha\gamma}\delta_{\beta\delta} + \delta_{\alpha\delta}\delta_{\beta\gamma} - \frac{2}{D}\delta_{\alpha\beta}\delta_{\gamma\delta}\right). \tag{6.6}$$

For general group-theoretical material concerning the isotropy of tensors with discrete symmetries in the context of lattice gases, we refer the reader to reference 39. Crucial observations for obtaining the two- and three-dimensional Navier-Stokes equations are the isotropy of pairwise symmetrical tensors for the triangular FHP lattice in two dimensions and the face-centered-hypercubic (FCHC) lattice in four dimensions, and thus also for the pseudo-four-dimensional three-dimensional model. We give now elementary proofs of these results.

In two dimensions, it is convenient to consider $T_{\alpha\beta\gamma\delta}$ as a linear map from the space E of two-by-two real symmetrical matrices into itself:

$$T: \quad A_{\alpha\beta} \mapsto T_{\alpha\beta\gamma\delta} A_{\gamma\delta}. \tag{6.7}$$

A basis of the space E is formed by the matrices P_1, P_2, and P_3, associated with the orthogonal projections onto the x_1-axis and onto two other directions at $2\pi/3$ and $4\pi/3$. In this representation, an arbitrary E-matrix may be written as

$$A = \chi_1 P_1 + \chi_2 P_2 + \chi_3 P_3, \tag{6.8}$$

and T becomes a three-by-three matrix T_{ab}, $(a, b = 1, 2, 3)$. The key observation is that the hexagonal group (rotations by multiples of $\pi/3$) becomes the permutation group of P_1, P_2, and P_3. Thus, T_{ab} is invariant under arbitrary permutations of the coordinates, i.e., is of the form

$$T_{ab} = \phi \operatorname{diag}_{ab}(1,1,1) + \chi 1_{ab}, \tag{6.9}$$

where $\operatorname{diag}_{ab}(1,1,1)$ is the diagonal matrix with entries one, 1_{ab} is the matrix with all entries equal to one, and ϕ and χ are arbitrary scalars. From equation (6.8), we have

$$\operatorname{tr}(A) = \chi_1 + \chi_2 + \chi_3, \tag{6.10}$$

where tr denotes the trace. We also note that

$$P_1 + P_2 + P_3 = (3/2)I, \tag{6.11}$$

where I is the identity (check it for the unit vectors of the x_1 and x_2 axis). Using equations (6.10) and (6.11), we can rewrite equation (6.9) as

$$T : \quad A \mapsto \phi A + \frac{3}{2}\chi \operatorname{tr}(A)I. \tag{6.12}$$

Reverting to tensor notations, this becomes

$$T_{\alpha\beta\gamma\delta} = \frac{\phi}{2}\left(\delta_{\alpha\gamma}\delta_{\beta\delta} + \delta_{\alpha\delta}\delta_{\beta\gamma}\right) + \frac{3\chi}{2}\delta_{\alpha\beta}\delta_{\gamma\delta}, \tag{6.13}$$

which is obviously isotropic.

We turn to the four-dimensional case, using the FCHC model of section 2.3. Invariance under permutations of coordinates and reversal of any coordinate implies that the most general possible form for $T_{\alpha\beta\gamma\delta}$ is

$$T_{\alpha\beta\gamma\delta} = \phi\delta_{\alpha\beta}\delta_{\beta\gamma}\delta_{\gamma\delta} + \chi\left(\delta_{\alpha\gamma}\delta_{\beta\delta} + \delta_{\alpha\delta}\delta_{\beta\gamma}\right) + \psi\delta_{\alpha\beta}\delta_{\gamma\delta}. \tag{6.14}$$

The χ and ψ terms are already isotropic. The vanishing of ϕ is a consequence of the invariance of the velocity set under the symmetry Σ with respect to the hyperplane $x_1 + x_2 + x_3 + x_4 = 0$, that is,

$$x_\alpha \mapsto x_\alpha - \sigma, \quad \sigma = \frac{1}{2}\sum_\alpha x_\alpha. \tag{6.15}$$

Indeed, consider the vector $v_\alpha = (2,0,0,0)$. Contracting the ϕ term four times with v_α, we obtain 16ϕ; the image of v_α under Σ is $w_\alpha = (1,-1,-1,-1)$, which contracted four times with the ϕ term gives 4ϕ. Thus, invariance requires $\phi = 0$, which proves isotropy.

We return to the general D-dimensional case, assuming isotropy. Substituting equation (6.6) into the macrodynamical momentum equation (5.17), we obtain

$$\partial_t\left(\rho u_\alpha\right) + \partial_\beta\left(\rho g(\rho)u_\alpha u_\beta\right) + \partial_\alpha\left(c_s^2\rho\left(1 - g(\rho)\frac{u^2}{c^2}\right)\right)$$

$$= \partial_\beta\left[\left(\nu_c(\rho) + \nu_p\right)\left(\partial_\alpha(\rho u_\beta) + \partial_\beta(\rho u_\alpha) - \frac{2}{D}\delta_{\alpha\beta}\partial_\gamma(\rho u_\delta)\right)\right]$$

$$+ O(\epsilon u^3) + O(\epsilon^2 u^2) + O(\epsilon^3 u), \tag{6.16}$$

with

$$g(\rho) = \frac{D}{D+2}\frac{b-2\rho}{b-\rho}, \quad c_s^2 = \frac{c^2}{D},$$

$$\nu_c(\rho) = -\frac{bc^4}{D(D+2)}\psi(\rho), \quad \nu_p = -\frac{c^2}{2(D+2)}. \tag{6.17}$$

Note that $g(\rho)$ appearing in equation (6.17) is not the same as $G(\rho)$ introduced in equation (4.15). Note also that $\psi(\rho)$, which was introduced in section 5, is still to be determined (see section 8).

We have now recovered macroscopic isotropy; equation (6.16) is very closely related to the fluid dynamical momentum (Navier-Stokes) equations. We postpone all further remarks to the next section.

7. Fluid dynamical regimes

Let us rewrite the macrodynamical equations for mass and momentum, derived in the previous sections in a compact form which brings out their similarities with the equations of fluid dynamics:

$$\partial_t \rho + \partial_\beta (\rho u_\beta) = 0, \tag{7.1}$$

$$\partial_t (\rho u_\alpha) + \partial_\beta P_{\alpha\beta} = \partial_\beta S_{\alpha\beta} + O(\epsilon u^3) + O(\epsilon^2 u^2) + O(\epsilon^3 u). \tag{7.2}$$

The momentum-flux tensor $P_{\alpha\beta}$ and the viscous stress tensor $S_{\alpha\beta}$ are given by

$$P_{\alpha\beta} = c_s^2 \rho \left(1 - g(\rho)\frac{u^2}{c^2}\right) \delta_{\alpha\beta} + \rho g(\rho) u_\alpha u_\beta, \tag{7.3}$$

and

$$S_{\alpha\beta} = \nu(\rho) \left(\partial_\alpha(\rho u_\beta) + \partial_\beta(\rho u_\alpha) - \frac{2}{D}\delta_{\alpha\beta}\partial_\gamma(\rho u_\gamma)\right)$$

$$\nu(\rho) = \nu_c(\rho) + \nu_p, \tag{7.4}$$

where $g(\rho)$, c_s^2, ν_c, and ν_p are defined in equation (6.17). Their values for the FHP-I and FCHC models are given below:

$$g(\rho) = \frac{3-\rho}{6-\rho}, \quad c_s^2 = \frac{1}{2}, \quad \nu_c(\rho) = -\frac{3}{4}\psi(\rho), \quad \nu_p = -\frac{1}{8}, \quad \text{for FHP-I}$$

$$g(\rho) = \frac{4}{3}\frac{12-\rho}{24-\rho}, \quad c_s^2 = \frac{1}{2}, \quad \nu_c(\rho) = -4\psi(\rho), \quad \nu_p = -\frac{1}{6}, \quad \text{for FCHC.}$$

$$\tag{7.5}$$

Various remarks are now in order. When the velocity **u** is very small, the momentum-flux tensor reduces to a diagonal pressure term $p\delta_{\alpha\beta}$ with the pressure given by the "isothermal" relation

$$p = c_s^2 \rho. \tag{7.6}$$

From this, we infer that the *speed of sound* should be c_s, namely $1/\sqrt{2}$ for FHP-I and FCHC.

The momentum-flux tensor in the "real world" is $P_{\alpha\beta} = p\delta_{\alpha\beta} + \rho u_\alpha u_\beta$. This form is a consequence of Galilean invariance, which allows one to relate thermodynamic equilibria with vanishing and nonvanishing mean velocities. The lattice gas momentum-flux tensor (7.3) with nonvanishing velocity differs by an additive term in the pressure and a multiplicative density-dependent factor $g(\rho)$ in the advection term. We will see later in this section how Galilean invariance can nevertheless be recovered.

Equation (7.4) is the stress-strain relation for a Newtonian fluid having kinematic viscosity $\nu_c + \nu_p$ and *vanishing bulk viscosity* [64]. The traceless character of $S_{\alpha\beta}$ (which implies this vanishing of the bulk viscosity) comes from the traceless character of $Q_{i\alpha\beta}$, defined by equation (4.15); this result would be upset by the presence of rest particles such as exist in the models FHP-II and III (see Appendix E). The kinematic viscosity has two contributions. One is the "collision viscosity" ν_c, not yet determined, which depends on the details of the collisions and is positive (see section 8). The other one is the "propagation viscosity" ν_p, which is *negative* and does not involve the collisions. The presence of such a negative propagation viscosity is an effect of the lattice discreteness [42].

The general strategy by which standard fluid dynamical equations are derived from equations (7.1) and (7.2) is to rescale the space, time, and velocity variables in such a way as to make undesirable terms irrelevant as $\epsilon \to 0$. Three different regimes will be considered in the following subsections. They correspond respectively to sound propagation, sound propagation with slow damping, and incompressible (Navier-Stokes) fluid dynamics.

7.1 Sound propagation

Consider a weak perturbation of the equilibrium solution with density ρ_0 and velocity zero. We write

$$\rho = \rho_0 + \rho'. \tag{7.7}$$

In a suitable limit, we expect that the only relevant terms in equations (7.1) and (7.2) will be[9]

$$\partial_t \rho' + \rho_0 \nabla \cdot \mathbf{u} = 0$$

$$\rho_0 \partial_t \mathbf{u} + c_s^2 \nabla \rho' = 0. \tag{7.8}$$

Formally, this regime is obtained by setting

$$\mathbf{r} = \epsilon^{-1} \mathbf{r}_1, \quad t = \epsilon^{-1} t_1, \quad \rho' = \epsilon^a \rho'_1, \quad \mathbf{u} = \epsilon^a \mathbf{U}, \quad a > 0. \tag{7.9}$$

It is then straightforward to check that the leading order terms take the form of equations (7.8) (in the rescaled variables). Eliminating $\mathbf{u}$ in equation (7.8), we obtain the scalar wave equation

$$\frac{\partial^2}{\partial t^2} \rho' - c_s^2 \nabla^2 \rho' = 0. \tag{7.10}$$

In other words, density and velocity perturbations with amplitudes $o(1)$ on temporal and spatial scales $O(\epsilon)$ propagate as sound waves with speed c_s.[10] Since the present regime of undamped sound waves involves only tensors of second order, it also applies to the HPP model.

---

[9] From here on, we use vector notation whenever possible.
[10] We have used here the Landau $O()$ and $o()$ notation.

7.2 Damped sound

Another regime includes the viscous damping term, so that instead of equation (7.8), we should have

$$\partial_t \rho' + \rho_0 \nabla \cdot \mathbf{u} = 0$$

$$\rho_0 \partial_t \mathbf{u} + c_s^2 \nabla \rho' = \rho_0 \nu(\rho_0) \left(\nabla^2 \mathbf{u} + \frac{D-2}{D} \nabla \nabla \cdot \mathbf{u} \right). \tag{7.11}$$

To obtain this regime, we proceed as in section 7.1 and include an additional time $t_2 = \epsilon^2 t$. Furthermore, in the scaling relation (7.9) we now require $a > 1$, that is, $\mathbf{u}$ and ρ' should be $o(\epsilon)$; otherwise, the nonlinear term also becomes relevant. Note that the damping is now on a time scale $O(\epsilon^{-2})$. Since propagation and damping are on time scales involving different powers of ϵ, it is not possible to describe them in a single equation without mixing orders.

7.3 Incompressible fluid dynamics: the Navier-Stokes equations

It is known that many features of low Mach number[11] flows in an ordinary gas can be described by the incompressible Navier-Stokes equation

$$\partial_t \mathbf{u} + \mathbf{u} \cdot \nabla \mathbf{u} = -\nabla p + \nu \nabla^2 \mathbf{u}$$

$$\nabla \cdot \mathbf{u} = 0. \tag{7.12}$$

In the "real world", the incompressible Navier-Stokes equation can be derived from the full compressible equations, using a Mach number expansion. There are some fine points in this expansion for which we refer the interested reader to reference 65. Ignoring these, the essential observation is that, to leading order, density variations become irrelevant everywhere except in the pressure term; the latter becomes slaved to the nonlinear term by the incompressibility constraint.

Just the same kind of expansion (with the same difficulties) can be applied to lattice gas dynamics. We start from equations (7.1) and (7.2) and freeze the density by setting it equal to the constant and uniform value ρ_0 everywhere except in the pressure term, where we keep the density fluctuations. We also ignore all higher-order terms $O(\epsilon^3 u)$, etc. This produces the following set of equations:

$$\rho_0 \partial_t \mathbf{u} + \rho_0 g(\rho_0) \mathbf{u} \cdot \nabla \mathbf{u} = -c_s^2 \nabla (p' - p(0) g(p_0) \frac{u^2}{c^2}) + \rho_0 \nu(\rho_0) \nabla^2 \mathbf{u}$$

$$\nabla \cdot \mathbf{u} = 0. \tag{7.13}$$

---

[11]The Mach number is the ratio of a characteristic flow velocity to the speed of sound.

The resulting equations (7.13) differ from equation (7.2) only by the presence of the factor $g(\rho_0)$ in front of the advection term $\mathbf{u} \cdot \nabla \mathbf{u}$. As it stands, equation (7.13) is not Galilean invariant. This, of course, reflects the lack of Galilean invariance at the lattice level. Similarly, the vanishing of $g(\rho_0)$ when the density per cell $d = \rho_0/b$ is equal to 1/2, i.e., for equal mean numbers of particles and holes, reflects a duality-invariance of the lattice gas without counterpart in the "real world" (see end of section 4.2). However, as soon as $d < 1/2$, it is straightforward to reduce equation (7.13) to the true Navier-Stokes equations (7.12); it suffices to rescale time and viscosity:

$$t \to \frac{t}{g(\rho_0)}, \quad \nu \to g(\rho_0)\nu. \tag{7.14}$$

Now we show that there is actually a rescaling of variables which reduces the macrodynamical equations to the incompressible Navier-Stokes equations. We set

$$\mathbf{r} = \epsilon^{-1}\mathbf{r}_1, \quad t = \frac{1}{g(\rho_0)}\epsilon^{-2}T, \quad \mathbf{u} = \epsilon \mathbf{U},$$

$$(p' - p(0))g(p_0)\frac{u^2}{c^2}) = \frac{\rho_0 g(\rho_0)}{c_s^2}\epsilon^2 P', \quad \nu = g(\rho_0)\nu'. \tag{7.15}$$

Thus, all the relevant terms are $O(\epsilon^2)$ in equation (7.1) and $O(\epsilon^3)$ in equation (7.2). The higher-order terms in the r.h.s. of equation (7.2) are $O(\epsilon^4)$ or smaller. In this way, we obtain to leading order (∇_1 denotes the gradient with respect to $\mathbf{r}_1$)

$$\partial_T \mathbf{U} + \mathbf{U} \cdot \nabla_1 \mathbf{U} = -\nabla_1 P' + \nu' \nabla_1^2 \mathbf{U}$$

$$\nabla_1 \cdot \mathbf{U} = 0, \tag{7.16}$$

which are exactly the incompressible Navier-Stokes equations.

Various comments are now made. The expansion leading to equation (7.16) is a large-scale and low Mach number expansion (the former is here inversely proportional to the latter). It also follows from the scaling relations (7.15) that the Reynolds number is kept fixed. It is not possible within our framework to have an asymptotic regime leading to nonlinear compressible equations at finite Mach number. Indeed, the speed of sound is here a *finite* fraction of the particle speed, and it is essential that the macroscopic velocity be small compared to particle speed, so as not to be contaminated by higher-order nonlinearities. It is noteworthy that models can be constructed having many rest particles (zero-velocity) with arbitrarily low speed of sound.

In a pure Navier-Stokes context, the non-Galilean invariance at the microscopic level is not a serious difficulty; as we have seen, Galilean invariance is recovered macroscopically, just by rescaling the time variable. However, when the models discussed here are generalized to include, for example, multi phase flow or buoyancy effects, a more serious problem may arise because the advection term of scalar quantities, such as chemical concentrations or temperature, involves usually a factor $g(\rho)$ different from that of the nonlinear advection term in the Navier-Stokes equations. Various solutions to this problem have been proposed [48,66].

There is a variant of our formalism, leading also to the incompressible Navier-Stokes equations, but in terms of the *mass current* $\mathbf{j} = \rho\mathbf{u}$ rather than the velocity $\mathbf{u}$. The analog of equation (7.13) (without rescaling) is then

$$\partial_t\mathbf{j} + \frac{g(\rho_0)}{\rho_0}\mathbf{j}\cdot\nabla\mathbf{j} = -c_s^2\nabla\rho' + \nu(\rho_0)\nabla^2\mathbf{j}$$

$$\nabla\cdot\mathbf{j} = 0. \tag{7.17}$$

Since $\mathbf{j}$ and $g(\rho_0)/\rho_0$ change sign under duality, equation (7.17) brings out duality-invariance.[12] A more decisive advantage of the $\mathbf{j}$-representation is that it gives a better approximation to the *steady state* Navier-Stokes equations when the Mach number is only moderately small. This is because in the steady state the continuity equation implies exactly $\nabla\cdot\mathbf{j} = 0$.

In three dimensions, when we use the pseudo-four-dimensional FCHC model, there are three independent space variables $\mathbf{r} = (x_1, x_2, x_3)$, but four velocity components:

$$\mathbf{U}_f = (\mathbf{U}, U_4) = (U_1, U_2, U_3, U_4). \tag{7.18}$$

The four-velocity $\mathbf{U}_f$ satisfies the four-dimensional Navier-Stokes equations with no x_4-dependence. Thus, the three-velocity $\mathbf{U}$ satisfies the three-dimensional Navier-Stokes equations (7.16),[13] while U_4 satisfies (note that the pressure term drops out)

$$\partial_T U_4 + \mathbf{U}\cdot\nabla_1 U_4 = \nu'\nabla_1^2 U_4. \tag{7.19}$$

This is the equation for a *passive scalar* with unit Schmidt number (ratio of viscosity to diffusivity).

Finally, we refer the reader to Appendix D for the inclusion of body forces in the Navier-Stokes equations.

---

[12]In the u representation, duality-invariance is broken because we have decided to work with the velocity of *particles* rather than with that of *holes*.

[13]Since the velocity set of the pseudo-four-dimensional model is the same as in four dimensions, isotropy is ensured for all fourth-order tensors depending only on the velocity set. Thus, the nonlinear term has the correct isotropic form. The viscous term is isotropic within the Boltzmann approximation (see section 8.2); otherwise, deviations from isotropy are expected to be small [2].

8.   The viscosity

All the macroscopic equations derived in section 7 have a universal form which does not depend on the details of collisions. The kinematic shear viscosity ν, which we will henceforth call the viscosity, does not possess this universality. Transport coefficients such as the viscosity characterize the linear response of equilibrium solutions to small externally imposed perturbations. It is known in statistical mechanics that the relaxation (or dissipation) of external perturbations is connected to the fluctuations at equilibrium via *fluctuation-dissipation* relations. Such relations have a counterpart for lattice gases. Two quite different approaches are known. In section 8.1, following a suggestion already made in [23], we present the "noisy" hydrodynamics viewpoint, in the spirit of Landau and Lifschitz [67,68]. Another approach, in the spirit of Kubo [69] and Green [70], using a Liouville equation formalism, may be found in reference 43. In section 8.2, we introduce the lattice analog of the Boltzmann approximation, which allows an explicit calculation of the viscosity. In section 8.3, we discuss some implications for the Reynolds numbers of incompressible flows simulated on lattice gases.

8.1   Fluctuation-dissipation relation and "noisy" hydrodynamics

We first explain the basic ideas in words. Spontaneous fluctuations at equilibrium involve modes of all possible scales. The fluctuations of very large scales should have their dynamics governed by the macroscopic equations derived in sections 5 through 7. Such fluctuations are also expected to be very weak, so that *linear* hydrodynamics should apply. Large-scale spontaneous fluctuations are constantly regenerated, and in a random manner; this regeneration is provided by a random force (noise) term which can be identified and expressed in terms of the fluctuating microscopic variables. If this random force has a short correlation-time (i.e., small compared to the life-time of the large-scale fluctuations under investigation), then each large-scale mode v has its dynamics governed by a Langevin equation.[14] It follows that the variance $\langle v^2 \rangle$ can be expressed in terms of the damping coefficient γ (related to the viscosity) and of the time-correlation function of the random force. Alternatively, the variance $\langle v^2 \rangle$ can be calculated from the known one-time equilibrium properties. Identification gives the viscosity in terms of equilibrium time-correlation functions. This is the general program that we now carry out for the special case of lattice gases. We restrict ourselves to equilibrium solutions with zero mean velocity.

   We will use in this section the following notation. The density ρ and the mass current j are no longer given by their expressions (3.24) in terms of the mean populations; instead, they are defined in terms of the fluctuating Boolean field

---

[14]For the case of lattice gases, we will actually obtain a finite difference equation.

$$\rho(t_*, \mathbf{r}_*) = \sum_i n_i(t_*, \mathbf{r}_*), \qquad \mathbf{j}(t_*, \mathbf{r}_*) = \sum_i \mathbf{c}_i n_i(t_*, \mathbf{r}_*). \tag{8.1}$$

We denote by $\tilde{n}_i$ the fluctuating part of the Boolean field, defined by

$$n_i(t_*, \mathbf{r}_*) = d + \tilde{n}_i(t_*, \mathbf{r}_*), \tag{8.2}$$

where d is the density per cell.

We introduce *meso-averaged* fields by taking *spatial* averages over a a distance ϵ^{-1}.[15] These will be denoted by angular brackets with the subscript ma. The meso-averages of n_i, ρ, and $\mathbf{j}$ are denoted $\bar{n}_i$, $\bar{\rho}$, and $\bar{\mathbf{j}}$ respectively. Locally, the equilibrium relation (4.14) should hold approximately for the meso-averaged populations. We thus write

$$\bar{n}_i = \frac{\bar{\rho}}{b} + \frac{D}{c^2 b} \bar{\mathbf{j}} \cdot \mathbf{c}_i + \delta_i + \bar{n}_i^{(1)}(t_*, \mathbf{r}_*). \tag{8.3}$$

δ_i represents the (still unknown) input from non-hydrodynamic fluctuations; $\bar{n}_i^{(1)}$ is the contribution analogous to $\epsilon N_i^{(1)}$ in equation (5.1), arising from the gradients of meso-averages. Note that in equation (8.3) we dropped contributions nonlinear in the mass current; indeed, we should be able to determine the viscosity from just linear hydrodynamics.[16]

We now derive the equations for noisy hydrodynamics. As usual, we start from the microscopic conservation relations (3.16) and (3.17) and we take their meso-averages:

$$\sum_i [\bar{n}_i(t_* + 1, \mathbf{r}_* + \mathbf{c}_i) - \bar{n}_i(t_*, \mathbf{r}_*)] = 0, \tag{8.4}$$

$$\sum_i \mathbf{c}_i [\bar{n}_i(t_* + 1, \mathbf{r}_* + \mathbf{c}_i) - \bar{n}_i(t_*, \mathbf{r}_*)] = 0. \tag{8.5}$$

Substituting equatino (8.3) into equation (8.5), we obtain

$$\frac{1}{b} \sum_i \mathbf{c}_i [\bar{\rho}(t_*+1, \mathbf{r}_*+\mathbf{c}_i) - \bar{\rho}(t_*, \mathbf{r}_*)] + \frac{D}{c^2 b} \sum_i \mathbf{c}_i \mathbf{c}_i \cdot [\bar{\mathbf{j}}(t_*+1, \mathbf{r}_*+\mathbf{c}_i) - \bar{\mathbf{j}}(t_*, \mathbf{r}_*)]$$

$$+ \sum_i \mathbf{c}_i [\bar{n}_i^{(1)}(t_* + 1, \mathbf{r}_* + \mathbf{c}_i) - \bar{n}_i^{(1)}(t_*, \mathbf{r}_*)] = \mathbf{f}(t_*, \mathbf{r}_*), \tag{8.6}$$

where

$$\mathbf{f}(t_*, \mathbf{r}_*) = -\sum_i \mathbf{c}_i [\delta_i(t_* + 1, \mathbf{r}_* + \mathbf{c}_i) - \delta_i(t_*, \mathbf{r}_*)] \tag{8.7}$$

is the random force. Using equations (8.1) through (8.5), we can also write (to leading order in gradients)

---

[15] More precisely, by dropping spatial Fourier components with wavenumber $k > \epsilon$.
[16] This is not exactly true in two dimensions as we will see below.

$$f(t_*, \mathbf{r}_*) =$$

$$\left\langle \frac{1}{c^2 b} \sum_{ij} \left(c^2 c_i + D c_i \cdot c_j \, c_i \right) \left[\tilde{n}_j(t_* + 1, \mathbf{r}_* + c_i) - \tilde{n}_j(t_* + 1, \mathbf{r}_* + c_j) \right] \right\rangle_{ma}$$

$$(8.8)$$

The l.h.s. of equation (8.6) is expanded in powers of gradients (i.e., of ϵ), as we did in section 5. However, we keep finite differences rather than derivatives in time because of the presence of the rapidly varying random force. Since we only want to identify the shear viscosity (the bulk viscosity is zero), it suffices to extract the solenoidal part of the hydrodynamical equation. For this and other reasons, it is better to work in Fourier space. We define the (spatial) Fourier transform of the fluctuating Boolean field by

$$\tilde{n}_i(t_*, \mathbf{r}_*) = \sum_{\mathbf{k}} e^{i\mathbf{k} \cdot \mathbf{r}_*} \hat{n}(t_*, \mathbf{k}), \tag{8.9}$$

where the components of $\mathbf{k}$ are multiples of 2π divided by the lattice periodicities in the various directions. We similarly define $\hat{\mathbf{j}}$ and $\hat{\mathbf{f}}$, the Fourier transforms of the mass current and the random force. Their solenoidal parts, projection on the hyperplane perpendicular to $\mathbf{k}$, are denoted $\hat{\mathbf{j}}_\perp$ and $\hat{\mathbf{f}}_\perp$.

To leading order in k, we obtain from equation (8.8) using equation (2.5)

$$\hat{\mathbf{f}}_\perp(t_*, \mathbf{k}) = -\sum_j i\mathbf{k} \cdot c_j \left(c_j - \frac{c_j \cdot \mathbf{k}\mathbf{k}}{k^2} \right) \hat{n}_j(t_* + 1, \mathbf{k}). \tag{8.10}$$

The meso-averaging is just the restriction that $k < \epsilon$. Fourier transforming equation (8.6) and taking the solenoidal part, we obtain for small k

$$\hat{\mathbf{j}}_\perp(t_* + 1, \mathbf{k}) - \hat{\mathbf{j}}_\perp(t_*, \mathbf{k}) + \nu k^2 \hat{\mathbf{j}}_\perp(t_*, \mathbf{k}) = \hat{\mathbf{f}}_\perp(t_*, \mathbf{k}). \tag{8.11}$$

This is our discrete Langevin equation. Note that ν is the (total) viscosity $\nu = \nu_c + \nu_p$. In principle, we must expand to second order in k to obtain the viscous terms, but we could as well have written the l.h.s of equation (8.11) a priori, since we want to use equation (8.11) to *determine* the viscosity. It is straightforward to solve the linear finite-difference equation (8.11). From the solution, we calculate the variance of $\hat{\mathbf{j}}_\perp$ and obtain, when the viscous damping time $1/(\nu k^2)$ is large compared to the correlation time of the random force

$$\left\langle \left| \hat{\mathbf{j}}_\perp(t_*, \mathbf{k}) \right|^2 \right\rangle = \frac{1}{2\nu k^2} \sum_{t_* = -\infty}^{t_* = +\infty} \left\langle \mathbf{f}_\perp(t_*, \mathbf{k}) \cdot \mathbf{f}_\perp^*(t_*, \mathbf{k}) \right\rangle \tag{8.12}$$

where the asterisk denotes complex conjugation. The variance of $\hat{\mathbf{j}}_\perp$ can also be calculated directly using equation (8.1) and

$$\langle \tilde{n}_i(t_*, \rho_*) \tilde{n}_j(t_*, 0) \rangle = \langle \tilde{n}_i^2 \rangle \delta_{ij} \delta_{\rho_*}$$

$$\langle \tilde{n}_i^2 \rangle = \langle n_i^2 \rangle - \langle n_i \rangle^2 = d - d^2, \tag{8.13}$$

where δ_{ρ_*} denotes a Kronecker delta in the spatial separation ρ_*. We obtain

$$\left\langle \left| \hat{\jmath}_\perp(t_*, \mathbf{k}) \right|^2 \right\rangle = \frac{1}{V} bc^2 d(1-d) \frac{D-1}{D}, \tag{8.14}$$

where V denotes the total number of lattice points in the periodicity volume. Thus, the l.h.s. of equation (8.12) is k-independent. We evaluate the r.h.s of equation (8.12) in the limit $k \to 0$, using equation (8.10). We skip some intermediate steps in which we (i) use the stationariness of the fluctuations at equilibrium, (ii) use the isotropy of second- and fourth-order symmetrical tensors, (iii) interchange the $k \to 0$ limit and the infinite summation over t_*.[17] Identifying the two expressions (8.12) and (8.14), we obtain for the viscosity

$$\nu = \frac{D}{2(D-1)(D+2)} \frac{1}{bc^2} \frac{1}{d(1-d)} \frac{1}{V}$$

$$\sum_{t_*=-\infty}^{t_*=+\infty} \sum_{ij\alpha\beta} Q_{i\alpha\beta} Q_{j\alpha\beta} \left\langle \hat{n}_i(t_*, 0) \hat{n}_j^*(0, 0) \right\rangle$$

$$= \frac{D}{2(D-1)(D+2)} \frac{1}{bc^2} \frac{1}{d(1-d)} \sum_{t_*=-\infty}^{t_*=+\infty}$$

$$\sum_{\rho_* \in \mathcal{L}} \sum_{ij\alpha\beta} Q_{i\alpha\beta} Q_{j\alpha\beta} \left\langle \tilde{n}_i(t_*, \rho_*) \tilde{n}_j(0, 0) \right\rangle \tag{8.15}$$

with

$$Q_{i\alpha\beta} = c_{i\alpha} c_{i\beta} - \frac{c^2}{D} \delta_{\alpha\beta}. \tag{8.16}$$

This completes the fluctuation-dissipation calculation of the viscosity. A consequence of the Fourier-space representation (the upper half of equation (8.15)) is the positivity of the viscosity; indeed, the viscosity is, within a positive factor, the time-summation of the autocorrelation of $\sum_i Q_{i\alpha\beta} \hat{n}_i(t_*, 0)$.

Several comments are now in order. It is easily checked that the $t_* = 0$ contribution to the viscosity (lower part of equation (8.15) is $c^2/(2(D+2))$, that is, just the opposite of the "propagation viscosity" ν_p introduced in section 7. The viscosity is the sum of the collision viscosity ν_c and ν_p. Using the identity

$$\sum_{t_*=-\infty}^{t_*=+\infty} Z(t_*) = 2 \sum_{t_*=0}^{t_*=+\infty} Z(t_*) - Z(0), \tag{8.17}$$

---

[17]This is equivalent to assuming that the viscosity is finite, see below.

(for an even function $Z(t_*)$), we find that ν_c has a representation similar to (8.15) (lower part), with an additional factor of 2 and the summation over t_* extending only from 0 to ∞. We thereby recover an expression derived in reference 43, using a discrete variant of the Green-Kubo formalism. It is reassuring to have two completely different derivations of the viscosity, since we consider our fluctuation-dissipation derivation somewhat delicate.

It is of interest that the fluctuation-dissipation derivation gives directly the (total) viscosity. This suggests that the splitting into collision and propagation viscosities is an artifact of our multi-scale formalism.

There is no closed form representation of the correlation function $\langle \tilde{n}_i(t_*, \rho_*) \tilde{n}_j(0,0) \rangle$, except for short times. However, (8.15) is a good starting point for a Monte-Carlo calculation of the viscosity [43].

In our derivation, we have dropped all contributions from *nonlinear* terms in the mass current j. Is this justified? If we reinstate the nonlinear terms, we obtain, for the solenoidal part of the meso-averaged mass current, the Navier-Stokes equations (7.17) of section 8 with the additional random force given in the Fourier representation by equation (8.10). On macroscopic scales, this force may be considered as δ-correlated in time. Its spectrum follows, for small k, a k^{D+1} power-law.[18] The Navier-Stokes equations with this kind of power-law forcing is one of the few problems in nonlinear statistical fluid mechanics which can be systematically analyzed by renormalization group methods [71,72]. For $D > 2$, the nonlinear term is irrelevant for small k so that our calculation of the viscosity is legitimate. At the "crossover" dimension $D = 2$, the nonlinear term becomes "marginal"; it produces a renormalization of the viscosity which is then logarithmically scale-dependent. Thus, in the limit of infinite scale-separation, the viscosity becomes infinite in two dimensions. This is an instance of the known divergence of transport coefficients in two-dimensional statistical mechanics [68,73]. Alternatively, the divergence of the viscosity in two dimensions can be viewed as due to the presence of a "long-time-tail", proportional to $t_*^{-D/2}$, in the correlation function appearing in equation (8.15). Attempts have been made to observe long-time-tails and scale-dependence of the viscosity in Monte-Carlo simulations of lattice gas models [8,23,43,44]. This is not easy because (i) the effects show up only at very long times (or large scales) and may then be hidden by Monte-Carlo noise (insufficient averaging), and (ii) the effects should get weaker as the number b of cells per node increases (see end of section 8.2).

Finally, the noisy hydrodynamics formalism can be used to estimate to what extent the microscopic noise contaminates the hydrodynamic macroscopic signal. Estimates, assuming the signal to be meso-averaged in space and time, have been made in the context of fully developed incompressible two- and three-dimensional turbulence.[19] It has been found that in two dimensions noise is relevant only at scales less than the dissipation scale,

---

[18] A factor k^2 comes from the average squared Fourier amplitude and another factor k^{D-1} from the D-dimensional volume element.

[19] Note that in the incompressible case, only solenoidal noise is relevant.

while in three dimensions this happens only far out in the dissipation range [14].

8.2 The Lattice Boltzmann approximation

Explicit calculation of transport coefficients can be done for lattice gases, using the Boltzmann approximation. In this approximation, one assumes that *particles entering a collision process have no prior correlations*. The microdynamical formalism of section 3.1 is particularly well suited for deriving what we will call the *lattice Boltzmann equation*. We take the ensemble average of equation (3.9). The Boolean variables n_i become the mean populations N_i. The average of the collision function Δ_i can be completely factorized, thanks to the Boltzmann approximation. We obtain

$$N_i(t_* + 1, \mathbf{r}_* + \mathbf{c}_i) = N_i(t_*, \mathbf{r}_*) + \Delta_i^{\text{Boltz}}$$

$$\Delta_i^{\text{Boltz}} = \sum_{s,s'} (s_i' - s_i) A(s \to s') \prod_j N_j^{s_j} (1 - N_j)^{(1-s_j)}. \qquad (8.18)$$

Here, all the N_j's are evaluated at t_* and $\mathbf{r}_*$. The $A(s \to s')$'s, the transition probabilities introduced in section 2.4, are the averages of the Boolean transition variables $\xi_{ss'}$. Note that the (Boltzmann) *collision function* Δ_i^{Boltz} vanishes at equilibrium.

The Boltzmann approximation in ordinary gases is associated with low density situations, when the mean-free path is so large that particles entering a collision come mostly from distant uncorrelated regions. The Boltzmann approximation for a lattice gas appears to have a very broad validity, not particularly restricted to low densities.[20] We will come back to the matter at the end of this section.

Our lattice Boltzmann equation (8.18) is a *finite difference* equation. There is a differential version of it, obtained by Taylor-expanding the finite differences to first order, namely

$$\partial_t N_i + \mathbf{c}_i \cdot \nabla N_i = \Delta_i^{\text{Boltz}} \qquad (8.19)$$

where Δ_i^{Boltz} is defined as in equation (8.18). Boltzmann equations of the form (8.19) have been extensively studied as discrete velocity approximations to the ordinary Boltzmann equation [15-17,19]. The (differential) Boltzmann formalism has been applied to various lattice gas models [35,39]. This formalism correctly captures all hydrodynamic phenomena involving only first-order derivatives. Indeed, for these, we have seen that only the equilibrium solutions matter, and the latter are completely factorized. Diffusive phenomena involve second-order derivatives. Hence, the *propagation viscosities* (see section 7), which are an effect of lattice-discreteness, are not

---

[20]Even at low densities, the Boltzmann approximation may not be valid. Indeed, without effectively changing the dynamics, we can reduce the density by arbitrary large factors by having the particles initially located on a sub-lattice with some large periodicity; these are, however, pathologically unstable configurations.

captured by the (differential) Boltzmann equation. At low densities, where collision viscosities dominate over propagation viscosities, the discrepancy is irrelevant.

We do not intend to engage in extended discussions of the consequences of the lattice Boltzmann equation because most of the derivation of the hydrodynamical equations is independent of this approximation. There are, however, two important results which follow from the lattice Boltzmann equation. The first concerns the irreversible approach to equilibrium. It is derived by adapting an H-theorem formalism to the fully discrete context (see Appendix F by Hénon).

The second result is an explicit derivation of the viscosity. From the Boltzmann equation, this is usually done by a Chapman-Enskog formalism [75,76] (see also Gatignol's monography, [17]). This formalism is easily adapted to the lattice Boltzmann equation [77]. With the general multi-scale formalism of sections 5 through 7, we have already covered a substantial fraction of the ground. Furthermore, an alternative derivation which stays completely at the microscopic level is presented in this volume by Hénon, who also discusses consequences of his explicit viscosity-formula [42]. We will therefore be brief.

The problem of the viscosity amounts to finding the coefficient ψ relating the gradient of the mass current ρu to the first-order perturbation $N_i^{(1)}$ of the mean population, through (see equation (5.14) of section 5)

$$N_i^{(1)} = \psi Q_{i\alpha\beta}\partial_{1\alpha}(\rho u_\beta)$$
$$Q_{i\alpha\beta} = c_{i\alpha}c_{i\beta} - \frac{c^2}{D}\delta_{\alpha\beta}. \tag{8.20}$$

We start from equation (5.1) with $N_i^{(0)}$ given by equation (4.14). We substitute into the lattice Boltzmann equation (8.18) and identify the terms $O(\epsilon)$. For this, we Taylor-expand finite differences to first order, use equations (5.6) and (5.7) to express time-derivatives in terms of space-derivatives, and ignore all terms beyond the linear ones in the velocity. We obtain

$$\frac{D}{bc^2}Q_{i\alpha\beta}\partial_{1\alpha}(\rho u_\beta) = \sum_j A_{ij}N_j^{(1)}. \tag{8.21}$$

Here,

$$A_{ij} = \left[\frac{\partial \Delta_i^{\text{Boltz}}}{\partial N_j}\right]_{N_i = \rho/b} \tag{8.22}$$

is the *linearized collision* matrix, evaluated at the zero-velocity equilibrium, which can be expressed in compact form as [42]

$$A_{ij} = -\frac{1}{2}\sum_{ss'}(s_i - s_i')A(s \to s')d^{p-1}(1 - d)^{b-p-1}(s_j - s_j'),$$

$$p = \sum_i s_i. \tag{8.23}$$

We eliminate $N_i^{(1)}$ between equations (8.20) and (8.21), to obtain

$$\left[\frac{D}{bc^2}Q_{i\alpha\beta} - \psi\sum_j A_{ij}Q_{j\alpha\beta}\right]\partial_{1\alpha}(\rho u_\beta) = 0. \tag{8.24}$$

This should hold for arbitrary gradients of the mass current. Thus, the quantity between square brackets vanishes. This means that, for any (α, β), $Q_{i\alpha\beta}$, considered as a vector with components labeled by i, is an eigenvector of the linearized collision matrix with eigenvalue $D/(bc^2\psi)$; a direct proof of this may be derived from the $\mathcal{G}$-invariance. From equation (8.24), we can easily calculate ψ; the simplest method is to multiply the vanishing square bracket by $Q_{i\alpha\beta}$ and sum over i, α, and β. If, in addition, we assume the isotropy of fourth-order tensors, we can use equation (6.17) to obtain a closed-form expression for the collision viscosity

$$\nu_c = -\frac{c^2}{D+2}\frac{\sum_{i\alpha\beta}Q_{i\alpha\beta}^2}{\sum_{ij\alpha\beta}Q_{i\alpha\beta}A_{ij}Q_{j\alpha\beta}}. \tag{8.25}$$

In Appendix E, we give explicit formulae calculated from equation (8.25) for the viscosities of the FHP models (including those with rest particles which require minor amendements of our formalism).

We finally address the question of the validity of the lattice Boltzmann equation. Comparisons of the viscosities obtained from simulations [25,29,31,33] or Monte-Carlo calculations [77] with the predictions of the lattice Boltzmann approximation suggest that the validity of the latter is not limited to low densities. We know that equilibrium solutions are factorized and that transport coefficients can be calculated with arbitrarily weak macroscopic gradients. However, this cannot be the basis for the validity of the Boltzmann approximation: a weak *macroscopic* gradient implies that the probability of changing the state of a given node from its equilibrium value is small; but when such a change takes place, it produces a strong *microscopic* perturbation in its environment. Otherwise, there would be no (weak) divergence of the viscosity in two dimensions; indeed, the Boltzmann approximation does not capture noise-induced renormalization effects (see end of section 8.1). A more likely explanation of the success of the lattice Boltzmann approximation may be that it is the leading order in some kind of $1/b$ expansion, where b is the number of velocity cells at each node. At the moment, we can only support this by the following heuristic argument. Deviations from Boltzmann require correlations between particles entering a collision. The latter arise from previous collisions;[21] when b is large, the weight pertaining to such events ought to be small.

8.3 The Reynolds number

Knowing the kinematic shear viscosity in terms of the density and the collision rules, we can calculate the Reynolds number associated to a large-scale flow.

---

[21] Collisions produce correlations whenever the particles are not exactly at equilibrium.

A natural unit of length is the lattice constant (distance of adjacent nodes), which has been taken equal to one for the two-dimensional HPP and FHP models. The four-dimensional FCHC model has a lattice constant of $\sqrt{2}$, but its three-dimensional projected version, the pseudo-four-dimensional FCHC model, resides on a cubic lattice which has also unit lattice constant. The time necessary for microscopic information to propagate from one node to its connecting neighbors defines a natural unit of time. We then have a natural unit of velocity: the speed necessary to travel the lattice constant (or the projected lattice constant for the pseudo-four-dimensional model) in a unit time. In these units, the characteristic scale and velocity of the flow will be denoted by ℓ_0 and u_0.

The standard definition of the Reynolds number is

$$R = \frac{\text{characteristic scale} \times \text{characteristic velocity}}{\text{kinematic shear viscosity}}. \tag{8.26}$$

In deriving the Navier-Stokes equations in section 7.3, we rescaled space, time, velocity, pressure, and viscosity (cf. equation (7.15)). The rescaling of space (by ϵ) and of velocity (by ϵ^{-1}) cancel in the numerator of equation (8.26). The rescaled viscosity is $\nu'(\rho_0) = \nu(\rho_0)/g(\rho_0)$. Hence, the Reynolds number is

$$R = \ell_0 u_0 \frac{g(\rho_0)}{\nu(\rho_0)}. \tag{8.27}$$

In order to operate in an incompressible regime, the velocity u_0 should be small compared to the speed of sound c_s. The latter is model-dependent: $c_s = 1/\sqrt{2}$ for FHP-I and FCHC, $c_s = \sqrt{3/7}$ for FHP-II and FHP-III (see section 7 and Appendix E). Let us therefore re-express the Reynolds number in terms of the Mach number

$$M = \frac{u_0}{c_s}. \tag{8.28}$$

We obtain

$$R = M\ell_0 R_*(\rho_0), \tag{8.29}$$

where

$$R_*(\rho_0) = \frac{c_s g(\rho_0)}{\nu(\rho_0)} \tag{8.30}$$

contains all the *local* information.

In flow simulations using lattice gases, it is of interest to operate at the density which maximizes R_*. Let us work this out for the simplest case of FHP-I. For the viscosity, we use the lattice Boltzmann value given in Appendix E. We have

$$g(\rho_0) = \frac{1}{2}\frac{1-2d}{1-d}, \quad \nu(\rho_0) = \frac{1}{12d(1-d)^3} - \frac{1}{8}, \quad d = \frac{\rho_0}{6}. \tag{8.31}$$

Here, d is the mean density per cell. Substituting in equation (8.30), we find that

$$R_*^{\text{max}} = \max R_* = 0.387, \quad \text{for} \quad d = d_{\text{max}} = 0.187. \qquad (8.32)$$

Results for FHP-II and FHP-III are given in Appendix E. Note that a gain of about a factor 6 is achieved in going from FHP-I to FHP-III, because the latter includes many more collisions. For the pseudo-four-dimensional FCHC model there is work in progress on the optimization of collisions. It is already known that R_*^{max} is at least 6.4 [78].

High Reynolds number incompressible turbulent flows have a whole range of scales. The smallest effectively excited scale is called the *dissipation scale* and denoted ℓ_d. It is then of interest to find how many lattice constants are contained in ℓ_d, since this will determine how effective lattice gases are in simulating high Reynolds number flows [1,36]. For this, let ℓ_0 denote the integral scale of the flow. Between ℓ_0, ℓ_d, and the Reynolds number R, there is the following relation

$$\frac{\ell_d}{\ell_0} = C R^{-m}. \qquad (8.33)$$

$m = 1/2$ in two dimensions and $m = 3/4$ in three dimensions; C is a dimensionless constant not given by theory. In two dimensions, equation (8.33) is a consequence of the Batchelor-Kraichnan [79,80] phenomenological theory of the enstrophy cascade, which is well supported by numerical simulations [81]. In three dimensions, equation (8.33) follows from the Kolmogorov [82] phenomenological theory of the energy cascade, which is well supported[22] by experimental data [83]. Using equations (8.29) and (8.33) and assuming that R_* has its maximum value R_*^{max}, we obtain

$$\ell_d = C \left(M R_*^{\text{max}}\right)^{-\frac{1}{2}} \ell_0^{\frac{1}{2}} = C \left(M R_*^{\text{max}}\right)^{-1} R^{\frac{1}{2}} \quad \text{in 2-D}, \qquad (8.34)$$

and

$$\ell_d = C \left(M R_*^{\text{max}}\right)^{-\frac{3}{4}} \ell_0^{\frac{1}{4}} = C \left(M R_*^{\text{max}}\right)^{-1} R^{\frac{1}{4}} \quad \text{in 3-D}. \qquad (8.35)$$

In all cases, we see that $\ell_d \to \infty$ as $R \to \infty$, but more slowly in three than in two dimensions. We are thus assured that at high Reynolds numbers the separation of scale between the lattice constant and ℓ_d necessary for hydrodynamic behavior is satisfied. Having it too well satisfied may however be a mixed blessing, as stressed in reference 36. Indeed, in hydrodynamic simulations using lattice gases, it is not desirable to have too much irrelevant microscopic information. We note that ℓ_0 appears in equations (8.34) and (8.35) with a larger exponent in the two-dimensional case; thus, the above mentioned problem is most severe for large lattices in two dimensions.

---

[22]Small intermittency corrections which would slightly increase the exponent m cannot be ruled out.

The highest Reynolds number which can be simulated by lattice gas methods in three dimensions can be estimated as follows. We take $M = 0.3$, a Mach number at which compressibility effects can be safely ignored [84]; we take the maximum known value $R_*^{max} = 6.4$ for the FCHC model, and we take $\ell_0 = 10^3$, a fairly large value which implies a memory requirement of at least 24 gigabits; from equation (8.29), we find that the maximum Reynolds number is about two thousand. It is of interest both in two and three dimensions to try to decrease the viscosity, thereby increasing R_*^{max}. Note that it is not correct to infer from dimensional analysis that necessarily R_*^{max} must be $O(1)$. R_*^{max} is very much a function of the *complexity* of collisions. For example, by going from FHP-I to FCHC (which can also be projected down to two dimensions), R_*^{max} increases more than sixteen times.

9. Conclusion

In statistical mechanics, there are many instances where two models, microscopically quite different, have the same large-scale properties. For example, the Ising model and a real Ferromagnet have presumably the same large-scale critical behavior. Similarly, the lattice gases studied in this paper, such as FHP and FCHC, are macroscopically indistinguishable from real fluids. This provides us with an attractive alternative to the traditional simulations of fluid mechanics. In lattice gas simulations, we just manipulate bits representing occupation of microscopic cells. The physical interpretation need not be in terms of particles moving and colliding. The idea can clearly be extended to include processes such as chemical reactions or multi-phase flow [53-57]. An open question is whether there are cellular automata implementations of processes which in the real world do not have a discrete microscopic origin, such as propagation of e.m. waves. More generally, what are the PDEs which can be *efficiently* implemented on cellular automata? We emphasize *efficiently*, because there are always brute force implementations: replace derivatives by finite differences on a regular grid and use finite floating point truncations of the continuous fields. The result may be viewed as a cellular automaton, but one in which there is no "bit democracy", insofar as there is a rigid hierarchical order between the bits.

Our derivation of hydrodynamics from the microdynamics leaves room for improvement. A key assumption made in section 4.1 may be formulated as follows. Among the invariant measures of the microdynamical equations, only the completely factorized ones (which play the role here of the microcanonical ensemble) are relevant in the limit of large lattices. On a finite lattice with deterministic and invertible updating rules, we expect that there are many other invariant measures. Indeed, phase space is a finite set and updating is a permutation of this set; it is thus unlikely that there should be a closed orbit going through all points. So, we do not expect the discrete equivalent of an ergodic theorem. Anyway, ergodic results should be irrelevant. On the one hand, on an $L \times L$ lattice with b bits per node, its takes 2^{bL^2} updates to visit all configurations (if they are accessible). On the other hand, we know (from simulations) that local equilibrium is achieved in a few updates and global equilibrium is achieved on a diffusive time scale (approximately L^2). We believe that, on large lattices, the factorized equilibrium distributions constitute some kind of "fixed point" to which there is rapid convergence of the iterated Boolean map defined by the microdynamical equations of section 3.1. Understanding this process should clarify the mechanism of irreversibility in lattice gases and, eventually, in real gases.

Acknowledgements

Many colleagues have contributed to this work by their suggestions as well as by their questions. They include V. Arnold, H. Cabannes, R. Caflisch, P. Clavin, A. Deudon, G. Doolen, R. Gatignol, F. Hayot, M. Hénon, S. Kaniel, R. Kraichnan, J. Lebowitz, D. Levermore, N. Margolus, J. L. Oneto, S. Orszag, H. Rose, J. Searby, Z. She, Ya. Sinai, T. Toffoli, G. Vichniac, S. Wolfram, V. Yakhot, and S. Zaleski. This work was supported by European Community grant ST-2J-029-1-F, CNRS-Los Alamos grant PICS "Cellular Automata Hydrodynamics", and DOE grant KC-04-01-030.

Appendix A.   Basic symmetries of HPP, FHP, and FCHC models

We show that the models HPP, FHP, and FCHC, introduced in section 2, satisfy the symmetry assumptions (1) through (4) of section 2.4. Assumptions (1) and (2) are obvious for all three models. Let us consider (3) and (4) successively for the three models.

HPP

Let us take the x_1 axis in the direction of the vector c_1. The isometry group $\mathcal{G}$ of the velocity set is generated by permutations of the x_1 and x_2 coordinates and reversals of any of them. Clearly, any two vectors c_i and c_j can be exchanged by some isometry, so that assumptions (3) holds. Consider a particular vector, say, c_1. The subgroup $\mathcal{G}_1$, leaving c_1 invariant reduces to the identity and reversal of x_2; this implies parts (a) and (b) of assumption (4).

FHP

Let us take the x_1 axis in the direction of c_1. The isometry group $\mathcal{G}$ is now generated by rotations of $\pi/3$ and reversal of the x_2 coordinate. Assumption (3) is obvious. The subgroup $\mathcal{G}_1$ reduces again to the identity and the reversal of x_2, so that (4) follows.

FCHC

The FCHC lattice was defined in section 2.3 with explicit reference to coordinates x_1, x_2, x_3, and x_4. In this coordinate system, the velocity set is formed of

$$(\pm1,\pm1,0,0), \quad (\pm1,0,\pm1,0), \quad (\pm1,0,0,\pm1)$$
$$(0,\pm1,\pm1,0), \quad (0,\pm1,0,\pm1), \quad (0,0,\pm1,\pm1). \tag{A.1}$$

By the orthonormal change of variables

$$\begin{pmatrix} y_1 \\ y_2 \\ y_3 \\ y_4 \end{pmatrix} = \frac{1}{\sqrt{2}} \begin{pmatrix} 1 & 1 & 0 & 0 \\ -1 & 1 & 0 & 0 \\ 0 & 0 & 1 & 1 \\ 0 & 0 & -1 & 1 \end{pmatrix} \begin{pmatrix} x_1 \\ x_2 \\ x_3 \\ x_4 \end{pmatrix}, \tag{A.2}$$

the velocity set becomes

$$(\pm\sqrt{2},0,0,0), \quad (0,\pm\sqrt{2},0,0), \quad (0,0,\pm\sqrt{2},0), \quad (0,0,0,\pm\sqrt{2}),$$
$$(\pm\frac{1}{\sqrt{2}},\pm\frac{1}{\sqrt{2}},\pm\frac{1}{\sqrt{2}},\pm\frac{1}{\sqrt{2}}). \tag{A.3}$$

The isometry group $\mathcal{G}$ is generated by permutations and reversals of the x_α coordinates and by the symmetry with respect to the hyperplane $x_1 + x_2 + x_3 + x_4 = 0$, which is conveniently written in terms of y_α coordinates as

$$\Sigma: \quad (y_1,y_2,y_3,y_4) \mapsto (-y_3,y_2,-y_1,y_4). \tag{A.4}$$

Assumption (3) is obvious in any of the coordinate systems. As for assumption (4), let us consider the subgroup $\mathcal{G}_1$ leaving invariant, say, the vector with y_α coordinates $(0,0,0,1/\sqrt{2})$. The restriction of $\mathcal{G}_1$ to the hyperplane $y_4 = 0$ is generated by the identity, permutations, and reversals of y_1, y_2, and y_3. Assumptions (a) and (b) follow readily.

Appendix B.   Symmetry-related properties

Using assumptions (1) through (4) of section 2.4, we prove properties **P1** through **P6**.

P1   *Parity-invariance.* The set of velocity vectors is invariant under space-reversal.

   Indeed, on a Bravais lattice, vectors connecting neighboring nodes come in opposite pairs.

P2   Any set of i-dependent vectors $v_{i\alpha}$, which is $\mathcal{G}$-invariant, is of the form $\lambda c_{i\alpha}$.

   We write v_i as the sum of its projection on c_i and of a vector perpendicular to c_i. This decomposition being $\mathcal{G}$-invariant, the latter vector vanishes by (4a).

P3 Any set of i-dependent tensors $t_{i\alpha\beta}$, which is $\mathcal{G}$-invariant, is of the form $\lambda c_{i\alpha} c_{i\beta} + \mu \delta_{\alpha\beta}$.

To the tensors $t_{i\alpha\beta}$, we associate the linear operators $T_i : x_\alpha \mapsto t_{i\alpha\beta} x_\beta$. $\mathcal{G}$-invariance means that the T_i's commute with any lattice isometry leaving c_i invariant. We now write the $\mathcal{G}$-invariant decomposition

$$T_i = P_i T_i P_i + (I - P_i) T_i P_i + P_i T_i (I - P_i) + (I - P_i) T_i (I - P_i), \quad \text{(B.1)}$$

where I is the identity in $\mathbf{R}^D$ and P_i is the orthogonal projection on c_i. The second operator in equation (B.1), applied to an arbitrary vector $\mathbf{w}$, gives

$$(I - P_i) T_i P_i \mathbf{w} = \frac{\mathbf{w} \cdot \mathbf{c}_i}{c^2} (I - P_i) T_i \mathbf{c}_i. \quad \text{(B.2)}$$

The vectors $(I - P_i) T_i \mathbf{c}_i$ are $\mathcal{G}$-invariant and orthogonal to c_i, and thus vanish by (4a). The third operator in B.1 vanishes for similar reasons (use the $\mathcal{G}$-invariance of the transposed of the T_i's). The fourth operator in B.1 is, by (4b) proportional to I_i, the identity in the subspace orthogonal to c_i. Since $I = I_i + P_i$, the proof is completed.

We mention that we obtained **P3** by trying to formalize a result used by Hénon [42] in deriving a closed-form viscosity formula.

P4 *Isotropy of second-order tensors.* Any $\mathcal{G}$-invariant tensor $t_{\alpha\beta}$ is of the form $\mu\delta_{\alpha\beta}$.

This is a special case of **P3**, when there is no i-dependence.

P5 Any $\mathcal{G}$-invariant third-order tensor vanishes.

This follows from **P1** (parity invariance).

P6 *Velocity moments.* Odd-order velocity moments vanish. The second-order velocity moment is given by

$$\sum_i c_{i\alpha} c_{i\beta} = \frac{bc^2}{D} \delta_{\alpha\beta}. \quad \text{(B.3)}$$

The vanishing of odd-order moments is a consequence of **P1**. Equation (B.3) follows from **P4** and the Identity

$$\sum_i c_{i\alpha} c_{i\alpha} = bc^2. \quad \text{(B.4)}$$

Appendix C.   Equilibrium solutions

We prove the

Lemma 1. *The following statements are equivalent:*

1. *The N_i's are a solution of*

$$\prod_j N_j^{s'_j}(1-N_j)^{(1-s'_j)} =$$

$$\sum_s A(s \to s') \prod_j N_j^{s_j}(1-N_j)^{(1-s_j)}, \quad \forall s'. \qquad (C.1)$$

2. *The N_i's are a solution of the set of b equations*

$$\Delta_i(N) \equiv$$

$$\sum_{ss'}(s'_i - s_i)A(s \to s') \prod_j N_j^{s_j}(1-N_j)^{(1-s_j)} = 0, \quad \forall i. \qquad (C.2)$$

3. *The N_i's are given by the Fermi-Dirac distribution*

$$N_i = \frac{1}{1 + \exp(h + \mathbf{q} \cdot \mathbf{c}_i)}, \qquad (C.3)$$

where h is an arbitrary real number and $\mathbf{q}$ is an arbitrary D-dimensional vector.

Proof: (1) implies (2).

We multiply equation (C.1) by s'_i and sum over all states s' to obtain

$$\sum_{s'} s'_i \prod_j N_j^{s'_j}(1-N_j)^{(1-s'_j)} = \sum_{ss'} s'_i A(s \to s') \prod_j N_j^{s_j}(1-N_j)^{(1-s_j)}.(C.4)$$

In the l.h.s. of equation (C.4), we change the dummy variable s' into s and decorate it with a factor $A(s \to s')$, summed over s', which is one by normalization of probability. Transferring everything into the r.h.s., we obtain equation (C.2). Note that the l.h.s of equation (C.2) resembles the "collision function" Δ_i of section 3.1 (equation (3.9)), but is evaluated with the mean populations instead of the Boolean populations n_i. The relation $\Delta_i = 0$ expresses that there is no change in the mean populations under collisions. ∎

Proof: (2) implies (3).

We define

$$\check{N}_i \equiv \frac{N_i}{1 - N_i}, \qquad (C.5)$$

$$\Pi \equiv \prod_j (1 - N_j).$$

(C.6)

Equation (C.2) may be written

$$\Delta_i / \Pi = \sum_{ss'} (s'_i - s_i) A(s \to s') \prod_j \check{N}_j^{s_j} = 0.$$

(C.7)

We now make use of a trick employed in proving H-Theorems in discrete velocity models (see [17], p. 29). We multiply equation (C.7) by $\log \check{N}_i$, sum over i, and use

$$\sum_i (s'_i - s_i) \log \check{N}_i = \log \frac{\prod_j \check{N}_j^{s'_j}}{\prod_j \check{N}_j^{s_j}},$$

(C.8)

to obtain

$$\sum_{ss'} A(s \to s') \log \left(\frac{\prod_j \check{N}_j^{s'_j}}{\prod_j \check{N}_j^{s_j}} \right) \prod_j \check{N}_j^{s_j} = 0.$$

(C.9)

Semi-detailed balance $(\sum_s A(s \to s') = \sum_{s'} A(s \to s') = 1)$ implies that

$$\sum_{ss'} A(s \to s') \left(\prod_j \check{N}_j^{s_j} - \prod_j \check{N}_j^{s'_j} \right) = 0.$$

(C.10)

Combining equations (C.9) and (C.10), we obtain

$$\sum_{ss'} A(s \to s') \left[\log \left(\frac{\prod_j \check{N}_j^{s'_j}}{\prod_j \check{N}_j^{s_j}} \right) \prod_j \check{N}_j^{s_j} + \prod_j \check{N}_j^{s_j} - \prod_j \check{N}_j^{s'_j} \right] = 0.$$

(C.11)

We make use of the relation $(x > 0,\ y > 0)$

$$y \log \frac{x}{y} + y - x = - \int_x^y \log \frac{t}{x} dt \le 0,$$

(C.12)

equality being achieved only when $x = y$. The l.h.s. of equation (C.11) is a linear combination of expressions of the form (C.12) with nonnegative weights $A(s \to s')$. For it to vanish, we must have

$$\prod_j \check{N}_j^{s_j} = \prod_j \check{N}_j^{s'_j}, \quad \text{whenever } A(s \to s') \ne 0.$$

(C.13)

This is equivalent to

$$\sum_i \log(\check{N}_i)(s'_i - s_i) A(s \to s') = 0 \quad \forall s, s'.$$

(C.14)

Equation (C.13) means that $\log \check{N}_i$ is a collision invariant. We now use assumption (5) of section 2.4, concerning the absence of spurious invariants, to conclude that

$$\log \check{N}_i = -(h + \mathbf{q} \cdot \mathbf{c}_i), \tag{C.15}$$

which is the most general collision invariant (a linear combination of the mass invariant and of the D momentum invariants). Reverting to the mean populations $N_i = \check{N}_i/(1 + \check{N}_i)$, we obtain (C.3).

Proof: (3) implies (1).

Equation (C.3) implies

$$\sum_j \log(\check{N}_j)(s_j - s'_j) = 0, \quad \text{whenever } A(s \to s') \neq 0. \tag{C.16}$$

This implies

$$\sum_s A(s \to s') \left(\prod_j \check{N}_j^{(s_j - s'_j)} - 1 \right) = 0. \tag{C.17}$$

Using semi-detailed balance, this may be written as

$$1 = \sum_s A(s \to s') \frac{\prod_j \check{N}_j^{s_j}}{\prod_j \check{N}_j^{s'_j}}. \tag{C.18}$$

Reverting to the N_j's, we obtain equation (C.1). This completes the proof of the equivalence lemma. ∎

Appendix D.   Inclusion of body-forces

Using the same notation as in section 7.3, we wish to obtain a Navier-Stokes equation with a body-force $\mathbf{f}$, that is

$$\partial_T \mathbf{U} + \mathbf{U} \cdot \nabla_1 \mathbf{U} = -\nabla_1 P' + \nu' \nabla_1^2 \mathbf{U} + \mathbf{f}$$

$$\nabla_1 \cdot \mathbf{U} = 0. \tag{D.1}$$

The force $\mathbf{f}$ may depend on space and time and can be velocity-independent (case I; e.g. gravity) or linear in the velocity $\mathbf{U}$ (case II; e.g. Coriolis force). The idea is to introduce a bias in the transition rules so as to give a net momentum input. Since all the terms in the Navier-Stokes momentum equation are $O(\epsilon^3)$ and the hydrodynamic velocity is $O(\epsilon)$ (before rescaling), the bias should be $O(\epsilon^3)$ for case I and $O(\epsilon^2)$ for case II.

We give now the modified form of the microdynamical equation (3.9) appropriate for body-forces. We introduce, in addition to the Boolean (transition) variables $\xi_{ss'}$ of section 3.1, the Boolean variables $\xi'_{ss'}$ such that

$$\langle \xi'_{ss'} \rangle = B(s \to s'). \tag{D.2}$$

The $B(s \to s')$'s are a set of transition probabilities associated to the body-force; they satisfy normalization

$$\sum_{s'} B(s \to s') = 1, \tag{D.3}$$

and mass conservation

$$\sum_i (s_i' - s_i) B(s \to s') = 0, \quad \forall s, s'. \tag{D.4}$$

They do not satisfy momentum conservation, semi-detailed balance and $\mathcal{G}$-invariance. The $\xi'_{ss'}$'s are chosen independently at each discrete time and node and the $B(s \to s')$'s may depend on space and time; further constraints will be given below. We also need a Boolean variable ς which acts as a *switch*: when $\varsigma = 0$ the force is off and the usual transition rules apply. The mean of ς is given by

$$\langle \varsigma \rangle = \rho_0 g(\rho_0) \epsilon^n$$

$$n = 3 \quad \text{case I}, \qquad n = 2 \quad \text{case II.} \tag{D.5}$$

This will take care of the scaling factors arising from the change of variables (7.15). The modified microdynamical equation is now

$$n_i(t_* + 1, \mathbf{r}_* + \mathbf{c}_i) = n_i + \Delta_i(n)$$

$$\Delta_i(n) =$$

$$\sum_{s,s'} (s_i' - s_i) \left((1 - \varsigma) \xi_{ss'} + \varsigma \xi'_{ss'} \right) \prod_j n_j^{s_j} (1 - n_j)^{(1-s_j)}. \tag{D.6}$$

Let us evaluate the body-force resulting from the insufficient additional ξ' term. For this, we multiply by $\mathbf{c}_i$ and average over the equilibrium distribution; deviations from equilibrium arising from hydrodynamic gradients are irrelevant. We ignore the ς-factor since it just provides the scaling factor.

We begin with case I. The average is then evaluated over the zero-velocity equilibrium distribution with density per cell d; we obtain

$$\mathbf{f} = \sum_{s,s',i} \mathbf{c}_i (s_i' - s_i) B(s \to s') \left(\frac{d}{1-d} \right)^p (1-d)^b, \quad p = \sum_j s_j \tag{D.7}$$

where b is the number of cells per node. Equation (D.7) is the additional constraint on the $B(s \to s')$'s for case I. If $\mathbf{f}$ is space- and/or time-dependent, so are the $B(s \to s')$'s. It is easy to check that for any given vector $\mathbf{f}$, there exist Boolean transition variables $\xi'_{ss'}$ of mean $B(s \to s')$ satisfying equation (D.7). When $\mathbf{f}$ is in the direction of a particular velocity vector, say $\mathbf{c}_{i_0}$, we can flip particles with velocity $-\mathbf{c}_{i_0}$ into particles with velocity $\mathbf{c}_{i_0}$ whenever this is possible, while leaving all other particles unchanged. This is done with a probability dependent on the amplitude of the force. Other directions of the force are handled by superposition.

We turn to case II. We wish to obtain a force of the form

$$f_\alpha = C_{\alpha\beta} U_\beta \tag{D.8}$$

where $C_{\alpha\beta}$ is a D-dimensional matrix. When the velocity $\mathbf{U}$ vanishes, the body-force should also vanish; this requires

$$\sum_{s,s',i} \mathbf{c}_i (s_i' - s_i) B(s \to s') \left(\frac{d}{1-d}\right)^p (1-d)^b = 0, \qquad p = \sum_j s_j. \tag{D.9}$$

With nonvanishing velocity, we must use the corresponding equilibrium populations given to relevant order by (cf. equation (4.14))

$$N_i = d + \frac{dD}{c^2} c_{i\alpha} u_\alpha. \tag{D.10}$$

Here, we have used the unscaled velocity $\mathbf{u}$. Below, however, we will use $\mathbf{U}$ since the scaling factor is taken care of by the Boolean switch ς. Using (D.10) in (D.6), we find that the average momentum imparted by $\xi_{ss'}'$ transitions is to leading order linear in $\mathbf{U}$. Identifying with equation (D.8), we find that the $B(s \to s')$'s must satisfy the following constraints

$$C_{\alpha\beta} = \frac{D}{c^2} (1-d)^{b-1} \sum_{s,s',i} c_{i\alpha}(s_i' - s_i) B(s \to s') \left(\frac{d}{1-d}\right)^p \sum_j s_j c_{j\beta},$$

$$p = \sum_j s_j. \tag{D.11}$$

Equations (D.9) and (D.11) are the additional constraints on the $B(s \to s')$'s for case II.

As an illustration, consider the case of the pseudo-four-dimensional FCHC model with a Coriolis force $2\Omega \wedge \mathbf{U}$, where Ω is in the x_3-direction. A possible implementation for the $\xi_{ss'}'$ transitions is through rotation by $\pi/2$ around the x_3-axis of those particles having their velocity perpendicular to this axis (with a probability dependent on Ω).

Appendix E.   Catalog of results for FHP models

The purpose of this appendix is to summarize all known analytic results for the FHP models, including the models II and III which have rest particles. Adapting the theory to cases with at most one rest particle is quite straightforward if one includes the rest-particle velocity, namely vector zero. Our derivations made extensive use of properties P1 to P6 of section 2.4. With rest particles, P1, P2, P4, and P5 are unchanged. In P3, λ and μ have usually different values for *moving* and *rest* particles. P6 becomes

$$\sum_i c_{i\alpha} c_{i\beta} = \frac{(b-1)c^2}{D} \delta_{\alpha\beta}, \tag{E.1}$$

	FHP-I	FHP-II	FHP-III
ρ_0	$6d$	$7d$	$7d$
c_s	$\frac{1}{\sqrt{2}}$	$\sqrt{\frac{3}{7}}$	$\sqrt{\frac{3}{7}}$
g	$\frac{1}{2}\frac{1-2d}{1-d}$	$\frac{7}{12}\frac{1-2d}{1-d}$	$\frac{7}{12}\frac{1-2d}{1-d}$
ν	$\frac{1}{12}\frac{1}{d(1-d)^3}-\frac{1}{8}$	$\frac{1}{28}\frac{1}{d(1-d)^3}\frac{1}{1-4d/7}-\frac{1}{8}$	$\frac{1}{28}\frac{1}{d(1-d)}\frac{1}{1-8d(1-d)/7}-\frac{1}{8}$
ς	0	$\frac{1}{98}\frac{1}{d(1-d)^4}-\frac{1}{28}$	$\frac{1}{98}\frac{1}{d(1-d)}\frac{1}{1-2d(1-d)}-\frac{1}{28}$
$R_*^{\max}$	0.387	1.08	2.22
$d_{\max}$	0.187	0.179	0.285

Table 1: Analytic results for three FHP models

where b is still the number of bits, so that $b-1$ is the number of particles *moving* with speed c.

In Table 1 below, we give results in terms of the mean density per cell d for the following quantities: the mean density ρ_0, the coefficient $g(\rho_0)$ rescaling the nonlinear term in the Navier-Stokes equation (see for example equation (7.13)), the kinematic shear viscosity ν, the kinematic bulk viscosity ς, the maximum value $R_*^{\max}$ of the coefficient R_* appearing in the Reynolds number (see equation (8.29)), and $d_{\max}$, the density at which the Reynolds number is maximum. The viscosities ν and ς are calculated within the lattice Boltzmann approximation (see section 8.2). $\rho_0\varsigma$ is the dynamic bulk viscosity; when it does not vanish, as is the case with rest particles, equation (7.11) becomes

$$\partial_t\rho' + \rho_0\nabla\cdot\mathbf{u} = 0$$

$$\rho_0\partial_t\mathbf{u} + c_s^2\nabla\rho' = \rho_0\nu\left(\nabla^2\mathbf{u} + \frac{D-2}{D}\nabla\nabla\cdot\mathbf{u}\right) + \rho_0\varsigma\nabla\nabla\cdot\mathbf{u}. \qquad (\text{E.2})$$

Appendix F.   An H-theorem for lattice gases[23]

F.1   Notation and basic equations

We number from 1 to b the cells at a given node (b is the number of different velocity vectors). It is not necessary that the velocity moduli are equal. Also, it will not be necessary to specify any symmetry for the lattice or for the collision rules. Finally, we will not make use of the conservation of the

---

[23]by M. Hénon, Observatoire de Nice.

number of particles or of the momentum, so that the proof is applicable to lattices where these conservation laws are violated.

We write $s_i = 1$ if particle i is present in the *input state*, 0 if it is absent. An input state is thus defined by $s = (s_1, \ldots, s_b)$. The number of distinct input states is 2^b.

We call $P(s)$ the probability of an input state s. We have

$$\sum_s P(s) = 1. \tag{F.1}$$

We call N_i the probability that particle i is present. We have

$$N_i = \sum_s s_i P(s), \qquad 1 - N_i = \sum_s (1 - s_i) P(s). \tag{F.2}$$

We define in the same way s_i', $s' = (s_1', \ldots, s_b')$, $P'(s')$, N_i' for the *output state*.

We call $A(s \to s')$ the probability that an input state s is changed into an output state s' by the collision. We have

$$P'(s') = \sum_s P(s) A(s \to s'). \tag{F.3}$$

We have, of course,

$$\sum_{s'} A(s \to s') = 1, \tag{F.4}$$

where the sum is over all output states. We will assume that the collision rules obey *semi-detailed balance*, i.e., that we have also

$$\sum_s A(s \to s') = 1. \tag{F.5}$$

F.2   Local theorem

Lemma 1. *If $f(x)$ is a convex function $(d^2 f/dx^2 > 0)$, then*

$$\sum_{s'} f[P'(s')] \leq \sum_s f[P(s)]. \tag{F.6}$$

Proof: From general properties of convex functions, we have

$$f\left[\frac{\sum_s q(s) P(s)}{\sum_s q(s)}\right] \leq \frac{\sum_s q(s) f[P(s)]}{\sum_s q(s)}, \tag{F.7}$$

where the $q(s)$ are arbitrary positive or zero coefficients. Taking $q(s) = A(s \to s')$, with s' given, and using equations (F.3) and (F.5), we obtain

$$f[P'(s')] \leq \sum_s A(s, s') f[P(s)]. \tag{F.8}$$

Summing over s' and using equation (F.4), we obtain equation (F.6). ∎

Lemma 2. *The following inequality holds:*

$$\sum_{s'} P'(s') \ln P'(s') \leq \sum_{s} P(s) \ln P(s). \qquad (F.9)$$

Proof: We apply Lemma 1 with $f(x) = x \ln x$. ∎

Lemma 3. *The following inequality holds:*

$$\sum_{s} P(s) \ln P(s) \geq \sum_{i=1}^{b} [N_i \ln N_i + (1 - N_i) \ln(1 - N_i)]. \qquad (F.10)$$

The equality holds if and only if

$$P(s_1, \ldots, s_b) = \prod_{i=1}^{b} N_i^{s_i} (1 - N_i)^{1-s_i}. \qquad (F.11)$$

Proof:[24] The right-hand side of equation (F.10) can be written, using equation (F.2):

$$\sum_{i=1}^{b} \sum_{s} [s_i P(s) \ln N_i + (1 - s_i) P(s) \ln(1 - N_i)], \qquad (F.12)$$

or

$$\sum_{s} P(s) \ln \left[\prod_{i=1}^{b} N_i^{s_i} (1 - N_i)^{1-s_i} \right]. \qquad (F.13)$$

Therefore, equation (F.10) can also be written

$$\sum_{s} P(s) \ln \left[\frac{\prod_{i=1}^{b} N_i^{s_i} (1 - N_i)^{1-s_i}}{P(s)} \right] \leq 0. \qquad (F.14)$$

We have, for any x:

$$\ln x \leq x - 1, \qquad (F.15)$$

where the equality holds only if $x = 1$. Therefore,

$$\ln \left[\frac{\prod_{i=1}^{b} N_i^{s_i} (1 - N_i)^{1-s_i}}{P(s)} \right] \leq \frac{\prod_{i=1}^{b} N_i^{s_i} (1 - N_i)^{1-s_i}}{P(s)} - 1. \qquad (F.16)$$

Multiplying this by $P(s)$ and summing over s, we obtain the desired result. ∎

The relation (F.11) corresponds to the Boltzmann approximation (independence of input particles).

---

[24] inspired by reference 85.

Local H-theorem

If the collision rules satisfy semi-detailed balance, and in the Boltzmann approximation, the following inequality holds:

$$\sum_{i=1}^{b}[N_i' \ln N_i' + (1 - N_i') \ln(1 - N_i')]$$

$$\leq \sum_{i=1}^{b}[N_i \ln N_i + (1 - N_i) \ln(1 - N_i)].) \tag{F.17}$$

Proof: From Lemma 3, we have

$$\sum_{s} P(s) \ln P(s) = \sum_{i=1}^{b}[N_i \ln N_i + (1 - N_i) \ln(1 - N_i)]. \tag{F.18}$$

Combining with Lemma 2:

$$\sum_{s'} P'(s') \ln P'(s') \leq \sum_{i=1}^{b}[N_i \ln N_i + (1 - N_i) \ln(1 - N_i)]. \tag{F.19}$$

Finally, applying Lemma 3 to the N_i''s and the P''s, we obtain equation (F.17). ∎

We remark that both conditions of the theorem are necessary; one can easily find counterexamples if one or the other is not satisfied. Consider, for instance, a node of the HPP lattice with probabilities before collision: $P(1,0,1,0) = 1/2$, $P(0,1,0,0) = 1/2$. We have: $N_1 = 1/2$, $N_2 = 1/2$, $N_3 = 1/2$, $N_4 = 0$; the Boltzmann approximation is not satisfied. We take the usual HPP collision rules. The probabilities after collision are then $P'(0,1,0,1) = 1/2$, $P'(0,1,0,0) = 1/2$. From this, we deduce $N_1' = 0$, $N_2' = 1$, $N_3' = 0$, $N_4' = 1/2$, and it can be immediately verified that the left-hand member of equation (F.17) is larger than the right-hand member.

Similarly, let us modify the collision rules and keep only one kind of collision: $(1,0,1,0)$ gives $(0,1,0,1)$, but not conversely. Semi-detailed balance is not satisfied. Take for instance $N_1 = N_2 = N_3 = N_4 = 1/2$. We assume that the Boltzmann approximation holds; therefore, $P(s) = 1/16$ for all s. We deduce $P'(1,0,1,0) = 0$; $P'(0,1,0,1) = 2/16$; $P'(s') = 1/16$ for the other s'; $N_1' = N_3' = 7/16$, $N_2' = N_4' = 9/16$; and here again the inequality (F.17) is violated.

F.3   Global theorem

First we sum equation (F.17) over all lattice nodes. We obtain a sum over all cells at all lattice nodes; their total number will be denoted by r:

$$\sum_{j=1}^{r}[N'^{(j)} \ln N'^{(j)} + (1 - N'^{(j)}) \ln(1 - N'^{(j)})]$$

$$\leq \sum_{j=1}^{r} [N^{(j)} \ln N^{(j)} + (1 - N^{(j)}) \ln(1 - N^{(j)})]. \tag{F.20}$$

Next we remark that this sum is invariant under propagation. We can therefore extend the theorem to an arbitrary number of time steps, and we obtain (with the same hypotheses as for the local theorem):

Global H-theorem

The function

$$\sum_{j=1}^{r} [N^{(j)} \ln N^{(j)} + (1 - N^{(j)}) \ln(1 - N^{(j)})] \tag{F.21}$$

is non-increasing as the lattice gas evolves.

Appendix F.1   Interpretation in terms of information theory

Consider a probability distribution over ν possible cases: $p_1, \ldots, p_\nu$. The associated information is

$$\log_2 \nu + \sum_{i=1}^{\nu} p_i \log_2 p_i. \tag{F.22}$$

This information has a minimal value 0 if all cases have the same probability: $p_1 = \cdots = p_\nu = 1/\nu$. It has a maximal value $\log_2 \nu$ if one of the p_i is 1 while the others are 0, i.e., for a deterministic choice between the ν cases.

We come back to lattices. $P(s)$ represents a probability distribution on 2^b cases, and therefore an information

$$b + \sum_{s} P(s) \log_2 P(s). \tag{F.23}$$

Thus, Lemma 2 expresses the following property: if semi-detailed balance is satisfied, then the information contained in the P can only remain constant or decrease in a collision.

From the P's, we can compute the N_i's by the formulas (F.2), but the converse is not generally true; in other words, the P's contain more information than the N_i's. Lemma 3 expresses this fact.

In the particular case of the Boltzmann approximation, the particles are considered as independent, and therefore, the P's contain no more information than the N_i's. We have then the equality in equation (F.10).

The proof of the local H-theorem can therefore be interpreted as follows: (i) initially the N_i's are given; this represents a given information; (ii) we compute the corresponding P's in the Boltzmann approximation; the information does not change; (iii) we compute the collision and obtain the P''s; the information decreases or stays constant; (iv) we compute the N''s from the P''s; here again, the information decreases or stays constant.

References

[1] U. Frisch, B. Hasslacher, and Y. Pomeau, *Phys. Rev. Lett.*, **56** (1986) 1505.

[2] D. d'Humières, P. Lallemand, and U. Frisch, *Europhys. Lett.*, **2** (1986) 291.

[3] J. von Neumann, *Theory of Self-Reproducting Automata* (Univ. of Illinois press, 1966).

[4] E. R. Berlekamp, J. H. Conway, and R. K. Guy, *Winning Ways for Your Mathematical Plays*, **2** (Academic Press, 1984).

[5] S. Wolfram, *Reviews of Modern Physics*, **55** (1983) 601.

[6] S. Wolfram, *Theory and Applications of Cellular Automata*, (World Scientific, 1986).

[7] Y. Pomeau, *J. Phys.*, **A17** (1984) L415.

[8] N. Margolus, T. Toffoli, and G. Vichniac, *Phys. Rev. Lett.*, **56** (1986) 1694.

[9] N. Margolus and T. Toffoli, "Cellular automata machines", submitted to *Complex Systems* (1987).

[10] W. D. Hillis, *The Connection Machine*, (MIT Press, 1985).

[11] J. P. Boon and S. Yip, *Molecular Hydrodynamics*, (McGraw-Hill, 1980).

[12] D. C. Rapaport and F. Clementi, *Phys. Rev. Lett.*, **57** (1986) 695.

[13] E. Meiburg, *Phys. Fluids*, **29** (1986) 3107.

[14] L. Hannon, private communications 1986.

[15] J. E. Broadwell, *Phys. Fluids*, **7** (1964) 1243.

[16] S. Harris, *Phys. Fluids*, **9** (1966) 1328.

[17] R. Gatignol, "Théorie Cinétique des Gaz à Répartition Discrète des Vitesses", *Lecture Notes in Physics*, **36** (Springer, Berlin, 1975).

[18] R. Gatignol, "The hydrodynamic description for a discrete velocity model of a gas", *Complex Systems*, **1** (1987) 708.

[19] H. Cabannes, *Mech. Res. Comm.*, **12** (1985) 289.

[20] R. Caflisch, "Asymptotics of the Boltzmann equation and fluid dynamics", lecture notes from CISM summer school on *Kinetic Theory and Gas Dynamics*, (Udine, Italy, June 1986).

[21] L. P. Kadanoff and J. Swift, *Phys. Rev.*, **165** (1968) 310.

[22] J. Hardy, Y. Pomeau, and O. de Pazzis, *J. Math. Phys.*, **14** (1973) 1746.

[23] J. Hardy, O. de Pazzis, and Y. Pomeau, *Phys. Rev.*, **A13** (1976) 1949.

[24] J. Hardy and Y. Pomeau, *J. Math. Phys.*, **13** (1972) 1042.

[25] D. d'Humières, P. Lallemand, and T. Shimomura, "An experimental study of lattice gas hydrodynamics", Los Alamos Report LA-UR-85-4051 (1985).

[26] D. d'Humières, Y. Pomeau, and P. Lallemand, *C. R. Acad. Sci. Paris II*, **301** (1985) 1391.

[27] J. Salem and S. Wolfram, in *Theory and Applications of Cellular Automata*, S. Wolfram. ed. (World Scientific, 1986) 362.

[28] D. d'Humières and P. Lallemand, *C. R. Acad. Sci. Paris II*, **302** (1986) 983.

[29] D. d'Humières and P. Lallemand, *Physica*, **140A** (1986) 326.

[30] D. d'Humières and P. Lallemand, *Helvetica Physica Acta*, **59** (1986) 1231.

[31] D. d'Humières and P. Lallemand, "Numerical simulations of hydrodynamics with lattice gas automata in two dimensions", *Complex Systems*, **1** (1987) 598.

[32] M. Hénon, "Isometric collision rules for the four-dimensional FCHC lattice gas", *Complex Systems*, **1** (1987) 475.

[33] J. P. Rivet, "Simulation d'écoulements tridimensionnels par la méthode des gaz sur réseaux: premiers résultats", submitted to *C. R. Acad. Sci. Paris* (1987).

[34] A. Clouqueur and D. d'Humières, "RAP1, a cellular automaton machine for fluid dynamics", *Complex Systems*, **1** (1987) 584.

[35] J. P. Rivet and U. Frisch, *C. R. Acad. Sci. Paris II*, **302** (1986) 267.

[36] V. Yakhot and S. Orszag, *Phys. Rev. Lett.*, **56** (1986) 169.

[37] V. Yakhot, B. Bayly, and S. Orszag, *Phys. Fluids*, **29** (1986) 2025.

[38] U. Frisch and J. P. Rivet, *C. R. Acad. Sci. Paris II*, **303** (1986) 1065.

[39] S. Wolfram, *J. Stat. Phys.*, **45** (1986) 471.

[40] F. Hayot, "Viscosity in lattice gas automata", *Physica D*, **26** (1987) 210.

[41] F. Hayot, "Higher order corrections to Navier-Stokes in lattice gas automata", *Complex Systems*, **1** (1987) 752.

[42] M. Hénon, "Viscosity of a lattice gas", *Complex Systems*, **1** (1987) 762.

[43] J. P. Rivet, "Green-Kubo formalism for lattice gas hydrodynamics and Monte-Carlo evaluation of shear viscosities", *Complex Systems*, **1** (1987) 838.

[44] L. Kadanoff, G. McNamara, and G. Zanetti, "Size-dependence of the shear viscosity for a two-dimensional lattice gas", preprint Univ. Chicago (1987).

[45] D. Tarnowski, *La Recherche*, **174** (1986) 272.

[46] D. d'Humières, P. Lallemand, and Y. Pomeau, *bull. Soc. Franç. Phys.*, **60** (1986) 14.

[47] L. Kadanoff, *Physics Today*, (September 1986) 7.

[48] C. Burges and S. Zaleski, *Complex Systems*, **1** (1987) 31.

[49] D. H. Rothmann, "Modeling seismic P-waves with cellular automata", preprint, MIT Earth Resources Lab (1986).

[50] D. Montgomery and G. Doolen, "Two cellular automata for plasma computations", *Complex Systems*, **1** (1987) 830.

[51] H. Chen and W. Matthaeus, "Cellular automaton formulation of passive scalar dynamics", *Phys. Fluids*, **30** (1987) 1235.

[52] H. Chen and W. Matthaeus, "A new cellular automaton model for magnetohydrodynamics", *Phys. Rev. Letters*, **58** (1987) 1865.

[53] A. de Masi, P. A. Ferrari, and J. L. Lebowitz, *Phys. Rev. Lett.*, **55** (1985) 1947.

[54] J. L. Lebowitz, *Physica*, **140A** (1986) 232.

[55] A. de Masi, P. A. Ferrari, and J. L. Lebowitz, "Reaction-Diffusion equations for interacting particle systems", *Physica*, in press.

[56] P. Clavin, D. d'Humières, P. Lallemand, and Y. Pomeau, *C. R. Acad. Sci. Paris II*, **303** (1986) 1169.

[57] P. Clavin, P. Lallemand, Y. Pomeau, and J. Searby, "Simulation of free boundaries in flow systems: a new proposal based on lattice-gas models", *J. Fluid Mech.*, in press.

[58] B. M. Boghosian and C. D. Levermore, *Complex Systems*, **1** (1987) 17.

[59] D. Levermore, private communication (1986).

[60] N. Ashcroft and D. Mermin, *Solid State Physics*, (Holt-Saunders International Editions, 1976).

[61] G. Doolen, private communication (1986).

[62] S. Chapman and T. G. Cowling, *The Mathematical Theory of Non-Uniform Gases*, (Cambridge Univ. Press, 1939).

[63] G. Uhlenbeck and G. Ford *Lectures in Statistical Mechanics*, (American Math. Soc., 1963).

[64] L. Landau and E. Lifschitz, *Fluid Mechanics*, (Pergamon 1981).

[65] A. Majda, *Compressible Fluid Flow and Systems of Conservation Laws in Several Space Variables*, (Springer Verlag, 1984).

[66] D. d'Humières, P. Lallemand, and J. Searby, "Numerical experiments in lattice gases: mixtures and Galilean invariance", *Complex Systems*, **1** (1987) 632.

[67] L. Landau and E. Lifschitz, *Zh. Eksp. Teor. Fiz.*, **32** (1975) 918; *Sov. Phys. JETP*, **5** (1957) 512.

[68] Y. Pomeau and P. Résibois, *Phys. Rep.*, **19C** (1975) 64.

[69] R. Kubo, *J. Phys. Soc. Jpn.*, **7** (1957) 439.

[70] H. S. Green, *J. Math. Phys*, **2** (1961) 344.

[71] D. Forster, D. R. Nelson, and M. J. Stephen, *Phys. Rev.*, **A16** (1977) 732.

[72] J. D. Fournier and U. Frisch, *Phys. Rev.*, **A28** (1983) 1000.

[73] B. J. Alder and T. E. Wainwright, *Phys. Rev.*, **A1** (1970) 18.

[74] V. Yakhot, S. Orszag, and U. Frisch, unpublished (1986).

[75] S. Chapman, *Philos. Trans. Roy. Soc. London*, Ser. **A216** (1916) 279; **217** (1917) 115.

[76] D. Enskog, *Kinetische Theorie der Vorgänge in Mässig Verdünnten Gasen*, dissertation, Uppsala (1917).

[77] J. P. Rivet, *Gas sur Réseaux*, internal report, Observatoire de Nice (1986).

[78] M. Hénon, private communication (1986).

[79] G. K. Batchelor, *Phys. Fluids (Suppl. 2)*, **12** (1969) 233.

[80] R. H. Kraichnan, *Phys. Fluids*, **10** (1967) 1417.

[81] M. E. Brachet, M. Meneguzzi, and P. L. Sulem, *Phys. Rev. Lett.*, **57** (1986) 683.

[82] A. W. Kolmogorov, *C. R. Acad. Sci. USSR*, **30** (1941) 301,538.

[83] A. S. Monin and A. M. Yaglom, *Statistical Fluid Mechanics*, **2**, J. L. Lumley, ed. (MIT Press, 1975) revised and augmented edition of the Russian original *Statisticheskaya Gidromekhanika*, (Nauka, Moscow 1965).

[84] T. Passot and A. Pouquet, "Numerical simulations of compressible homogeneous fluids in the turbulent régime", *J. Fluid Mech.*, in press.

[85] R. M. Fano, *Transmission of Information*, (Wiley, 1961) 43.

K. Diemer,* K. Hunt,† S. Chen,† T. Shimomura,† and G. Doolen†

*Physics Dept. University of California Santa Cruz, Santa Cruz, CA 95064 and †Los Alamos National Laboratory, Center for Nonlinear Studies and Theoretical Division, Los Alamos, NM 87545

Density and Velocity Dependence of Reynolds Numbers for Several Lattice Gas Models

Analytic calculations of shear viscosity η and a reduced Reynolds number R_* for eleven different 2-D and six different 3-D lattice gas hydrodynamic models are compared. The mean free path is calculated for ten of the 2-D models. Only models with one non-zero speed are considered. All analytic derivations of η have ignored the velocity dependence of η. For comparison with the analytic results, Reynolds numbers for several 2-D models are calculated, using the microscopic definition of shear viscosity, from very low velocity 2-D Couette flow simulations. A similar check, using the definition of mean free path, is done for one of the 2-D models. Viscosity for one model is calculated using a computer simulation to demonstrate a velocity dependence and thus give a rough idea of the accuracy of these analytic calculations of the viscosity. The velocity dependence of the mean free path is calculated analytically, compared with computer calculations, and shown to be strongly anisotropic.

Lattice Gas Methods for Partial Differential Equations, SFI SISOC,
Eds. Doolen et al., Addison-Wesley Publishing Co., 1990 **137**

1. INTRODUCTION

It has been shown recently that the 2-D and 3-D Navier-Stokes equations can be simulated on a particular class of cellular automata, called lattice gas models.[1,4,5,12] A detailed account can be found in Frisch.[5] In such lattice gas flow simulations, it is useful to choose collision rules which minimize the viscosity and thereby maximize the Reynolds number, because the computer calculation time is proportional to R_*^3 in two dimensions and R_*^4 in three dimensions.[10] While it is not practical to search the space of all possible collision rules to determine the optimal table, a comparison of several models gives an estimate of how these quantities depend on collision rules. Analytical results for shear viscosity η for eleven different 2-D and six different 3-D lattice gas models are given below. All viscosity calculations done here use a single speed algorithm developed by Hénon[7,8] or an extension of it for models including rest particles . The viscosity is derived in the Boltzmann approximation from a straightforward algebraic calculation which depends on the choice of collision rules. Agreement is obtained with the analytic results of d'Humières and Lallemand,[2] obtained using a Chapman-Enskog expansion and a Taylor series in velocity to obtain η.

In section 2 of this paper, we establish the notation to be used in this paper. Section 3 summarizes Hénon's method for calculating the shear viscosity for 2-D hexagonal lattice gas simulations. The mean free path λ and R_* are also given for this case. Section 4 generalizes the results for R_* and η for arbitrary dimension and for the inclusion of rest particles. Section 5 presents analytic results for η, R_*, and λ for a number of 2-D models. The kinematic viscosity is shown to be nonlinearly dependent on mean free path. Section 6 presents analytic results for η and R_* for a number of 3-D models. The limitations of these analytic results are discussed in section 7. The viscosities calculated in sections 5 and 6 are velocity independent. Computer simulation results from one model show a definite velocity dependence. The computer simulations were also run for several models at very low velocities to compare with the velocity-independent analytic results. Section 7 also includes an analytic calculation of the velocity dependence of the mean free path, which is verified with measurements from a computer simulation. Concluding remarks are given in section 8.

2. NOTATION

The notation is chosen to be consistent, for the most part, with Hénon's.[8] The velocities are $c_i = (c_{i1}, \cdots, c_{iD})$, with i running over all possible velocities on the lattice (including rest particles) and D equal to the dimension of the lattice. Both the lattice and the set of velocity vectors must satisfy certain symmetry conditions, (see Frisch[5] or Hénon[8] for details). The number of possible velocities at a node is n, the number of possible non-zero velocities is b. 2-D models are run on a hexagonal lattice, 3-D models are run on a 4-D face-centered hypercubic lattice (or its dual) and then projected down to three dimensions. The input state is s, the output state is s', and the transition probability between the two states is represented as $A(s \rightarrow s')$. Each state is specified by a set of parameters $s = (s_1, \cdots, s_i, \cdots, s_n)$. The presence or absence of a particle at a lattice site with a velocity c_i is indicated by a value of 1 or 0, respectively, for the parameter s_i. The total number of particles at a given site is p. Density ρ is given in terms of *mass/node*, with unit mass particle defined to be m. The density ρ', in terms of mass/"volume", is determined from ρ by $\rho' = f\rho$. The conversion factor f is proportional to l^{-D}, where l is the lattice spacing. For all hexagonal models, f is $(2l^{-2})/\sqrt{3}$. The time step is defined to be τ. The speed of sound is c_s. The angle between c_i and c_j is defined to be θ_{ij}. We define a reduced Reynolds number, $R_* = R/Ml_o$, where M is the Mach number and l_o is the characteristic scale of the flow expressed in units of l. This reduced Reynolds number is especially useful in comparing the models because computer calculation time varies as a power of R_*, as mentioned above.

3. SUMMARY OF THE HÉNON'S ALGORITHM FOR VISCOSITY

To clarify the method used by the present authors to compute viscosity and R_*, a summary of Hénon's algorithm for the 6-bit model is shown below. More extensive treatments of this subject can be found in Hénon,[8] Frisch,[5] and Dubrelle.[3] Viscosity and R_* can also be calculated using an adaptation of the Chapman-Enskog expansion for lattice gasses. Such treatments can be found in d'Humières[2] and Wolfram.[13]

The gas density is defined by

$$\rho'(\mathbf{x}) = fm \sum_{i=1}^{6} N_i(\mathbf{x}), \qquad (3.1)$$

where N_i is the probability of finding a particle arriving at node x with velocity c_i. The mean velocity $\mathbf{u}$ at a node is defined by

$$\rho'\mathbf{u}(\mathbf{x}) = fm \sum_{i=1}^{6} N_i(\mathbf{x})\mathbf{c}_i. \qquad (3.2)$$

The conventions for the directions associated with each N_i are listed in Table 1, where c_{i1} are the x_1-direction cosines and c_{i2} are the x_2-direction cosines.

Consider an approximately homogeneous state (i.e., $\rho(\mathbf{x}) = \rho$), in which the velocity distribution is assumed to be isotropic, then all the N_i are equal to a constant value d,

TABLE 1 Conventions for the Directions Associated with Each N_i

i	mass	c_{i1}	c_{i2}
1	1	$1/2$	$\sqrt{3}/2$
2	1	$-1/2$	$\sqrt{3}/2$
3	1	-1	0
4	1	$-1/2$	$-\sqrt{3}/2$
5	1	$1/2$	$-\sqrt{3}/2$
6	1	1	0

$$N_i = \frac{\rho'}{6fm} = d.$$

Hénon considers a shear flow perturbed slightly from isotropy:

$$N_i = d + \nu_i(\mathbf{x}),$$

with

$$\nu_i(\mathbf{x}) \ll 1.$$

From conservation of mass, one obtains:

$$\sum_{i=1}^{6} \nu_i = 0.$$

Consider the small shear flow:

$$u_1 = Tx_2, \qquad u_2 = 0.$$

Hénon uses the ansatz that

$$\nu_i(\mathbf{x}) = k_i x_2 + \epsilon_i,$$

and obtains the analytic expressions given below for k_i and ϵ_i. In the Boltzmann approximation, the probability of input state s is

$$P_s = \prod_{j=1}^{6} N_j{}^{s_j}(1 - N_j)^{s_j}.$$

Equivalently,

$$N_i = \sum_s s_i P_s.$$

The probability of a particle leaving node x with velocity c_i is then

$$N_i' = \sum_s \sum_{s'} s_i' A(s \to s_i') P_s,$$

where $A(s \to s_i')$ is the probability of a transition from state s to s'. Using semi-detailed balance,

$$N_i' - N_i = \sum_s \sum_{s'} (s_i' - s_i) A(s \to s_i') P_s.$$

Expanding to first order in ν_i,

$$\nu_i' - \nu_i = \sum_s \sum_{s'} (s_i' - s_i) A(s \to s_i') d^{p-1} (1-d)^{5-p} \sum_{j=1}^{6} s_j \nu_j,$$

where p is the number of particles in state s.

The basic propagation equation is

$$N_i'(\mathbf{x}, t) = N_i(\mathbf{x} + \tau c_i, t + \tau),$$

where τ is the time step (usually chosen to be 1). With the assumption of steady state,

$$N_i'(\mathbf{x}) = N_i(\mathbf{x} + \tau c_i),$$

one obtains

$$\nu_i'(\mathbf{x}) - \nu_i(\mathbf{x}) = \tau k_i c_{i2}.$$

A set of equations for k_i and a set for ϵ_i are obtained by equating the collision and propagation equations equal to each other and collecting terms proportional to and independent of x_2, respectively. The k_i and ϵ_i must also satisfy the following constraint equations (from conservation of mass and momentum),

$$\sum_{i=1}^{6} k_i = 0, \qquad \sum_{i=1}^{6} c_{i1} k_i = 6dT, \qquad \sum_{i=1}^{6} c_{i2} k_i = 0,$$

$$\sum_{i=1}^{6} \epsilon_i = 0, \qquad \sum_{i=1}^{6} c_{i1} \epsilon_i = 0, \qquad \sum_{i=1}^{6} c_{i2} \epsilon_i = 0.$$

The solutions to these 12 equations are

$$k_i = \frac{2dT}{c^2} c_{i1},$$

and

$$\epsilon_i = -\Delta \frac{2\tau dT}{c^2} c_{i1} c_{i2},$$

where

$$\Delta \equiv \frac{1}{2\sum_s \sum_{s'} (s_i - s_i') A(s - s') d^{p-1}(1-d)^{5-p} \sum_j s_j \cos^2 \theta_{ij}}.$$

The parameter Δ has the physical interpretation that collisions damp the anisotropy by a factor of $1/\Delta$ each time step.

To calculate the viscosity, the momentum flux Q across a plane parallel to the flow is required.

$$Q = \eta T l \tau = m(N_4(x)c_{41} + N_5(x)c_{51} - N_1(x + \tau c_1)c_{11} - N_2(x + \tau c_2)c_{21}) \quad (3.3)$$

Using the definition of N_i, the shear viscosity is

$$\eta(\rho) = \frac{m}{2T\tau^2}\left[(k_2 + k_5 - k_1 - k_4)x_2 - (\epsilon_2 + \epsilon_5 - \epsilon_1 - \epsilon_4) - (k_1 - k_2)\frac{\sqrt{3}l}{2}\right],$$

Substituting in the results for k_i and ϵ_i from above, one obtains,

$$\eta(\rho) = \frac{md\sqrt{3}}{\tau}\left[\Delta - \frac{1}{2}\right]. \quad (3.4)$$

The kinematic viscosity, ν, is $\eta/(\rho f)$. The reduced Reynolds number, R_*, is defined by:

$$R_* = c_s g(\rho)/\nu,$$

where

$$c_s = 1/\sqrt{2}, \quad \text{and} \quad g(\rho) = \frac{1}{2}\frac{(1-2d)}{(1-d)}.$$

c_s is the sound speed and $g(\rho)$ is the coefficient of the $\mathbf{u} \cdot \nabla \mathbf{u}$ term in the Navier-Stokes equation.

To calculate the mean free path, we first calculate the average number of collisions per time step per particle. This number is the the sum over all collisions of the number of particles involved in the collision, p, times the probability of a collision per time step, $A(s \to s')(1 - \delta_{ss'})/\tau$, times the probability of a given state s being occupied, $d^p(1-d)^{6-p}$. The mean free path is then just the velocity of the particle divided by the average number of collisions per time step per particle:

$$\lambda = \frac{c\tau}{\sum_s \sum_{s'} pA(s \to s')(1 - \delta_{ss'})d^p(1-d)^{6-p}}. \quad (3.5)$$

Note that this isotropic λ is calculated in the limit that the mean flow velocity goes to zero. For finite velocities, the mean free path is not isotropic.

4. GENERALIZATION OF HENON'S ALGORITHM

The generalization of these equations to higher dimensions[5] and to the inclusion of a rest particles[1] is:

$$\eta = \frac{\rho f \tau c^2}{2(D+2)} \left(\frac{1+2\mu_7}{1-2\mu_7} \right),$$

$$R_* = \frac{c_s g(\rho)}{\nu}$$

(4.1)

with the following definitions,

$$\rho = nd,$$

$$c_s = \left(\frac{c^2 b}{Dn} \right)^{\frac{1}{2}},$$

$$g(\rho) = \frac{D}{D+2} \left(\frac{n}{b} \right) \left(\frac{1-2d}{1-d} \right),$$

$$\mu_7 \equiv \frac{D}{2(D+1)bc^4} \sum_{s} \sum_{s'} A(s \to s') d^{p-1} (1-d)^{(b-p-1)} \sum_{\alpha} \sum_{\beta} Y_{\alpha\beta}(s) Y_{\alpha\beta}(s'),$$

$$Y_{\alpha\beta}(s) \equiv \sum_{i=1}^{n} s_i \left(c_{i\alpha} c_{i\beta} - \frac{|c_i|^2}{D} \right)_{\alpha\beta}.$$

5. 2-D ANALYTICAL RESULTS

We categorize collision rules using the following terms: minimal, minimum w, or collision-saturated rules, deterministic or nondeterministic rules, chiral or symmetric rules, presence or absence of rest particles, and presence or absence of collisions that create or destroy rest particles. Minimal rules mean the smallest set of acceptable rules that can be used. Such rules conserve mass, momentum, and energy, but have the minimum number of known spurious conservation laws. (The total number of spurious conservation laws is unknown.) An example of such a law would be particle number conservation along a lattice line. Collision-saturated rules mean the maximum possible collisions are used; that is, no output state is the same as an input state if there exists a different state that conserves mass and momentum. In the minimum w model, the set of velocity vectors in the output state is deduced

[1]These formula can be obtained from those of Dubrelle[3] in the following limits (basically removing the violation of semi-detailed balance)—$[d_* A^+ + (1 - d_*) A^-] \to [A(s \to s')]$ and $[d_*] \to [d]$. The variable p must also be reinterpreted as the total number of particles at the lattice site rather than just the number of moving particles.

from those of the input state by an isometry chosen so as to preserve the momentum and minimize the value of w defined below:

$$w = \frac{1}{D(D+2)} \left(\sum_\alpha \sum_{\beta > \alpha} (a_{\alpha\beta} + a_{\beta\alpha})^2 + (\sum_\alpha a_{\alpha\alpha})^2 + 2 \sum_\alpha a_{\alpha\alpha}^2 \right),$$

where $a_{\alpha\beta}$ are the components of the matrix representation of the isometry. Roughly speaking, this number measures the average correlation between an input state s and the corresponding output state s' obtained by that particular isometry. When w is minimized, the viscosity is minimized. The method is described in detail in Hénon.[7] Deterministic implies that for a given input state, the output state is always the same. For a nondeterministic model, the output state is chosen randomly from states that have the same mass and momentum as the input state. Chiral describes rules obtained by a preferred direction of rotation of the input state. Chirality implies a universe with a definite handedness. Non-chiral rules have the equal numbers of clockwise and counterclockwise rotations within each group of input states with the same mass. All the nondeterministic collision rules are non-chiral. All deterministic models, except the 7-Bit Model C, are made as non-chiral as possible, given that some of the mass groups in the deterministic rules have an odd number of states, making a collision-saturated, one-to-one mapping of input states to output states inherently chiral. Model C is a completely chiral version of model B.

The analytical results for η and R_* are quoted in Eqs. 5.1–5.34 for the eleven different collision tables listed in appendix A. Graphs of R_* vs. reduced ρ are shown in Figures 1–3, and the maximum value of R_* (and the density at which it is maximal) for each model is given in the Table 2. The mean free path was evaluated in the small velocity limit (Eq. 3.6), for the 6- and 7-bit models, and plotted in Figures 4 and 5. The mean free path was also plotted versus the kinematic viscosity in Figures 6 and 7 to show the deviation from linearity. The reduced density varied over a range of values for these two plots. Reference points are plotted at $d = 0.2$, $d = 0.5$, and $d = 0.8$.

TABLE 2 2-D Maximum $R_*(\rho)$ Values

Model	ρ	$R_*(\rho)$
6-bit Minimal Nondeterministic	1.12	0.387
6-bit Minimum w Nondeterministic	1.81	1.08
6-bit Collision-Saturated Nondeterministic	1.92	1.32
6-bit Collision-Saturated Deterministic	1.92	1.32
7-bit Minimal Nondeterministic	1.25	1.08
7-bit Collision-Saturated Nondeterministic A	1.99	2.22
7-bit Collision-Saturated Deterministic A	1.99	2.22
7-bit Nondeterministic B	2.14	2.65
7-bit Deterministic B	2.14	2.65
7-bit Deterministic C (Chiral)	2.14	2.65
8-bit Minimal Nondeterministic	1.39	1.11

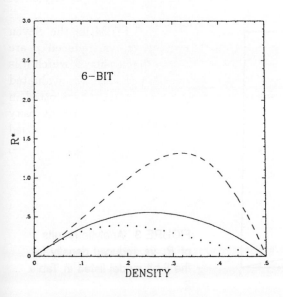

FIGURE 1 Analytic results of R_* vs. reduced density for the 6-bit models listed in Table 2. The dotted line corresponds to the minimal model. The solid line corresponds to the minimum w model. The dashed line corresponds to the collision-saturated models.

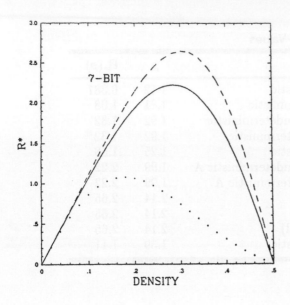

FIGURE 2 Analytic results of R_* vs. reduced density for the 7-bit models listed in Table 2. The dotted line corresponds to the minimal model. The solid line corresponds to the collision-saturated models A. The dashed line corresponds to the models B and C.

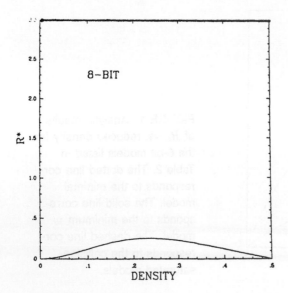

FIGURE 3 Analytic results of R_* vs. reduced density for the 8-bit model listed in Table 2.

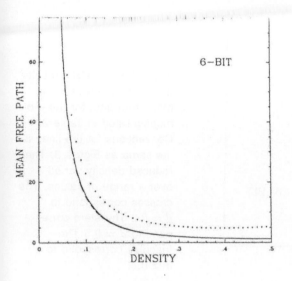

FIGURE 4 Analytic results of mean free path vs. reduced density for the 6-bit models listed in Table 2. Conventions for the lines are the same as Figure 3.

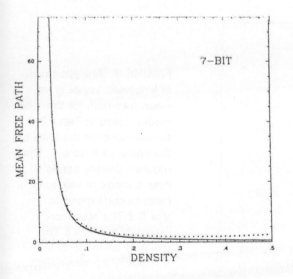

FIGURE 5 Analytic results of mean free path vs. reduced density for the 7-bit models listed in Table 2. Conventions for the lines are the same as Figure 3.

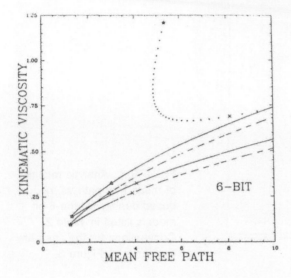

FIGURE 6 Analytic results of kinematic viscosity vs. mean free path for the 6-bit models listed in Table 2. Conventions for the lines is the same as Figure 3. The reduced density varied over a range of values. The crosses correspond to $d = 0.2$ The stars correspond to $d = 0.5$ The triangles correspond to $d = 0.8$.

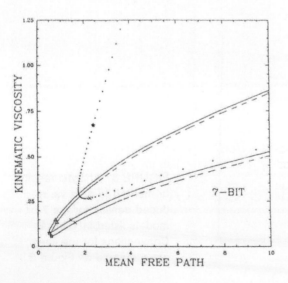

FIGURE 7 Analytic results of kinematic viscosity vs. mean free path for the 7-bit models listed in Table 2. Conventions for the lines are the same as Figure 4. The reduced density varied over a range of values. The crosses correspond to $d = 0.2$ The stars correspond to $d = 0.5$ The triangles correspond to $d = 0.8$.

η AND R_* FOR THE 6-BIT MINIMAL NONDETERMINISTIC MODEL [2]

$$\eta = \frac{\rho f \tau c^2}{4}\left(\frac{1}{2\mu} - \frac{1}{2}\right) \tag{5.1}$$

$$\mu \equiv \frac{3d}{2}(1-d)^3 \tag{5.2}$$

$$R_* = \frac{2\sqrt{2}}{\tau c}\left(\frac{1-2d}{1-d}\right)\left(\frac{\mu}{1-\mu}\right) \tag{5.3}$$

$$\rho = 6d \tag{5.4}$$

$$\lambda = \frac{c\tau}{6d^2(1-d)^4 + 6d^3(1-d)^3} \tag{5.5}$$

η AND R_* FOR THE 6-BIT MINIMUM W NONDETERMINISTIC MODEL

$$\eta = \frac{\rho f \tau c^2}{4}\left(\frac{1}{2\mu} - \frac{1}{2}\right) \tag{5.6}$$

$$\mu \equiv \frac{3d}{2}(1-d)(1+d(1-d)) \tag{5.7}$$

$$R_* = \frac{2\sqrt{2}}{\tau c}\left(\frac{1-2d}{1-d}\right)\left(\frac{\mu}{1-\mu}\right) \tag{5.8}$$

$$\rho = 6d \tag{5.9}$$

$$\lambda = \frac{c\tau}{6d^2(1-d)^4 + 33d^3(1-d)^3 + 12d^4(1-d)^2} \tag{5.10}$$

η AND R_* FOR THE 6-BIT COLLISION-SATURATED NONDETERMINISTIC MODEL AND THE 6-BIT COLLISION-SATURATED DETERMINISTIC MODEL

$$\eta = \frac{\rho f \tau c^2}{4}\left(\frac{1}{2\mu} - \frac{1}{2}\right) \tag{5.11}$$

$$\mu \equiv \frac{3}{2}d(1-d)^3 + 6d^2(1-d)^2 + \frac{3}{2}d^3(1-d) \tag{5.12}$$

$$R_* = \frac{2\sqrt{2}}{\tau c}\left(\frac{1-2d}{1-d}\right)\left(\frac{\mu}{1-\mu}\right) \tag{5.13}$$

$$\rho = 6d \tag{5.14}$$

$$\lambda = \frac{c\tau}{6d^2(1-d)^4 + 36d^3(1-d)^3 + 12d^4(1-d)^2} \tag{5.15}$$

[2]This is referred to as Model 1 in d'Humières[2] and as FHP-I in Frisch.[5]

η AND R_* FOR THE 7-BIT MINIMAL NONDETERMINISTIC MODEL [3]

$$\eta = \frac{\rho f \tau c^2}{4}\left(\frac{1}{2\mu} - \frac{1}{2}\right) \tag{5.16}$$

$$\mu \equiv \frac{7}{2}d(1-d)^3(1 - \frac{4d}{7}) \tag{5.17}$$

$$R_* = \frac{\sqrt{21}}{12\tau c}\left(\frac{1-2d}{1-d}\right)\left(\frac{8\mu}{1-\mu}\right) \tag{5.18}$$

$$\rho = 7d \tag{5.19}$$

$$\lambda = \frac{c\tau}{30d^2(1-d)^5 + 15d^3(1-d)^4 + 8d^4(1-d)^3} \tag{5.20}$$

η AND R_* FOR THE 7-BIT COLLISION-SATURATED NONDETERMINISTIC MODEL A[4]

$$\eta = \frac{\rho f \tau c^2}{4}\left(\frac{1}{2\mu} - \frac{1}{2}\right) \tag{5.21}$$

$$\mu \equiv \frac{7}{2}d(1-d)(1 - \frac{8d}{7}(1-d)) \tag{5.22}$$

$$R_* = \frac{\sqrt{21}}{12\tau c}\left(\frac{1-2d}{1-d}\right)\left(\frac{8\mu}{1-\mu}\right) \tag{5.23}$$

$$\rho = 7d \tag{5.24}$$

$$\lambda = \frac{c\tau}{30d^2(1-d)^5 + 69d^3(1-d)^4 + 92d^4(1-d)^3 + 75d^5(1-d)^2} \tag{5.25}$$

η AND R_* FOR THE 7-BIT NONDETERMINISTIC MODEL B, THE 7-BIT DETERMINISTIC MODEL B, AND THE 7-BIT DETERMINISTIC MODEL C (CHIRAL)

$$\eta = \frac{\rho f \tau c^2}{4}\left(\frac{1}{2\mu} - \frac{1}{2}\right) \tag{5.26}$$

$$\mu \equiv \frac{7}{2}d(1-d)(1 - \frac{6d}{7}(1-d)) \tag{5.27}$$

$$R_* = \frac{\sqrt{21}}{12\tau c}\left(\frac{1-2d}{1-d}\right)\left(\frac{8\mu}{1-\mu}\right) \tag{5.28}$$

$$\rho = 7d \tag{5.29}$$

$$\lambda = \frac{c\tau}{30d^2(1-d)^5 + 51d^3(1-d)^4 + 68d^4(1-d)^3 + 75d^5(1-d)^2} \tag{5.30}$$

[3]This is referred to as Model 2 in d'Humières[2] and as FHP-II in Frisch.[5]

[4]This is referred to as Model 3 in d'Humières[2] and as FHP-III in Frisch.[5] and the 7-Bit Collision-Saturated Deterministic Model A

η AND R_* FOR THE 8-BIT MINIMAL NONDETERMINISTIC MODEL [5]

$$\eta = \frac{\rho f \tau c^2}{4} \left(\frac{1}{2\mu} - \frac{1}{2} \right) \tag{5.31}$$

$$\mu \equiv \frac{7}{2} d(1-d)^3 \left(1 - \frac{4d}{7} \left(1 - \frac{d(1-d)}{d^2 + (1-d)^2} \right) \right) \tag{5.32}$$

$$R_* = \frac{1}{12\tau c} \left(\frac{3(d^2 + (1-d)^2)^2}{7(d^2 + (1-d)^2)^2 + 4d(1-d)} \right)^{1/2} \left(7 + \frac{2d}{d^2 + (1-d)^2} \right) \times$$
$$\left(\frac{1-2d}{1-d} \right) \left(\frac{8\mu}{1-\mu} \right) \tag{5.33}$$

$$\rho = d \left(7 + \frac{2d}{d^2 + (1-d)^2} \right) \tag{5.34}$$

6. 3-D ANALYTIC RESULTS

The state at a node in the face-centered hypercubic (FCHC) single-speed models can be represented by 14, 18, or 24 bits.[1,11,12]

The 24-bit model requires a sublattice of the 4-D integral lattice (i.e., the nodes are all points with integral coordinates (x, y, z, w)). The sublattice consists of all integral points for which the coordinate sum is even. In the sublattice, all points of least nonzero distance from the origin are used, these being the 24 vectors: $(\pm 1, \pm 1, 0, 0)$ and permutations. The random model is one where the output state is chosen randomly from all input states with the same mass and momentum as the output state. The method for choosing the collision rules for the minimum w model is the same as described above for the 2-D models. In the partially optimized collision table,[9] the input and output states were chosen to minimize the shear viscosity given below, using Hénon's earliest heuristic optimization. Hénon[7] casts the viscosity into a form that is more easily optimized.

The 18-bit model is run on the same sublattice. In order to reduce the computational burden, the following useful device called married pairs, first proposed in Rothaus,[11] is used. Whenever at a site there is a particle at a site headed in the direction (x, y, z, w), there is also one headed in the direction $(x, y, z, -w)$. This reduces the bits required per site to 18.

The 14-bit model is run on a different 4-D sublattice than the above 18- and 24-bit models. Essentially the dual of the sublattice considered above, it consists of points all of whose coordinates are even or all of whose coordinates are odd.

---

[5]This is referred to as Model 4 in d'Humières.[2] The collision table in appendix B corrects what appears to be a typographical error in the table listed in d'Humières[2] so that it matches the verbal description of the model therein.

TABLE 3 3-D Maximal R* Values

Model	ρ	$R_*(\rho)$
14-bit Minimal Married Pair	1.32	0.0876
14-bit Random Collision Saturated Married Pair	1.32	0.0877
18-bit Minimal Married Pair	1.60	0.131
18-bit Random Collision Saturated Married Pair	2.00	0.094
24-bit Random Collision Saturated[f]	0.17	2
24-bit Minimum Collision Saturated[f]	0.20	1.74
24-bit Collision Saturated Partially Optimized	7.64	6.21

The vectors considered in this instance are $(\pm2,0,0,0)$, $(0,\pm2,0,0)$, $(0,0,\pm2,0)$, $(0,0,0,\pm2)$, $(\pm1,\pm1,\pm1,\pm1)$. With the restriction of married pairs in the fourth dimension, the state at a node can be represented by 15 bits. However, since the effect on the evolution of the automaton arising from the pair of directions $(0,0,0,\pm2)$, if these are not used in any scattering laws, is irrelevant and does not create any unwanted conservation laws in the 3-D section $w = 0$, the number of bits needed to describe the state at a site can be reduced to 14. More detailed information on the minimal 14-bit and 18-bit married pair collision tables is contained in appendix B.

The general analytic formulas for η and R_* for FCHC lattices is quoted in Eqs. 6.1–6.4. Due to the size of the collision tables (up to approximately 16 Megabytes), these formulas are evaluated numerically rather than analytically. Graphs of R_* vs. reduced ρ (for the models computed by the present authors) are shown in Figures 8–10, and the maximum value of R_* is given in Table 3.

$$\eta = \frac{\rho f \tau c^2}{6}\left(\frac{3}{4\mu} - \frac{1}{2}\right) \tag{6.1}$$

$$\mu \equiv \sum_s \sum_{s'} (s_i - s_i')Ad^{p-1}(1-d)^{n-p-1}\sum_j s_j \cos^2\theta_{ij} \tag{6.2}$$

$$R_* = \frac{2}{\tau c}\left(\frac{1-2d}{1-d}\right)\left(\frac{4\mu}{3-2\mu}\right) \tag{6.3}$$

$$\rho = (\text{number of bits} + \text{number of married pairs})d \tag{6.4}$$

Hénon has shown that with the inclusion of rest particles, the value of R_* can be increased to 8.46 with one rest particle per node, to 10.22 with two, and at least 10.71 with three.[6]

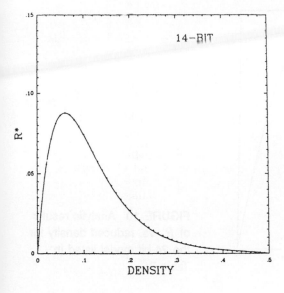

FIGURE 8   Analytic results of R_* vs. reduced density for the 14-bit models listed in Table 3. The solid line corresponds to the minimal model. The dotted line corresponds to the collision-saturated model.

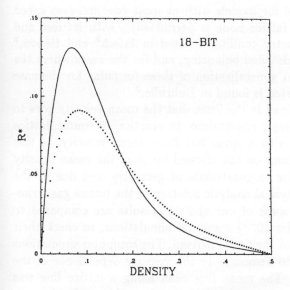

FIGURE 9   Analytic results of R_* vs. reduced density for the 18-bit models listed in Table 3. The solid line corresponds to the minimal model. The dotted line corresponds to the collision-saturated model.

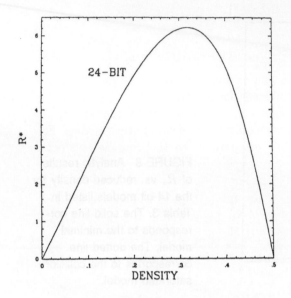

FIGURE 10 Analytic results of R_* vs. reduced density for the 24-bit model listed in Table 3.

7. LIMITATIONS OF HÉNON'S ALGORITHM

The formulae of section 4 are valid for models with at most one non-zero speed (more than one rest particle at a lattice node is permitted) , with lattices and collision rules satisfying the symmetry conditions listed in Frisch[5] and Hénon,[8] with collision rules satisfying semi-detailed balancing, and for the cases where the Boltzman approximation is valid. A generalization of these formulae for the case where semi-detailed balance is violated is found in Dubrulle.[3]

All the above results were obtained in the limit that the mean velocity goes to zero, i.e., systems not far from isotropic equilibrium. In practice, of course, lattice gas computer simulations are run with a small but finite mean velocity. In this section, we first discuss the limitations on the allowed range of the mean density ρ and the mean flow velocity u due to constraints of geometry and due to the requirement of the existence of a physical analytic solution to the lattice gas fermi-dirac distribution function. Next, some of our analytic results are compared to results from very low velocity (order 10^{-3}) computer simulations, to check their accuracy in the parameter range that they were derived. The computer simulations were run at higher velocities to demonstrate the velocity dependence of the viscosity and the mean free path. The mean free path along a lattice line was calculated analytically to show the (anisotropic) velocity dependence of viscosity.

A. ANALYTIC SOLUTIONS FOR THE LATTICE GAS DISTRIBUTION FUNCTION

In the low velocity limit, the equilibrium fermi-dirac distribution

$$N_i = \frac{1}{e^{h(n,u)+q(n,u)\cdot c_i} + 1}$$

can be written in the form,

$$N_i = \frac{1}{e^{\alpha+\beta u\cdot c_i} + 1}, \tag{7.1}$$

where α and β are dependent only on the mean density and the magnitude of the mean velocity, and are independent of θ_i, the angle between u and c_i. The coefficients α and β can be determined from the two relations:

$$n = \sum_i N_i \tag{7.2}$$

$$nu = \sum_i N_i c_i, \tag{7.3}$$

where n is the mean density per node, u is the mean velocity, and c_i is the velocity of a single particle at the lattice site.

For the FHP models, these equations can be solved analytically for a mean velocity u that is halfway between two adjacent lattice directions. Substituting in this particular u into Eqs. 7.2 and 7.3, we obtain

$$n = \frac{2}{e^{\alpha+\frac{\sqrt{3}}{2}u} + 1} + \frac{2+R}{e^{\alpha} + 1} + \frac{2}{e^{\alpha-\frac{\sqrt{3}}{2}u} + 1}, \tag{7.4}$$

and

$$nu = \frac{\sqrt{3}}{e^{\alpha+\frac{\sqrt{3}}{2}\beta u} + 1} - \frac{\sqrt{3}}{e^{\alpha-\frac{\sqrt{3}}{2}\beta u} + 1}, \tag{7.5}$$

where R is the number of particles with $u = 0$ allowed at a lattice site. We define $H \equiv e^{\alpha}$, $Q \equiv e^{\frac{\sqrt{3}}{2}\beta u}$, and $v \equiv \frac{2}{\sqrt{3}}u$. We then have

$$n = \frac{2}{1+HQ} + \frac{2+R}{1+H} + \frac{2}{1+HQ^{-1}},$$

and

$$nv = \frac{2}{1+HQ} - \frac{2}{1+HQ^{-1}}.$$

Adding and subtracting these two equations gives two independent linear equations for Q .

$$Q = -\frac{(H+1)(nv+n) - 4H - 6 - R}{(H^2+H)(nv+n) - 2H - HR}$$

and
$$Q = -\frac{(H^2 + H)(n\mathrm{v} - n) - 2H + HR}{(H + 1)(n\mathrm{v} - n) + 4H + 6 + R}.$$

Equating these two expressions for Q and factoring out $(1+H)$ yields the following cubic equation for H:

$$(1 - \mathrm{v}^2)n^2 H^3 + ((1 - \mathrm{v}^2)n^2 - 4n - 2nR)H^2 + ((\mathrm{v}^2 - 1)n^2 + 8n - 12$$
$$+ (R+4)R)H + ((\mathrm{v}^2 - 1)n^2 + 12n - 36 - (R - 2n + 12)R) = 0 \tag{7.6}$$

This equation has one real positive root for low v. In order for the N_i's to be between zero and one, both H and Q must be positive. From the solution to Eq. 7.6 for $H = 0$, one can see that H is positive for $n < ((6 + R)\sqrt{3})/(\sqrt{3} + 2u)$. From the mass conservation equation, one can see that Q is positive for $n < \sqrt{3}/u$. Thus, a physical solution exists only for the region below the two curves($H = 0$ and $Q = 0$). Figure 11 plots these two curves for the 6 bit model. H and Q can then be substituted back into N_i.

The dotted line in Figure 11 shows the geometrical constraints on the maximum velocity (in any direction, not just halfway between adjacent lattice vectors) as a function of density. By geometrical constraints, we mean the existence of a maximum velocity due to the limitation of a finite number of velocity vectors at each site. To calculate this, the mean density ρ was distributed among the different lattice directions such that the mean momentum was maximized. That is, the state was made as anisotropic as possible with the only constraint being that the density does not exceed one in any direction. The cusps in the diagram are caused by that constraint—note that they occur when the density is a multiple of one. The

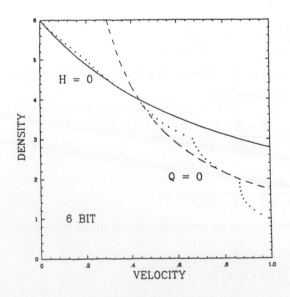

FIGURE 11 The area below both the solid and dashed lines is the range of mean densities and mean flow speeds for which there exists a physical analytic solution to the 6-bit fermi-dirac lattice gas distribution as described in section 7A. The mean flow velocity is halfway between adjacent lattice directions. The dotted line shows the maximum mean flow speed (in any direction) as a function of mean density due to geometrical constraints, also described in section 7A.

geometrical constraints for a mean velocity halfway between adjacent lattice vector yields a somewhat smaller maximum velocity than that shown in Figure 11.

A similar calculation can be done for the FCHC lattice. First, set $\mathbf{u} = (u/2, u/2, u/2, u/2)$, then Eqs 7.2 and 7.3 become:

$$n = \frac{6}{e^{\alpha+\beta u}+1} + \frac{12+R}{e^{\alpha}+1} + \frac{6}{e^{\alpha-\beta u}+1}, \tag{7.7}$$

and

$$n\frac{u}{2} = \frac{3}{e^{\alpha+\beta u}+1} - \frac{3}{e^{\alpha-\beta u}+1}. \tag{7.8}$$

We define $H \equiv e^{\alpha}$ and $Q \equiv e^{\beta u}$. (Note this is slightly different from the previous definition of Q.) We then have

$$n = \frac{6}{1+HQ} + \frac{12+R}{1+H} + \frac{6}{1+HQ^{-1}},$$

and

$$nu = \frac{6}{1+HQ} - \frac{6}{1+HQ^{-1}}.$$

Adding and subtracting these two equations gives two independent linear equations for Q.

$$Q = -\frac{(H+1)(nu+n) - 12H - 24 - R}{(H^2+H)(nu+n) - 12H - HR}$$

and

$$Q = -\frac{(H^2+H)(nu-n) + 12H + HR}{(H+1)(nu-n) + 12H + 24 + R}.$$

Equating these two expressions for Q obtained from these independent linear equations and factoring out $(1+H)$ yields the following cubic equation for H

$$(1-u^2)n^2 H^3 + ((1-u^2)n^2 - 24n - 2nR)H^2 + ((u^2-1)n^2$$
$$+24n + (R+24)R)H + ((u^2-1)n^2 + 48n \tag{7.9}$$
$$-576 - (R-2n+48)R) = 0.$$

This equation also has one real positive root for low u, as did Eq. 7.6. The requirement that N_i is between zero and one yields the constraints $n < (24+R)/(1+u)$ and $n < 6/u$, which are plotted in Figure 12, for the 24 bit model. H and Q can then be substituted back into N_i. In Figure 12, the dotted line shows the geometrical constraints, calculated in the same way as the six bit model.

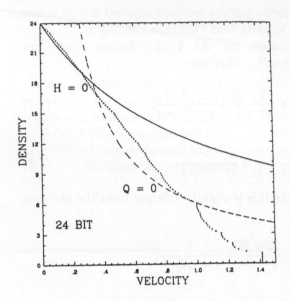

FIGURE 12 The area below both the solid and dashed lines is the range of mean densities and mean flow speeds for which there exists a physical analytic solution to the 24-bit fermi-dirac lattice gas distribution as described in section 7A. The mean flow velocity is in the (+1,+1,+1,+1) direction. The dotted line shows the maximum mean flow speed (in any direction) as a function of mean density due to geometrical constraints, also described in section 7A.

B. VELOCITY DEPENDENCE OF VISCOSITY AND MEAN FREE PATH

The plot in Figure 13 of R_* vs. reduced density was calculated from a 2-D lattice gas simulation, where one boundary was held fixed and the opposite side was held at a constant velocity of order 0.25. The two sides perpendicular to this had periodic boundary conditions. Each side of the lattice was 4096 nodes. The simulation was run for 50,000 time steps. Three different sets of 7 bit collision rules were used. The viscosity was calculated locally across the row next to the bottom row of the lattice. The mean flow velocity for that row was of order 10^{-3}. The number of particles crossing a given row with each of the different velocities was counted.

From this data the momentum flux across the row was calculated. The shear rate was calculated from the difference in average velocity of the rows directly above and below the row being considered. From Eq. (3.3) , the viscosity can be calculated directly. R_* was calculated from Eq. (4.1). The data points in Figure 13 correspond to results from these computer simulations. For comparison, the velocity-independent analytic results are also plotted in Figure 13.

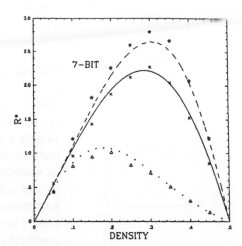

FIGURE 13 Comparison of values of R_* from analytic calculations (in the limit that the mean flow velocity u goes to zero) and from numerical simulations as described in section 7B (u is of order 10^{-3}). The dotted, solid and dashed lines are the analytical results and the triangles, crosses, and stars are the computer results. The dotted line and the triangles correspond to the minimal model. The solid line and the crosses correspond to the collision-saturated nondeterministic model A. The dashed line and the stars correspond to the nondeterministic model B.

The mean free path was also calculated from a zero mean velocity 2-D computer simulation to verify our velocity-independent analytic results. The size of the lattice was 128×128. Periodic boundary conditions were used. The 6-bit minimal collision rules were used. The simulation was run for 40,000 time steps, keeping track of the number of two- and three-body collisions over the course of the run. To calculate the mean free path, the particle velocity c is divided by the average number of collisions per time step. The average number of collisions per particle per time step is: (Number of two-body collisions / 2 + Number of three-body collisions / 3) / (Total number of time steps). The data from this simulation is plotted in Figure 14, along with the analytic result.

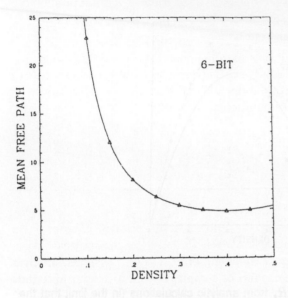

FIGURE 14 Comparison of values of mean free path λ from analytic calculations (in the limit that the mean flow velocity u goes to zero) and from numerical simulations as described in section 7B (u is of order 10^{-3}). The 6-bit minimal rules were used. The solid line is the analytical result and the triangles are the computer results.

The plot in Figure 15 of η vs. velocity shows that η does depend slightly on velocity. The data is from a computer simulation similar to the one described for Figure 13. In this case, however, the viscosity was calculated locally over each row of the lattice, each row being, of course, at a different velocity. The collision rules were the 7-bit minimal ones described above.

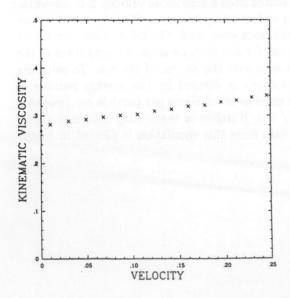

FIGURE 15 Shear viscosity vs. mean flow velocity. The crosses are data points from the computer simulation described in section 7B. The 7-bit minimal rules were used. The dot shows the analytic result for the limit that the mean flow velocity goes to zero. The density ρ for both cases is 1.2.

In section 3, we calculated (Eq. 3.5) a velocity-independent mean free path, averaged over all directions. For a finite mean flow velocity, we expect the mean free path to become anisotropic, due to the lack of Galilean invariance on the microscopic scale. To give an example, if the mean flow velocity is along a lattice direction, the mean free path becomes for maximum flow speeds essentially infinite along that direction, since each particle along that direction moves out of the way of the one behind it each time step. To illustrate this effect, we calculated the mean free path along the lattice lines, i.e., how far does the average particle move along one of the six lattice directions before colliding. Notice that this is a slightly different definition of mean free path than what was used in section 5. Here we are only looking at a limited region of space (six lines), whereas the definition in section 5 involved an average over all directions each time step.

To do this calculation, we will make use of the analytic N_i's obtained in section 7A, for a mean flow velocity halfway between adjacent lattice directions. First, we introduce some notation. The subscript f (for forward) will be used to denote quantities associated with lattice directions that have a positive component along the direction of the mean flow. The subscript p will be used for directions that are entirely perpendicular to the mean flow. The subscript b will be used for directions that are opposite to the mean flow direction. Thus we have the following correspondence between this notation and the notation of Table 1:

$$N_f = N_1 = N_2 = \frac{1}{1 + HQ}$$

$$N_p = N_3 = N_6 = \frac{1}{1 + H}$$

$$N_b = N_4 = N_5 = \frac{1}{1 + (H/Q)}$$

To calculate the mean free path λ_f along a forward lattice line, we need to first calculate the probability of a collision, P_{cf}, given the presence of a particle going in the forward direction at a lattice site. For the 6-bit minimal collision rules, this is simply the probability of a two- or three-body collision with the substitution that the probability of the presence of a particle going forward in the cell is 1 rather than N_f:

$$P_{cf} = N_b(1 - N_f)(1 - N_p)^2(1 - N_b) + N_b N_p(1 - N_f)(1 - N_p)(1 - N_b)$$

The probability that a particle goes a distance of nl, where l is the lattice spacing, before undergoing a collision is $(1 - P_{cf})^{n-1} P_{cf}$. To calculate λ_f, we sum over all sites along the path the product of the collision probability at that site times the distance traveled:

$$\lambda_f = \sum_{n=1}^{\infty} nl(1 - P_{cf})^{n-1} P_{cf} = \frac{l}{P_{cf}}$$

Similarly, we have

$$\lambda_p = \frac{l}{P_{cp}}$$

where

$$P_{cp} = N_p(1 - N_f)^2(1 - N_b)^2 + N_b N_f(1 - N_f)(1 - N_p)(1 - N_b)$$

and

$$\lambda_b = \frac{l}{P_{cb}}$$

where

$$P_{cb} = N_f(1 - N_f)(1 - N_p)^2(1 - N_b) + N_f N_p(1 - N_f)(1 - N_p)(1 - N_b)$$

These three λ's do not reduce, in the zero velocity limit, to those of section 5 because of the difference in definition as discussed above. The solid lines in Figures 16–18 show these three λ's vs. velocity for several different densities. The triangles in the plots are measurements of these λ's from a 2-D constant mean flow velocity lattice gas computer simulation. At each time step $(t = n\tau)$, the percentage of particles, in each lattice direction that have not collided and scattered out of their original $(t = 0)$ direction, was measured. This percentage is the probability that a particle has gone a distance nl along the lattice direction without a collision. In other words, this percentage is the value of the normalized mean free path distribution function for that particular distance. Figure 19 shows the distribution functions for the forward, perpendicular directions. The decay is initially exponential. The sum of the normalized distribution times l is the mean free path along that lattice direction. As expected, the mean free path is significantly longer in the forward direction than in the backward direction.

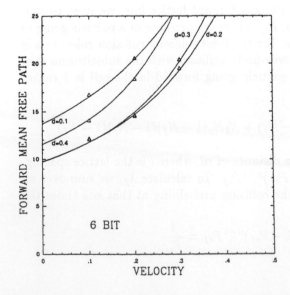

FIGURE 16 Solid lines correspond to analytic results for mean free path along the forward direction (with respect to the mean velocity) vs. velocity. The model used is the minimal 6-bit model (only two- and three-body collisions without spectators). The different lines correspond to different values of reduced density d, in steps of 0.1. The triangles mark measurements from computer simulations.

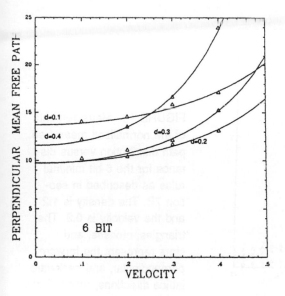

FIGURE 17 Solid lines correspond to analytic results for mean free path along the perpendicular direction (with respect to the mean velocity) vs. velocity. The model used is the minimal 6-bit model (only two and three body collisions without spectators). The different lines correspond to different values of reduced density d, in steps of 0.1. The triangles mark measurements from computer simulations.

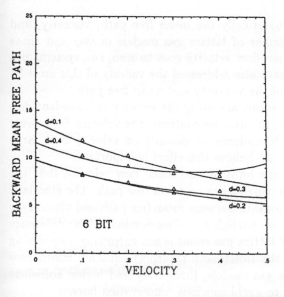

FIGURE 18 Solid lines correspond to analytic results for mean free path along the backward direction (with respect to the mean velocity) vs. velocity. The model used is the minimal 6-bit model (only two- and three-body collisions without spectators). The different lines correspond to different values of reduced density d, in steps of 0.1. The triangles mark measurements from computer simulations.

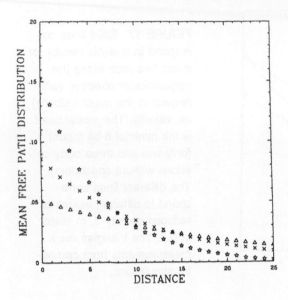

FIGURE 19   Measurements of the normalized mean free path distribution versus distance for the 6-bit minimal rules as described in section 7B. The density is 1.2 and the velocity is 0.3. The triangles, crosses, and stars represent the forward, perpendicular, and backward lattice directions, respectively.

8. CONCLUSIONS

In this paper we have calculated analytically the mean free path, viscosity, and reduced Reynolds number for a number of lattice gas models in two and three dimensions, in the limit that the mean flow velocity goes to zero, i.e., systems not far from isotropic equilibrium. We have also addressed the validity of this limit by examining the velocity dependence of the viscosity and mean free path.

Most lattice gas computer simulations are using the velocity-independent definition of viscosity and R_*. However, in these simulations, the velocity field never goes to zero everywhere. Since the dependence of viscosity on velocity is so large ($\tilde{}10\%$) even for moderate velocities, we believe this effect is important.

We have shown analytically, in the limit of zero mean flow velocity, that the viscosity has, in general, a nonlinear dependence on mean free path. The standard hydrodynamic result of a linear relationship between mean free path and viscosity is valid only for small densities, where two particle collisions dominate. The difference between the standard result and our lattice gas result is not surprising considering the standard result is derived from the Boltzmann transport equation which uses only binary collisions, and in lattice gas models, three-body and higher collisions are needed to obtain a high R_* and to avoid spurious conservation laws.

When we relaxed the velocity independence requirement for the mean free path, we found an anisotropic mean free path along the different directions (along the lattice lines) relative to the mean flow. This result strongly suggest an anisotropic viscosity for lattice gas models.

R_*'s are still useful for comparing different sets of collision rules and evaluating which one would be best for a particular computer simulation use. In order to use the lattice gas models to simulate turbulent flows, one needs to increase R_* or decrease viscosity as much as possible. One of the 2-D models in this paper obtained a R_* value of 2.8, measured from a computer simulation, which is bigger than the other models of its type, (single-speed models satisfying semi-detailed balance), currently in use. For studying turbulent flow in three dimensions, it is clear a well-optimized version of the 24-bit model is needed. However, there are situations where a 14-bit model might be useful. One such candidate is 3-D flow through porous media and oil recovery simulations, which require very complicated boundaries but only very low Reynolds number flow of the order 1. The 14-bit model, with its reduced complexity of collision rules (compared with the 4-megaword rule table needed for a reasonably high R_* for the 24-bit model) and reduced memory requirements due to fewer bits per state, should be able to run faster and for larger universes.

APPENDIX A - 2-D COLLISION TABLES

The following notation, from d'Humières,[2] is used to describe the collision tables. In all of the following, c, τ, and l have been set equal to 1. Table 5 gives all possible configurations of the 6-bit models, listing only the cases such that $j_x \geq j_y \geq 0$ where $j_x = 2c_x$ and $j_y = 4c_y/\sqrt{3}$ so integral indices can be used. Tables 6 and 7 list half the possible configurations of the 7- and 8-bit models, respectively, where the remaining configurations can be obtained by duality—replacing particles with holes and vice-versa. The first three columns give the total mass and momentum. The fourth column shows the different states, the legal collisions exchanging states appearing within the same row. The configurations are represented as binary numbers using the bit conventions listed in Table 4. Configurations that can be obtained by the application of the symmetry group are not explicitly listed, but the number of such states that can be obtained from each listed state is given in the last column. The collision probability is written $A_I(j \rightarrow k)$ where I is an index representing the mass and momentum as given in the first three columns of the table and j and k are the positions of the input and output states in the rows of the table. The number of states in a row (i.e., the number of possible states with the same mass and momentum) is q_I.

TABLE 4 Notation for all hexagonal models

bit	i	mass	c_{i1}	c_{i2}
0	1	1	1/2	$\sqrt{3}/2$
1	2	1	−1/2	$\sqrt{3}/2$
2	3	1	−1	0
3	4	1	−1/2	$-\sqrt{3}/2$
4	5	1	1/2	$-\sqrt{3}/2$
5	6	1	1	0
6	7	1	0	0
7	8	2	0	0

TABLE 5 List of Configurations for 6-Bit Tables

m	j_x	j_y	configurations			degeneracies
0	0	0	000000			1
1	2	0	100000			6
2	0	0	100100	010010	001001	1
2	2	0	010001			6
2	3	2	100001			6
3	0	0	101010	010101		1
3	2	0	110010	101001		6
3	4	0	110001			6
4	3	2	110011			6
4	2	0	110101			6
4	0	0	011011	101101	110110	1
5	2	0	111011			6
6	0	0	111111			1

TABLE 6 List of Configurations for 7-Bit Tables

m	j_x	j_y	configurations					degeneracies
0	0	0	0000000					1
1	0	0	1000000					1
1	2	0	0100000					6
2	0	0	0100100	0010010	0001001			1
2	2	0	1100000	0010001				6
2	3	2	0100001					6
3	0	0	1100100	1010010	1001001	0010101	0101010	1
3	2	0	1010001	0110010	0101001			6
3	3	2	1100001					6
3	4	0	0110001					6

TABLE 7 List of Configurations for 8-Bit Tables

m	j_x	j_y	configurations				degeneracies
0	0	0	00000000				1
1	0	0	01000000				1
1	2	0	00100000				6
2	0	0	10000000	00100100	00010010	00001001	1
2	2	0	01100000	00010001			6
2	3	2	00100001				6
3	0	0	11000000	01100100	01010010	01001001	
			00010101	00101010			1
3	2	0	10100000	01010001	00110010	00101001	6
3	3	2	01100001				6
3	4	0	00110001				6
4	0	0	10100100	10010010	10001001	01010101	1
			01101010	00110110	00101101	00011011	
4	2	0	11100000	10010001	01110010	01101001	
			00110101				6
4	3	2	10100001	00110011			6
4	4	0	01110001				6

6-BIT MINIMAL NONDETERMINISTIC MODEL

$$A_I(j \rightarrow k) = \delta_{jk} \qquad \forall\, I \notin \{200, 300\}$$
$$A_{200}(j \rightarrow k) = \frac{1 - \delta_{jk}}{2} \qquad j, k \in \{1, 2, 3\}$$
$$A_{300}(j \rightarrow k) = 1 - \delta_{jk} \qquad j, k \in \{1, 2\}$$

6-BIT MINIMUM W NONDETERMINISTIC MODEL

$$A_I(j \rightarrow k) = \delta_{jk} \qquad \forall\, I \notin \{200, 300, 320, 400\}$$
$$A_{200}(j \rightarrow k) = \frac{1 - \delta_{jk}}{2} \qquad j, k \in \{1, 2, 3\}$$
$$A_{300}(j \rightarrow k) = \frac{1 - \delta_{jk}}{2} \qquad j, k \{1, 2\}$$
$$A_{320}(j \rightarrow k) = 1 - \delta_{jk} \qquad j, k \{1, 2\}$$
$$A_{400}(j \rightarrow k) = \frac{1 - \delta_{jk}}{2} \qquad j, k \in \{1, 2, 3\}$$

6-BIT COLLISION SATURATED NONDETERMINISTIC MODEL

$$
\begin{aligned}
A_I(j \to k) &= \delta_{jk} & \forall\, I &\notin \{200, 300, 320, 400\} \\
A_{200}(j \to k) &= \tfrac{1-\delta_{jk}}{2} & j, k &\in \{1, 2, 3\} \\
A_{300}(j \to k) &= 1 - \delta_{jk} & j, k &\in \{1, 2\} \\
A_{320}(j \to k) &= 1 - \delta_{jk} & j, k &\in \{1, 2\} \\
A_{400}(j \to k) &= \tfrac{1-\delta_{jk}}{2} & j, k &\in \{1, 2, 3\}
\end{aligned}
$$

6-BIT COLLISION SATURATED DETERMINISTIC MODEL

$$
\begin{aligned}
A_I(j \to k) &= \delta_{jk} & \forall\, I \notin \{200, 300, 320, 400\} \\
A_{200}(1 \to 2) &= A_{200}(2 \to 3) = A_{200}(3 \to 1) = 1 \\
A_{300}(1 \to 2) &= A_{300}(2 \to 1) = 1 \\
A_{320}(1 \to 2) &= A_{320}(2 \to 1) = 1 \\
A_{400}(1 \to 3) &= A_{400}(2 \to 1) = A_{400}(3 \to 2) = 1
\end{aligned}
$$

7-BIT MINIMAL NONDETERMINISTIC MODEL

$$
\begin{aligned}
A_I(j \to k) &= \delta_{jk} & \forall\, I &\notin \{200, 220, 300, 400\} \\
A_{200}(j \to k) &= \tfrac{1-\delta_{jk}}{2} & j, k &\in \{1, 2, 3\} \\
A_{220}(j \to k) &= 1 - \delta_{jk} & j, k &\in \{1, 2\} \\
A_{300}(j \to k) &= \tfrac{1-\delta_{jk}}{2} & j, k &\in \{1, 2, 3\} \\
A_{300}(j \to k) &= 1 - \delta_{jk} & j, k &\in \{4, 5\} \\
A_{300}(j \to k) &= A_{300}(k \to j) = 0 & j &\in \{1, 2, 3\},\ k \in \{4, 5\} \\
A_{400}(j \to k) &= \delta_{jk} & j, k &\in \{1, 2, 3\} \\
A_{400}(j \to k) &= 1 - \delta_{jk} & j, k &\in \{4, 5\} \\
A_{400}(j \to k) &= A_{400}(k \to j) = 0 & j &\in \{1, 2, 3\},\ k \in \{4, 5\}
\end{aligned}
$$

7-BIT COLLISION SATURATED NONDETERMINISTIC MODEL A

$$
\begin{aligned}
A_I(1 \to 1) &= 1 & \forall\, I \text{ such that } q_I = 1 \\
A_I(j \to k) &= \tfrac{1-\delta_{jk}}{q_I - 1} & \forall\, I \text{ such that } 1 < q_I < 4,\ j, k \in \{1, \cdots, q_I\} \\
A_{300}(j \to k) &= \tfrac{1-\delta_{jk}}{2} & j, k \in \{1, 2, 3\} \\
A_{300}(j \to k) &= 1 - \delta_{jk} & j, k \in \{4, 5\} \\
A_{300}(j \to k) &= A_{300}(k \to j) = 0 & j \in \{1, 2, 3\},\ k \in \{4, 5\} \\
A_{400}(j \to k) &= \tfrac{1-\delta_{jk}}{2} & j, k \in \{1, 2, 3\} \\
A_{400}(j \to k) &= 1 - \delta_{jk} & j, k \in \{4, 5\} \\
A_{400}(j \to k) &= A_{400}(k \to j) = 0 & j \in \{1, 2, 3\},\ k \in \{4, 5\}
\end{aligned}
$$

7-BIT COLLISION-SATURATED DETERMINISTIC MODEL A

$$A_I(1 \to 1) = 1 \qquad\qquad \forall\, I \text{ such that } q_I = 1$$
$$A_{200}(1 \to 3) = A_{200}(2 \to 1) = A_{200}(3 \to 2) = 1$$
$$A_{220}(1 \to 2) = A_{220}(2 \to 1) = 1$$
$$A_{300}(1 \to 2) = A_{300}(2 \to 3) = A_{300}(3 \to 1) = 1$$
$$A_{300}(4 \to 5) = A_{300}(5 \to 4) = 1$$
$$A_{320}(1 \to 3) = A_{320}(2 \to 1) = A_{320}(3 \to 2) = 1$$
$$A_{400}(1 \to 2) = A_{400}(2 \to 3) = A_{400}(3 \to 1) = 1$$
$$A_{400}(4 \to 5) = A_{400}(5 \to 4) = 1$$
$$A_{420}(1 \to 3) = A_{420}(2 \to 1) = A_{420}(3 \to 2) = 1$$
$$A_{500}(1 \to 3) = A_{500}(2 \to 1) = A_{500}(3 \to 2) = 1$$
$$A_{520}(1 \to 2) = A_{520}(2 \to 1) = 1$$

7-BIT NONDETERMINISTIC MODEL B

$$A_I(1 \to 1) = 1 \qquad\qquad \forall\, I \text{ such that } q_I = 1$$
$$A_I(j \to k) = \frac{1 - \delta_{jk}}{q_I - 1} \qquad \forall\, I \in \{200, 220, 500, 520\},\ j, k \in \{1, \cdots, q_I\}$$
$$A_{300}(j \to k) = \frac{1 - \delta_{jk}}{2} \qquad j, k \in \{1, 2, 3\}$$
$$A_{300}(j \to k) = 1 - \delta_{jk} \qquad j, k \in \{4, 5\}$$
$$A_{300}(j \to k) = A_{300}(k \to j) = 0 \qquad j \in \{1, 2, 3\},\ k \in \{4, 5\}$$
$$A_{320}(1 \to 1) = 1$$
$$A_{320}(j \to k) = 1 - \delta_{jk} \qquad j, k \in \{2, 3\}$$
$$A_{320}(j \to k) = A_{320}(k \to j) = 0 \qquad j \in \{1\},\ k \in \{2, 3\}$$
$$A_{400}(j \to k) = \frac{1 - \delta_{jk}}{2} \qquad j, k \in \{1, 2, 3\}$$
$$A_{400}(j \to k) = 1 - \delta_{jk} \qquad j, k \in \{4, 5\}$$
$$A_{400}(j \to k) = A_{400}(k \to j) = 0 \qquad j \in \{1, 2, 3\},\ k \in \{4, 5\}$$
$$A_{420}(1 \to 1) = 1$$
$$A_{420}(j \to k) = 1 - \delta_{jk} \qquad j, k \in \{2, 3\}$$
$$A_{420}(j \to k) = A_{420}(k \to j) = 0 \qquad j \in \{1\},\ k \in \{2, 3\}$$

7-BIT DETERMINISTIC MODEL B

$$A_I(1 \to 1) = 1 \qquad\qquad \forall\, I \text{ such that } q_I = 1$$
$$A_{200}(1 \to 3) = A_{200}(2 \to 1) = A_{200}(3 \to 2) = 1$$
$$A_{220}(1 \to 2) = A_{220}(2 \to 1) = 1$$
$$A_{300}(1 \to 2) = A_{300}(2 \to 3) = A_{300}(3 \to 1) = 1$$
$$A_{300}(4 \to 5) = A_{300}(5 \to 4) = 1$$
$$A_{320}(1 \to 1) = A_{320}(2 \to 3) = A_{320}(3 \to 2) = 1$$
$$A_{400}(1 \to 2) = A_{400}(2 \to 3) = A_{400}(3 \to 1) = 1$$
$$A_{400}(4 \to 5) = A_{400}(5 \to 4) = 1$$
$$A_{420}(1 \to 1) = A_{420}(2 \to 3) = A_{420}(3 \to 2) = 1$$
$$A_{500}(1 \to 3) = A_{500}(2 \to 1) = A_{500}(3 \to 2) = 1$$
$$A_{520}(1 \to 2) = A_{520}(2 \to 1) = 1$$

7-BIT DETERMINISTIC MODEL C (CHIRAL)

$$A_I(1 \to 1) = 1 \qquad\qquad \forall\ I \text{ such that } q_I = 1$$

$$A_{200}(1 \to 3) = A_{200}(2 \to 1) = A_{200}(3 \to 2) = 1$$

$$A_{220}(1 \to 2) = A_{220}(2 \to 1) = 1$$

$$A_{300}(1 \to 3) = A_{300}(2 \to 1) = A_{300}(3 \to 2) = 1$$

$$A_{300}(4 \to 5) = A_{300}(5 \to 4) = 1$$

$$A_{320}(1 \to 1) = A_{320}(2 \to 3) = A_{320}(3 \to 2) = 1$$

$$A_{400}(1 \to 2) = A_{400}(2 \to 3) = A_{400}(3 \to 1) = 1$$

$$A_{400}(4 \to 5) = A_{400}(5 \to 4) = 1$$

$$A_{420}(1 \to 1) = A_{420}(2 \to 3) = A_{420}(3 \to 2) = 1$$

$$A_{500}(1 \to 2) = A_{500}(2 \to 3) = A_{500}(3 \to 1) = 1$$

$$A_{520}(1 \to 2) = A_{520}(2 \to 1) = 1$$

8-BIT MINIMAL NONDETERMINISTIC MODEL

$$A_I(j \to k) = \delta_{jk} \qquad\qquad \forall\ I \notin \{200, 220, 300, 320, 400, 420, 500, 600\}$$

$$A_{200}(1 \to 1) = 1$$

$$A_{200}(j \to k) = \frac{1 - \delta_{jk}}{2} \qquad j, k \in \{2, 3, 4\}$$

$$A_{200}(1 \to j) = 0 \qquad j \in \{2, 3, 4\}$$

$$A_{220}(j \to k) = 1 - \delta_{jk} \qquad j, k \in \{1, 2\}$$

$$A_{300}(1 \to 1) = 1$$

$$A_{300}(j \to k) = \frac{1 - \delta_{jk}}{2} \qquad j, k \in \{2, 3, 4\}$$

$$A_{300}(j \to k) = 1 - \delta_{jk} \qquad j, k \in \{5, 6\}$$

$$A_{300}(1 \to j) = 0 \qquad j \in \{2, 3, 4, 5, 6\}$$

$$A_{300}(j \to k) = A_{300}(k \to j) = 0 \qquad j \in \{2, 3, 4\},\ k \in \{5, 6\}$$

$$A_{320}(j \to k) = 1 - \delta_{jk} \qquad j, k \in \{1, 2\}$$

$$A_{320}(j \to k) = \delta_{jk} \qquad j, k \in \{3, 4\}$$

$$A_{320}(j \to k) = A_{320}(k \to j) = 0 \qquad j \in \{1, 2\},\ k \in \{3, 4\}$$

$$A_{400}(j \to k) = \frac{1 - \delta_{jk}}{2} \qquad j, k \in \{1, 2, 3\}$$

$$A_{400}(j \to k) = 1 - \delta_{jk} \qquad j, k \in \{4, 5\}$$

$$A_{400}(j \to k) = \delta_{jk} \qquad j, k \in \{6, 7, 8\}$$

$$A_{400}(j \to k) = A_{400}(k \to j) = 0 \qquad j \in \{1, 2, 3\},\ k \in \{4, 5, 6, 7, 8\}$$

$$A_{400}(j \to k) = A_{400}(k \to j) = 0 \qquad j \in \{4, 5\},\ k \in \{6, 7, 8\}$$

$$A_{420}(j \to k) = 1 - \delta_{jk} \qquad j, k \in \{1, 2\}$$

$$A_{420}(j \to k) = \delta_{jk} \qquad j, k \in \{3, 4, 5\}$$

$$A_{420}(j \to k) = A_{420}(j \to k) = 0 \qquad j \in \{1, 2\},\ k \in \{3, 4, 5\}$$

$$A_{500}(j \to k) = \delta_{jk} \qquad j, k \in \{1, 2, 3\}$$

$$A_{500}(j \to k) = 1 - \delta_{jk} \qquad j, k \in \{4, 5\}$$

$$A_{500}(j \to k) = \frac{1 - \delta_{jk}}{2} \qquad j, k \in \{6, 7, 8\}$$

$$A_{500}(j \to k) = A_{500}(k \to j) = 0 \qquad j \in \{1, 2, 3\},\ k \in \{4, 5, 6, 7, 8\}$$

$$A_{500}(j \to k) = A_{500}(k \to j) = 0 \qquad j \in \{4, 5\},\ k \in \{6, 7, 8\}$$

$$A_{600}(j \to k) = \delta_{jk} \qquad j, k \in \{1, 2, 3, 4\}$$

$$A_{600}(j \to k) = 1 - \delta_{jk} \qquad j, k \in \{5, 6\}$$

$$A_{600}(j \to k) = A_{600}(k \to j) = 0 \qquad j \in \{1, 2, 3, 4\},\ k \in \{5, 6\}$$

APPENDIX B - 3-D MINIMAL MARRIED PAIR COLLISION TABLES

18-BIT MODEL

For the minimal table, collisions involve solely exchanging pairs of particles. There may be more than two particles in an input state but only two get changed, the rest of the particles being just "spectators." The pairs are chosen from the groups listed in table 8 which lists the pairs in terms of hexadecimal numbers where the notation for the bits is listed in table 9. Each line in the table represents a different scattering group. Exchange takes place only within groups, not between them. The transition probability for a given input state to go into a given output state is simply $(1/(\text{number of possible output states for that input state}))$.

TABLE 8 Binary scattering groups for 18-bit models

00003	0000c	00030	000c0	00300	00c00
00005	00050	01000			
00009	00500	04000			
00090	00900	10000			
0000a	000a0	02000			
00006	00a00	08000			
00060	00600	20000			

TABLE 9 Notation for 18-bit married pair models

i	bit	mass	c_{ix}	c_{iy}	c_{iz}	c_{iw}
1	0	1	1	1	0	0
2	1	1	−1	−1	0	0
3	2	1	1	−1	0	0
4	3	1	−1	1	0	0
5	4	1	1	0	1	0
6	5	1	−1	0	−1	0
7	6	1	1	0	−1	0
8	7	1	−1	0	1	0
9	8	1	0	1	1	0
10	9	1	0	−1	−1	0
11	10	1	0	1	−1	0
12	11	1	0	−1	1	0
13	12	1	1	0	0	1
		1	1	0	0	−1
14	13	1	−1	0	0	1
		1	−1	0	0	−1
15	14	1	0	1	0	1
		1	0	0	0	−1
16	15	1	0	−1	0	1
		1	0	−1	0	−1
17	16	1	0	0	1	1
		1	0	0	1	−1
18	17	1	0	0	−1	1
		1	0	0	−1	−1

14-BIT MODEL This is similar to the 18-bit model except some collisions exchange four particles are used in addition to pair exchange. Again, more than two or four particles maybe present at a site but only two or four particles, respectively, are exchanged. The transition probability for a given input state to go into a given output state is again (1/(number of possible output states for that input state)). The groups that can be exchanged are listed in table 10 using notation in table 11. Again, each line in the scattering group table corresponds to a different group.

TABLE 10 Scattering Groups
for 14-Bit Models

Binary Scattering Groups		
0003	000c	0030

Quaternary Scattering Groups	
2040	000f
1080	000f
0900	000f
0600	000f
2040	0033
1080	0033
0900	0033
0600	0033
2040	003c
1080	003c
0900	003c
0600	003c

TABLE 11 Notation for 14-Bit Married Pair Models

i	bit	mass	c_{ix}	c_{iy}	c_{ix}	c_{iw}
1	0	1	$\sqrt{2}$	0	0	0
2	1	1	$-\sqrt{2}$	0	0	0
3	2	1	0	$\sqrt{2}$	0	0
4	3	1	0	$-\sqrt{2}$	0	0
5	4	1	0	0	$\sqrt{2}$	0
6	5	1	0	0	$-\sqrt{2}$	0
7	6	1	$1/\sqrt{2}$	$1/\sqrt{2}$	$1/\sqrt{2}$	$1/\sqrt{2}$
		1	$1/\sqrt{2}$	$1/\sqrt{2}$	$1/\sqrt{2}$	$-1/\sqrt{2}$
8	7	1	$1/\sqrt{2}$	$1/\sqrt{2}$	$-1/\sqrt{2}$	$1/\sqrt{2}$
		1	$1/\sqrt{2}$	$1/\sqrt{2}$	$-1/\sqrt{2}$	$-1/\sqrt{2}$
9	8	1	$1/\sqrt{2}$	$-1/\sqrt{2}$	$1/\sqrt{2}$	$1/\sqrt{2}$
		1	$1/\sqrt{2}$	$-1/\sqrt{2}$	$1/\sqrt{2}$	$-1/\sqrt{2}$
10	9	1	$1/\sqrt{2}$	$-1/\sqrt{2}$	$-1/\sqrt{2}$	$1/\sqrt{2}$
		1	$1/\sqrt{2}$	$-1/\sqrt{2}$	$-1/\sqrt{2}$	$-1/\sqrt{2}$
11	10	1	$-1/\sqrt{2}$	$1/\sqrt{2}$	$1/\sqrt{2}$	$1/\sqrt{2}$
		1	$-1/\sqrt{2}$	$1/\sqrt{2}$	$1/\sqrt{2}$	$-1/\sqrt{2}$
12	11	1	$-1/\sqrt{2}$	$1/\sqrt{2}$	$-1/\sqrt{2}$	$1/\sqrt{2}$
		1	$1/\sqrt{2}$	$1/\sqrt{2}$	$-1/\sqrt{2}$	$-1/\sqrt{2}$
13	12	1	$-1/\sqrt{2}$	$-1/\sqrt{2}$	$1/\sqrt{2}$	$1/\sqrt{2}$
		1	$-1/\sqrt{2}$	$-1/\sqrt{2}$	$1/\sqrt{2}$	$-1/\sqrt{2}$
14	13	1	$-1/\sqrt{2}$	$-1/\sqrt{2}$	$-1/\sqrt{2}$	$1/\sqrt{2}$
		1	$-1/\sqrt{2}$	$-1/\sqrt{2}$	$-1/\sqrt{2}$	$-1/\sqrt{2}$

ACKNOWLEDGMENTS

We are grateful to H. D. Chen for valuable discussions. This work is supported by the U.S. Department of Energy.

REFERENCES

1. d'Humières, D., P. Lallemand, and U. Frisch (1986), "Lattice Gas Models for 3-D Hydrodynamics," *Europhys. Lett.* **2**(4), 291–297.
2. d'Humières, D., and P. Lallemand (1987), "Numerical Simulations of Hydrodynamics with Lattice Gas Automata in Two and Three Dimensions," *Complex Systems* **1**, 598–631.
3. Dubrulle, B. (1988), "Method of Computation of the Reynolds Number for Two Models of Lattice Gas Involving Violation of Semi-Detailed Balance," *Complex Systems* **2**, 577–609.
4. Frisch, U., B. Hasslacher, and Y. Pomeau (1986), "Lattice-Gas Automata for the Navier-Stokes Equation," *Phys. Rev. Lett.* **56**, 1505–1508.
5. Frisch, U., D. d'Humières, B. Hasslacher, P. Lallemand, Y. Pomeau, and J. Rivet (1987), "Lattice Gas Hydrodynamics in Two and Three Dimensions," *Complex Systems* **1**, 648–707.
6. Frisch, U., private communication.
7. Hénon, M. (1987), "Isometric Collision Rules for the Four-Dimensional FCHC Lattice Gas," *Complex Systems* **1**, 475–494.
8. Hénon, M. (1987), "Viscosity of a Lattice Gas," *Complex Systems* **1**, 763–789.
9. Hénon, M., private communication.
10. Orszag, S., and V. Yakhot (1986), "Reynolds Number Scaling of Cellular-Automation Hydrodynamics," *Phys. Rev. Lett.* **56**, 1691–1693.
11. Rothaus, 0., unpublished report.
12. Shimomura, T., G. Doolen, B. Hasslacher, and C. Fu (1987), "Calculations Using Lattice Gas Techniques," *Los Alamos Science* **15**, 201–210.
13. Wolfram, S. (1986), "Cellular Automata Fluids I," *J. Stat. Phys.* **45**, 471–526.

Michel Hénon
C.N.R.S., Observatoire de Nice, France, B.P. 139, 06003 Nice Cedex, France

Viscosity of a Lattice Gas

This paper originally appeared in *Complex Systems* (1987), Volume 1, pages 763–789.

ERRATA

The last part of equation (B.1) should read

$$=\frac{D}{2(D-1)nc^4}\sum_\alpha\sum_\beta$$
$$\sum_i\left(c_{i\alpha}c_{i\beta}-\frac{c^2}{D}\delta_{\alpha\beta}\right)\sum_j\left(c_{j\alpha}c_{j\beta}-\frac{c^2}{D}\delta_{\alpha\beta}\right)\sum_s s_is_jd^{p-1}(1-d)^{n-p-1}.$$
$$(B.1)$$

Complex Systems **1** (1987) 763–789

Viscosity of a Lattice Gas

Michel Hénon
C.N.R.S., Observatoire de Nice, France, B.P. 139, 06003 Nice Cedex, France

Abstract. The shear viscosity of a lattice gas can be derived in the Boltzmann approximation from a straightforward analysis of the numerical algorithm. This computation is presented first in the case of the Frisch-Hasslacher-Pomeau two-dimensional triangular lattice. It is then generalized to a regular lattice of arbitrary dimension, shape, and collision rules with appropriate symmetries. The viscosity is shown to be positive. A practical recipe is given for choosing collision rules so as to minimize the viscosity.

1. Introduction

1.1 Goal

For computational efficiency, the collision rules in a lattice gas automaton should be chosen so as to make the shear viscosity as small as possible [1]. For a given set of rules, the viscosity can be estimated through numerical simulations [2,3]. It would be much more convenient, however, to have an explicit formula through which the viscosity could be computed directly from the lattice rules. Here, I present such a formula, which is applicable when the following conditions are satisfied:

1. there is a single population of particles (all velocities have the same modulus);

2. the lattice and the collision rules are "sufficiently symmetrical", in a sense which will be made more precise below (section 3);

3. the Boltzmann approximation is valid (the probabilities of arrival of particles at a node from different directions can be assumed independent);

4. the system is not far from isotropic equilibrium (low Mach number);

5. the only quantities conserved by collisions are the number of particles and the momentum; and

6. the collisions satisfy semi-detailed balancing.

It can be shown from the formula that the viscosity is always positive. Moreover, from the structure of the formula, one can derive a simple rule for the optimization of individual collisions.

1.2   Method

The formula will be derived by a straightforward analysis of the lattice rules; apart from the basic definition of shear viscosity, nothing is borrowed from the classical theories of fluid dynamics or statistical mechanics. The derivation consists of the following steps:

1. Define a steady, homogeneous state, with linear shear velocity field.

2. Write equations for collisions.

3. Write equations for propagation.

4. Solve these equations to determine the detailed structure of the steady state.

5. Compute the momentum flux and the viscosity.

In order to show the argument more clearly, the computation is first presented in detail for a simple case: the triangular lattice introduced by Frisch, Hasslacher, and Pomeau [4] (*FHP lattice*). The five steps are described in sections 2.1 to 2.5, then the computation is generalized to arbitrary lattices in sections 3.1 to 3.5.

1.3   The FHP lattice

We recall briefly the definition of the FHP lattice. We consider a plane triangular lattice, populated by particles. At any given time, a particle is in a given node and has a velocity pointing towards one of the six neighbor nodes, with a given modulus c. At a given node, no two particles can have the same velocity (exclusion principle). Evolution proceeds in two alternating phases: (i) *collisions*: particles arriving at a node "collide" and may change their velocities, according to definite *collision rules*; and (ii) *propagation*: each particle moves to the next node in the direction of its velocity. Several variations are possible concerning the collision rules; here we will use the original rules as defined in [4], i.e. only binary head-on collisions and triple collisions (see also [1], figures 4a and b). Note that each collision preserves the number of particles and the momentum. Note also the overall symmetry of the rules: they are invariant under any rotation or symmetry which preserves the lattice.

1.4   Notation

The velocities are $c_i = (c_{i1}, c_{i2})$, with $i = 1$ to 6 (figure 1). The position of a node is $x = (x_1, x_2)$. $N_i(x)$ is the probability of a particle *arriving* at node x with velocity c_i. (In general, N_i should depend on time as well as on position; here, however, we will limit our attention to a steady state). $N_i'(x)$ is the probability of a particle *leaving* node x with velocity c_i. We will use physical units rather than dimensionless units in order to avoid any ambiguity in the use of the final formula: l is the link length, τ is

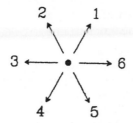

Figure 1: The six velocities of the FHP lattice.

the propagation time, and m is the mass of a particle. Thus, the velocity modulus is $c = l/\tau$, and the number of nodes per unit area is $f = 2/(\sqrt{3}\,l^2)$.

2.   Computation of the viscosity for the FHP lattice

2.1   Steady state

The gas density is

$$\rho(\mathbf{x}) = fm \sum_{i=1}^{6} N_i(\mathbf{x}). \tag{2.1}$$

We will consider a homogeneous state: $\rho(\mathbf{x})$ has a constant value ρ.

In the simplest case of an isotropic velocity distribution, all N_i are equal to the same constant value d:

$$N_i(\mathbf{x}) = \frac{\rho}{6fm} = d. \tag{2.2}$$

We will consider a state which does not differ too much from isotropy:

$$N_i(\mathbf{x}) = d + \nu_i(\mathbf{x}), \qquad N_i'(\mathbf{x}) = d + \nu_i'(\mathbf{x}), \tag{2.3}$$

with

$$\nu_i(\mathbf{x}), \nu_i'(\mathbf{x}) \ll 1. \tag{2.4}$$

We have

$$\sum_{i=1}^{6} \nu_i = 0. \tag{2.5}$$

All computations will be made to first order in the ν_i.

The mean velocity $\mathbf{u}$ at a node $\mathbf{x}$ is defined by

$$\rho\mathbf{u}(\mathbf{x}) = fm \sum_{i=1}^{6} N_i(\mathbf{x})\mathbf{c}_i. \tag{2.6}$$

We will consider a state in which $\mathbf{u}$ is a shear flow:

$$u_1 = Tx_2, \qquad u_2 = 0, \tag{2.7}$$

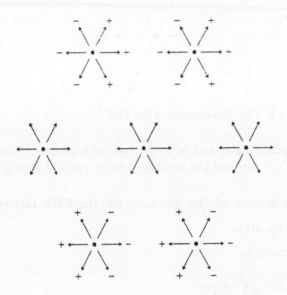

Figure 2: Deviations from equilibrium at seven neighboring nodes.

where T is a given constant. The steady state is thus defined by two constants, ρ and T. Our objective is to compute its detailed structure, i.e., the values of the $N_i(\mathbf{x})$.

It seems natural to assume that each ν_i is also a linear function of x_2:

$$\nu_i(\mathbf{x}) = k_i x_2 + \epsilon_i, \tag{2.8}$$

where the k_i and ϵ_i ($i = 1$ to 6) are 12 constants to be determined. The interpretation of the constants k_i is straightforward. Figure 2 represents the velocities at seven neighboring nodes. The central horizontal line corresponds to $x_2 = 0$: the mean velocity is $\mathbf{u} = 0$ as shown by equation (2.7). Therefore, we do not expect any systematic first-order deviation from equilibrium on that line. In the upper horizontal line, x_2 is positive, and the mean velocity $\mathbf{u}$ has a positive horizontal constant. Therefore, we expect an increase of the populations N_i for the velocities $i = 1, 5, 6$ lying in the right half-plane, and a decrease for the velocities $i = 2, 3, 4$ lying in the left half-plane, as indicated by the $+$ and $-$ signs (which are the signs of the ν_i). In the lower horizontal line, we have the opposite effect. For any particular velocity $\mathbf{c}_i$, then, we have a linear variation of ν_i with the ordinate x_2.

The interpretation of the ϵ_i terms is more subtle; they result from the combination of shear and propagation. Consider the situation one propagation step after figure 2. The central node receives an under-average number of particles along directions 1 and 4, and an over-average number

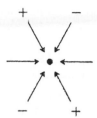

Figure 3: Shear-induced anisotropy.

along directions 2 and 5 (figure 3). Thus, a *shear-induced* *anisotropy* is created. This anisotropy is represented by the ϵ_i terms.

Substitution of equation (2.8) in the mass and momentum equations (2.1) and (2.6) gives three equations for the ϵ_i:

$$\sum_{i=1}^{6} \epsilon_i = 0, \qquad \sum_{i=1}^{6} \epsilon_i c_{i1} = 0, \qquad \sum_{i=1}^{6} \epsilon_i c_{i2} = 0, \tag{2.9}$$

and three equations for the k_i:

$$\sum_{i=1}^{6} k_i = 0, \qquad \sum_{i=1}^{6} c_{i1} k_i = 6dT, \qquad \sum_{i=1}^{6} c_{i2} k_i = 0. \tag{2.10}$$

These equations are not sufficient to determine the ϵ_i and the k_i. To complete the determination, we must write that the state of the system is invariant under collisions plus propagation. So we first establish equations for these two mechanisms.

2.2   Collisions

In this section, we consider a given node $\mathbf{x}$; to simplify, we will omit the coordinate $\mathbf{x}$ and write N_i for $N_i(\mathbf{x})$, etc.

A particular collision is defined by an *input state*, defined as a subset of the n velocities, and an *output state*, similarly defined. The input state is conveniently defined by a collection of n numbers:

$$s = (s_1, \ldots, s_6), \tag{2.11}$$

where $s_i = 1$ if velocity c_i is present in the input state, $s_i = 0$ otherwise. Similarly, we define

$$s' = (s'_1, \ldots, s'_6) \tag{2.12}$$

for the output state. We write a collision as $C = (s; s')$. Each collision C has an associated probability, which we write

$$A(C) \qquad \text{or} \qquad A(s; s'). \tag{2.13}$$

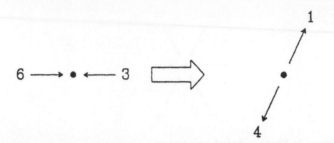

Figure 4: Example of a collision.

As an example, for the collision represented on figure 4, we have

$$s = (0,0,1,0,0,1), \qquad s' = (1,0,0,1,0,0), \tag{2.14}$$

and the associated probability is, under the classical FHP rules:

$$A(0,0,1,0,0,1;1,0,0,1,0,0) = A(s;s') = 1/2. \tag{2.15}$$

It will be convenient to formally define a *collision* for all s and s', even if the collision rules do not provide for any actual transition from s to s'; in that case, we simply write $A(s;s') = 0$. There are $2^6 = 64$ possible input states s, and 64 possible output states s'. Therefore, $A(s;s')$ can be written as a 64×64 matrix. In this way, the whole set of collision rules is neatly encoded into a single matrix. It is a very sparse matrix: most cases correspond to forbidden transitions, i.e. $A = 0$ (in particular, all cases where the number of particles or the momentum of s and s' do not agree). Note also that for an input state s which does not change, there is $A(s;s') = 1$ for $s' = s$, and $A(s;s') = 0$ for $s' \neq s$.

The sum of all A corresponding to a given input state s must of course be 1; so for any s we have

$$\sum_{s'} A(s;s') = 1. \tag{2.16}$$

We will assume that the symmetric relations are also satisfied, i.e. for any s', there is

$$\sum_{s} A(s;s') = 1. \tag{2.17}$$

This is called *semi-detailed balancing*.

In what follows, we will frequently have to sum over all collisions, and we will use the various notations

$$\sum_C A \mathcal{F} = \sum_{\int} \sum_{\int'} A(\int;\int') \mathcal{F} =$$

$$\sum_{s_1=0}^{1} \cdots \sum_{s_6=0}^{1} \sum_{s_1'=0}^{1} \cdots \sum_{s_6'=0}^{1} A(s_1, \ldots, s_6; s_1', \ldots, s_6') \mathcal{F}. \qquad (2.18)$$

where $\mathcal{F}$ is any expression.

For any given collision $(s; s')$ with non-zero probability, the number of particles p and the momentum $\mathbf{q}$ must be preserved. In other words, if $A(s; s') \neq 0$, then

$$p = \sum_{i=1}^{6} s_i = p' = \sum_{i=1}^{6} s_i', \qquad (2.19)$$

and

$$\mathbf{q} = \sum_{i=1}^{6} s_i \mathbf{c}_i = \mathbf{q}' = \sum_{i=1}^{6} s_i' \mathbf{c}_i. \qquad (2.20)$$

We use now the Boltzmann approximation and we write that the probability of an input state s is

$$\prod_{j=1}^{6} N_j^{s_j} (1 - N_j)^{1-s_j}. \qquad (2.21)$$

Therefore, the probability of having a particle with velocity $\mathbf{c}_i$ in the output state is:

$$N_i' = \sum_C s_i' A \prod_{j=1}^{6} N_j^{s_j} (1 - N_j)^{1-s_j}. \qquad (2.22)$$

If we sum the probabilities of all input states which contain $\mathbf{c}_i$, we obtain N_i:

$$N_i = \sum_s s_i \prod_{j=1}^{6} N_j^{s_j} (1 - N_j)^{1-s_j} \qquad (2.23)$$

or, using equation (2.16)

$$N_i = \sum_C s_i A \prod_{j=1}^{6} N_j^{s_j} (1 - N_j)^{1-s_j}. \qquad (2.24)$$

Therefore,

$$N_i' - N_i = \sum_C (s_i' - s_i) A \prod_{j=1}^{6} N_j^{s_j} (1 - N_j)^{1-s_j}. \qquad (2.25)$$

If $p = 0$, there is $s_i = s_i' = 0$ for all i, and the corresponding term in the right-hand side of equation (2.25) vanishes. Similarly, the term corresponding to $p = 6$ vanishes. Therefore, we will consider in equation (2.25) and in all subsequent equations (which derive from it) that the summation is to be made only on collisions with

$$1 \le p \le n - 1. \tag{2.26}$$

Substituting $N_i = d + \nu_i$, $N'_i = d + \nu'_i$, and developing to first order in the ν_i, we obtain

$$\nu'_i - \nu_i = \sum_C (s'_i - s_i) A d^p (1-d)^{6-p} \left[1 + \frac{1}{d(1-d)} \sum_{j=1}^{6} (s_j - d)\nu_j \right]. \tag{2.27}$$

The first term can be written, using equations (2.19), (2.16), and (2.17)

$$\sum_C s'_i A d^{p'} (1-d)^{6-p'} - \sum_C s_i A d^p (1-d)^{6-p} =$$

$$\sum_{s'} s'_i d^{p'} (1-d)^{6-p'} - \sum_s s_i d^p (1-d)^{6-p}. \tag{2.28}$$

This vanishes since the two terms represent the same summation. Using also equation (2.5), we obtain the fundamental *collision equation*:

$$\nu'_i - \nu_i = \sum_C (s'_i - s_i) A d^{p-1} (1-d)^{5-p} \sum_{j=1}^{6} s_j \nu_j. \tag{2.29}$$

As an example, consider the case $i = 1$. $s'_i - s_i$ is different from zero in six cases:

$$s = (1,0,0,1,0,0), \qquad s' = (0,1,0,0,1,0), \qquad p = 2, \qquad A = 1/2;$$

$$s = (1,0,0,1,0,0), \qquad s' = (0,0,1,0,0,1), \qquad p = 2, \qquad A = 1/2;$$

$$s = (0,1,0,0,1,0), \qquad s' = (1,0,0,1,0,0), \qquad p = 2, \qquad A = 1/2;$$

$$s = (0,0,1,0,0,1), \qquad s' = (1,0,0,1,0,0), \qquad p = 2, \qquad A = 1/2;$$

$$s = (1,0,1,0,1,0), \qquad s' = (0,1,0,1,0,1), \qquad p = 3, \qquad A = 1;$$

$$s = (0,1,0,1,0,1), \qquad s' = (1,0,1,0,1,0), \qquad p = 3, \qquad A = 1.$$

$$\tag{2.30}$$

Therefore, the collision equation is

$$\nu'_1 - \nu_1 = \frac{1}{2} d(1-d)^3 (\nu_2 + \nu_5 + \nu_3 + \nu_6 - 2\nu_1 - 2\nu_4)$$

$$+ d^2 (1-d)^2 (\nu_2 + \nu_4 + \nu_6 - \nu_1 - \nu_3 - \nu_5) \tag{2.31}$$

The other equations ($i = 2, \ldots, 6$) are deduced by rotation of indices.

2.3   Propagation

The probability of arrival of a particle at a node equals the probability of leaving the previous node:

$$N_i(\mathbf{x} + \tau c_i, t + \tau) = N_i'(\mathbf{x}, t). \tag{2.32}$$

Since we assume a steady state, this equation reduces to

$$N_i(\mathbf{x} + \tau c_i) = N_i'(\mathbf{x}). \tag{2.33}$$

Substituting the expressions (2.3) for N and N' and using (2.8), we obtain the fundamental *propagation equation*:

$$\nu_i(\mathbf{x}) - \nu_i'(\mathbf{x}) = -\tau k_i c_{i2}. \tag{2.34}$$

2.4   Computation of the steady state

Combining the collision and propagation equations, (2.29) and (2.34), and substituting (2.8), we obtain a set of six equations:

$$\sum_C (s_i' - s_i) A d^{p-1} (1 - d)^{5-p} \sum_{j=1}^{6} s_j (k_j x_2 + \epsilon_j) - \tau k_i c_{i2} = 0$$

$$(i = 1, \ldots, 6). \tag{2.35}$$

We use now these equations, together with equations (2.9) and (2.10), to determine the steady state. We consider first the terms proportional to x_2 in equation (2.35), which give six equations:

$$\frac{1}{2} d(1 - d)^3 (k_2 + k_5 + k_3 + k_6 - 2k_1 - 2k_4)$$

$$+ d^2 (1 - d)^2 (k_2 + k_4 + k_6 - k_1 - k_3 - k_5) = 0 \tag{2.36}$$

and five similar equations deduced by circular permutation. This is a set of linear homogeneous equations. But in fact, there are only three independent equations, because of the conservation of mass and momentum. Therefore, equation (2.36) has non-zero solutions, which have the general form

$$k_i = K_0 + K_1 c_{i1} + K_2 c_{i2} \tag{2.37}$$

where K_0, K_1, K_2 are arbitrary constants. Making use now of equation (2.10), we find that the K_i are uniquely determined and the k_i are

$$k_i = \frac{2dT}{c^2} c_{i1}. \tag{2.38}$$

We remark that k_i is proportional to the horizontal component of c_i; this agrees with the intuitive description of figure 2.1.

We consider next the terms independent of x_2 in equation (2.35), which give after substitution of the solution (2.38):

$$\frac{1}{2}d(1-d)^3(\epsilon_2 + \epsilon_5 + \epsilon_3 + \epsilon_6 - 2\epsilon_1 - 2\epsilon_4)$$

$$+d^2(1-d)^2(\epsilon_2 + \epsilon_4 + \epsilon_6 - \epsilon_1 - \epsilon_3 - \epsilon_5) - \frac{2d\tau T}{c^2}c_{11}c_{12} = 0 \qquad (2.39)$$

and five similar equations. These are six linear inhomogeneous equations. Again, there are only three independent equations. Combining with equation (2.9), we have a system of six independent equations to solve.

It will be helpful to inquire into the physical meaning of equation (2.39). Figures 2 and 3 showed that propagation produces a deficit of particles in directions 1 and 4 and an excess in directions 2 and 5. Thus, it tends to produce an anisotropy in the ϵ_i. On the other hand, since the collision rules are symmetrical, they should tend to damp out any anisotropy. Thus, the first two terms in equation (2.39) represent the damping of the anisotropy by collisions, while the last term represents the excitation of the anisotropy by propagation. The equation is satisfied when these two processes are in equilibrium.

This suggests the following conjecture: the equilibrium anisotropy, represented by the ϵ_i, should be proportional to the excitation term in equation (2.39):

$$\epsilon_i \propto -c_{i1}c_{i2}. \qquad (2.40)$$

Looking at figure 1, we see that this can be written

$$\epsilon_1 = -\epsilon, \qquad \epsilon_2 = \epsilon, \qquad \epsilon_3 = 0,$$
$$\epsilon_4 = -\epsilon, \qquad \epsilon_5 = \epsilon, \qquad \epsilon_6 = 0 \qquad (2.41)$$

where ϵ is a constant. The equations (2.9) are satisfied. Substituting into equation (2.39), we find that the conjecture is true: the equations are satisfied if we take

$$\epsilon = \frac{\tau T}{2\sqrt{3}\,c^2}(1-d)^{-3} \qquad (2.42)$$

and the solution is

$$\epsilon_i = -\frac{2\tau T}{3c^2}(1-d)^{-3}c_{i1}c_{i2}. \qquad (2.43)$$

The steady state is thus completely determined. It is given by equations (2.3), (2.8), (2.38), (2.43):

$$N_i = d + \frac{2dT}{c^2}c_{i1}x_2 - \frac{2\tau T}{3c^2}(1-d)^{-3}c_{i1}c_{i2}. \qquad (2.44)$$

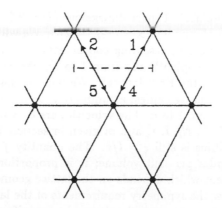

Figure 5: Computation of the momentum flux across a segment (shown as a dashed line).

2.5   Momentum flux

In view of the spatial periodicity of the lattice, it will be sufficient to compute the horizontal momentum Q transferred downwards across a horizontal segment of length l, situated above a node $\mathbf{x}$, during one propagation time τ (figure 5). The shear viscosity η is then defined by

$$Q = \eta T l \tau. \tag{2.45}$$

Contributions to Q come from four possible particle motions, corresponding to velocity directions 1, 2, 4, 5, as indicated on the figure. For instance, there is a probability $N_4(\mathbf{x})$ that a horizontal momentum mc_{41} will be transferred downwards, etc. Summing these contributions, we obtain

$$Q =$$

$$m[N_4(\mathbf{x})c_{41} + N_5(\mathbf{x})c_{51} - N_1(\mathbf{x} + \tau c_1)c_{11} - N_2(\mathbf{x} + \tau c_2)c_{21}]. \tag{2.46}$$

Equating this to equation (2.45) and substituting the values (2.44) found for the steady state, we finally obtain the shear viscosity for the FHP lattice:

$$\eta = \rho \tau c^2 \left[\frac{1}{12d(1-d)^3} - \frac{1}{8} \right]. \tag{2.47}$$

The first term agrees with the result derived for small d in reference 5. Good agreement is found with the numerical simulation of reference 2.

3.   Computation of the viscosity for a general lattice

We generalize now the computation to arbitrary lattices (subject to the conditions stated in the introduction).

We call D the number of space dimensions ($D = 2$ for the FHP lattice). Thus, the position of a node is $\mathbf{x} = (x_1, \ldots, x_D)$. We will use Greek letters α, β, ... for subscripts representing coordinates; a summation on one of these subscripts will implicitly run from 1 to D. We call n the number of velocities ($n = 6$ for the FHP lattice). We will use Latin letters i, j for subscripts representing velocities; a summation on one of these subscripts will implicitly run from 1 to n. The velocities are $\mathbf{c}_i$, with components $c_{i\alpha}$. The definitions of N_i, N_i', l, τ, and m given in section 1.4 are unchanged. The velocity modulus is still $c = l/\tau$. The quantity f is now defined as the number of nodes per unit volume; it is proportional to l^{-D}, with a numerical coefficient which depends on the lattice geometry.

We specify now the symmetry requirements of the lattice. For the purpose of the present paper, we will only need to specify symmetries for the set of n velocities $V = \{\mathbf{c}_1, \ldots, \mathbf{c}_n\}$.

1. *The set V must be isotropic to fourth order.* By this, we mean the following. For a given integer r, we define a tensor of order r

$$B_{\alpha_1 \ldots \alpha_r} = \sum_i c_{i\alpha_1} \ldots c_{i\alpha_r}. \tag{3.1}$$

We require that up to order $r = 4$ these tensors are isotropic (i.e., their components are invariant in any rotation of the coordinate axes). As is easily shown, the tensors are then given by [6]:

$$B_\alpha = \sum_i c_{i\alpha} = 0,$$

$$B_{\alpha\beta} = \sum_i c_{i\alpha} c_{i\beta} = \frac{nc^2}{D} \delta_{\alpha\beta},$$

$$B_{\alpha\beta\gamma} = \sum_i c_{i\alpha} c_{i\beta} c_{i\gamma} = 0,$$

$$B_{\alpha\beta\gamma\varsigma} = \sum_i c_{i\alpha} c_{i\beta} c_{i\gamma} c_{i\varsigma} = \frac{nc^4}{D(D+2)} (\delta_{\alpha\beta}\delta_{\gamma\varsigma} + \delta_{\alpha\gamma}\delta_{\beta\varsigma} + \delta_{\alpha\varsigma}\delta_{\beta\gamma}) \tag{3.2}$$

where $\delta_{\alpha\beta}$ is the Kronecker symbol.

2. All velocities should be in a sense interchangeable. This is formalized as follows. We call G the group of the isometries of the D-dimensional space which map V on itself. Then, *for any two velocities $\mathbf{c}_i$ and $\mathbf{c}_j$, there exists an isometry of G which maps $\mathbf{c}_i$ on $\mathbf{c}_j$.*

3. We consider now a particular velocity $\mathbf{c}_i$. In addition to the symmetry around the origin, expressed by (1) above, the set V must also exhibit a definite symmetry around $\mathbf{c}_i$, which we now specify. First, we decompose each velocity $\mathbf{c}_j$ as

$$c_j = c_{j\parallel} + c_{j\perp} , \tag{3.3}$$

where $c_{j\parallel}$ is parallel to c_i and $c_{j\perp}$ is perpendicular to c_i. We call G_i the subgroup of G consisting of all isometries in which c_i is invariant. For a given j, consider the set V_{ij} of all velocities $c_{I(j)}$ which can be obtained from c_j by an isometry $I \in G_i$. Their parallel components are equal:

$$c_{I(j)\parallel} = c_{j\parallel}. \tag{3.4}$$

The symmetry condition will be that *the set of perpendicular components $c_{I(j)\perp}$ is isotropic to second order in the hyperplane perpendicular to c_i.*

To translate this condition into equations, it will be convenient to temporarily redefine the coordinate system in such a way that the x_1 axis is parallel to c_i. We have then $c_{j\parallel} = (c_{j1}, 0, \ldots, 0)$, $c_{j\perp} = (0, c_{j2}, \ldots, c_{jD})$, and the condition is

$$
\begin{aligned}
\sum_{I \in G_i} c_{I(j)\alpha} &= 0, & (\alpha = 2, \ldots, D), \\
\sum_{I \in G_i} c_{I(j)\alpha} c_{I(j)\beta} &= R\delta_{\alpha\beta}, & (\alpha, \beta = 2, \ldots, D),
\end{aligned}
\tag{3.5}
$$

where R is the same for all α, β. (Note that we have summed over the isometries $I \in G_i$ rather than over the elements of V_{ij}. This is easily shown to be equivalent. Consider the subgroup G_{ij} of G_i made of the isometries which leave c_i and c_j invariant. For any velocity $c_k \in V_{ij}$, the isometries which map c_j into c_k form a left coset of G_{ij}; their number is $|G_{ij}|$. Thus, summing over $I \in G_i$, we obtain each velocity the same number of times).

The constant R can be computed as follows. We call θ_{ij} the angle between c_i and c_j. Then,

$$c_{I(j)1} = c_{j1} = c \cos\theta_{ij} \tag{3.6}$$

and

$$\sum_{\alpha=2}^{D} c_{I(j)\alpha}^2 = c^2 \sin^2\theta_{ij}. \tag{3.7}$$

Summing over $I \in G_i$ and using (3.2b), we obtain

$$R = \frac{|G_i| c^2 \sin^2\theta_{ij}}{D-1}. \tag{3.8}$$

Finally, we specify the symmetry requirements for the collision rules. They are quite simple:

4. *Collision rules must be invariant under any isometry of G. In other words, for any collision $(s; s')$ and for any isotropy $I \in G$, there is*

$$A[I(s); I(s')] = A(s; s'),$$ (3.9)

where $I(s)$ has the obvious meaning.

These conditions are satisfied by the FHP lattice, and also by the face-centered-hypercubic lattice (FCHC) introduced by d'Humières, Lallemand, and Frisch [7,1]. For the HPP lattice [1,8–10], however, relation (3.2d) is not satisfied.

3.1 Steady state

Equations (2.1) to (2.6) are still valid, with 6 replaced by n. For an isotropic velocity distribution, instead of (2.2) we have

$$N_i(\mathbf{x}) = \frac{\rho}{nfm} = d.$$ (3.10)

We consider a state in which the mean velocity field is an arbitrary linear function of the position $\mathbf{x}$: instead of equation (2.7), we postulate the more general form

$$u_\alpha = \sum_\beta T_{\alpha\beta} x_\beta \qquad (\alpha = 1, \ldots, D).$$ (3.11)

The $T_{\alpha\beta}$ are the components of a tensor, which is the velocity gradient:

$$T_{\alpha\beta} = \frac{\partial u_\alpha}{\partial x_\beta}.$$ (3.12)

ν_i is then assumed to be also a linear function of coordinates: instead of equation (2.8), we have

$$\nu_i = \sum_\beta k_{i\beta} x_\beta + \epsilon_i$$ (3.13)

where the $k_{i\beta}$ and the ϵ_i are constants to be determined.

The equations (2.9) and (2.10) become

$$\sum_i \epsilon_i = 0, \qquad \sum_i \epsilon_i c_{i\alpha} = 0,$$ (3.14)

and

$$\sum_i k_{i\beta} = 0, \qquad \sum_i c_{i\alpha} k_{i\beta} = ndT_{\alpha\beta}.$$ (3.15)

Note that α and β can take all values in these equations. Thus, equation (3.14) represents a total of $D + 1$ equations, and equation (3.15) represents a total of $D(D + 1)$ equations.

3.2 Collisions

The collision equation is almost the same as equation (2.29), the only difference being that 6 is replaced by n:

$$\nu_i' - \nu_i = \sum_C (s_i' - s_i) A d^{p-1} (1-d)^{n-p-1} \sum_j s_j \nu_j. \tag{3.16}$$

3.3 Propagation

The propagation equation (2.34) becomes

$$\nu_i(\mathbf{x}) - \nu_i'(\mathbf{x}) = -\tau \sum_\alpha k_{i\alpha} c_{i\alpha}. \tag{3.17}$$

3.4 Computation of the steady state

Combining these equations and substituting (3.13), we obtain a set of n equations:

$$\sum_C (s_i' - s_i) A d^{p-1} (1-d)^{n-p-1} \sum_j s_j \left(\sum_\beta k_{j\beta} x_\beta + \epsilon_j \right) - \tau \sum_\alpha k_{i\alpha} c_{i\alpha} = 0$$

$$(i = 1, \ldots . n). \tag{3.18}$$

Considering first the terms linear in x, we obtain a set of Dn linear homogeneous equations for the $k_{i\beta}$. Mass and momentum are conserved; we assume that the collision rules have been so chosen that there is no other conserved quantity. Then, only $D(n-D-1)$ of these equations are independent, and solutions have the form

$$k_{i\beta} = K_{0\beta} + \sum_\alpha K_{\alpha\beta} c_{i\alpha}, \tag{3.19}$$

where the $K_{0\beta}$ and $K_{\alpha\beta}$ are arbitrary constants. Using the relations (3.15) and the symmetry conditions (3.2), we find that these constants are uniquely determined and the $k_{i\beta}$ are

$$k_{i\beta} = \frac{Dd}{c^2} \sum_\alpha c_{i\alpha} T_{\alpha\beta}. \tag{3.20}$$

Substituting equation (3.20) in equation (3.17), we obtain

$$\nu_i(\mathbf{x}) - \nu_i'(\mathbf{x}) = -\frac{\tau Dd}{c^2} \sum_\alpha \sum_\beta T_{\alpha\beta} c_{i\alpha} c_{i\beta}. \tag{3.21}$$

Summing over i, and taking into account the conservation of particle number, we obtain

$$\sum_\alpha T_{\alpha\alpha} = 0. \tag{3.22}$$

Thus, the tensor $T_{\alpha\beta}$ cannot be entirely arbitrarily chosen: its trace must be zero. This is a consequence of the fact that the density has been assumed constant in space and time.

We consider next the terms independent of $\mathbf{x}$ in equation (3.18). We obtain a set of n linear inhomogeneous equations:

$$\sum_C (s_i' - s_i) A d^{p-1} (1-d)^{n-p-1} \sum_j s_j \epsilon_j - \frac{\tau Dd}{c^2} \sum_\alpha \sum_\beta T_{\alpha\beta} c_{i\alpha} c_{i\beta} = 0$$

$$(i = 1, \ldots, n). \tag{3.23}$$

Only $n - D - 1$ of them are independent. Combining with equation (3.14), we have a system of n equations to solve for the n unknowns ϵ_j. It might seem at first view that this system cannot be explicitly solved in the general case. However, we can again conjecture that the anisotropy (represented by the ϵ_i) is proportional to the propagative excitation, i.e., that

$$\epsilon_i = -\lambda \frac{\tau Dd}{c^2} \sum_\alpha \sum_\beta T_{\alpha\beta} c_{i\alpha} c_{i\beta} \tag{3.24}$$

where λ is a constant. It turns out that the conjecture is true and that (3.24) is indeed a solution of the equations, with λ given by

$$\frac{1}{\lambda} = \frac{D}{D-1} \sum_C (s_i - s_i') A d^{p-1} (1-d)^{n-p-1} \sum_j s_j \cos^2 \theta_{ij}, \tag{3.25}$$

where i is an arbitrarily chosen direction and θ_{ij} is the angle between velocities c_i and c_j. A detailed proof of this result is given in Appendix A. The steady state is now fully determined. It is given by equations (2.3), (3.13), (3.20), and (3.24):

$$N_i = d + \frac{Dd}{c^2} \sum_\alpha \sum_\beta T_{\alpha\beta} c_{i\alpha} x_\beta - \lambda \frac{\tau Dd}{c^2} \sum_\alpha \sum_\beta T_{\alpha\beta} c_{i\alpha} c_{i\beta}. \tag{3.26}$$

The non-dimensional factor λ has a simple physical meaning. We define the post-collision anisotropy terms ϵ_i' by

$$\nu_i' = \sum_\beta k_{i\beta} x_\beta + \epsilon_i'. \tag{3.27}$$

From equations (3.21) and (3.24), we deduce

$$\epsilon_i' = (1-\lambda) \frac{\tau Dd}{c^2} \sum_\alpha \sum_\beta T_{\alpha\beta} c_{i\alpha} c_{i\beta}, \tag{3.28}$$

or

$$\epsilon_i' = \left(1 - \frac{1}{\lambda}\right) \epsilon_i. \tag{3.29}$$

In other words, collisions damp a fraction $1/\lambda$ of the anisotropy.

We remark also that equation (3.26) can be written

$$N_i = d + \frac{Dd}{c^2} \sum_\alpha \sum_\beta T_{\alpha\beta} c_{i\alpha} (x_\beta - \lambda\tau c_{i\beta}), \tag{3.30}$$

which could lead to an interpretation of λ as a *mean-free path*, expressed in link lengths.

3.5 Momentum flux

We compute the momentum flux across a "surface" of dimension $D-1$, of measure σ, centered in x, perpendicular to a unit vector v. This surface is assumed to be large in relation to link length: $\sigma \gg l^{D-1}$. Links parallel to c_i which cross the surface correspond to destination nodes which lie in a volume $\sigma\tau c_i \cdot v$. The number of these nodes is $\sigma\tau f c_i \cdot v$. The average position of a destination node is $x + \tau c_i/2$; therefore, the probability of existence of a particle on a link crossing the surface is on the average

$$N_i(x + \frac{1}{2}\tau c_i) = N_i(x) + \frac{1}{2}\tau c_i \cdot \nabla N_i = N_i(x) + \frac{1}{2}\tau k_i \cdot c_i, \tag{3.31}$$

according to equation (3.13). (Note that this could be written more symmetrically: $[N_i(x) + N_i'(x)]/2$.) Thus, the number of particles crossing the surface in the direction v during the time τ is

$$\sigma\tau f(c_i \cdot v) \left[N_i(x) + \frac{1}{2}\tau k_i \cdot c_i\right]. \tag{3.32}$$

Each particle carries a momentum mc_i. Finally, we sum on i and we divide by τ and σ to obtain the desired flux per unit surface:

$$fm \sum_i c_i(c_i \cdot v) \left[N_i(x) + \frac{1}{2}\tau k_i \cdot c_i\right]. \tag{3.33}$$

Taking for v the axis directions, we obtain the components of the tensorial flux:

$$F_{\gamma\varsigma} = fm \sum_i c_{i\gamma} c_{i\varsigma} \left[N_i(x) + \frac{1}{2}\tau k_i \cdot c_i)\right]. \tag{3.34}$$

We substitute the values of N_i and k_i given by equations (3.26) and (3.20):

$$F_{\gamma\varsigma} =$$

$$fm \sum_i c_{i\gamma} c_{i\varsigma} \left[d + \frac{Dd}{c^2} \sum_\alpha \sum_\beta T_{\alpha\beta} c_{i\alpha} x_\beta - \frac{\tau Dd}{c^2} \left(\lambda - \frac{1}{2}\right) \sum_\alpha \sum_\beta T_{\alpha\beta} c_{i\alpha} c_{i\beta}\right].$$

$$\tag{3.35}$$

Using equations (3.2) and (3.10), we obtain

$$F_{\gamma\varsigma} =$$

$$\frac{\rho c^2}{D}\delta_{\gamma\varsigma} - \frac{\rho\tau c^2}{D+2}\left(\lambda - \frac{1}{2}\right)\sum_\alpha\sum_\beta T_{\alpha\beta}(\delta_{\alpha\beta}\delta_{\gamma\varsigma} + \delta_{\alpha\gamma}\delta_{\beta\varsigma} + \delta_{\alpha\varsigma}\delta_{\beta\gamma}) \quad (3.36)$$

or, using equation (3.22):

$$F_{\gamma\gamma} = \qquad \frac{\rho c^2}{D} - \frac{2\rho\tau c^2}{D+2}\left(\lambda - \frac{1}{2}\right)T_{\gamma\gamma},$$

$$F_{\gamma\varsigma} = \quad -\frac{\rho\tau c^2}{D+2}\left(\lambda - \frac{1}{2}\right)(T_{\gamma\varsigma} + T_{\varsigma\gamma}) \qquad (\gamma \neq \varsigma). \tag{3.37}$$

Equation (3.37b) shows that the shear velocity is

$$\eta = \frac{\rho\tau c^2}{(D+2)}\left(\lambda - \frac{1}{2}\right). \tag{3.38}$$

Note that the kinematic viscosity η/ρ can be interpreted, as in classical gases, as the product of the particle velocity c by a mean-free path. This mean-free path equals the link length $l = \tau c$ multiplied by the dimensionless constant $(\lambda - 1/2)/(D+2)$.

Substituting equation (3.25), we obtain the explicit formula

$$\eta = \frac{\rho\tau c^2}{D+2}\left[\frac{D-1}{D\sum_C(s_i - s_i')Ad^{p-1}(1-d)^{n-p-1}\sum_j s_j\cos^2\theta_{ij}} - \frac{1}{2}\right] \tag{3.39}$$

i in this formula is an arbitrarily chosen direction.

4.   Minimization of the viscosity

We show now that the formula (3.39) can be written in a different way, involving a sum of squared quantities. As a consequence, the viscosity is always positive. The new formula provides also a practical recipe for choosing collision rules which minimize the viscosity.

The right-hand side of equation (3.25) is independent of the chosen i. We can therefore sum over i and divide by n:

$$\frac{1}{\lambda} = \frac{D}{(D-1)n}\sum_C Ad^{p-1}(1-d)^{n-p-1}\sum_i(s_i - s_i')\sum_j s_j\cos^2\theta_{ij}. \tag{4.1}$$

This can be written

$$\begin{aligned}\frac{1}{\lambda} &= \frac{D}{(D-1)nc^4}\sum_C Ad^{p-1}(1-d)^{n-p-1}\\ &\quad \sum_i(s_i - s_i')\sum_j s_j\sum_\alpha c_{i\alpha}c_{j\alpha}\sum_\beta c_{i\beta}c_{j\beta}\\ &= \frac{D}{(D-1)nc^4}\sum_C Ad^{p-1}(1-d)^{n-p-1}\sum_\alpha\sum_\beta\\ &\quad \sum_i(s_i - s_i')c_{i\alpha}c_{i\beta}\sum_j s_j c_{j\alpha}c_{j\beta}.\end{aligned} \tag{4.2}$$

We define

$$X_{\alpha\beta} = \sum_i s_i c_{i\alpha} c_{i\beta}. \tag{1.3}$$

The tensor $X_{\alpha\beta}$ is the *second-order momentum* of the input state. Note that it is symmetrical: $X_{\alpha\beta} = X_{\beta\alpha}$. Similarly,

$$X'_{\alpha\beta} = \sum_i s'_i c_{i\alpha} c_{i\beta} \tag{4.4}$$

is the second-order momentum of the output state. Equation (4.2) becomes

$$\frac{1}{\lambda} = \frac{D}{(D-1)nc^4} \sum_C A d^{p-1} (1-d)^{n-p-1} \sum_\alpha \sum_\beta (X_{\alpha\beta} - X'_{\alpha\beta}) X_{\alpha\beta}. \tag{4.5}$$

The trace of $X_{\alpha\beta}$ is

$$\sum_\alpha X_{\alpha\alpha} = \sum_\alpha \sum_i s_i c_{i\alpha}^2 = \sum_i s_i c^2 = pc^2. \tag{4.6}$$

We split $X_{\alpha\beta}$ into isotropic and anisotropic parts:

$$X_{\alpha\beta} = \frac{pc^2}{D} \delta_{\alpha\beta} + Y_{\alpha\beta}. \tag{4.7}$$

The tensor $Y_{\alpha\beta}$ has a null trace:

$$\sum_\alpha Y_{\alpha\alpha} = 0. \tag{4.8}$$

Similarly, we write

$$X'_{\alpha\beta} = \frac{pc^2}{D} \delta_{\alpha\beta} + Y'_{\alpha\beta}. \tag{4.9}$$

Substituting into equation (4.5) and going over to the more detailed collision notation (see equation (2.18)), we obtain

$$\frac{1}{\lambda} = \frac{D}{(D-1)nc^4} \sum_s \sum_{s'} A(s;s') d^{p-1} (1-d)^{n-p-1} \sum_\alpha \sum_\beta (Y_{\alpha\beta} - Y'_{\alpha\beta}) Y_{\alpha\beta}.$$

We introduce the quantities

$$\mu_1 = \frac{D}{2(D-1)nc^4} \sum_s \sum_{s'} A(s;s') d^{p-1} (1-d)^{n-p-1} \sum_\alpha \sum_\beta Y_{\alpha\beta}^2,$$

$$\mu_2 = \frac{D}{2(D-1)nc^4} \sum_s \sum_{s'} A(s;s') d^{p-1} (1-d)^{n-p-1} \sum_\alpha \sum_\beta Y_{\alpha\beta}'^2, \tag{4.10}$$

$$\mu_3 = \frac{D}{4(D-1)nc^4} \sum_s \sum_{s'} A(s;s') d^{p-1} (1-d)^{n-p-1} \sum_\alpha \sum_\beta (Y_{\alpha\beta} - Y'_{\alpha\beta})^2,$$

$$\mu_4 = \frac{D}{4(D-1)nc^4} \sum_s \sum_{s'} A(s;s') d^{p-1} (1-d)^{n-p-1} \sum_\alpha \sum_\beta (Y_{\alpha\beta} + Y'_{\alpha\beta})^2.$$

Equation (4.10) can be written

$$\frac{1}{\lambda} = \mu_1 - \mu_2 + 2\mu_3, \tag{4.11}$$

or

$$\frac{1}{\lambda} = 3\mu_1 + \mu_2 - 2\mu_4. \tag{4.12}$$

μ_1 and μ_2 are computed in Appendix B; their values are simply

$$\mu_1 = \mu_2 = \frac{1}{2}. \tag{4.13}$$

From equations (4.12) and (4.13), we then have

$$\frac{1}{2\lambda} = \mu_3 = 1 - \mu_4. \tag{4.14}$$

Substituting in equation (3.38), we obtain

$$\eta = \frac{\rho \tau c^2}{2(D+2)} \frac{\mu_4}{1 - \mu_4}. \tag{4.15}$$

μ_3 and μ_4 cannot be negative, as shown by their expressions (4.11c) and (4.11d). Therefore, from equation (4.15):

$$0 \leq \mu_4 \leq 1. \tag{4.16}$$

It follows that *the viscosity η is positive of zero.* μ_4 is a dimensionless number, lying between 0 and 1, which characterizes the viscosity of the lattice gas; it might be called the *viscosity index.*

The limiting case $\mu_4 = 1$ corresponds to an infinite viscosity. It requires $\mu_3 = 0$, or, as shown by equation (11c), $Y'_{\alpha\beta} = Y_{\alpha\beta}$ for all collisions: the second-order momentum must be invariant in collisions. This happens in particular in the trivial case where there are no proper collisions (i.e., the velocities remain unchanged during the collision phase).

The other limiting case, $\mu_4 = 0$, is more interesting: It corresponds to zero viscosity. For this, we must have for every collision:

$$Y_{\alpha\beta} + Y'_{\alpha\beta} = 0. \tag{4.17}$$

This rule has a simple geometrical interpretation. Let us call *total state* the sum of the input and output states; the rule is then: *for every collision, the second-order momentum of the total state must be isotropic.* (Note that "sum" is taken in the algebraic sense: if a velocity exists both in the input and in the output state, it must be counted twice in the computation of the second-order momentum). A collision which satisfies this condition will be called a *perfect collision.*

In practice, not all collisions can be perfect. In particular, when there is only one velocity ($p = 1$), the output state is necessarily identical to the input state since the first-order momentum must be conserved. On the other hand, the second-order momentum is anisotropic. The quantity

(4.18) is non-zero. We can therefore strengthen our result. *The viscosity is always positive.*

We remark that for any input state with $p = n/2$ (i.e., with exactly one half of the incoming links occupied by particles) a perfect collision rule can be devised as follows:

1. Take the dual of the input state (i.e. replace particles by "holes" and vice versa);

2. Effect a symmetry with respect to the origin. In other words, call $\sigma(i)$ the index corresponding to the opposite velocity (i.e., $c_{\sigma(i)} = -c_i$); then, the rule is: $s'_{\sigma(i)} = 1 - s_i$. It is easily seen that the number of particles and the first-order momentum are preserved, while the second-order momentum of the total state is isotropic. For instance, in the FHP lattice, we obtain the rules for triple collisions shown in figure 6. In the first two cases, the velocities are unchanged. The third case is the "head-on collision with spectator" [11].

In order to minimize the viscosity, one must first sort the states into subsets according to particle number and momentum, and then pair the states so as to minimize the quantities (4.18). Roughly speaking, each collision should replace the input state by an output state whose second-order momentum is as much as possible "symmetrical with respect to isotropy." Work is in progress to apply this recipe to the four-dimensional, 24-velocity lattice gas proposed by d'Humières, Lallemand, and Frisch [1,7]. Present results indicate that this fine-tuning of the collision rules can lower the viscosity by a factor between 3 and 4 with respect to a collision algorithm in which the output state is randomly chosen (among all states having the same particle number and momentum as the input state). More generally, it seems likely that the larger the number n of velocities, the more closely the limit $\eta = 0$ can be approached.

Appendix A.

We prove here that equation (3.24), with λ given by (3.25), is a solution of the equations (3.14) and (3.23). The first part is simple: using equations (3.2) and (3.22), we immediately find that the equations (3.14) are verified.

We introduce the abbreviated notation

$$Q_i = \sum_\alpha \sum_\beta T_{\alpha\beta} c_{i\alpha} c_{i\beta}. \tag{A.1}$$

Substituting equation (3.24) in (3.23), we find that the equations to be satisfied are

$$Q_i = \lambda \sum_C (s_i - s'_i) A d^{p-1} (1-d)^{n-p-1} \sum_j s_j Q_j$$

$$(i = 1, \ldots, n). \tag{A.2}$$

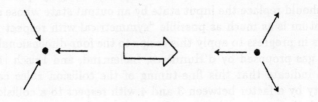

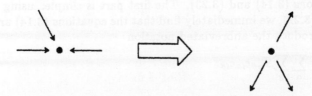

Figure 6: "Perfect" rules for triple collisions in the FHP lattice.

We consider now a particular value of i, and the subgroup G_i of G consisting of all isometries in which c_i is invariant. We divide the collisions into equivalence classes according to the following rule: two collisions belong to the same class W if there exists an isometry of G_i which maps one into the other. More precisely, two collisions $(s; s')$ and $(\tilde{s}; \tilde{s}')$ belong to the same class if there exists an isometry $I \in G_i$ which maps s into $\tilde{s}$ and s' into $\tilde{s}'$.

We decompose now the sum over collisions in equation (A.1): we sum first over the members of each class W, and then over the classes:

$$Q_i = \lambda \sum_W \sum_{C \in W} (s_i - s_i') A d^{p-1} (1-d)^{n-p-1} \sum_j s_j Q_j. \tag{A.3}$$

Because of symmetry assumption (4) (see beginning of section 3), A is constant inside a class W. s_i, s_i', p are also constant inside a class, and therefore equation (A.3) can be written

$$Q_i = \lambda \sum_W (s_i - s_i') A d^{p-1} (1-d)^{n-p-1} \sum_{C \in W} \sum_j s_j Q_j. \tag{A.4}$$

Consider a particular class W and a particular collision $C^* = (s^*; s^{prime*}) \in W$. We call G_i^* the subgroup of G_i made of isometries which leave C^* invariant. For any collision $C \in W$, the isometries which map C^* into C form a left coset of G_i^*; their number is $|G_i^*| = |G_i|/|W|$. So, instead of summing on collisions in W, we can sum on isometries in G_i:

$$\sum_{C \in W} \mathcal{F}(C) = \frac{|W|}{|G_i|} \sum_{I \in G_i} \mathcal{F}[I(C^*)], \tag{A.5}$$

where $\mathcal{F}$ is an arbitrary function. Thus, equation (A.4) becomes

$$Q_i = \lambda \sum_W (s_i - s_i') A d^{p-1} (1-d)^{n-p-1} \frac{|W|}{|G_i|} \sum_{I \in G_i} \sum_j s_j^* Q_{I(j)}. \tag{A.6}$$

We have used here the fact that each input velocity of C is the image of an input velocity of C^*. Substituting equation (A.1) in the right-hand side and changing the summation order, we have

$$Q_i = \lambda \sum_W (s_i - s_i') A d^{p-1} (1-d)^{n-p-1} \frac{|W|}{|G_i|} \sum_j s_j^* \sum_\alpha \sum_\beta T_{\alpha\beta} \sum_{I \in G_i} c_{I(j)\alpha} c_{I(j)\beta}. \tag{A.7}$$

We take a coordinate system such that the x_1 axis is parallel to c_i. Then $c_{I(j)1} = c_{j1}$ and equation (A.7) takes the form

$$Q_i = \lambda \sum_W (s_i - s_i') A d^{p-1} (1-d)^{n-p-1} \frac{|W|}{|G_i|} \sum_j s_j^*$$

$$\left[T_{11} c_{j1}^2 |G_i| + c_{j1} \sum_{\beta=2}^{D} T_{1\beta} \sum_{I \in G_i} c_{I(j)\beta} \right.$$

$$+c_{j1}\sum_{\alpha=2}^{D}T_{\alpha 1}\sum_{I\in G_i}c_{I(j)\alpha}+\sum_{\alpha=2}^{D}\sum_{\beta=2}^{D}T_{\alpha\beta}\sum_{I\in G_i}c_{I(j)\alpha}c_{I(j)\beta}\Bigg]. \tag{A.8}$$

Using equations (3.5), (3.6), and (3.8), we obtain

$$Q_i = \lambda\sum_W(s_i - s_i')Ad^{p-1}(1-d)^{n-p-1}|W|\sum_j s_j^*$$
$$\left[T_{11}c^2\cos^2\theta_{ij} + \sum_{\alpha=2}^{D}T_{\alpha\alpha}\frac{c^2\sin^2\theta_{ij}}{D-1}\right] \tag{A.9}$$

or, using equation (3.22)

$$Q_i =$$

$$\lambda\sum_W(s_i - s_i')Ad^{p-1}(1-d)^{n-p-1}|W|\sum_j s_j^* T_{11}c^2\frac{D\cos^2\theta_{ij}-1}{D-1}. \tag{A.10}$$

Summing over all collisions $C^* \in W$, dividing by $|W|$, and dropping the asterisks, we obtain

$$Q_i =$$

$$\lambda\sum_W(s_i - s_i')Ad^{p-1}(1-d)^{n-p-1}\sum_{C\in W}\sum_j s_j T_{11}c^2\frac{D\cos^2\theta_{ij}-1}{D-1}. \tag{A.11}$$

From equation (A.1), we find

$$Q_i = T_{11}c^2. \tag{A.12}$$

We can then divide both sides by Q_i and we obtain

$$1 = \lambda\sum_C(s_i - s_i')Ad^{p-1}(1-d)^{n-p-1}\sum_j s_j\frac{D\cos^2\theta_{ij}-1}{D-1}. \tag{A.13}$$

We remark that the tensor $T_{\alpha\beta}$ has entirely disappeared: the equation (A.13) depends only on the collision rules. As a consequence, because of symmetry assumptions (2) and (4) (section 3), this equation is the same for all directions i. The conjecture is verified, and we have found a solution for the probabilities N_i.

The rightmost term $1/(D-1)$ in equation (A.13) can be deleted because it gives a null contribution; this is easily seen by summing over i. We obtain then for λ the equation (3.25).

Appendix B.

We evaluate here μ_1 and μ_2 defined by equations (4.11a) and (4.11b). μ_1 does not depend on s'; therefore, its value is, according to equation (2.16):

$$\mu_1 = \frac{D}{2(D-1)nc^4} \sum_s d^{p-1}(1-d)^{n-p-1} \sum_\alpha \sum_\beta \left[X_{\alpha\beta} - \frac{pc^2}{D}\delta_{\alpha\beta} \right]^2 =$$

$$\frac{D}{2(D-1)nc^4} \sum_s d^{p-1}(1-d)^{n-p-1} \sum_\alpha \sum_\beta$$

$$\left[\sum_i s_i\left(c_{i\alpha}c_{i\beta} - \frac{c^2}{D}\delta_{\alpha\beta} \right) \right]\left[\sum_j s_j\left(c_{j\alpha}c_{j\beta} - \frac{c^2}{D}\delta_{\alpha\beta} \right) \right]$$

$$= \frac{D}{2(D-1)nc^4} \sum_\alpha \sum_\beta$$

$$\sum_i \left(c_{i\alpha}c_{i\beta} - \frac{c^2}{D}\delta_{\alpha\beta} \right) \sum_j \left(c_{j\alpha}c_{j\beta} - \frac{c^2}{D}\delta_{\alpha\beta} \right) \sum_s s_i s_j d^{p-1}(1-d)^{n-p-1}. \quad \text{(B.1)}$$

If $i = j$, the sum over s is

$$\sum_s s_i d^{p-1}(1-d)^{n-p-1} = \sum_{s_1=0}^{1} \cdots \sum_{s_n=0}^{1} s_i d^{p-1}(1-d)^{n-p-1}, \quad \text{(B.2)}$$

or, using $p = \sum_k s_k$:

$$\frac{1}{d(1-d)} \sum_{s_1=0}^{1} d^{s_1}(1-d)^{1-s_1} \cdots \sum_{s_n=0}^{1} d^{s_n}(1-d)^{1-s_n} s_i = \frac{1}{1-d} \quad \text{(B.3)}$$

(the sum over s_i gives d; the sums over s_k for $k \neq i$ give 1).

If $i \neq j$, the sum over s in equation (1) is

$$\sum_s s_i s_j d^{p-1}(1-d)^{n-p-1} = \frac{d}{1-d}, \quad \text{(B.4)}$$

by a similar computation. Equation (1) becomes

$$\mu_1 = \frac{D}{2(D-1)nc^4} \sum_\alpha \sum_\beta \left(\frac{1}{1-d}\sum_i \sum_{j=i} + \frac{d}{1-d}\sum_i \sum_{j\neq i} \right)$$

$$\left(c_{i\alpha}c_{i\beta} - \frac{c^2}{D}\delta_{\alpha\beta} \right)\left(c_{j\alpha}c_{j\beta} - \frac{c^2}{D}\delta_{\alpha\beta} \right). \quad \text{(B.5)}$$

The summations over i and j can also be written

$$\left(\sum_i \sum_{j=i} + \frac{d}{1-d}\sum_i \sum_j \right). \quad \text{(B.6)}$$

The second term vanishes, and there remains

$$\mu_1 = \frac{D}{2(D-1)nc^4} \sum_\alpha \sum_\beta \sum_i \left(c_{i\alpha} c_{i\beta} - \frac{c^2}{D} \delta_{\alpha\beta} \right)^2. \tag{B.7}$$

For $\alpha = \beta$, the sum over i is, using equation (3.2):

$$\sum_i c_{i\alpha}^4 - 2\frac{c^2}{D} \sum_i c_{i\alpha}^2 + n\frac{c^4}{D^2} = \frac{3nc^4}{D(D+2)} - \frac{nc^4}{D^2}, \tag{B.8}$$

and for $\alpha \neq \beta$:

$$\sum_i c_{i\alpha}^2 c_{i\beta}^2 = \frac{nc^4}{D(D+2)}. \tag{B.9}$$

Finally, we sum over α and β, and equation (B.5) reduces to

$$\mu_1 = \frac{1}{2}. \tag{B.10}$$

We consider now the quantity μ_2. It is independent of s. Using the assumption of semi-detailed balancing (2.17), we can sum over s and eliminate A. A computation similar to the above one gives then

$$\mu_2 = \frac{1}{2}. \tag{B.11}$$

Acknowledgements

My thanks go to Uriel Frisch, who with unfailing patience educated me on the subject of lattice gases. I have also benefited from discussions with D. d'Humières, J. L. Oneto, Y. Pomeau, and J. P. Rivet.

References

[1] U. Frisch, D. d'Humières, B. Hasslacher, P. Lallemand, Y. Pomeau, and J. P. Rivet, "Lattice gas hydrodynamics in two and three dimensions", *Complex Systems*, **1** (1987) 648.

[2] D. d'Humières and P. Lallemand, "Lattice gas automata for fluid mechanics", *Physica*, **140A** (1986) 337.

[3] D. d'Humières and P. Lallemand, *Complex Systems*, **1** (1987) 598.

[4] U. Frisch, B. Hasslacher, and Y. Pomeau, "Lattice-gas automata for the Navier-Stokes equation", *Phys. Rev. Lett.*, **56** (1986) 1505.

[5] J. P. Rivet and U. Frisch, "Automates sur gaz de réseau dans l'approximation de Boltzmann", *C. R. Acad. Sci. Paris*, **302** (1986) 267.

[6] S. Wolfram, "Cellular automaton fluids 1: Basic theory", *J. Stat. Phys.*, **45** (1986) 471.

[7] D. d'Humières, P. Lallemand, and U. Frisch, "Lattice gas models for 3D hydrodynamics", *Europhys. Lett.*, **2** (1986) 291.

[8] J. Hardy and Y. Pomeau, "Thermodynamics and hydrodynamics for a modelled fluid", *J. Math. Phys.*, **13** (1972) 1042.

[9] J. Hardy, Y. Pomeau, and O. de Pazzis, "Time evolution of a two-dimensional model system: I. Invariant states and time correlation functions", *J. Math. Phys.*, **14** (1973) 1746.

[10] J. Hardy, O. de Pazzis, and Y. Pomeau, "Molecular dynamics of a classical lattice gas: Transport properties and time correlation functions", *Phys. Rev. A*, **13** (1976) 1949.

[11] D. Levermore, private communication.

Computer Hardware Papers

A. Despain,† C. E. Max,† G. Doolen,‡ and B. Hasslacher‡
†JASON, Mitre Corporation and ‡ Los Alamos National Laboratory

Prospects for a Lattice-Gas Computer

A two-day workshop was held in June of 1988 to discuss the feasibility of designing and building a large computer dedicated to lattice-gas cellular automata. The primary emphasis was on applications for modeling Navier-Stokes hydrodynamics. The meeting had two goals: (1) to identify those theoretical issues which would have to be addressed before the hardware implementation of a lattice-gas machine would be possible; and (2) to begin to evaluate alternative architectures for a dedicated lattice-gas computer. This brief paper contains a summary of the main issues and conclusions discussed at the workshop.

INTRODUCTION

This note summarizes a two-day workshop held in La Jolla, California in June of 1988. The workshop was co-sponsored by the Center for Nonlinear Studies at the Los Alamos National Laboratory, and by the JASON group, Mitre Corporation. The purpose of the workshop was to identify, define, and begin to resolve substantive issues which must be addressed before a lattice-gas computer can be implemented in hardware.

The workshop attendees were:

- George Adams, Purdue University
- Gary Doolen, Los Alamos National Laboratory
- Paul Frederickson, NASA Ames, RIAC project
- Castor Fu, Stanford University
- Brosl Hasslacher, Los Alamos National Laboratory
- Fung F. Lee, Stanford University
- Norman Margolus, MIT Laboratory for Computer Science
- Tsutomu Shimomura, Los Alamos National Laboratory
- Tom Toffoli, MIT Laboratory for Computer Science

and the following members of the JASON group:

- Kenneth Case, University of California at San Diego
- Alvin Despain, University of California at Berkeley
- Freeman Dyson, Institute for Advanced Study
- Michael Freedman, University of California at San Diego
- Claire Max, Lawrence Livermore National Laboratory
- Oscar Rothaus, Cornell University

The primary emphasis of the workshop was on the use of cellular automata for simulations of three-dimensional incompressible Navier-Stokes hydrodynamics. Within this context, there are two types of applications for which a special-purpose computer might offer important potential advantages over conventional numerical hydrodynamics techniques implemented on general-purpose supercomputers:

1. Studies of flows with complex boundary conditions. For example, one might look at a boundary layer and study various techniques that have been suggested for drag-reduction and boundary-layer modification.

2. Studies of three-dimensional incompressible flows at high Reynolds numbers. These could include studies of the onset of fluid turbulence, free-boundary problems (such as ship wakes and drag), or the combination of hydrodynamics and simple chemical-reaction systems.

The issues discussed at the workshop fall into three general categories: theory, computer simulation, and hardware.

THEORETICAL ISSUES

The most prevalent use of cellular automata for modeling hydrodynamics has been the so-called lattice gas. In this approach, one follows the motions of many individual particles which interact via given collision laws at fixed lattice sites. The individual particles are allowed to have only one (or at most a few) discrete speeds relative to the grid of lattice sites. The hydrodynamic limit is regained by averaging over a large number of these discrete particles, to obtain the first few moments of their distribution function.

In two spatial dimensions, the properties of possible sets of collision rules for the particles and lattice geometries for the sites are now reasonably well understood. There are two practical ways to represent a given rule set: via a look-up table which enumerates all the possible incoming and outgoing configurations, or via an algorithm or computation which generates the rules anew at each timestep and each collision site. However, in three spatial dimensions, the possible rule sets are far more complicated, and there are many unsolved questions regarding appropriate collision rules and their efficient execution. For maximum efficiency, a special-purpose lattice-gas computer should probably contain a hard-wired implementation of a particular rule set. However, the general consensus at the workshop was that there is not yet a sufficient understanding of rule sets that have been proposed for three spatial dimensions to settle upon an optimum one for hardware implementation.

Important issues that remain to be solved concerning collision rules for three-dimensional hydrodynamics are the following:

1. What is the "best" rule set to use for modeling three-dimensional hydrodynamics?

   a. How does the choice of this "best" rule set change with the type of application one wants to solve? For example, are some rule sets better for studies of boundary-layer effects or free-boundary problems, while others are optimum for studying the onset of turbulence at high Reynolds number?

   b. How can rules be "tuned" to get optimum results for given problem parameters? For example, how can one optimize for high Reynolds number, or for specific types of boundary conditions?

2. Rules for lattice gases representing three-dimensional hydrodynamics tend to be very complicated. One way to implement them computationally is to use look-up tables, but these become very large. If there are n bits at each lattice site, then there are 2^n table entries. For example, the 24-bit model requires 16 million entries. How can this large number of rules be reduced by "factoring" or "grouping" them, to reduce the size of the rule representation in the look-up table? What is the fundamental dimension of the rule set?

3. In several proposed rule sets, one has to choose whether the same collisio will always have the same outcome, or whether one will implement a randomization process within the rule set to "mix up" the collision outcomes. The addition

of an explicit randomization procedure is expensive computationally. Under what circumstances can one rely on the inherently high frequency of particle collisions to achieve randomization, so that it does not have to be explicitly included in the rule engine?

4. A related question concerns the desirability of adding a "collision bit" to the algorithm. This is an additional bit determining whether a particle will or will not undergo a collision at the next lattice site that it reaches, if all other conditions for a collision at that site are satisfied. If all particles undergo collisions whenever they can (no collision bit), one obtains a more "collisional" rule set, leading to the potential for attaining higher Reynolds numbers. Are there circumstances in which a less collisional rule set would be desirable?

5. What advantages are there to using nonperiodic tiling or quasilattices for modeling three-dimensional hydrodynamics, as compared with the so-called four-dimensional schemes or other periodic tiling schemes?

6. Is there a lattice gas analogue for adaptive-mesh hydrodynamic techniques, so that greater spatial resolution can be achieved in regions where it is needed? Can sub-grid scaling rules be derived to extend the spatial resolution of the lattice gas method?

7. What physical laws or partial differential equations to the various rule sets represent? Can the differences between the Navier-Stokes equations and the lattice-gas implementations be systematically understood?

   a. Under what conditions (limits on the Mach number, particle density, Reynolds number) does the lattice-gas model with a given rule set reduce to three-dimensional Navier-Stokes hydrodynamics?

   b. Given a set of physical constraints, can an algorithm be developed that will systematically generate a corresponding lattice-gas rule set?

   c. Each given rule set implies a particular functional form for the viscosity as a function of density. Given that the density is nearly constant in space for incompressible flows with Mach numbers small compared to unity, does it matter whether or not the density dependence of the viscosity law is physical?

   d. It has been suggested that the nonphysical function $g(\rho)$ appearing in front of the $u \cdot \nabla u$ term in the momentum equation can be eliminated by using rules which include two or more discrete (nonzero) velocities. Is this generally valid? Under what conditions would it be desirable to use more than one particle velocity? What is the gain in accessible Reynolds number when additional speeds are allowed? Are there advantages to these schemes that would allow lattice gases to satisfy statistics other than Fermi statistics? (The latter prevail for most currently used rules.)

In addition to the above questions concerning rules for lattice gas representations of hydrodynamics, there are a set of issues involving extensions of the cellular automata methodology to other physical models:

1. Can hydrodynamics be modeled by using cellular automata particles to represent vorticity, in analogy with finite-difference vorticity-tracking algorithms? What range of Reynolds numbers could such a technique model?

2. How practical would it be to add some simple extensions to three-dimensional lattice-gas hydrodynamics? Some extensions that would be useful include gravity or other body-forces, two or more different fluid types, or simple chemistry, such as Maxwell's equations, would require a very different algorithmic approach.

ISSUES CONCERNING NUMERICAL SIMULATIONS

The workshop participants felt that there was a need to develop "benchmark" simulation problems. These would consist of a few canonical two- and three-dimensional hydrodynamics problems, for which the numerical results of various lattice-gas models could be compared with each other, with conventional hydrodynamics simulation, and with experimental results. This is particularly important because of the fact that different lattice gas rule sets may represent different approximations to the Navier-Stokes equations (i.e., they may approach the Navier-Stokes equations in different asymptotic limits).

A parallel effort should be made to compare lattice-gas simulation results with standard analytic solutions to the Navier-Stokes equations, in cases where these are known. Possible examples are channel flow, pipe flow, Pouseille flow, Couette flow, and so forth. This has been done to a limited extent for two-dimensional lattice-gas models, but three-dimensional applications have not yet been well studied.

A different type of test of lattice-gas algorithms was thought to be important as well. One should perform the standard numerical test of increasing the grid resolution, while holding fixed all of the "physical" parameters describing the problem. The goal would be to check that the higher-resolution result is identical to that obtained with lower numerical resolution.

A final numerical simulation issue thought to be important by the workshop participants concerns how to generate adequate graphical visualizations of the results of a three-dimensional lattice-gas simulation. It was pointed out that the amount of data storage needed for a three-dimensional simulation at high Reynolds number will be very high. Therefore, thought must be given to how to integrate input-output and graphical display within the process of numerical computation itself. For the types of physical problems which one wants to address using lattice gases, it may not be adequate to obtain graphical displays of the results based entirely upon post-processing.

ISSUES CONCERNING THE HARDWARE PERFORMANCE OF A LATTICE-GAS COMPUTER

In order to focus on the issue of hardware design for a lattice-gas machine, a set of performance measures was chosen. The idea was to outline hypothetical specifications for hardware components, so that when different candidate architectures were compared with each other, they would all be making the same assumptions about the capabilities of commonly used hardware components.

Table 1 gives a rough overview of the capabilities of VLSI technology today and in five years. Using these characteristics, which are of course only approximate, one can outline the characteristics and performance of various architectures for a lattice-gas supercomputer.

There appears to be a practical limit on the total number of chips it is plausible to include in a supercomputer. Existing supercomputers have about a third of a million chips. Workshop participants hypothesized that in the future one might build supercomputers with up to a million chips. Since the total number of lattice points required for a lattice-gas computation of high-Reynolds-number, three-dimensional hydrodynamics is much larger than a million, one is led to a design in which many cellular automata lattice points are placed on each chip.

The next hardware issue is how to implement the set of collision rules. Since only a few rule sets for three spatial dimensions have been studied to date, the workshop participants felt that it was premature to choose a specific rule set for implementation in hardware. It was suggested that even after more three-dimensional rules have been studied, it would be desirable to leave flexibility in the choice of rules for the lattice-gas supercomputer. There are two reasons for this choice. First, a new and better set of collision rules for hydrodynamics might be invented at any

TABLE 1 VLSI Technology (CMOS)

	Today	In Five Years
	1 cm^2 active area	1 cm^2
	200 pins	400 pins
	1 Mbit DRAM	4 Mbit
	50K "random" transistors	200K
	10 nsec internal clock (on-chip communications)	1 nsec
	80 nsec external drive (off-chip communications)	8 nsec

time; and second, one may at some later point want to use the lattice-gas computer to study other physical models such as the mixing of two different gases, or hydrodynamics with simple chemistry.

Flexibility in the choice of rule set would have the most straightforward implementation if collision rules were executed via look-up tables. In that case, one could feed an alternative look-up table into the computer when one wanted to change rules. The difficulty with this approach is that the rules suggested to date for hydrodynamics in three spatial dimensions would require very large look-up tables, with the disadvantages that the tables would use up large amounts of memory and would be slow to compute collisions.

Thus there is a lot to be gained by understanding the symmetries underlying each proposed rule set, so that the look-up table can be collapsed into a considerably smaller amount of memory space.

The second method of implementing a rule set is to design a computational engine in hardware that would re-calculate the rules "on the fly" for each collision. The advantage of this technique relative to a look-up table will depend on the details of the chosen rule set. At present for three spatial dimensions, it seems that the look-up table approach is preferable from the point of view of speed and feasibility. In addition, the hardware rule-engine is less flexible than a table look-up approach, unless a software layer can be added to customize the rule engine for a choice of several different rule sets. Since many lattice site reside on each chip, and since off-chip communications are slower than those that remain on-chip, it seems desirable to locate on each chip the look-up tables or hardware rule engines which calculate the collision outcomes. This avoids the time delays which would occur if one had to go off-chip to calculate collision outcomes. If there are many lattice points on a chip, one may want to have many "computational nodes" on each chip. (Here a "computational node" is defined to be a look-up table or a rule engine for calculating collision outcomes.) This would avoid the time delays inherent in updating all of the lattice sites on a given chip sequentially.

Thus one must choose how to trade off the number of lattice sites which can be stored on a chip with the number of "computational nodes" that will fit on a chip. The results of this trade-off will probably vary with the specific type of rule set chosen, since the size and complexity of the "computational node" and the number of bits required for a lattice site will in general vary. In the example shown in the following table, it was decided to allocate half of the chip space to lattice sites and half to "computational nodes."

With the above discussion as background, the workshop arrived at the target performance characteristics of a hypothetical lattice-gas computer in Table 2.

TABLE 2 Target Performance Parameters

Problem Definition:

three-dimensional incompressible hydrodynamics;
flexible boundary conditions; some flexibility in the
rule set; if possible, high input-output rate

Hardware Aspects:

512 lattice sites per "computational node"
64 "computational nodes" per chip
32,000 lattice sites per chip
about a million chips total
3×10^{10} lattice sites total ($> 10^{11}$ within 5 years
10 nsec update rate (on-chip)
about 64×10^8 site updates per chip per second
about 6×10^{15} site updates/sec total
($> 6 \times 10^{16}$ site updates/sec within 5 years)

CONCLUSIONS

The workshop in La Jolla produced a considerable amount of enthusiasm about the potential of a dedicated lattice-gas computer. Preliminary estimates based on the above performance numbers suggest that such a machine could surpass the present performance of a general-purpose supercomputer such as the CRAY II (3×10^7 site updates/sec) by a factor of about 10^8 and possibly considerably more. Of course, the target machine could be expensive; first-of-a-kind supercomputers can cost from tens to a hundred million dollars. The cost of this machine is proportional to the number of chips, so a change in speed by a factor of 10 would result in a factor of 10 change in cost.

In view of the combination of large cost and high scientific potential for such a machine, it will be imperative to proceed along two parallel paths: (1) refinement of the theoretical understanding of cellular automata rules and lattices in three spatial dimensions, and (2) building of intermediate-scale hardware implementations of dedicated cellular automata computational engines, so as to gain expertise in the practical areas of architecture trade-offs and implementations. A promising example of such an intermediate engine is the CAM-8 machine recently proposed by Margolus and Toffoli, which would deliver 2×10^{10} site updates per second with 16 bits per site. Progress along both of these paths will be necessary in order to learn how to best exploit the potential of a dedicated lattice-gas supercomputer.

Norman Margolus and Tommaso Toffoli
MIT Laboratory for Computer Science, Cambridge, MA

Cellular Automata Machines

The advantages of an architecture optimized for cellular automata simulations are so great that, for sufficiently large-scale CA experiments, it becomes absurd to use any other kind of computer.

In this article we discuss cellular automata machines in general, give some illustrative examples of the use of an existing machine, and then describe a much more ambitious architecture (which is under development) suitable for extensive three-dimensional simulations.

1. INTRODUCTION

The focus of the research conducted by the Information Mechanics Group at the MIT Laboratory for Computer Science has been the study of the physical bases of computation, and the computational modeling of physics-like systems. Much of this research has involved reversible models of computation and cellular automata (CA).

In 1981, the frustrating inefficiency of conventional computer architectures for simulating and displaying cellular automata became a serious obstacle to our experimental studies of reversible cellular automata. Even using conventional components, it was clear that several orders of magnitude in performance could be gained by devising hardware which would take advantage of the predictability and locality of the updating process.

Our first prototype was a sequential machine which scanned a two-dimensional array of cells, producing new states for the cells fast enough and in the right order so that it could keep up with the beam of an ordinary raster-scan television monitor. After a few years of experimentation and refinement, arrangements were made for a version of our machine, namely CAM-6, to be produced commercially, so that it would be available to the general research community.[2,13,14]

The existence of even such small-scale CAMs (Cellular Automata Machines) has already had a direct impact on the subject of CA simulations of fluid mechanics. In informal studies of gas-like models, we found one that Yves Pomeau had previously investigated—the HPP gas.[8][1] According to Pomeau, seeing his CA running on our machine made him realize that what had been conceived primarily as a *conceptual* model could indeed be turned, by using suitable hardware, into a *computationally accessible* model, and stimulated his interest in finding CA rules which would provide better models of fluids.[6]

In fact (as we shall see below) the advantages of an architecture optimized for CA simulations are so great that, for sufficiently large-scale CA experiments, it becomes absurd to use any other kind of computer.

2. TRULY MASSIVE COMPUTATION

Cellular automata constitute a general paradigm for massively parallel computation. In CA, size and speed are decoupled—the speed of an individual cell is not constrained by the total size of the CAM. Maximum size of a CAM is limited not by any essential feature of the architecture, but by economic considerations alone. Cost goes up essentially linearly with the size of the machine, which is indefinitely extendable.

These properties of CAMs arise principally from two factors. First, in conventional computers, the cycle time of the machine is constrained by the finite propagation speed of light—the universal speed limit. The length of signal paths in the computer determines the minimum cycle time, and so there is a conflict between speed and size. In CA, cells only communicate with spatially adjacent neighbors, and so the length of signal paths is inherently independent of the number of cells in the machine. Size and speed are decoupled.

---

[1]Pomeau's result was brought to our attention by Gérard Vichniac, then working with our group.

Second, this locality permits a modular architecture: there are no addressing or speed difficulties associated with simply adding on more cells. As you add cells, you also add processors. Whether your module of space contains a separate processor for each cell or time-shares a few processors over many cells is just a technological detail. What is essential is that adding more cells doesn't increase the time needed to update the entire space—since you always add associated processors at a commensurate rate. For the foreseeable future, there are no practical technological limits on the maximum size of a simulation achievable with a fixed CAM architecture.

The reason that CA can be realized so efficiently in hardware can ultimately be traced back to the fact that they incorporate certain fundamental aspects of physical law, such as locality and parallelism. Thus the structure of these computations maps naturally onto physical implementations. It is of course exactly this same property of being physics-like that makes CA a natural tool for physical modeling (e.g., fluid behavior). Von Neumann-architecture machines emulate the way we consciously think: a single processor that pays attention to one thing at a time. CA emulate the way nature works: local operations happening everywhere at once. For certain physical simulations this latter approach seems very attractive.

3. A PROCESSOR IN EVERY CELL?

In order to maintain the advantages of locality and parallelism, CAMs should be constructed out of modules, each representing a "chunk" of space. The optimal ratio of processors to cells within each module is a compromise dictated by factors such as

- technological and economic constraints,
- the relative importance of speed versus simulation size,
- the complexity and variability of processing at each cell,
- the importance of three-dimensional simulations,
- I/O and inter-module communications needs, and
- a need for analysis capabilities of a less local nature than the updating itself.

Just to give an idea of one extreme at the fine-grained end of the spectrum, consider a machine having a separate processor for each cell, and some simple two-dimensional cellular-automaton rule built in.[2] We estimate that, with integrated-circuit technology, a machine consisting of 10^{12} cells and having an update cycle of 100 pico-seconds for the entire space will be technologically feasible within 10 years. If the same order of magnitude of hardware resources contemplated for this CAM (using the same technology) were assembled as a serial computer with a single

[2]This approach does not necessarily restrict one to a single specific application. There are simple universal rules (cf. LOGIC in [14]) which can be used to simulate any other two-dimensional rule in a local manner.

processor, the machine might require seconds rather than pico-seconds to complete a single updating of all the cells.

There are serious technological problems which must be overcome before three-dimensional machines of this maximally parallel kind will be feasible. The immediate difficulty is that our present electronic technologies are essentially two dimensional, and massive interconnection of planar arrays (or "sheets") of cells in a third dimension is difficult. In the short term, this problem can be addressed by time-sharing relatively few processors over rather large groups of cells on each sheet; this allows interconnections between sheets to also be time-shared. The architectures of the CAMs built by our group make use of this idea.

A more fundamental problem which will eventually limit the size of CAMs is heat dissipation: heat generation in a truly three-dimensional CAM will be proportional to the number of cells, and thus to the volume of the array, while heat removed must all pass through the surface of this volume. Since virtually dissipationless computation seems to be (at least in principle) possible,[1,5] this may essentially be a technological problem.

4. AN EXISTING CAM

CAM-6 is a cellular automata machine based on the decision that each space-module should have few processors and many cells. In addition to drastically reducing the number of wires needed for interconnecting modules (even in two dimensions), this allows a great deal of flexibility in each processor while still maintaining a good balance between hardware resources devoted to processing and those devoted to the storage of state-variables (i.e., cell states).

Each CAM-6 module contains 256K bits of cell-state information and eight 4K-bit look-up tables which are used as processors. Both cell-state memory and the processors are ordinary memory chips, similar to those found in any personal computer. The rest of the machine consists of a few dozen garden-variety TTL chips, and one other small memory chip used for buffering cell data as it is accessed. All of this fits on a card that plugs into a personal computer (we used an IBM-PC, because of its ubiquity) and gives a performance, in many interesting CA experiments, comparable to that of a CRAY-1.[3]

The architecture which accomplishes this is very simple.

Cell-state memory is organized as 65,536 cells in a 256×256 array, with 4 bits of state in each cell. The cell states are mapped as pixels on a CRT monitor. To

[3]For the simulation of extremely simple CA rules, without any simultaneous analysis or display processing, any computer equipped with raster-op hardware will be able to perform almost as fast as CAM-6, since this CAM is really just a specialized raster-op processor. These computers will not be able to compete as the processing becomes more sophisticated, or as we add more modules to simulate a bigger space without any slowdown.

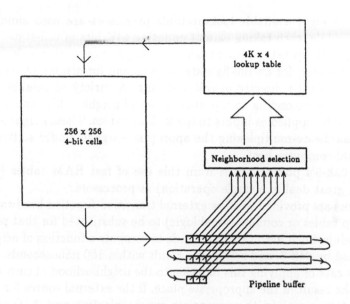

FIGURE 1   As the 4 planes are scanned, a stream of 4-bit cell values flow through a pipeline-buffer. From this buffer, 9 cell values at a time are available for use as neighbors. Of these 36 bits, up to 12 are sent to the lookup table, which produces a new 4-bit cell value.

achieve this effect, all 4 bits of a cell are retrieved in parallel (with the array being scanned sequentially in a left-to-right, top-to-bottom order). The timing of this scan is arranged to coincide with the framing format of a normal raster-scan color monitor—cell values are displayed as the electron beam scans across the CRT. Thus a complete display of the space occurs 60 times per second.

Such a memory-mapped display is very common in personal computers. What we add (see Figure 1) is the following: as the data streams out of the memory in a cyclic fashion, we do some buffering (with a pipeline that stretches over a little more than two scan lines) so that all the values in a 3×3 window (rather than a single cell at a time) are available simultaneously. We send the center cell of this window to the color monitor, to produce the display as discussed above. Subsets of the 36 bits of data contained in this window (and certain other relevant signals) are applied to the address lines of look-up tables: the resulting 4 output bits are inserted back in memory as the new state of the center cell. In essence, the set of neighbor values is used as an index into a table, which contains the appropriate responses for each possible neighborhood case. Even when a new cell state has been computed, the above-mentioned buffering scheme preserves the cell's current state as long as it is needed as a neighbor of some other cell still to be updated, so that every sixtieth of a second an updating of the entire space is completed exactly as if the transition function had been applied *to all cells in parallel.*

Four of the eight available look-up-table processors are used simultaneously within each module, each taking care of updating 64K bits of cell-state. The other four *auxiliary* look-up tables can be used, in conjunction with a color-map table and an event-counter, for on-the-fly data analysis and for display transformations. They can also be used directly in cell updating. A variety of neighborhoods are available, each corresponding to a particular set of neighbor bits and other useful signals that can be applied as inputs to the look-up tables. These neighborhoods are achieved by hardware-multiplexing the appropriate signals under software control of the personal-computer host.

Most of CAM-6's power derives from this use of fast RAM tables (which can accomplish a great deal in a single operation) as processors.

Connectors are provided to allow external transition-function hardware (such as larger look-up tables or combinational logic) to be substituted for that provided on the CAM-6 module. Such hardware only needs to compute a function of neighborhood values supplied by CAM-6, and settle on a result within 160 nanoseconds. The CAM-6 module takes care of applying this function to the neighborhood of each cell in turn and storing the result in the appropriate place. If the external source for a new cell-value is a video camera (with appropriate synchronization and A/D conversion), then CAM-6 can be used for real-time video processing.

The connectors also allow external signals to be brought into the module as neighbors, allowing the output of an external random number generator, or signals from other CAM-6 modules, to be used as arguments to the transition function. When several modules are used together, they all run in lockstep, updating corresponding cell positions simultaneously. Three-dimensional simulations can be achieved by having each module handle a two-dimensional slice, and *stacking* the slices by connecting neighbor signals between adjacent slices.

The hardware resources and usage of CAM-6 are discussed in more detail in the book *Cellular Automata Machines: a new environment for modeling*.[14] For illustrative purposes a few of the physical modeling examples discussed in this book will be surveyed in the next section.

5. PHYSICAL MODELING WITH CAM-6

CAM-6 (simply "CAM" in this section) is a general-purpose cellular automata machine. It is intended as a laboratory for experimentation, a vehicle for communication of results, and a medium for real-time demonstration.

The experiments illustrated in this section were performed with a single CAM module, with no external hardware attached.

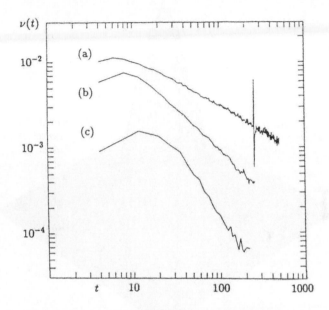

FIGURE 2   Time-correlation function $\nu(t)$ for HPP–GAS (a), TM–GAS (b), and
FHP–GAS (c).

TIME CORRELATIONS Figure 2 shows the results of some time-correlation experiments that made use of CAMs event counter.[11] In these simulations, two copies of the same system were run simultaneously, each using half of the machine. Corresponding cells of the two systems were updated at the same moment. Each run was begun by initializing both systems with identical cell values, and then holding one of the systems fixed while updating the other a few times. The systems were then updated in parallel for several thousand steps, with a constant time-delay between the two versions of the same system. Velocity-velocity autocorrelations were accumulated by comparing the values of corresponding cells as they were being updated, and sending the results of the comparisons to a counter that was read by the host computer between steps. In addition to time-correlations, space and space-time correlations could similarly be accumulated simply by introducing a spatial shift between the two systems before beginning to accumulate correlations. The three time-correlation plots refer to three different lattice gases; each data point represents the accumulation of over a billion comparisons. The whole experiment entailed accumulating about 3/4 of a trillion comparisons, and took about two-and-a-half days to run.

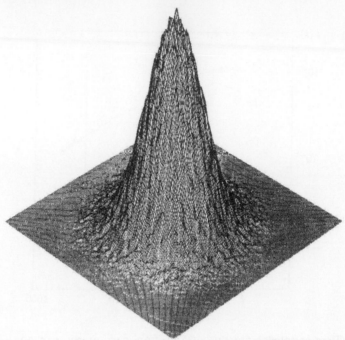

FIGURE 3 Histogram of $P(x, y; t)$—the probability that a particle of TM-GAS will be found at x, y at time t—as determined by a long series of simulation runs on CAM.

SELF-DIFFUSION Figure 3 is a histogram showing the probability that a particle of the TM-GAS lattice gas[14] started at the origin of coordinates will be found at a position (x, y) after some fixed number of steps (1024 steps in this case).[4] The data was accumulated by "marking" one of the particles (using a different cell value for it than for the rest, but not changing its dynamics) and then using the auxiliary look-up tables in combination with the event counter to track its collisions, and hence its movements. For each (x, y) value the height of the plot indicates the number of runs in which the particle ended up at that point.

Though such an experiment requires a massive amount of computation, the essential results of each run can be saved in a condensed form (as a string of collision data for a single particle) for post-analysis. In this way, a single experiment can be used for studying various kinds of correlations.

[4]This experiment was conducted by Andrea Califano.

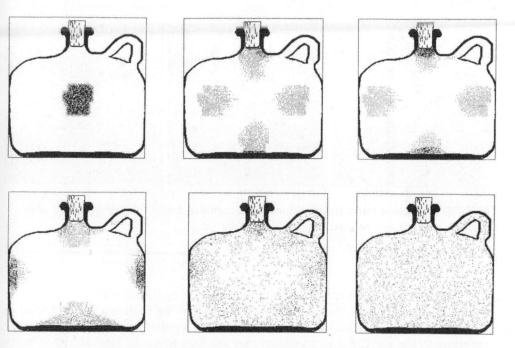

FIGURE 4 Expansion of a TM-GAS cloud in a vacuum. Repeated collisions between particles and with container's walls eventually lead to thorough thermalization.

THERMALIZATION Figure 4 shows the expansion of a clump of particles of TM-GAS. In this experiment one bit of state within each cell is devoted to indicating whether or not that cell contains a piece of the wall; this bit represents a boundary-condition *parameter* of the simulation, and doesn't change with time. Other state information in each cell is used to simulate the moving gas. Cells which don't border on a wall follow the TM-GAS rule (similar to the better known HPP-GAS rule[8]). Near a wall, the rule is modified so that particles are reflected. An arbitrary boundary can be simulated simply by drawing it—here we've drawn a jug. Initially it is evident that there are only four directions of travel available to the particles, but as the gas equilibrates this microscopic detail becomes invisible.

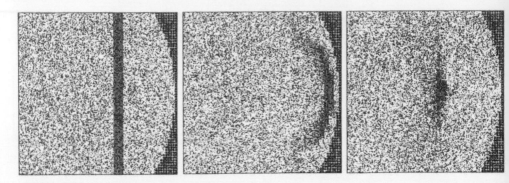

FIGURE 5   A plane pulse traveling towards a concave mirror (a) is shown right after the reflection (b) and approaching the focal point (c).

REFLECTION AND REFRACTION   Figure 5 shows exactly the same kind of simulation as Figure 4, but with a different initial condition. Here we've drawn a wall shaped as a concave mirror, and illustrate reflection of a density enhancement which is initially traveling to the right. For compactness, we use here a special kind of high-density nondissipative wave (a "soliton") that this rule supports (on a slightly larger scale, such phenomena can of course be demonstrated with ordinary near-equilibrium "acoustic" waves).

In a similar experiment, Figure 6 shows the refraction of a wave by a lens. As before, we draw our obstacle by reserving one bit of each cell's state as a spatial parameter denoting whether the cell is inside or outside the lens. Particles outside the lens follow a lattice-gas rule. Inside the lens, this rule is modified so that particles travel only half as fast as outside (this is accomplished simply by having the particles

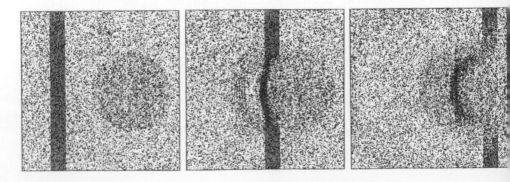

FIGURE 6   Refraction and reflection patterns produced by a spherical lens.

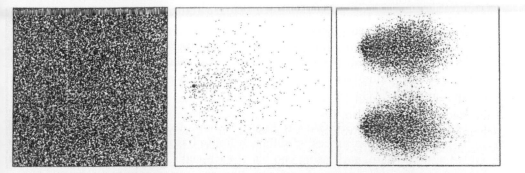

FIGURE 7 (a) The direction of drift is invisible if the fluid has uniform density. (b) Markers ejected by a smokestack diffuse in the fluid. (c) On a larger-scale simulation, the streamlines start becoming visible.

move only during half of the steps). Rules that depend on time in such a manner are provided for in CAMs hardware by supplying "pseudo-neighbor" signals that can be seen simultaneously by every cell as part of its neighborhood, and can be changed between steps under software control.

TRACING A FLOW Figure 7 illustrates an experiment in which smoke is used to trace the flow of a lattice gas. Frame (a) shows a lattice gas with a net drift to the right—this is not evident if we don't color the particles to indicate their velocities. Frame (b) shows the diffusion of particles released from a single point. This source is implemented in the same manner as the mirrors and lenses discussed previously—we mark the cells that are to be sources, and follow a different rule there. The "smoke" particles released from this source are colored differently from the other particles; however, the dynamics is "color-blind," and treats them just as ordinary gas particles. By looking only at these diffusing smoke particles, one can immediately see their collective net drift. Frame (c) shows the same phenomenon as (b), but using a space 16 times larger (1024×1024 rather than 256×256). Since the width of the diffusion pattern is proportional to $\sqrt{t}$, whereas the net distance a particle drifts is proportional to t, the drift becomes more and more evident as the scale is increased.

The larger cellular automaton shown in that last frame was simulated by a single CAM module,[5] using a technique called *scooping*. The 1024×1024 array of cells resides in the host computer's memory, and CAMs internal 256×256 array is used as a cache: this is loaded with a portion of the larger array, updated for a couple of dozen steps, and then stored back; the process is repeated on the next portion, until all of the larger array has been updated. Since scooping entails some overhead (data must be transferred between main memory and cache, and data

---

[5]This experiment was conducted by Tom Cloney.

at the edges of the cache—where some of the neighbors are not visible—must be recomputed in a later scoop), the effective cell-update rate drops somewhat, but to no worse than about half of CAMs normal rate. A similar technique can be used for three-dimensional simulations with a single CAM (this works particularly well with partitioning rules[10,14]).

DIFFUSION-LIMITED AGGREGATION Figure 8 shows two stages in the growth of a dendritic structure by a process of diffusion-limited aggregation.[16] There are three coupled systems here, each using one bit of each cell's state. The first system is a lattice gas with a 50% density of particles. This gas is used only as a "thermal bath" to drive the diffusion of particles in a second system. The contents of the cells in this second system are randomly permuted in a local manner that depends on the thermal bath. The third system is a growing cluster started from a seed consisting of a single particle: whenever a particle of the diffusing system wanders next to a piece of the cluster, the particle is transferred to the cluster system, where it remains frozen in place. Owing to this capture process, there will be fewer diffusing particles near the growing cluster than away from it, and the net diffusion flow is directed toward the cluster. Most of the new arrivals get caught on the periphery of the cluster, giving rise to a dendritic pattern.

ISING SPIN SYSTEMS Figure 9 contains two views of a deterministic Ising dynamics[3,9,12,15]: both frames correspond to a single configuration of spins. The one on the left shows the spins themselves, the one on the right illustrates the use of CAMs auxiliary tables to display in real time a *function* of the system's state rather

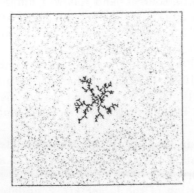 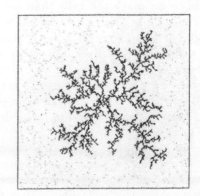

FIGURE 8 Dendritic growth by diffusion-limited aggregation. The process was started from a one-cell seed in the middle, and with a 10% density of diffusing particles.

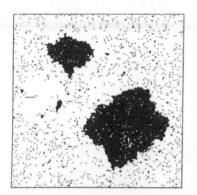 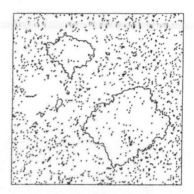

FIGURE 9   (a) A typical spin configuration; (b) the same configuration, but displaying the energy rather than the spins.

than the *state* itself—in this case, the bond energy. One can watch the motion of this energy (which is a conserved quantity and thus obeys a continuity equation) while the evolution is taking place; one can run space-time correlation experiments on either magnetization or energy, etc. By using a heat bath (as in the preceding aggregation model) one can also implement canonical Ising models. Figure 10 plots the magnetization in such a model versus the Monte Carlo acceptance probability.[6] Techniques which allow CAM itself to generate (in real time) the finely tunable random numbers needed to implement the wide range of acceptance probabilities used in this experiment are discussed in Toffoli and Margolus.[14] The actual method used in the experiment plotted here involved using a second CAM machine for this purpose and taking advantage of an instant-shift hardware feature that happens to be present in CAM-6; this feature is central to the design of CAM-8.

OTHER PHENOMENA   Other physical phenomena for which CAM-6 models are provided in Toffoli and Margolus[14] include nucleation, annealing, erosion, genetic drift, fractality, and spatial reactions analogous to the Zhabotinsky reaction. A number of models which are interesting for the study of the physics of computation are also given, including a reversible cellular-automaton model of computation and some models of asynchronous computation. These examples were developed to illustrate a variety of techniques for using CAM-6; they may also serve to clarify what we mean when we call this device a general-purpose cellular automata machine.

[6]This experiment was conducted by Charles Bennett.

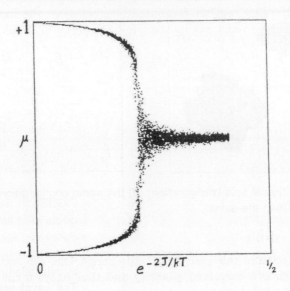

FIGURE 10 Magnetization μ in the canonical-ensemble model, versus the Monte Carlo acceptance probability. Note the sharp transition at the critical temperature T_{crit}.

6. CAM-8

CAM-6 makes it practical to perform with a personal computer scientifically useful cellular automata experiments that would otherwise require a supercomputer. Nevertheless it is much too small for many investigations, particularly ones involving the large-scale simulation of physical systems. Furthermore, CAM-6 is very limited in its choice of neighborhoods, in the number of bits available within each cell, and in the number of dimensions it can simulate. For these reasons, we have designed a much larger and more flexible machine (CAM-8) based on our experience with our earlier machines.[7]

The CAM-8 architecture makes it practical to simulate a fully parallel synchronous updating involving *billions* of 16-bit cells configured as a multidimensional

[7]This design is still evolving somewhat; what will be presented here is the machine as we conceive of it today.

rectangular array—of course, only three of these dimensions are indefinitely scalable. This is accomplished by using thousands of look-up-table processors, each shared among a million cells.

Each million-cell CAM-8 module updates its portion of the cell space 25 times a second; thus regardless of how many cells it may contain, the entire space can always be updated 25 times per second. If speed is more important than size, we can have each processor service fewer cells and the simulation will run correspondingly faster. For example, if each processor services only a thousand cells (rather than a million), then a space one thousandth as big can be updated about 25,000 times per second.

In addition to extending the size and speed of accessible CA computations by several orders of magnitude beyond that reached by CAM-6, CAM-8 is also much more flexible. In a single updating step, every cell in CAM-8 "sees" a neighborhood consisting of 16 bits selected from any of the few *million* nearest cells. These 16-bits are used as an index into a 64K-word look-up table, to produce a new 16-bit cell value for the given cell. Thus the rule applied to this rather general neighborhood can be arbitrarily chosen with absolutely no loss in speed. By using a sequence of steps, *arbitrarily large neighborhoods and cell sizes can be achieved.* The updating rule on these larger cells is implemented as a sequence of arbitrary 16-input/16-output functions.

Just as CAM-6 was controlled by a "host" computer, the many parameters of CAM-8's operation are managed by a host. Such matters as the size and shape of the cell array being used, the neighborhood used at each step, initialization of cells and look-up tables, real-time display and analysis, and error recovery, are all under the control of a small supervisory computer that acts as the intermediary and agent of the user.

As few as one of the million-cell CAM-8 modules can be made into a minimal machine capable of performing 25 million cell updates per second. Such a machine could be integrated into a personal computer host much as CAM-6 was, at a similar cost, allowing small-scale experimentation with models which could then be transferred without change to larger machines. This kind of compatibility should greatly enhance the scientific value of both large and small machines.

6.1 BASIC STRUCTURAL ELEMENTS

CAM-8's speed comes from its parallelism: the machine is made out of ordinary commodity RAM chips, driven at full memory bandwidth, plus some rather simple "glue" logic which will almost all go into a *controller-chip*, several copies of which will be associated with each million cell module. We feel that this restriction to inexpensive memory is important, since it should make it economically feasible to build CAM-8 machines which can perform CA experiments that involve hundreds of billions, or even trillions, of cell updates per second.

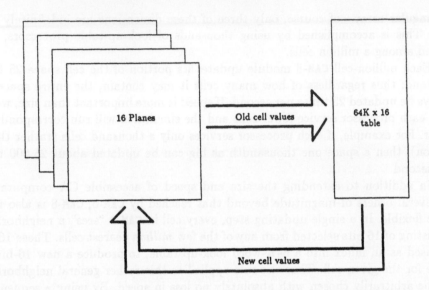

FIGURE 11   A space module of CAM-8, containing a cell array and a lookup table. As the million-cell array is scanned, a stream of 16-bit values are sent as addresses to a 64K×16 lookup table—the 16-bit results are put back into the array, as new cell values.

The atomic unit of the CAM-8 architecture is a *space module* which contains one million 16-bit cells along with the machinery for updating them. A complete CAM-8 machine consists of one or more such modules, plus associated packaging, provisions for power and clock distribution, cooling, host interface, and display circuitry.

The space module of CAM-8 consists of two more elementary modules: a *table module* and a *cell module*. The table module consists simply of a few static RAM chips, which together form two 64K×16 look-up tables (16 inputs and 16 outputs) each with a 30ns access time. The cell module consists of 16 *bit modules*, each of which handles one bit-slice of the 16-bit cell memory. Each bit-module in turn consists of two chips: a megabit dynamic RAM chip, and a controller chip.[8]

It is the controller chip which really embodies the CAM-8 architecture: everything else is just ordinary RAM chips! The controllers are interfaced to the host via a control bus of moderate width, and mediate the transmission of plane and table data to or from the host. Counters on the controller chips are the primary resource used for data analysis. These chips are also responsible for controlling the order in which data is retrieved from their associated DRAM chip during cell updating, look after putting a bit of the updated cell back into memory, and communicate

[8]This same controller chip will support larger DRAM chips as they become available.

with controllers in neighboring modules in order to tie the machine together into a single uniform space.

The basic operation of a space module is rather simple: every 40ns, the cell module outputs the value of some 16-bit cell, presenting it to the table module. This table module uses the value as a pointer into a look-up table and returns the corresponding 16-bit entry. This updated result is presented to the input of the cell module, which puts it back into the cell we started with. In Figure 11, we illustrate this basic operation for cells arranged in a two-dimensional array (as they were in CAM-6). Thus each cell is updated as a function of itself alone. How then do we bring neighboring cell values into the picture?

6.2 NEIGHBORHOODS

The most significant architectural difference between CAM-6 and CAM-8 lies in the way that neighbors are assembled for simultaneous application to a look-up table.

CAM-6 was designed primarily for running CA which employ traditional neighborhood formats, such as the "Moore" and "von Neumann" neighborhoods, in which one cell is updated as a function of more than one cell. Since cells within each CAM-6 module are processed sequentially, new cell values cannot simply replace old values if the updating is to result in the same state that a simultaneous updating would produce—the old values must be retained as long as they may be needed in computing the new state of some cell. Because of this CAM-6 requires some buffering of cell values—neighborhood values sent to the lookup table are taken from this buffer (see Figure 1). For a 3×3 neighborhood, CAM-6 requires a 515-bit long buffer (2 lines plus 3 bits).

CAM-8 takes as its primary neighborhood format *partitioning cellular automata* —a format wherein space is subdivided into disjoint subsets of cell bits. Lattice gas models follow this format: each site is updated independently of all the others, and then data is transferred between sites. With this format, sites can be updated in any order, and the new values can immediately replace the old ones—no buffering such as was done in CAM-6 is needed. This format has a simpler hardware realization than traditional formats, and allows an enormous range of neighbor choices (as will be explained below). The traditional formats can still be obtained rather efficiently by mapping them into the partitioning scheme: each million-cell module can run the 3×3 Moore neighborhood on a space of 8 billion 1-bit cells at a rate of over 65 million cell updates per second (see section 6.6).

Thus a CAM-8 step actually consists of two parts: an updating of all elements of the current partition, and a regrouping of data bits to form a new partition. The elements of the partition are just the 16-bit cells, each of which is updated by applying its value to a look-up table and storing the 16-bit result back into the cell. The partition is changed by shuffling bits between cells—how this is done is at the heart of CAM-8's design.

We simply take advantage of the fact that the elementary "chunk" of CAM-8's space is much larger than a single cell: it contains an array of a million cells. Since each of the 16 bit-arrays that constitute this cell-array are scanned independently, *we can change the grouping of bits that appear together as a single cell by simply changing the relative order of scanning of the different bit-arrays.*

6.3 SCANNING THE CELL ARRAY

CAM-8 derives much of its power from its flexibility in scanning the bit-arrays that constitute the cell array. All of our neighbor gathering and in fact all of our ability to control the size and shape of the cell array rests on the relative coordination of the separate bit-array scans.

To avoid complications, in this section we will restrict our attention to the behavior of a single space module within which the space is *wrapped around:* opposite edges of the cell array are logically adjacent (i.e., "periodic" boundary conditions). We will postpone until the next section a discussion of how an array of such modules are "glued" together to form a uniform space.

Suppose that we want to have our space module treat its cells as a two-dimensional array of size 1024×1024. We can picture our cell bits as forming 16 bit-planes (as in Figure 11). Each bit-module handles one of these bit-planes, and so it is a simple matter to have a given plane shift relative to the others. Assume for simplicity that all bit-planes are scanned from left to right and from top to bottom, and that all bit-modules start their scan at the upper left corner of the array. Then, to shift a given plane up by 100 positions we simply start the scan for that plane 100 rows down, and continue at the top when we reach the bottom. If all other bit-planes are scanned starting at the top, then the cell values that will be produced by the simultaneous action of the various bit-modules will correspond to the given plane shifted periodically, and the others left in place.

In terms of hardware, it is of course the controller chip within each bit-module which is responsible for scanning the bits of its bit-plane in an order corresponding to a version of the plane that is shifted (with wraparound) by the desired amount. By communicating with controller chips during a scan, the host sets up the order to be used for the next scan. Thus no time is stolen from the updating to accomplish relative shifting of bit-planes, and so we refer to them as *instant shifts*. In CAM-8, neighbors are gathered together by instant shifts.

In general, as long as there is some known default order in which cell bits will be scanned for a bit-array that has not been shifted, bit-modules can accomplish any desired shift of their bit-arrays by simply returning the appropriate bits in the appropriate order. This is true regardless of whether the cell array is organized in 1, 2, 3, or more dimensions. Thus the cell module may be configured with great flexibility as a multidimensional array of up to a million cells in two, three, or more dimensions. Cells that are not scanned during one step may be scanned in a later step; scanning a smaller array decreases the time taken for each update of the space.

6.4 GLUING MODULES TOGETHER

The only point about neighbor gathering that remains to be explained is how the instant-shift process works when many cell-modules are "glued" together edge-to-edge to form a larger space. What happens is that each module internally performs bit-array shifts exactly as described above, with wraparound occurring within each module. Because of this wraparound, bits that should have shifted out the side of one module and into the opposite side of the adjacent module have instead been reinjected into the opposite side of the *same* module. The positions of these bits relative to the edges of a module are exactly as they should be for a true shift: they are just in the wrong module. However, since all bit-modules process corresponding bits at the same moment, each of the bit-modules can produce a truly shifted output by simply replacing its own bit with that of a neighboring module (obtained via an interconnecting wire) when appropriate. Thus each cell module produces 16 truly-shifted outputs which together are sent to the lookup table to produce a result that is deposited back in that cell module.

For example, consider a billion-cell CAM-8 running in its 64K×16K two-dimensional configuration. Each of the 16 bit-planes in this configuration is distributed among 1024 bit-modules, each of which scans an area 1024×1024. Now suppose we want to shift one of the bit-planes 50 positions to the left. Each of the rows within each of the bit-modules is rotated (circularly shifted) 50 positions to the left by appropriately changing the order of accessing the cell memory. Each bit-module's controller chip will produce as a "glued" output a 1024×1024 window onto a portion of the complete shifted plane in the following way: the first 974 cell values of each row will come from the bit-module's own rotated data, while the last 50 values will be "borrowed" from the rotated data of the module to its right.

Vertical gluing of bit-planes is achieved in a similar fashion. That is, the controller chip first glues bit-modules together horizontally; the output of this gluing process is further multiplexed across vertically adjacent modules, yielding the final output. In this way, each module only needs to be connected (by a single bidirectional line[9]) to each of its four nearest-neighbor modules, and any shift of up to ±1024 positions horizontally, ±1024 vertically, or any combination of these can be accommodated. Thus any 16 bits (one from each plane) in a 2049×2049 region can be brought together and used as the neighbors to be jointly sent to the look-up tables.[10]

A shift-only step would entail sending the final glued *truly shifted* output produced by each cell module directly to be deposited within that module—without first passing it through the look-up table. Notice that by simply interleaving some

---

[9]Since all bit-modules belonging to the same bit-plane are shifted in the same manner, whenever one module needs to "borrow" a bit from the module to its right, then all the modules do: separate wires for information flowing in the two possible directions between modules are not needed.

[10]Such large neighborhoods are, for example, particularly useful in image processing. They may also be useful for modeling systems with a variety of time scales (e.g., several widely different ranges of particle speeds).

shift-only steps between normal shift-and-update steps, *arbitrarily wide neighborhoods can be achieved*. Since shifts of the full width of a space module can be accomplished in a single step, the speed penalty for using larger neighborhoods is not great.

Although our example above was two-dimensional, CAM-8 is actually organized as a three-dimensional array of space modules, each connected to its *six* nearest neighboring modules. This makes one further level of glue multiplexing available, allowing shifts of the entire space in three dimensions. For example, suppose that each cell module's data is organized as a three-dimensional $128 \times 128 \times 64$ array: any 16 bits (one from each bit-array) in a $257 \times 257 \times 129$ region could be brought together as part of a single updating step.

Inter-neighbor communication achieved by shifting bit-arrays is very natural for models such as lattice gases, which have particles that move in various directions and interact, and also for constructing reversible rules.[14] If for some reason we want to run a rule that uses a very wide conventional neighborhood (i.e., all bits coming from the same bit-array), this can always be implemented on CAM-8 by simply using a rule that sends the same value to all bits of the cell, and then using our normal shifting to gather neighbors. For conventional CA with small neighborhoods, there are usually much more efficient ways to embed them in the partitioning scheme (see section 6.6).

6.5 MAKING ENDS MEET

Given a three-dimensional array of space modules, we would like to be able to simulate systems that have no boundaries: we would like to be able to make the entire space periodic. Naively, this implies big bundles of wires going from each edge of the array to the opposite edge. Much worse, the time delay associated with such wires would be a characteristic of the architecture which depends on the size of the system: updating rate for the entire space would no longer be independent of system size.

In fact, we can do much better than this. Consider the chain of modules shown in Figure 12. In the top diagram we illustrate the naive scenario: one long wire is used to achieve wraparound. Note that we have numbered the modules for reference, and we have also shaded the right half of the chain. Next we show the same chain with the second half interwoven with the first to achieve wraparound with only local interconnections. Notice that this second arrangement can be obtained from the first without rotating or flipping any of the modules. Hence the same rearrangement can be performed along one entire dimension of a three-dimensional array of modules without affecting the connections in the other two dimensions. Thus the array can be wrapped around in all three dimensions using only local interconnections by simply repeating this construction for each of the dimensions in turn.

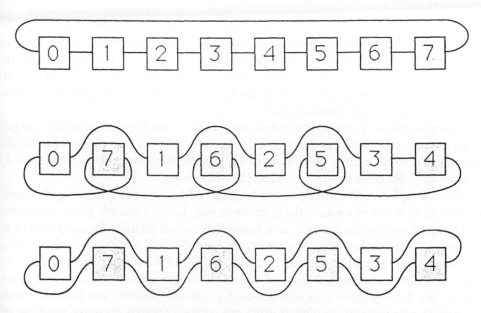

FIGURE 12 Achieving periodic wraparound of the cell space. A chain of numbered modules (a) is interwoven with itself to achieve wraparound (b). By changing positions of attachment, the second half of the chain is simplified (c).

In the bottom diagram in Figure 12 we have redrawn the preceding picture so that the shaded modules have each interchanged the point of attachment for right-neighbor and left-neighbor connections: this was not logically necessary, but it makes the maximum wirelength slightly shorter and the arrangement somewhat neater. It is clear from this last diagram that if we have *next-nearest neighbor interconnections* rather than nearest neighbor interconnections along each dimension, our three-dimensional array *can simulate a periodically wrapped around space.*

In a CAM-8 machine, we don't actually require distinct kinds of modules (corresponding to the different points of neighbor-wire attachment discussed above): we achieve exactly the same effect by having our controller chips internally exchange their usage of opposite wires, so that any module can play any of the 8 distinct roles (in terms of use of connecting wires) required by the next-nearest neighbor wraparound scheme in three dimensions.

6.6 INTERNAL DIMENSIONS

Although only three dimension can be arbitrarily extended by gluing together modules, additional *internal* dimensions are available.[11] Since periodic shifts along internal dimensions just involve a scan order choice (taken before the step is started), all hardware gluing for the external dimensions can be considered to take place after these shifts have already happened. Thus data can be shifted along all dimensions (internal and external) simultaneously.

For example, a typical four-dimensional configuration for a space module might be $64 \times 64 \times 64 \times 4$, where the 4 cubes are internally treated as a shallow fourth dimension. An $n \times n \times n$ array of modules would then handle a space $64n \times 64n \times 64n \times 4$, and the neighborhood would be a $129 \times 129 \times 129 \times 4$ region.[12]

Of course, internal dimensions can be regarded as a way of increasing the number of bits in a cell in a particularly uniform way. In the example given above, we could regard each space module as a three-dimensional $64 \times 64 \times 64$ array of 64-bit *supercells*. Now instead of 16 bit-arrays we have 64, and we can still choose the shift *separately* for each bit array, still with a $129 \times 129 \times 129$ neighborhood. To do this, we divide the complete scan of our space into 4 component scans—one for each of the 4 component-cells of our 64-bit supercell—and set the shifts for each component scan separately. This is often not enough, though, to make good use of these large cells, since our look-up tables only see one component cell at a time. What we typically do is use a longer sequence of scans, each with its own shifts and look-up table (we double-buffer the look-up tables, so that a new table can be prepared while the old one is being used). As we scan the array of supercells, each update involves the selection of a 16-bit subset of the 64 bits in a supercell. This selection makes use of shifts along the fourth dimension of our array in order to permute bits between the component cells of the supercell. Since this shift can also be set separately for each scan, and since the look-up tables can put their outputs in any desired bit-position within a cell, such a sequence of scans can implement *any function* on the larger cell.

Similarly *we can make our supercells arbitrarily large* (within the constraints of module memory limits) by an appropriate choice of array dimensions. Unfortunately, very large supercells become impractical because of the large number of steps needed to update them, and also because of table-changing overhead: as each module contains fewer but larger supercells our scans become shorter, leaving less time for preparing the table to be used by the next scan. Thus if supercells become too large we waste almost all of our time waiting between scans while table changing is completed. Of course this consideration doesn't apply when internal dimensions are used as true dimensions rather than as a general way of putting more bits in

---

[11] Our initial design restricts all internal dimensions to be a power of two in size. Thus for the million-cell array contained in one cell module, the data may be organized in up to 20 internal dimensions, in addition to the three external dimensions.

[12] In the case of lattice gases, it has already proven useful to construct rules in four dimensions in order for their three-dimensional projections to have full three-dimensional rotational symmetry.[7]

a cell, since in this case the tables aren't changed and updating is done in a single scan of the entire space.

As an example of the use of supercells, consider the problem of embedding a two-dimensional space of 1-bit cells employing the 3×3 Moore neighborhood into a two-dimensional CAM-8 space. To use our CAM-8 cells efficiently, we might decide to put 16 1-bit cells into a single CAM-8 cell, with each CAM-8 cell representing a 4×4 patch of the embedded space. By shifting bit planes we can move our patch around, and by updating the 4 middle cells in each patch, it seems that we can do a complete update of the space in 4 steps. There is, however, a problem: since the entire patch isn't updated simultaneously, we need to keep copies of all of the old cell values while we construct the new values. We can add the extra bits we need to each cell by making our system out of 32-bit supercells. It is noteworthy that *we still only require 4 scans* in order to construct the new cell values, since each update of a patch looks at exactly 16 of the 32 bits in the supercell. Two additional scans suffice to make a copy of the new state within each supercell (in preparation for the next step), giving us an overall performance of 16 Moore neighborhood 1-bit cell updates for every 6 CAM-8 cell updates, and each million cell CAM-8 module services a space of 8 million 1-bit cells.

6.7 INPUT AND OUTPUT

All CAM-8 I/O passes through and is mediated by the controller chips. Typically, most of the I/O involves communication with the control host in order to change CAM control parameters and data. There is, however, an additional high-bandwidth data bus, the *flywheel bus*, available for interfacing special purpose hardware (and for debugging purposes). The flywheel bus consists of the final glued outputs of the bit-modules, together with inputs to these same modules. *Every bit of cell memory* in the machine is available to be examined and modified once during every step. Depending on how CAM-8 is configured, flywheel-bus input bits can be ignored (in favor of internally generated new cell values), routed as inputs to the look-up tables, or sent directly to the bit-arrays.

Besides the two data lines (one for input and one for output) that it contributes to the flywheel bus, each bit-module also has a small number of control lines. Some of these lines are merged together into a *control bus* of moderate width that is used by a host computer to control the machine. Areas that can be accessed via the control bus include

- the lookup tables (two SRAM modules)
- the bit-array (a single DRAM chip)
- various registers located within the controller chip

the array-dimensions control registers
the scan control registers
the interconnect control registers
the I/O control registers

the error management registers
the table-address source select registers
the array-data source select registers
the event-counter source select registers
the event-counter buffer register

■   various status registers

Each bit-module is connected both to one data line and to one address line of a look-up table. During normal updating, the address line is fed sequentially with the glued output of the bit-module, and the values appearing on the data line are written sequentially into the bit-array as its new contents. This is, however, just one possible combination of table-address and array-data sources—by writing to a module's "source select" registers, any of the following three bits may be sent either as an address bit to the look-up table, as a data bit to be written directly into the bit-array, or as an event to be counted:

■   the glued output for this bit-module
■   one bit from the site address
■   one external bit, taken from

    the flywheel-bus input for this bit-module
    the wired "or" of one bit chosen by each bit-module
    a signal from an adjacent module (on a neighbor-wire)

Notice that the table output doesn't appear in this list—it can be included in the functions sent to the bit-array and event counter, but not in the function sent to the look-up table.

By appropriately controlling the sources both for table addresses and for array data we can, for example, run a step in which a constant value of 0 or 1 is sent to the table address while the array data is not affected—the bit-array can even be shifted during this step, since the table is not needed for this. Thus one can run steps during which one or more address bits of the table are host-selected constants, analogous to the "phase" bits[14] used by CAM-6. This allows one to split a lookup table into several subtables, to be used during consecutive steps without having to download new tables. Of course downloading new tables isn't a great problem as long as all the tables are identical (or there are only a few different kinds), since all table modules that should contain the same information can be written simultaneously, and double-buffering allows us to send the next table while the previous table is still being used.

Data is read from or written to either arrays or tables by the host in a similar manner: a stream of bits is sent to or from the bit-module associated with the data. For bit-arrays, the scan control registers are used not only during steps, but also to control where the data-bits sent by the host to the bit-module should go. For tables, each bit-module controls one bit of the address of a table, and is told by the host which bit of its internal address counter should be shown to its table, to control where data-bits go.

6.8 DATA ANALYSIS

Each bit-module contains an event counter that is used for real-time data analysis.

As mentioned above, any function of the current (glued) cell bit, the new value (from the table) for this bit, a bit of the site address, and an external bit, can be counted sequentially by the event counter as the bit-array is sequentially scanned.

For example, if a bit-array is being used to store a spatial parameter (such as an obstacle in a fluid-flow experiment), the associated table output is not needed for updating, and may be programmed for data analysis and counted. If there aren't enough such "free" tables, or if the analysis requires a different neighborhood than the updating, separate analysis steps may be interleaved between updating steps by rewriting tables.

The event counter can be used to perform local averaging by appropriately controlling the scan. For example, a large three-dimensional array can be updated as a collection of small cubes, with counts read immediately after each cube is updated (these cubes can be much smaller than the region serviced by each cell module).

Using an internal dimension, each module can be split into two halves, each of which runs exactly the same dynamics except that the initial configurations of cell data for the two halves differ in some respect. For example, if we are computing autocorrelations (see Section 5), the initial configurations in the two halves will be different versions of the same system, but with a space or a time shift between them (or both). We would like to accumulate a count of the differences between the two halves. This can be done by simply alternating steps during which the two systems are updated, with steps during which groups of corresponding bits from the two halves are gathered together (eight from one and eight from the other) as 16-bit values to be compared and counted. By performing the shift portion of the step either as part of the update, or as part of the gather for counting, the bits being compared will either be the outputs from the last update, or the inputs to the next update, whichever is desired.

Machines with more than one module can be split in a different manner which can also be used for computing autocorrelations, but which is useful for many other purposes. In particular, this new technique can play an important role in error detection and correction (see next section). This splitting requires extra flexibility in the use of intermodule connecting wires, but doesn't require any extra wires.[13]

To explain this new splitting method, we first recall that a CAM-8 step logically consists of two operations: a gathering of data by shifting, followed by an updating

[13] In the scheme discussed here, the *first-neighbor* connections between modules are time-multiplexed to serve first for shifting data, and then for transmitting comparison data between corresponding cells of two subsystems. If we added *second-neighbor wires* (first neighbors of first neighbors along the first-neighbor directions), then these wires could take over the data-shifting function while the first-neighbor wires were transmitting comparisons, allowing both operations to happen simultaneously. This would gain us a factor of up to two in speed for counting comparisons, but would more than double the number of wires running between modules of our machine, and so this is not planned for our initial version.

of individual cells. Although we have arranged to be able to do both operations simultaneously, there is nothing to prevent us from doing separate gather and update steps. In this case, during the update steps there would be no use made of the intermodule connections, since all of the data would already have been gathered into the correct module. Thus these wires would be free to be used for other purposes during update.

Now once again we use an internal dimension to split our space in half in preparation for computing autocorrelations. This time, however, instead of keeping corresponding cells of the two systems being compared in the two halves of a single module, we store the data for the second system in a shifted fashion, so that the partner of a given cell in the first system is always found in an adjacent module (the one to the right, say). Since, as we have already noted, separate gather-and-shift steps leave the wires connecting modules unused during the cell updating steps, we can arrange our scans so that pairs of corresponding cells are updated simultaneously in adjacent modules, and *use the interconnecting wires to pass data between the modules for an on-the-fly comparison.*

Using interconnecting wires along one dimension, we have thus coupled the two halves of a space divided in two. In a similar fashion, using the interconnections in all three directions, we can split the space into as many as 8 coupled subspaces.

6.9 ERROR HANDLING

CAM-8 is designed to be a scalable architecture, making it practical to build machines containing billions of cells and performing hundreds of billions or even trillions of updates each second. While there are no inherent architectural limits on how many space modules can be hooked together, there is a practical problem which grows as more and more modules are added, namely, error handling. Using built-in hardware consistency checks, it will be possible to discover and recover from hardware errors.

Every sequence of I/O transactions between the host and the space modules includes checksums to test for garbled transmission: in case of an error we retransmit.

To directly test look-up table integrity, we can simply have the host perform a verify-write of all tables, in which the old contents is read and compared with what the host is writing. Since tables are double buffered, this can be done without stopping the updating. A verify-checksum can also be accumulated on the *old* table while a new table is being written—this avoids the need to retransmit the old table in order to check it.

Cell-memory is tested by each bit-module during every step. Checksum bits reflecting the number of ones last written and their positions are compared to corresponding checksums performed on the data subsequently read. By having a relatively long checksum, we can make the chance of getting the correct checksum when there is a memory error very small.

Characteristics of the experiment can be used to help monitor or verify correctness. For example, conserved quantities can be periodically checked (using the event counters) to see that they haven't changed; reversible experiments can be run backwards as a test, to see if the initial state is recovered; for non-conserving irreversible rules, we can periodically count some aspect of the state (for example, the number of 1-bits in the system) and use this sequence of counts as a "signature" to be compared with a repeat run.

Hard errors, caused by bad components, can be tested for whenever any error is detected. Soft errors, in which memory bits are typically changed, are principally caused by alpha particles. Modern commercial memory chips, which constitute most of CAM-8, are inherently quite reliable: even with absolutely no provision for error correction, it should be possible to run a billion-cell machine for several days at a time without any errors. Thus for such a machine it may be perfectly practical in most cases to simply detect errors, and rerun an experiment if any occur. In fact for many experiments of a statistical nature, such as fluid flow past obstacles, we could even decide that rare errors in which a few particles are added or dropped don't matter.

For longer runs, or for large machines, if we want to guarantee exactly correct operation, it is probably most practical to use each machine as two correlated half-machines, both running the same experiment, using the adjacent-module splitting technique discussed in the previous section. Since the chance of two different space modules both experiencing a soft error during the same step is extraordinarily small (expected perhaps once in 10^{13} steps for a trillion-cell machine), we can assume that one out of every correlated pair of bit-modules will always be correct. Bits that were updated incorrectly are fixed by using data from the correct twin (cf. below), and incorrect tables are simply rewritten.

Given an error, there remains the problem of deciding which of the pair of correlated modules is incorrect. For data errors, we rely on the internal checksums maintained by the bit-modules to tell us which module to fix. Otherwise we make use of one further facility provided by the hardware in order to quickly and reliably find the error. Whenever table comparisons disagree, both the original contents of the cell where the error occurred and the updated value are latched by the controller chips. By examining this information (at the end of the step), the host can tell which of the modules was updated incorrectly.

6.10 DISPLAY

Being able to display the state of our system in real time provides important feedback as to whether or not everything is working as expected, and what parts of the system are doing something interesting that should be investigated more closely.

Two-dimensional display is not much of a problem for CAM-8, since this machine can provide its data in the correct format for a color monitor. This machine can even, if desired, scan its data in the correct format for an interlaced display: since each cell is updated independently of all others, the rows can be scanned in whatever order you choose.

For a complete two-dimensional display ($64k \times 16k$ for a billion-cell machine[14]), we could cover an enormous wall with thousands of color monitors, each of which would show a 1024×1024 patch (using a four-fold interlaced scan on a 15kHz monitor). Of course it would be more practical to use only one monitor (or just a few), and shift the data to move the window around.

Since all of the neighbors that would be used for a cell-update are available simultaneously, it is a simple matter to display a function of the neighborhood, rather than the neighborhood itself. For example, in a fluid-mechanics experiment you might want to show only the smoke particles that trace the flow.

Another display technique would be to let the host computer use the average data accumulated by counters (see Section 6.8) in order to produce a display on its own screen.

In more dimensions, how to display the cell data is less obvious. Clearly, though, once we decide how to display a section of the system in a convenient manner, then we can see any part of the system by appropriate shifts (in however many dimensions).

One method that allows CAM-8 to *always* produce some sort of a display is to map the million cell array serviced by each space module into a 1024×1024 square in the most natural manner you can think of, and then display this square array. For example, a $128 \times 128 \times 64$ configuration could be arranged as an 8×8 "sheet" of 128×128 "postage stamps." If we now scan the data in this flattened format (which just involves a particular choice of the order in which cells are updated), then it can be displayed on a 15kHz monitor, exactly as the 1024×1024 configuration that we discussed above for two-dimensional displays. If not all cells are being updated, then we can map the cell array into a pattern that fits *within* a 1024×1024 array, and display the unused areas as black areas on the monitor.[15] While we are displaying this black area, we can actually proceed to update the active region many times, so that for a very small active region, many hundreds of updates may occur between frames—thus we could watch a small area run at thousands of frames per second! If 256K cells or less are being processed, we can get a better display by mapping the cell array into a 512×512 matrix, and displaying this matrix on a 30kHz non-interlaced monitor as we scan it.

Enlarged views are useful for studying rules in action and for constructing initial states. Essentially any degree of expansion can be produced by taking advantage of CAM-8's scan flexibility to cause rescans of bits and rows.

---

[14] In order to realize a $64K \times 16K$ space without adding extra interconnections, we unfold CAM-8's basic $8 \times 8 \times 16$ array of space modules into a 64×16 sheet by appropriately programming the use of the existing interconnect wires.

[15] When we aren't locked to the framing format of a monitor, we don't waste time on these unused areas, and the number of complete updating steps per second therefore rises when we process smaller areas.

6.11 OTHER PROCESSORS

The heart of the CAM-8 architecture is the scanning/shifting/gluing technique embodied in the cell module; the processor is just a look-up table. Related machines which can take advantage of putting space directly into the hardware can be built by simply using a different processor.

Our decision to use 16-bit cells in CAM-8 was predicated on our use of look-up tables as processors: a 16-bit address was about the most we could handle by fast table look-up. If we abandoned table look-up and used other kinds of processors, cells could be much larger. In this case, our cell modules would simply consist of more bit-modules, with no change at all needed in the controller chips. Gluing and virtual shifts would work exactly as usual; only the cell size and the processor would be different.

In addition to larger cell sizes, the CAM-8 controller chip is actually designed to accommodate a pipelined processor with a much longer latency than a fast SRAM look-up table: although the processor must still finish a computation about every 30ns, this computation can have several stages.[16] For example, various compositions of moderate-size look-up tables, appropriately pipelined, could be used in many interesting cases to handle the updating of CA that require cells too large to treat with single lookup tables.[17] Field Programmable Gate Arrays (FPGAs), which are themselves close kin to cellular automata, could also serve as processors in this context.

As an extreme example of large cells and a complex processor built on the CAM-8 bit-module, we might make a machine in which each cell contains several 80-bit floating-point numbers, with a pipelined updating performed using commercial floating point chips. As part of each step, the floating-point numbers would be gathered together using CAM-8's instant shifts.[18]

CAM-8 is a rather general-purpose tool; much more highly parallel application-specific machines could be made using simple processors optimized for certain small sets of CA rules, such as two-dimensional lattice gas rules. Many such processors—perhaps thousands—could fit on a single custom integrated circuit. This leads to a problem when we want to interconnect chips. We can, however, borrow CAM-8's solution to the wiring problem: time multiplex each processor over several cells, in order to reduce the wire count. For example, if each processor services 64 cells, then the number of wires required to connect to a two-dimensional array of such processors is 1/8 of what it would be if each cell had its own processor. Within the range of acceptable wire counts, the CAM-8 method of time-multiplexing processors

---

[16]There is no reason why the controller chip couldn't be modified to accommodate very long pipelines, involving perhaps hundreds or thousands of stages, since the computation performed by each module is completely sequential, and the data access is perfectly predictable.

[17]Note that if we abandoned conventional memory chips, both cell access-rate and update-rate could be greatly increased by simply dividing each memory access into several short pipelined stages.

[18]Since all 80 bits of a floating-point number would tend to shift in the same direction, it might be more economical to use controllers that service more than one DRAM chip.

also allows us to carefully balance silicon area devoted to processing against area devoted to cell memory.

7. CONCLUSIONS

Cellular automata constitute a general paradigm for massively parallel computation by means of uniform arrays of simple processors. There is a natural match between this paradigm and the capabilities of parallel digital hardware: locality and scalability make it possible to build massively parallel machines with an astronomical computing rate and number of cells. To study the possibilities of the CA architecture, we have built a number of machines optimized for general purpose CA work, and plan to build machines which can handle *billions* of 16-bit cells, and *hundreds of billions* of cell updates each second. We expect that the availability of CA machines with even such modest performance will encourage scientists in many disciplines to develop CA algorithms, which will in turn provide additional impetus for the development of much more powerful CA machines.

ACKNOWLEDGMENTS

This research was supported by grants from the National Science Foundation, the Department of Energy, International Business Machines, and the Defense Advance Research Projects Agencies.

REFERENCES

1. Bennett, Charles (1973) "Logical Reversibility of Computation," *IBM J. Res. Develop.* **6**, 525–532.
2. Califano, Andrea, Norman Margolus and Tommaso Toffoli (1987), *CAM-6 User's Guide*; and Kenneth Porter, *CAM-6 Hardware Manual*, Systems Concepts, 55 Francisco St., San Francisco 94133.
3. Creutz, Michael (1986), "Deterministic Ising Dynamics," *Annals of Physics* **167**, 62–76.
4. Feynman, Richard (1982), "Simulating Physics with Computers," *Int. J. Theor. Phys.* **21**, 467–488.
5. Fredkin, Edward, and Tommaso Toffoli (1982), "Conservative Logic," *Int. J. Theor. Phys.* **21**, 219–253.
6. Frisch, Uriel, Brosl Hasslacher, and Yves Pomeau (1986), "Lattice-Gas Automata for the Navier-Stokes Equation," *Phys. Rev. Lett.* **56**, 1505–1508.
7. Frisch, Uriel, Dominique d'Humières, Brosl Hasslacher, Pierre Lallemand, Yves Pomeau, and Jean-Pierre Rivet (1987), "Lattice Gas Hydrodynamics in Two and Three Dimensions," *Complex Systems* **1**, 649–707.
8. Hardy, J., O. de Pazzis, and Yves Pomeau (1986), "Molecular Dynamics of a Classical Lattice Gas: Transport Properties and Time Correlation Functions," *Phys. Rev.* **A13**, 1949–1960.
9. Herrmann, Hans (1986), "Fast Algorithm for the Simulation of Ising Models," *J. Phys.* **45**, 145–151.
10. Margolus, Norman (1984), "Physics-Like Models of Computation," *Physica* **10D**, 81–95.
11. Margolus, Norman, Tommaso Toffoli, and Gérard Vichniac (1986), "Cellular-Automata Supercomputers for Fluid Dynamics Modeling," *Phys. Rev. Lett.* **56**, 1694–1696.
12. Pomeau, Yves (1984), "Invariant in Cellular Automata," *J. Phys.* **A17**, L415–L418.
13. Toffoli, Tommaso (1984), "CAM: A High-Performance Cellular-Automaton Machine," *Physica* **10D**, 195–204.
14. Toffoli, Tommaso, and Norman Margolus (1987), *Cellular Automata Machines—A New Environment for Modeling* (Cambridge: MIT Press).
15. Vichniac, Gérard (1984), "Simulating Physics with Cellular Automata," *Physica* **10D**, 96–115.
16. Witten, Thomas, and Leonard Sander (1981), *Phys. Rev. Lett.* **47**, 1400.

Andre Clouqueur and Dominique d'Humières
CNRS, Laboratoire de Physique de l'Ecole Normale Supérieure, 24 rue Lhomond, 75231 Paris Cedex 05, France

RAP1, A Cellular Automaton Machine for Fluid Dynamics

This paper originally appeared in *Complex Systems* (1987), Volume 1, pages 585-597.

Complex Systems 1 (1987) 585 597

RAP1, a Cellular Automaton Machine
for Fluid Dynamics

Andre Clouqueur
Dominique d'Humières
CNRS, Laboratoire de Physique de l'Ecole Normale Supérieure,
24 rue Lhomond, 75231 Paris Cedex 05, France

Abstract. RAP1 is a special purpose computer built to study lattice gas models. It allows the simulation of any model using less than 16 bits per node, and interactions restricted to first and second nearest neighbors on a 256 × 512 square lattice. The time evolution of the automaton is displayed in real time on a color monitor at a speed of 50 frames per second.

1. Introduction

The concept of cellular automata was introduced in the early fifties by von Neumann and Ulam [1] to study the behavior and the organization of complex systems. A cellular automaton (CA) is a set of *identical* processors located on a *regular lattice* and with *limited connections* with their neighbors. For each time step, the CA is described by the values of the states of all the processors. At time $t+1$, all the processors compute in parallel their new state as a given function of their state and those of the connected processors at time t. Wolfram [2] has shown that very simple one-dimensional CA with one-bit internal states may give extremely complicated behaviors, as soon as each cell is connected to its first- and second-nearest neighbors and its time evolution is given by a Boolean function chosen within the suitable set of Boolean functions of five Boolean variables. The CAM machines built at MIT by T. Toffoli [3] played a prominent part in the interest for CA during the last five years. These machines demonstrated that cheap, but very powerful, special-purpose computers can be built to study a wide class of CA. They have also shown the impact of direct visualization on the study of very complex phenomena.

During the same time, several attempts were done in the physicist community to find simple ways to describe and study the motion of a collection of interacting particles [4,5]. In the simplest model [5], the particles are constrained to move on a square lattice from a node to one of its nearest neighbors and to interact on the nodes only. However, this model was too simple to give realistic flows. Two years ago, Frisch, Hasslacher, and

Pomeau [6] found that the same kind of model on a triangular lattice leads to a more accurate approximation of a real gas. Using standard methods of statistical mechanics, they were able to show that the time evolution of this gas is described by the Navier-Stokes equation like most of the real fluids. Since, lattice gas models have received a lot of interest. This class of models, which first lead to Navier-Stokes equation are now applied to a wide range of systems from thermal effects to combustion phenomena [7]. Thus, lattice gas automata not only support the conjecture that cellular automata are able to simulate partial differential equations, but bring standard methods of physics to build and study a wide class of cellular automata. In this class, the ith bit of the processor states is viewed as a particle which can jump at each time step from one node of the lattice to its neighbor in the direction c_i. Thus, the time evolution is split in two substeps. During the first one, called the "collision step", each processor computes its new state as a function of its state at time t; during the second, the "propagation step", the bit i is moved from each processor to its neighbor in the direction c_i.

While the CAM machine is very well suited for the study of two-dimensional cellular automata deriving from the original study of Von Neumann and Ulam, in which the neighborhood relations are essential, its use to simulate the relative motion of bits of information needed for lattice gas models requires quite a subtle trick known as the Margolus neighborhood [3,8,9]: the nodes are no longer equivalent but are packed in two-by-two cells which become the new node of the automaton, thus decreasing by a factor of four the effective size of the lattice. The specific algorithm of the two-dimensional lattice gas models leads to a more practical architecture: the original two-dimensional lattice, made of $M \times N$ cells with b bits per cells, is also viewed as a three-dimensional lattice, made of b one-bit $M \times N$ planes. During the collision step, the two-dimensional structure is used, each cell computing its new state; during the propagation step, the ith plane is moved as a whole in the direction c_i with respect to a reference plane.

In section 2, the similarities and the differences of RAP1 with the raster displays and the CAM machines will be presented. Section 3 will be devoted to the description of the hardware implementation. Preliminary results of hydrodynamical simulations will be given in section 4.

2. Video architectures

The RAP1 project started at the beginning of October 1985, with the following constraints:

1. Versatility, to allow the machine to be used to study a large class of lattice gas models of physical interest.

2. Direct display of the automaton evolution, to remove a classical bottleneck of the simulations on general purpose computers, i.e., the

visualization of the results.

3. Fast completion in order to build the prototype before the subject drifts too far from its starting point. This point implied the use of the limited resources of the laboratory for this not scheduled project.

4. Possibility to extend the design to larger lattice sizes.

These constraints implied several technical choices. First, we ruled out the use of true parallel architecture and chose to take benefit of the CAM experience, that is, to use a serial implementation of the algorithm. A consequence of this choice and of the synchronization of the computation process with the visualization restricted the size of the lattice to 256 lines of 512 pixels, a size which can easily be displayed on a low-resolution color monitor. In the following subsection, the evolution from raster displays to the RAP1 architecture will be presented at the conceptual level to stress the basic similarities along with the main differences between raster displays, CAM, and RAP machines.

2.1   Raster dispays

The raster displays are the basis of both the CAM and RAP machines. In these displays, the image is stored in a memory which we will refer to as screen memory. This memory is serially read row by row, synchronously with the sweep of the horizontal lines of the screen; the content of each memory location gives the intensity of the corresponding dot location on the screen as shown figure 1a. The time is divided into frame corresponding to the display of a full screen. The beginning of each frame is marked by a VSYNC signal. Each frame is in turn divided into lines corresponding to the display of one horizontal line. The beginning of each line is marked by an HSYNC signal. A VBLANK signal selects the lines during which the screen memory rows are actually displayed, and a HBLANK signal sets the active part of the lines. The VBLANK and HBLANK signals reset row and dot counters respectively; these counters are then incremented by the HSYNC and dot clocks respectively to give the address of the pixel to be read in the screen memory. The value of the pixel is then sent to a color look-up table which feeds digital-to-analog converters (DACs) to control the intensity of the red, green, and blue inputs of the monitor. Today, the details of the hardware implementation of this basic scheme are handled by graphic display processors (GDPs), which provide most of the needed signals.

For a typical low-resolution European monitor, such as the one used for RAP1, the frame frequency is 50 Hz or 20 ms per frame, and each line is 64-μs long. For a 256 × 512 resolution, the dot clock frequency is 14 MHz or 71.4 ns per pixel, the HBLANK signal is 36.57-μs long and occurs 17.71 μs after the HSYNC signal, and the VBLANK signal is active from line 38 to 293.

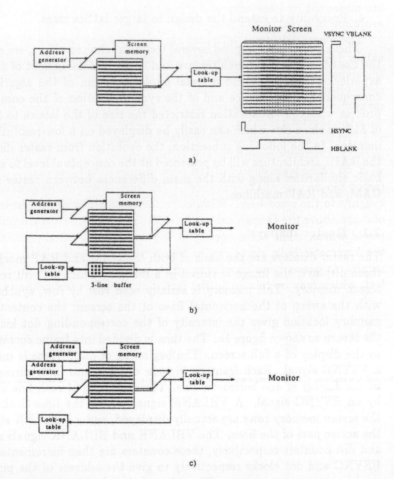

Figure 1: (a) Schematic of a raster display. (b) Schematic of the CAM machine. (c) Schematic of the RAP1.

2.2   The CAM machines

The first version of the CAM machine was completed in the early eighties at MIT and was one of the very first special-purpose computers designed to study physical problems. The description of one of its descendants, CAM-5, is given in reference 8, and the CAM-6 version is now commercially available along with a user manual and an extensive amount of software [9]. The basic idea of all these machines was to simulate two-dimensional cellular automata on a square lattice with the so-called von Neumann neighborhood: connections of each node to its first and second neighbors on a square lattice. The CAM machines were built around a raster display architecture. The simplest possible version is schematized in figure 1b. The screen memory is sequentially read row by row and sent to the display and to a three-line buffer which stores the data needed for a node and its eight neighbors. This buffer feeds a "computation" look-up table with nine inputs. The output of this table is then written back to the memory. In principle, this scheme can be used for multi-bit states; however, the size of the look-up table grows exponentially with the number of bits. Even for two-bit states, it is necessary to restrict either the number of bits interacting together or the number of available rules for the automaton, using some combinations of smaller tables. For example, the CAM-6 machine has four bits per state organized as two pairs of two bits, one of them interacting with its associated von Neumann neighborhood and the other alone. Up to ten bits per state can be obtained using the Margolus neighborhood [8], which has been used to study correlation functions of different lattice gas models [10].

2.3   The RAP1 machine

The RAP1 machine uses the same kind of basic architecture as the CAM machines use, but the neighborhood interactions are replaced by plane displacements (as in the lattice gas simulations [11] where time evolution of the automaton is split in two steps per frame). During the first step, the collision step, the value of the displayed pixel is sent to a computation look-up table, its input being written back to the screen memory as in the CAM machines; however, in the RAP, only the value of the pixel itself is used instead of the value of the pixel and of its neighbors. During the second step, the propagation step, the automon is viewed as made of a collection of one-bit two-dimensional planes, one for each bit of the internal state of the nodes. The displacement of bits from one node to one of its neighbors is done once by the corresponding translation of the whole plane, using an address generator per plane rather than only one for all the screen memory as is used for raster displays and CAM machines.

As for the CAM machines, the neighborhood relations can be implemented on the RAP machine. In this case, the information must be duplicated at each step as many times as there are relevant neighbors, then all these replica are moved in the corresponding direction. This process wastes part of the screen memory, a penalty which can be somewhat decreased when several nodes are packed in the same 16-bit pixel. For example, four cells of the Conway's game of life [12] can be packed in one pixel, saving a factor of two for the overhead of the duplication.

3.   RAP1 scheme

3.1   Collision step

In this section, we assume that the machine has only to compute its state at time $t+1$ as a function of its state at time t with no need for information coming from the neighborhood. This computation corresponds exactly to the collision step of the lattice gas automata. In this mode, the information stored in the screen memory are serially read, sent to the computation look-up table, and written back to the screen memory during the display window.

For that, we use video random access memory (VRAM), which are the combination of a standard 65536 bits random access memory (64k RAM)[1] and a shift register 256 bits long. This shift register can be loaded by the content of an arbitrary row or its content can be written back to an arbitrary row of the memory; both operations use only one memory access. Except for these two operations, the use of the RAM array and the shift register are completely decoupled. The content of the shift register can be serially clocked out, while an external signal is clocked in at the same time. The shift registers of the VRAM are designed to be easily cascaded, and the screen memory of the RAP1 machine is made of 16 planes 256×512, each plane being made of two VRAM TMS 4161. The RAMs corresponding to the left and right parts of the screen will be said to be in zone 0 and 1 respectively.

This VRAM allows a very simple architecture. The row counter sets the address of the row to be displayed, which is down loaded into the shift registers 3.43 μs before the HBLANK signal becomes active. The shift registers are then clocked out 512 times at a 14-MHz rate and the 16-bit output word is used as an address for 16 static random access memories (64k SRAM) to produce a new 16-bit word which is written back to the shift registers. During the same time, the output of the shift register is also sent to an other look-up table made of three $4k \times 4$ SRAMs to generate information for the color display. After the visualization window, the shift register contains in place the new line after processing and is written back to the corresponding row of the screen memory.

3.2   Horizontal shifts

The horizontal shifts needed to simulate the horizontal motion of planes are obtained through a slight modification of the previous scheme. If the

---

[1]The RAM is organized as 256 rows of 256 bits each.

shift register is clocked 511 times and written back to the memory, the net result of this operation is a global shift of the line toward the right of the screen. To define periodic boundary conditions in the horizontal direction, the new state of the point located on the left edge of the row must be computed and the result written directly into the row of the left memory (zone 0) at location 00. For that, the output of the look-up table must be stored in an additional register inserted in the computation loop after the look-up table. In a similar way, if the shift register is clocked 513 times and written back to the memory, the result is a shift of the line toward the left of the screen. After 513 clock pulses, the leftmost point is shifted out of the shift register. To avoid a new computation of its value, another register must be inserted in the computation loop just before the look-up table. For periodic boundary conditions, the content of this register is directly written into the row of the right memory (zone 1) at location FF.[2] Thus, two registers must be inserted in the computation loop introducing two additional time delays in the loop. These delays must be compensated for by two additional clock pulses requiring 513 pulses for right shifts, 514 pulses for in place computations, and 515 ones for left shifts.

In fact, the TMS 4161 memories require an even number of clock pulses to correctly write back the content of the shift register, and the previous scheme must be slightly corrected by insertion in the computation loop of a third register and a multiplexer allowing this new register to be bypassed. When this third register is bypassed and the shift registers are clocked 514 times, there is no horizontal shift. If the third register is inserted in the loop, 514 or 516 clock pulses give right or left shifts respectively. The final loop is represented in figure 2 and the processing of any line can be summarized as follows:

1. Download into the shift registers the row memories at address given by the row counter.

2. Repeat steps 3 to 5 514 times.

3. Clock the shift registers and the three additional registers.

4. Compute the new state through the look-up table.

5. Feedback the shift register.

6. Write back the shift registers to the row memories for the planes with right shift.

7. Fill directly location 00 of the row memories of zone 0 for the planes with right shift.

8. Write back the shift registers to the row memories for the planes without horizontal shift.

---

[2] Hexadecimal notation.

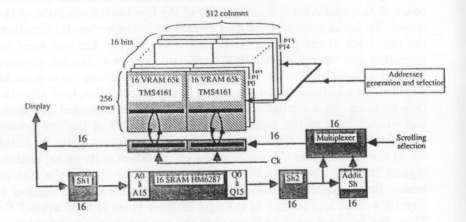

Figure 2: Detail of the computation loop of RAP1.

9. Clock twice the shift registers and the three additional registers, doing steps 3 to 5.

10. Write back the shift registers to the row memories for the planes with left shift.

11. Fill directly location FF of the row memories of zone 1 for the planes with left shift.

Steps 3 to 5 are done at the same time, using the pipelined architecture of the loop, at a 14-MHz rate, and steps 1 and 6 to 11 are done at a much slower rate during the blanking windows.

Since the underlying structure of memories is a square lattice, a trick is needed to map hexagonal lattices on them. The simplest way consists to use the vertical and horizontal directions along with one of the 45° directions, leading to very severe geometric distortions: a rectangular area in the ordinary plane becoming a diamond. A better solution is given in reference 11 in which the directions of the oblique links depend on the parity of the lines. This technique is implemented in RAP1 with two different horizontal shift masks: one for the even lines and one for the odd ones, selected by the least significant bit of the row counter.[3]

3.3 Vertical scrolling

While the horizontal shifts correspond to physical displacements inside the screen memory, the vertical scrollings correspond to virtual displacements.

---

[3]For future use, a similar posibility is provided for the vertical scrolling, but depending now on the frame parity.

For that, a frame counter is needed and the address generator is split in three pieces:

1. one for the planes without scrolling with the same address for the displayed line and the downloaded memory row,

2. one for the plane with an upward scrolling, the address of the memory row being now the address of the displayed line plus the content of the frame counter,[4] and

3. one for the plane with a downward scrolling, the address of the memory row being now the address of the displayed line minus the content of the frame counter.

All these operations are done modulo 256 and naturally implement periodic vertical boundary conditions.

In principle, every plane should have its own generator and a selection mechanism to select the model dependent scrolling direction. In fact, to reduce the number of components, the address selection is done by a multiplexing technique and only one address generator. For that, steps 1 and 6 through 11 of the previous section are divided into three substeps, one for each of the three kinds of addresses. A four-bit pattern is associated with these substeps. The direction of the plane motion is then fixed by a four-bit mask and a selector on each plane prevents the row address select (RAS) and the column address select (CAS) signals to be sent to the screen memories except when a match occurs between the substep pattern and the plane mask.

3.4 Hardware realization

The RAP1 machine is made of ten printed boards, with printed circuits for the regular connections like the address lines of the memory and wire wrapping for random connections:

1. Two $5V - 6A$ power supply boards with serial regulators.

2. One interface board connects the RAP1 to an IBM PC-compatible microcomputer through two 16-bit parallel lines with handshake. One set of lines is used to send to the RAP1 a 16-bit command word giving the next operating mode:

   (a) Write the plane direction patterns in one of the four-plane boards;

   (b) set the memory address of the following input-output operation (this step is needed once for consecutive memory addresses, any input-output operation incrementing an I/O address counter);

   (c) write a word in the computation look-up table;

   (d) write a word in the color look-up table;

---

[4]The address of the top and bottom lines on the screen are respectively 0 and FF.

(e) write a word in the screen memory;

(f) read a word from the screen memory;

(g) read the line counter;

(h) read the frame counter (two words);

(i) start the computation process;

(j) stop the computation process.

3. The second set of lines is used to exchange 16-bit words between the PC and the RAP1 during the input-output operations.

4. One board provides all the timing and the address signals. The heart of this board is a graphic display processor (GDP) EF9365 which provides all the synchronization signals along with the line address counter for the display.

5. One board contains the 16 HM6287 SRAMs for the computation look-up table and two of the 16-bit registers in the computation loop.

6. Four boards with four planes per board are used for the screen memory. Each board contains eight TMS 4161 VRAMs, the displacement masks for the four planes, the RAS and CAS selectors, and the third 16-bit register and the multiplexer in the computation loop.

7. Half of the last board is used for the display interface, while the second half is available for future use, like post-processing of the data. This color look-up table is made of three 4k × 4 HM6168 SRAMs and the 16 bits to 12 bits selection is done the following way:

(a) the 12 least significant bits for the blue look-up table,

(b) the 12 middle bits for the green look-up table,

(c) the 12 most significant bits for the red look-up table.

8. This configuration restricts the color coding somewhat, but allows substantial space saving. In most of the studied cases, a suitable choice of the plane meaning allows correct color codings.

4.   Lattice gas simulations with RAP1

The debugging of RAP1 was finished at the end of the first quarter of 1986, six months after we started the project. The next six months were used to write enough software on the host PC to allow friendly use by non-expert users. Most of the efforts were concentrated on the hexagonal lattice gas models, especially on model III of reference 11, which allow reasonable Reynolds numbers to be obtained for moderate lattice sizes. This model was implemented using two seven-bit collision tables as in reference 11 computed once. Then, a spreadsheet allows quick definitions of different look-up tables using the remaining nine bits to define obstacles, sink or

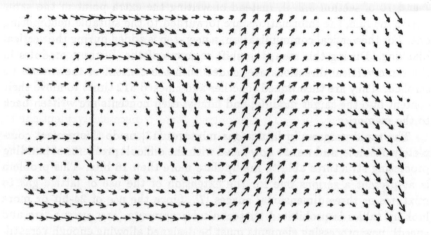

Figure 3: Von Karman street behind a flat plate obtained with RAP1. Local averages over 16 × 16 nodes are performed in the host IBM-PC computer from data retrieved from RAP1.

sources of particles, and so on. At present, we are using a configuration with seven bits for the particles: one to choose between the two possible collision tables, one for obstacles corresponding to "bounce-back" conditions, one for "absolute" sinks which destroy all the incoming particles, and the remaining for directed sources, one for each of the six directions of the triangular lattice. The existing software allows a very easy definition of flows around arbitrary obstacles with different "wind tunnel" conditions. The final geometry can be saved in some kind of library.

Despite the limited size of the RAP1 memory, several interesting hydrodynamics instabilities were observed: Kelvin-Helmholtz and von Karman instabilities. Figure 3 gives an example of the results obtained on RAP1 for a Reynolds number of about 100 and a Mach number of 0.5 around a flat plate, showing the shedding of vortices in the wake. This figure was obtained forty times faster than the corresponding flow of figure 1 of reference 13 obtained on a FPS164 for model II, clearly demonstrating the potentiality of this class of special-purpose computers for lattice gas simulations.

5.   Conclusion

When we started this project, we knew that the size of the RAP1 memory would be too small for realistic hydrodynamics simulations, but we thought that the sixteen bits would be sufficient for future applications. At this time, the project for the next stage is to build a second version with several RAP1 modules with slight modifications to allow the exchange of data between them. The horizontal communication is easily done through steps

7 and 10 of section 3.2, if, instead of writing the extra point in the same RAP1 module, the informations are forwarded to the suitable neighboring one. For the vertical communication, a new step is added during the vertical blanking. During this step, the shift registers of the different modules in the same vertical directions are loaded with the content of the row to be scrolled. They are then linked together and clocked 512 times to move their content into the target module, and finally, their contents are written back to the memories.

The rapid progress in lattice gas mixtures [13] made this project completely obsolete, and we are now faced with the difficult problems of building processing structures able to manipulate more than 16 bits. This problem is at present a serious one for the extension of the use of lattice gas to mixtures or three-dimensional spaces [7]. Since the use of 24-bit or more look-up tables is prohibited by technical arguments (cost versus size and speed), new processing elements must be designed allowing enough versatility for a moderate complexity. For two-dimensional lattice gas models, this goal can be partially reached, at least for models with less than 16 bits, if the full word describing a node is split into two parts: one for the particles themselves (16 for example) with a collision step computed with look-up tables and a second one to drive extra memories used as multiplexers? to implement obstacles, sinks, sources, body forces, random choices between the look-up tables of the different modules, and so on. A $1024 \times 1024 \times 24$ machine working at 50 Msites per second with these kind of computation tables is planned for the next stage and a second one four times as big and as fast soon after.

Acknowledgments

We thank P. Giron, who built the RAP1 machine; J. C. Bernard, S. Bortzmeyer, and V. Beliard, who wrote the software used for the simulations; and P. Lallemand and Y. Pomeau for helpful discussions.

References

[1] J. von Neumann, *Theory of Self-Reproducing Automata*, (Univ. of Illinois press, 1966).

[2] S. Wolfram, *Rev. Mod. Phys.*, **55** (1984) 96.

[3] T. Toffoli, *Physica*, **10D** (1984) 195.

[4] I. E. Broadwell, *Phys. of Fluids*, **7** (1964) 1243; R. Gatignol, "Théorie Cinétique des Gaz à Répartition Discrète de Vitesse", *Lecture Notes in Physics* (Springer-Verlag, Berlin, 1975).

[5] J. Hardy, and Y. Pomeau, *J. Math. Phys.*, **13** (1972) 1042; J. Hardy, O. de Pazzis and Y. Pomeau, *J. Math. Phys.*, **14** (1973) 1746; J. Hardy, O. de Pazzis and Y. Pomeau, *Phys. Rev.*, **A13** (1976) 1949.

[6] U. Frisch, B. Hasslacher and Y. Pomeau, *Phys. Rev. Lett.*, **56** (1986) 1505.

[7] P. Clavin, P. Lallemand, Y. Pomeau and G. Searby, "Simulation of Front Dynamics in Flows: a New Proposal Based on Lattice-Gas Models", *J. Fluid Mech.* (1986), in press. P. Clavin, D. d'Humières, P. Lallemand and Y. Pomeau, *C. R. Acad. Sci. Paris XII*, **303** (1986) 1169.

[8] N. Margolus, *Physica*, **10D** (1984) 81.

[9] T. Toffoli and N. Margolus, *Cellular Automata Machines: a New Environment for Modeling*, (M.I.T. press, 1986).

[10] N. Margolus, T. Toffoli, and G. Vichniac, *Phys. Rev. Lett.*, **56** (1986) 1694.

[11] D. d'Humières and P. Lallemand, "Numerical Simulations of Hydrodynamics with Lattice Gas Automata in Two Dimensions", *Complex Systems*, **1** (1987) 598.

[12] E. R. Berlekamp, J. H. Conway and R. K. Guy, *Winning Ways for Your Mathematical Plays*, **2** (Academic Press, 1984).

[13] D. d'Humières, Y. Pomeau and P. Lallemand, *C. R. Acad. Sci. Paris XII*, **301** (1985) 1391.

Hydrodynamic Studies and
Application Papers

Steven A. Orszag and Victor Yakhot
Applied and Computational Mathematics, Princeton University, Princeton, New Jersey 08544

Reynolds Number Scaling of Cellular-Automaton Hydrodynamics

This paper originally appeared in *Physical Review Letters*, Volume 56, Number 16, April 21, 1986, pp. 1691–1693.

We argue that the computational requirements for presently envisaged cellular-automaton simulations of continuum fluid dynamics are much more severe than for solution of the continuum equations.

PACS numbers: 47.10+g

It has been recently been suggested[1,2] that cellular automata (CAs) [defined as discretely and locally linked, finite- (and few-) state machines] may be an effective way to compute complex fluid flows. These automata have the advantage that they may be simply and perhaps inexpensively constructed with use of specially designed parallel hardware. With suitable interaction rules, it has been argued,[1,2] the space-time average kinetic behavior of the CA system follows the incompressible Navier-Stokes dynamical equations. While the Navier-Stokes equations for continuum fluids can be calculated efficiently on parallel-architecture machines, it is probably easier to make efficient use of the parallel architecture with CAs. In this Letter, we wish to point out that there are some considerations that require resolution before these methods can be considered to be a viable alternative to traditional continuum mechanical methods for high-Reynolds-number fluid dynamics.

Let us compare the resolution and work requirements for a CA simulation of a high-Reynolds-number flow with those of direct numerical solution of the incompressible Navier-Stokes equations. It is well known[3,4] that, at Reynolds number N_{Re}, the Kolmogorov and Batchelor-Kraichnan theories of three- and two-dimensional equilibrium range dynamics, respectively, predict that the range of excited scales is of order $N_{Re}^{3/4}$ and $N_{Re}^{1/2}$ and the computational work required to calculate a significant time in the evolution of large-scale flow structures is of order N_{Re}^3 and $N_{Re}^{3/2}$ in three and two dimensions, respectively.

The suggested evolution rules for CAs to reproduce hydrodynamic behavior are based on conservation laws of mass, momentum, and energy. Dissipation is modeled through the thermalization of coherent hydrodynamic modes. Therefore, the lattice resolution of the CA calculation must be much finer than that of the hydrodynamic simulation, the latter requiring the retention of only those degrees of freedom describing motions on scales of the dissipation range or larger. Thus, the lattice spacing a must be smaller than the dissipation scale η in the turbulent fluid.

We now discuss some conditions that CA models should satisfy to describe high-Reynolds-number fluid flows. We present three successively more restrictive arguments that show that η/a must grow rapidly with Reynolds number.

SIGNAL-TO-NOISE RATIO The hydrodynamic velocity in the CA simulation is calculated by subdividing the computational domain into cells with linear dimensions $\gg a$, averaging over the CAs within a (finite) cell, and smoothing (filtering) the resulting (noisy) velocity field. Thus, the hydrodynamic velocity at a point x is the (space-time) filtered velocity of the CAs in the cell C_x centered at x, $v_H(x) = \langle v(x) \rangle$, where the local velocity in C_x is

$$v(x) = \frac{1}{n} \sum_{i \in C_x} v_i, \tag{1}$$

where n is the number of occupied sites i within the cell. We assume that the possible velocity values at an occupied CA site are $v_i = \pm v_{th}$ where v_{th} is the constant (thermal) velocity over the CA grid. At low Mach numbers, $v_H \ll v_{th}$. In this case, the fluctuations in $v(x)$ are of order $n^{-1/2}v_{th}$. In order that the hydrodynamic velocity found in this way may be a good representation of the continuum hydrodynamics, it is necessary that the noise $n^{-1/2}v_{th}$ be small compared to the smallest significant hydrodynamic velocity. The smallest significant hydrodynamic velocity is the eddy velocity on scales of order of the dissipation scale η. In three dimensions, $\eta = O((\epsilon v^3)^{-1/4})$ and the eddy velocity on the scale of η is $v_\eta = O((\epsilon v)^{1/4})$. Here v is the viscosity and ϵ is the turbulent energy dissipation rate per unit mass. Thus, we require that the number of CA sites n within a cell of size η be at least

$$n \gg \frac{v_{th}^2}{(\epsilon v)^{1/2}}. \tag{2}$$

Since $\epsilon = O(U^3/L)$ where U is the large-scale rms fluctuating velocity and L is the associated large-scale length of these velocity fluctuations, we find that $n \gg N_{Re}^{1/2} M^2$, where $N_{Re} = UL/v$ is the Reynolds number and $M = U/v_{th}$ is the Mach number.[5] Since the number of cells of size η within a three-dimensional turbulent eddy of size L scales as $N_{Re}^{9/4}$, the overall number of CA sites must increase at least as $N_{Re}^{11/4}/M^2$.

Since the effective evolution time of the fluid system is L/U, while the time step on the CA lattice is a/v_{th}, it follows that CA simulation requires at least L/aM steps in time. Since the computational work for each site update is of order 1, it follows that the CA simulation requires at least of order $(N_{Re}/M)^{11/3}$ work.

In summary, the above signal-to-noise considerations suggest the following lower-bound estimates for the computer storage S and work W for CA simulations fo high-Reynolds-number, low-Mach-number flows (where, for reference, we include the corresponding estimates for the continuum Navier-Stokes equations): for CA (2D),

$$S = \frac{N_{Re}^{3/2}}{M^2}, \qquad W = \frac{N_{Re}^{9/4}}{M^4};$$

for Navier-Stokes (2D),[6]

$$S = N_{Re}, \qquad W = N_{Re}^{3/2};$$

for CA (3D),

$$S = \frac{N_{Re}^{11/4}}{M^2}, \qquad W = \left(\frac{N_{Re}}{M}\right)^{11/3};$$

for Navier-Stokes (3D),[7]

$$S = N_{Re}^{9/4}, \qquad W = N_{Re}^3.$$

UPPER BOUND FOR THE REYNOLDS NUMBER

A more stringent condition on the Reynolds-number dependence of the minimal number of lattice sites in a CA simulation of hydrodynamics is found as follows. If the discrete velocity of the CAs is $\pm v_{th}$ (again, the thermal velocity or sound speed on the CA lattice) and the lattice spacing is a, then the kinematic viscosity v on the lattice is at least of order va. For the CA to give a self-consistent hydrodynamic simulation, the viscosity determined on the "molecular" level must equal the viscosity governing the dissipation of the hydrodynamic modes. Thus the Reynolds number of the simulated fluid can be at most UL/v or ML/a. Since the number N of CA sites in the lattice is of order $(L/a)^d$, where d is the dimension of space, we obtain the result that N must be at least of order $(N_{Re}/M)^d$. As above, the CA simulation of the flow requires at least L/aM steps in time. It follows that the CA simulation requires at least of order $(N_{Re}/M)^d$ memory and of order N_{Re}^{d+1}/M^{d+2} work.

These estimates for storage S and work W based on lower bounds for the effective viscosity on the lattice are of order

$$S = \left(\frac{N_{\text{Re}}}{M}\right)^2, \qquad W = \frac{N_{\text{Re}}^3}{M^4};$$

for CA (2D);

$$S = N_{\text{Re}}, \qquad W = N_{\text{Re}}^{3/2}$$

for Navier-Stokes (2D);

$$S = \left(\frac{N_{\text{Re}}}{M}\right)^3, \qquad W = \frac{N_{\text{Re}}^4}{M^5}$$

for CA (3D); and

$$S = N_{\text{Re}}^{9/4}, \qquad W = N_{\text{Re}}^3$$

for Navier-Stokes (3D).

HYDRODYNAMIC FLUCTUATIONS The CA system will yield a self-consistent continuum hydrodynamic description only if the thermal energy fluctuations on hydrodynamic spatial scales are small compared to the energy of the hydrodynamic modes on corresponding length scales. If the "mass" of an occupied CA site m, its energy is $d/2mv_{\text{th}}^2$ in d space dimensions. Then the fluctuation in total thermal energy over a cell with n occupied CAs is $\sqrt{n}mv_{\text{th}}^2$. (We note that in a CA with velocity states $\pm v_{\text{th}}$, energy fluctuations are proportional to density fluctuations.) The corresponding hydrodynamic energy within a cell of size η is $\rho\eta^3 v_H^2$ where v_H is the hydrodynamic velocity and ρ is the hydrodynamic density. In three dimensions, the dissipation scale is η and the associated hydrodynamic velocity is v_η. Also, the relation between m and ρ is $nm = \rho\eta^3$. Thus, for thermal fluctuations to be small, we must require that $n \gg N_{\text{Re}}/M^4$. In two dimensions, the corresponding result is $n \gg N_{\text{Re}}^2/M^4$.

This argument shows that the storage and work required for a self-consistent hydrodynamic description using CAs is of order

$$S = \frac{N_{\text{Re}}^3}{M^4}, \qquad W = \frac{N_{\text{Re}}^{9/2}}{M^7}$$

for CA (2D);

$$S = N_{\text{Re}}, \qquad W = N_{\text{Re}}^{3/2}$$

for Navier-Stokes (2D);

$$S = \frac{N_{\text{Re}}^{13/4}}{M^4}, \qquad W = \frac{N_{\text{Re}}^{13/3}}{M^{19/3}}$$

for CA (3D); and

$$S = N_{\text{Re}}^{9/4}, \qquad W = N_{\text{Re}}^{3}$$

for Navier-Stokes (3D).

The CA models approximate fluids that are by their nature necessarily compressible. This means that an equation of state for pressure is needed. However, self-consistency requires that thermodynamic pressure fluctuations over dissipation scales be small. This latter condition leads to results identical to those just obtained by use of energy estimates. Indeed, it is known[8] that the rms thermodynamic pressure fluctuations in a volume η^3 are $(\rho k T c^2/\eta^3)^{1/2}$, where T is the temperature of the fluid and c is the sound speed. But the hydrodynamic pressure fluctuations over length scales of order η are of order $\rho(\epsilon v)^{1/2}$. Since $kT = mv_{\text{th}}^2$ and $c \approx v_{\text{th}}$, the previously given estimates apply.

We believe that these pessimistic estimates for high Reynolds numbers and low Mach numbers must be overcome before CAs can be an effective modeling tool for complex fluid flows. This can, in principle, be done by averaging over the shortest scales $a \ll \eta$ in order to reduce the number of degrees of freedom.[9] However, it seems that this renormalization can be useful (in the context of local, few-bit, parallel computations) only if it does not generate nonlocal, complex interactions in the set of basic rules defining CAs. Unfortunately, we do not now understand why this kind of "turbulence transport" modeling should be either easier or more successful on the CA lattice than for the continuum equations or for molecular dynamics.

While the above estimates for CA simulations of turbulence are quite pessimistic, there may be cases in which CA simulations of turbulence may be effective. In a turbulent boundary layer, the local Reynolds numbers is $O(1)$ in the viscous sublayer and is modest within the buffer layer. A CA model could be effective in these regions in the modeling of turbulent burst formation and evolution. However, this application requires the development of three-dimensional CA models and suitable techniques to match the outer regions of the flow.

We would like to thank Robert Kraichnan and Stephen Wolfram for discussions. This work was supported by the Office of Naval Research under Contracts No. N00014-82-C-0451 and No. N00014-85-K-0201, and by the Air Force Office of Scientific Research under Contract No. F49620-85-C-0026.

REFERENCES

1. Frisch, U., B. Hasslacher, and Y. Pomeau (1986), *Phys. Rev. Lett.* **56**, 1505.
2. Salem, J., and S. Wolfram, to be published.
3. Orszag, S. A. (1970), *J. Fluid Mech.* **41**, 363–386.
4. Herring, J. R., S. A. Orszag, R. H. Kraichnan, and D. G. Fox (1974), *J. Fluid Mech.* **66**, 417–444.
5. Notice that the Mach number is bounded in CA simulations of fluid dynamics. With allowed velocities of $\pm v_{th}$, $M \leq O(1)$, the upper limit being achieved when the CAs exhibit pure streaming, nonhydrodynamic motion.
6. The Batchelor-Kraichnan theory suggests that there are may be logarithmic corrections to these estimates.
7. Intermittency effects may slightly change these estimates.
8. Landau, L. D., and E. M. Lifschitz (1959), *Fluid Mechanics* (London: Pergamon), 529.
9. A kind of renormalization using "pseudovertices" is suggested in Ref. 1. The idea is to introduce vertices on subhydrodynamic scales that are not kept track of explicitly. If this idea can be successfully implemented, it would reduce the computational storage requirements given above but apparently not the computational work. We are grateful to C. H. Bennett for this comment.

Daniel H. Rothman & Jeffrey M. Keller
Department of Earth, Atmospheric, and Planetary Sciences, Massachusetts Institute of Technology, Cambridge, MA 02139

Immiscible Cellular-Automaton Fluids

This paper originally appeared in *Journal of Statistical Physics*, volume 52(3/4), August 8, 1988, pp. 1119–1127.

We introduce a new deterministic collision rule for lattice-gas (cellular-automaton) hydrodynamics that yields immiscible two-phase flow. The rule is based on a minimization principle and the conservation of mass, momentum, and particle type. A numerical example demonstrates the spontaneous separation of two phases in two dimensions. Numerical studies show that the surface tension coefficient obeys Laplace's formula.

Recently, Frisch, Hasslacher, and Pomeau (FHP)[1] introduced a discrete lattice-gas model for the numerical solution of the 2-D incompressible Navier-Stokes equations. In their model, space, time and the velocities of particles are discrete. Identical particles of equal mass populate a triangular lattice, obey simple collision rules, and travel to neighboring sites at each time step. Because the model is entirely discrete, and because the evolution of a site is determined by the state of the site and its nearest neighbors, the lattice gas is a cellular automaton.[2] Despite its simplicity, the macroscopic behavior of the lattice-gas automaton asymptotically approaches continuum flow. Since its introduction, this new model of fluid dynamics has not only been the subject of extensive theoretical and numerical studies,[3-7] but has also been extended to 3-D[8] and applied to a wide range of problems.[9-11]

Here we introduce a simple yet fundamental extension of the lattice gas that leads to immiscible two-phase flow with interfacial tension between fluid phases. In regions occupied by only a single phase, our 2-D model is (barring irrelevant details) identical to the FHP gas. When two phases occupy the same region, however, we apply a new collision rule that causes preferential grouping of like phases. To demonstrate the validity of our model, we provide empirical evidence that it correctly honors the physics of interfacial tension. Of course, the asymptotic arguments of FHP apply equally well to our model in regions of homogeneity.

The equations of motion for immiscible two-phase flow are given by the Navier-Stokes equations within each phase, and by boundary conditions at the interfaces between phases.[12,13] The incompressible Navier-Stokes equations are

$$\nabla \cdot \mathbf{u} = 0 \tag{1}$$

$$\rho \partial_t \mathbf{u} + \rho(\mathbf{u} \cdot \nabla)\mathbf{u} = -\nabla p + \mu \nabla^2 \mathbf{u}, \tag{2}$$

where ρ denotes the density, $\mathbf{u}$ the velocity, p the pressure, and μ the shear viscosity. Two boundary conditions govern the behavior of an interface. The first is the purely kinematical statement that the component of velocity locally normal to the boundary in each phase must equal the normal component of the interfacial velocity:

$$\mathbf{u}_1 \cdot \mathbf{n} = \mathbf{u}_2 \cdot \mathbf{n} = \mathbf{u}_{int} \cdot \mathbf{n}. \tag{3}$$

Here the subscripted velocities refer to the two phases and the interface; $\mathbf{n}$ is the unit normal to the interface. The second boundary condition is a dynamical description of the momentum flux across the interface. The requirement here is that the stress difference at the interface be balanced by surface tension. In 2-D, this equation is

$$\mathbf{T}_2 \cdot \mathbf{n} - \mathbf{T}_1 \cdot \mathbf{n} = (\sigma/R)\mathbf{n}, \tag{4}$$

where σ is the surface tension coefficient, R is the radius of curvature (considered positive when the center of curvature is on the side of the interface to which n points), and $\mathbf{T}_1$ and $\mathbf{T}_2$ denote the stress tensor $\mathbf{T} = -p\mathbf{I} + \mu(\nabla \mathbf{u} + (\nabla \mathbf{u})^{\mathbf{T}})$ in phases 1 and 2, respectively.

Of these four equations, the FHP gas, after a rescaling of variables and macroscopic averaging, models the first two. The idea of FHP is to set up a triangular lattice, with identical particles traveling between neighboring sites at each time step. Up to six particles, each with unit velocity, may reside at a site, but there may be at most one particle moving in each of the six possible directions. When particles meet at the same site, they obey collision rules that conserve mass (particle number) and momentum. Macroscopic fields are obtained by coarse-grain averaging in space and time; the correspondence of these macroscopic fields to Eqs. (1) and (2) is primarily due to the microscopic conservation of mass and momentum and the symmetries of the triangular lattice.

To model two-phase flow, one must not only satisfy Eqs. (1) and (2), but also the boundary conditions of Eqs. (3) and (4). Our two-phase model is built upon the FHP foundation. We define two kinds of particles, and refer to them by *color* –"red" and "blue."[7,9] In addition to conserving mass and momentum, however, collisions must now also conserve the number of reds (or blues) and encourage the preferential grouping of like colors.

We employ a version of the FHP model that includes zero-velocity "rest particles"[1,3,6]; there are thus 7 available velocities. The ith velocity vector is denoted by c_i; $c_0 = 0$ and c_1 through c_6 are unit vectors connecting neighboring sites on the triangular lattice. Red and blue particles may simultaneously occupy the same site, but not with the same velocity. The Boolean variable $r_i(x) \in \{0,1\}$ indicates the presence or absence of a red particle with velocity c_i at lattice site x; the variable $b_i(x) \in \{0,1\}$ plays the same role for a blue particle. The configuration at a site is thus completely described by the two 7-bit variables $r = \{r_i, i = 0,\ldots,6\}$ and $b = \{b_i, i = 0,\ldots,6\}$.

Cohesion in real liquids results from short-range intermolecular forces of attraction.[14] We model these short-range forces by allowing the particles at sites which are the nearest neighbors of site x to influence the configuration of particles at site x. Specifically, we define a local *color flux* and a local *color field*, and design collision rules such that the "work" performed by the flux against the field is minimized, subject to the constraints of mass, momentum, and color conservation.

The local color flux $q[r(x), b(x)]$ is defined to be the difference between the net red momentum and net blue momentum at site x:

$$q\left[r(x), b(x)\right] \equiv \sum_i c_i \left[r_i(x) - b_i(x)\right]. \qquad (5)$$

The local color field $f(x)$ is defined to be the direction-weighted sum of the differences between the number of reds and the number of blues at neighboring sites:

$$f(x) \equiv \sum_i c_i \sum_j \left[r_j(x + c_i) - b_j(x + c_i)\right]. \qquad (6)$$

The work W performed by the flux against the field is then

$$W(r, b) = -f \cdot q(r, b). \qquad (7)$$

Here we have incorporated repulsive forces between different colors in addition to attractive forces between like colors. The result of a collision, $r \to r'$, $b \to b'$, is determined by the solution to the minimization problem

$$W(r', b') = \min_{r'', b''} W(r'', b'') \qquad (8)$$

subject to the constraints of colored mass conservation

$$\sum_i r_i'' = \sum_i r_i, \qquad \sum_i b_i'' = \sum_i b_i, \qquad (9)$$

and colorblind momentum conservation

$$\sum_i \mathbf{c}_i(r_i'' + b_i'') = \sum_i \mathbf{c}_i(r_i + b_i). \tag{10}$$

If the solution is nonunique, the outcome of a collision is chosen with equal *a priori* weight from the set of configurations that satisfy Eqs. (8), (9) and (10). In the resulting configuration, particles of each color will be preferentially moving toward concentrations of like color and away from concentrations of unlike color. Once collisions have occurred, each particle moves one lattice unit in its direction of motion.

Figure 1 illustrates the nonequilibrium behavior of this two-fluid model. The initial configuration is a random mixture. The reduced density $d = \rho/7 = .75$; ρ is the average number of particles per site. There are 128^2 sites; boundaries are periodic, both horizontally and vertically. The initial distribution of particles is random, with reds and blues equally probable. A high density of red particles at a site is illustrated as white and a high density of blue particles as black; lower densities or mixtures of red and blue are illustrated by the appropriate shade of gray. No averaging has been performed.

The automaton quickly acts to smooth out all surfaces, producing a number of 2-D bubbles. Random motion eventually causes small bubbles to meet, producing even larger bubbles. The final equilibrium state is full separation, with plane horizontal interfaces.

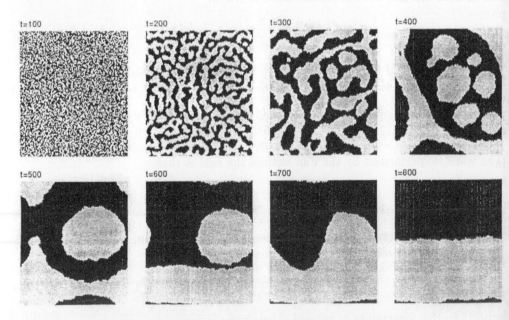

FIGURE 1 Nonequilibrium behavior of the two-phase automaton, illustrated at representative time steps. The initial configuration is a random mixture.

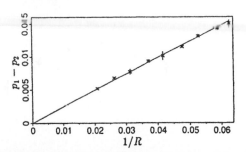

FIGURE 2 Numerical confirmation of Laplace's formula, Eq. (11). R is the radius of a 2-D bubble. The calculated points should each lie on the same horizontal line.

Quantitative tests show that the automaton is indeed modeling surface tension. We test the validity of Eq. (4), for the simplified case of no flow. This boundary condition is then Laplace's formula,[13]

$$p_1 - p_2 = \sigma/r. \tag{11}$$

Figure 2 is a plot of $R\Delta p$ vs. R, where R is the radium of a 2-D bubble and $\Delta p = p_1 - p_2$. Pressure is calculated from the equation of state $p = 3\rho/7$ given in Frisch et al.[3] The tests were performed by initializing a lattice of size $(4R)^2$ with a blue bubble of radius R in a sea of red, with $d = .75$ and $u = 0$ in both phases. The bubble maintains its gross shape; Δp is measured by computing the difference between the average density of sites occupied by only blue particles and the average density of sites occupied by only red particles. The typical density contrast is less than 1%. As with the real bubbles, these 2-D bubbles undergo free oscillations evident in plots of Δp vs. time. Here we have simply averaged Δp over 2000 time steps beginning at time step 500. For each R, this calculation is performed 4 times with different initial configurations. The graph shows the mean and standard deviation of the averages computed in the 4 runs, for each value of R, compared to the best-fitting horizontal line. As predicted by Eq. (11), $R\Delta p$ is invariant with R, except for statistical fluctuations. We conclude that the surface tension for $d = .75$ is $\sigma = .247 \pm .002$.

Figure 3 depicts the continuous variation of surface tension with density. Surface tension is low at low densities, partly because fewer particles provide fewer attractive forces, and also because fewer particles provide less choices in the minimization of W. At unusually high densities, the surface tension will also be low but surfaces do not break; σ vanishes at $d = 1$ because Δp must equal zero at full density.

Of the two boundary conditions, we have quantitatively studied only Eq. (4). Because we have observed the continuous flow of surfaces without rupture, we take the automaton's adherence to Eq. (3) to be self-evident for u measured from the lattice. However, the momentum equation, Eq. (2), is attained only after a rescaling of variables to attain Galilean invariance[1,3]; this rescaling does not apply to

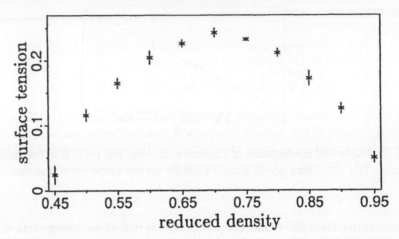

FIGURE 3 Numerical measurements of surface tension as a function of reduced density.

the velocity of the interface. Thus our model as it stands lacks Galilean invariance. D'Humières, Lallemand, and Searby[7] have shown, however, that the FHP collision-rules can be tuned to provide Galilean invariance without a rescaling of variables; their corrections should be applicable to our model as well.

We comment briefly on extensions of the model. Fluids of different viscosities may be modeled by either limiting the set of possible collisions in one fluid, or by decreasing the frequency of collisions in one fluid. Buoyant forces may be modeled using the techniques described in Burges and Zaleski.[9] Extensions to 3-D[8] are straightforward in principle. Flow of three or more fluids can be modeled by dispensing with the unified expressions for the flux and field given by Eqs. (5) and (6), and instead defining red fields, blue fields, and, say, green fields, and likewise for the fluxes. The (possibly weighted) sum of the work of each flux against the corresponding field of the same color would then be minimized. No repulsive forces would be modeled, but we do not consider them necessary.

We have presented a new model, based on discrete dynamics, for the solution of the flow equations for two immiscible fluids. We expect that the model will be useful for the study of a variety of two-fluid instability phenomena, possibly with increased computational efficiency over other methods[15] insofar as they exist. The model should be particularly useful for the study of viscous fingering[16] in microscopic models of porous media,[10] a problem for which no comprehensive numerical model has yet been found. The simplicity of this automaton for two-phase flow should thus provide not only a novel approach to the study of fluid mixtures, but also the ability to study computationally what has heretofore been accessible only by experiment.

ACKNOWLEDGMENTS

It is a pleasure to thank Stéphane Zaleski for numerous thought-provoking discussions. We would also like to thank A. Gunstensen for technical assistance, J. Berryman, B. Boghosian, and T. Madden for their advice and encouragement, and S. Cole, J. Dellinger, and R. Ottolini for providing graphics software. This work was supported in part by Army Research Office grant DAAG29-85-K-0226 and by the Department of Earth, Atmospheric, and Planetary Science, Massachusetts Institute of Technology.

REFERENCES

1. Frisch, U., B. Hasslacher, and Y. Pomeau (1986), *Phys. Rev. Lett.* **56**, 1505.
2. Wolfram, S. (1986), *Theory and Applications of Cellular Automata* (Singapore: World Scientific).
3. Frisch, U., D. d'Humières, B. Hasslacher, P. Lallemand, Y. Pomeau, and J-P. Rivet (1987), *Complex Systems* **1**, 648.
4. Wolfram, S. (1986), *J. Stat. Phys.* **45**, 471.
5. Kadanoff, L., G. McNamara, and G. Zanetti (1987), "From Automata to Fluid Flow: Comparisons of Simulation and Theory," *University of Chicago preprint*.
6. D'Humières, D., and P. Lallemand (1987), *Complex Systems* **1**, 598.
7. D'Humières, D., P. Lallemand, and G. Searby (1987), *Complex Systems* **1**, 632.
8. D'Humières, D., P. Lallemand, and U. Frisch (1987), *Europhys. Lett.* **2**, 291; Rivet, J-P. (1987), *C.R. Acad. Sci. Paris II* **305**, 751.
9. Burges, C., and S. Zaleski (1987), *Complex Systems* **1**, 31.
10. Rothman, D. (1988), *Geophysics* **53**, April 1988, in press.
11. See bibliography of 3 above.
12. Drew, D. A. (1983), *Ann. Rev. Fluid Mech.* **15**, 261.
13. Landau, L. D., and E. M. Lifshitz (1959), *Fluid Mechanics* (New York: Pergamon Press), 230; Batchelor, G. K. (1967), *An Introduction to Fluid Dynamics* (Cambridge: Cambridge University), 68.
14. Rowlinson, J. S., and B. Widom (1982), *Molecular Theory of Capillarity* (Oxford: Clarendon Press).
15. Aref, H. (1987), "Finger, Bubble, Tendril, Spike," *University of California, San Diego, preprint*.
16. Homsy, G. M. (1987), *Ann. Re. Fluid Mech.* **19**, 271.

Victor Yakhot, Bruce J. Bayly, and Steven A. Orszag
Applied and Computational Mathematics, Princeton University, Princeton, New Jersey 08544

Analogy between Hyperscale Transport and Cellular Automaton Fluid Dynamics

It is argued that the dynamics of a very large-scale (hyperscale) flow superposed on the stationary small-scale flow maintained by a force $\mathbf{f}(\mathbf{x})$ is analogous to the cellular automation hydrodynamics on a lattice having the same spatial symmetry as the force $\mathbf{f}$.

While real fluids consist of discrete particles, they can be regarded as continuous media at scales that are much lager than the typical intermolecular distance, and, on these scales, they can be described by the equations of continuum hydrodynamics. The equations are quite insensitive to the details of the molecular dynamics; the microscopic interactions affect only the viscosity coefficient. Microscopically dissimilar fluids can be described by the Navier-Stokes equations, although the microscopic properties of different fluids may be reflected in a very wide range of viscosity coefficients.

The lack of dependence of the hydrodynamics on the microscopic properties of the fluids is the basis for the recent interest in discrete approximations to molecular dynamics or Cellular Automata (CA's).[1-4] Cellular automata are discretely and locally linked, finite-state machines. The "molecules" in a CA fluid move in discrete steps over the lattice sites and interact according to a well-defined set of rules that typically conserve momentum and the total number of particles. The hydrodynamic

Lattice Gas Methods for Partial Differential Equations, SFI SISOC,
Eds. Doolen et al., Addison-Wesley Publishing Co., 1990   **283**

behavior of the CA fluid is given by the evolution of the average macroscopic properties of the system ("slow" modes). Some limitations of CA hydrodynamics have been discussed in Ref. 5.

The first lattice model of a fluid was introduced by Hardy, de Passis, and Pomeau (HPP).[1] Recently, new models, which are modifications of the HPP ideas, have led to simulations of two-dimensional fluid motions that appear to be compatible with experimental data.[2–4] Hydrodynamic equations for such a CA fluid can be derived using techniques based on the Chapman-Enskog expansion, as in the kinetic theory of gases. It has been found that the form of the continuum equations for a given CA fluid depends strongly on the symmetry properties of the lattice. In particular, the HPP lattice gas model based on the two-dimensional square lattice leads to anisotropic viscosity and anisotropic nonlinear terms in the resulting continuum dynamics. However, a regular hexagonal lattice, again in two dimensions, introduced by Frisch, Hasslacher, and Pomeau[2] is symmetric enough to produce the Navier-Stokes equations with isotropic viscosity and nonlinear terms. It has been pointed out by Wolfram[6] that in three dimensions none of the space-filling crystallographic lattices have sufficient symmetry to guarantee the isotropy of the corresponding hydrodynamic equations. However, icosahedral symmetry would produce isotropic viscosity and nonlinear terms.[6] Unfortunately, no periodic lattice has such symmetry (with the possible exception of some recently conjectured quasicrystal structures).[7]

Hydrodynamic equations are derived from the microscopic equations of motion by averaging over small scales. It is natural to pose the following problem: Let us consider a viscous fluid driven by a force which generates a stationary field v^l on the small-scale l. The equation of motion for the perturbation v^L defined on scales L which are much lager than the scale l of the basic flow can be derived by averaging over the small-scale velocity field v^l. The resulting equations describe *hyperscale hydrodynamics*. To see that this problem arises quite naturally, let us imagine a system of microscopic particles driven by an external force $\mathbf{f}$. The molecules participate in two kinds of motion, one related to thermodynamic noise and the other caused by the external force. Filtering out the smallest (thermodynamic) scales, one derives the Navier-Stokes equation subject to the external force. If we also average over the scales corresponding to the force $\mathbf{f}$, the resulting equation will not necessarily be the Navier-Stokes equation, but will rather be an equation describing the large-scale (hyperscale) notion that does not explicitly include the external force.

Some examples of hyperscale dynamics have been considered in Ref. 8. It has been shown that a system of square vortices (eddies) gives rise to the equation of motion for the velocity perturbation at large scales with an anisotropic viscosity:

$$\nu_\phi = \nu(1 + \frac{3}{8}\mathrm{Re}^2 - \frac{1}{2}\mathrm{Re}^2\sin 2\phi)\,, \tag{1}$$

where ν_ϕ is the effective viscosity in the direction $(\cos\phi, \sin\phi)$. In this and all subsequent equations, Re denotes the Reynolds number of the small-scale flow; the formulas quoted are the lowest-order nontrivial results of a hierarchy of successive smoothing approximations.[8] A plane parallel system of eddies also leads to an anisotropic viscosity.[8] However, a system of triangular vortices having hexagonal symmetry (invariance under rotation by 60 deg) generates an isotropic viscosity coefficient for the hyperscale motion:

$$\nu_1 = \nu(1 + \frac{3}{4}\mathrm{Re}^2) \ . \tag{2}$$

The analogy with cellular automata is striking: in two dimensions only the triangular lattice and triangular set of vortices produce an isotropic equation for the large-scale velocity fluctuations. Moreover, it has been shown by Sivashinsky[10] that hyperscale hydrodynamics is not Galilean invariant because the averaged Navier-Stokes equation with the forcing term is not. The same holds for the CA hydrodynamics: It has been pointed out[2] that the continuum equations following from cellular automaton models are not Galilean invariant as a result of the discrete lattice underlying the model.

The dynamics of hyperscale flows superposed on small-scale flows in three dimensions have been studied in Refs. 9 and 11. The analogs of steady cellular flows in two dimensions are the family of so-called Beltrami flows in there edimensions, defined by finite Fourier sums of the form

$$\mathbf{v}(\mathbf{x}) = \sum_{Q \epsilon S} A(Q)(\mathbf{n} + iQ \times \mathbf{n})e^{iQ \cdot x/1} \ . \tag{3}$$

Here S is a finite set of unit-magnitude vectors Q, $\mathbf{n}$ is a unit vector perpendicular to Q, and the complex amplitudes $A(Q)$ satisfy the reality condition $A(-Q) = A^*(Q)$. These flows, like two-dimensional cellular flows, are exact steady solutions of the inviscid fluid equations, and can be maintained in viscous fluid by the action of an externally imposed body force.

The form of the equation of motion for hyperscale flow on the small-scale flow (3) involves the fourth-rank tensor[9,11]

$$\mathcal{N}_{ijkl} = \sum_{(Q) \epsilon S} |\, AQ\,|^2 \left(\delta_{ik}\delta_{jl} - 2\delta_{ik}Q_jQ_l - \delta_{jl}Q_iQ_k - 2\delta_{ij}Q_iQ_jQ_kQ_l\right),$$

which is isotropic only if

$$\sum_{Q \epsilon S} |\, A(Q)\,|^2 \, Q_iQ_j = \lambda\delta_{ij},$$

$$\sum_{Q \epsilon S} |\, A(Q)\,|^2 \, Q_iQ_jQ_kQ_l = \mu(\delta_{ij}\delta_{kl} + \delta_{il}\delta_{kj} + \delta_{ik}\delta_{jl}),$$

for some constants λ and μ. Clearly, the more vectors Q we have in S, the better our chances of being able to sellect amplitudes $A(Q)$ of the corresponding components so as to obtain an isotropic tensor $\mathcal{N}$. We shall give some examples to illustrate the connection between the underlying small-scale flow structures and the resulting hyperscale dynamics.

The simplest Beltrami flow in the family (3) has only two Fourier components with the wave vectors $\pm Q_o = (\pm 1, 0, 0)$ and amplitudes $\mathcal{A}(\pm Q_o) = c/2$; $v^l = c$ $(0, \cos x/l, -\sin x/l)$. A hyperscale perturbation with the wave vector $(0, k, 0)$ then obeys the equation of motion

$$\dot{v}_1 = -\nu k^2 (1 - \frac{1}{2}\mathrm{Re}^2)v_1, \qquad \dot{v}_2 \equiv 0,$$

$$\dot{v}_3 = -\nu k^2 (1 - \frac{1}{2}\mathrm{Re}^2)v_3$$

The effective viscosity for the v_1 component is negative, and the hyperscale flow is therefore unstable, if the small-scale Reynolds number $\mathrm{Re} \equiv cl/\nu$ exceeds $\sqrt{2}$.

The so-called ABC flow is obtained when S consists of six wave vectors located at the vertices of an octahedron:

$$S = \{Q\} = \left\{ \begin{pmatrix} \pm 1 \\ 0 \\ 0 \end{pmatrix} \begin{pmatrix} 0 \\ \pm 1 \\ 0 \end{pmatrix} \begin{pmatrix} 0 \\ 0 \\ \pm 1 \end{pmatrix} \right\}, \tag{4}$$

with all amplitudes having the same modulus $\mid \mathcal{A}(Q) \mid = cl/\sqrt{12}$, where c is again the rms fluid velocity. Now the hyperscale equation takes the form

$$\frac{\partial}{\partial t}v_a = -\nu k^2 [\delta_{ab} + \mathrm{Re}^2 M_{ab}(\hat{\mathbf{k}})]v_b,$$

where M is a matrix that depends only on the unit vector $\hat{\mathbf{k}}$ in the direction of $\mathbf{k}$. The eigenvalues of M are all greater than or equal to zero, with equality occurring only if $\hat{\mathbf{k}}$ lies in one of the coordinate planes. The hyperscale flow is therefore stable although still somewhat anisotropic.

A more complex flow with smilar structure to the ABC flow can be obtained by augmenting the wave vector set by

$$S^1 = \left\{ \begin{pmatrix} \pm\frac{2}{3} \\ \pm\frac{2}{3} \\ \pm\frac{1}{3} \end{pmatrix} \begin{pmatrix} \pm\frac{1}{3} \\ \pm\frac{2}{3} \\ \pm\frac{2}{3} \end{pmatrix} \begin{pmatrix} \pm\frac{2}{3} \\ \pm\frac{1}{3} \\ \pm\frac{2}{3} \end{pmatrix} \right\}, \tag{5}$$

and assigning the corresponding Fourier modes amplitudes $\mathcal{A}(Q) = \lambda(c/\sqrt{12})$, $Q \epsilon S^1$. Here λ is a parameter: setting $\lambda = 0$ recovers the ABC flow with its anisotropic equation of motion, but if λ is raised to the special value of $9/(156)^{1/2}$ then the conditions for isotropy of the tensor $\mathcal{N}_{ijkl}$ are satisfied, and we obtain an isotropic equation of motion for the hyperscale modes. This example illustrates the fact that a small-scale flow can be constructed to have different large-scale properties from a discrete lattice gas with the same spatial symmetry group.

Going on to more and more symmetric flows, it turns out that *icosahedral* or *dodecahedral* symmetry in the small-scale flow gives exact isotropy to the hyperscale dynamics. For example, the wave vector set for the icosahedral flow is

$$S = \left\{ \frac{1}{s} \begin{pmatrix} \pm 1 \\ \pm \tau \\ 0 \end{pmatrix}, \frac{1}{s} \begin{pmatrix} 0 \\ \pm 1 \\ \pm \tau \end{pmatrix}, \frac{1}{s} \begin{pmatrix} \pm \tau \\ 0 \\ \pm 1 \end{pmatrix} \right\}, \tag{6}$$

where τ is the golden ration of $(1 + \sqrt{5})/2$ and $S = (5 + \sqrt{5})/2$. It is easily checked that the tensor $\mathcal{N}_{ijkl}$ is isotropic for this flow, provided that the amplitudes of the modes are all chosen equal. The hyperscale properties of the icosahedral flow appear to be indistinguishable from those of the exactly isotropic flow obtained as the limiting case of flows with more and more Fourier components distributed uniformly on the unit sphere.

It is interesting to observe that not only does the momentum equation for the hyperscale modes take the classical form, but so does the hyperscale diffusion equation The tensor that enters the correction for the effective diffusivity in the lowest smoothing approximation[8,9] is the sum of the "projection tensors"

$$T_{ij} = \sum_{Q \in S} (\delta_{ij} - Q_i Q_j),$$

which is isotropic for all the aforementioned flows except the simple flow with only two Fourier components. Indeed, it has been demonstrated independently of the smoothing theory that a passive contaminant disperses diffusively in the icosahedral and augmented cubic flows. Simulations of particles dispersion in these flows demonstrate that, for large times, almost all particles migrate away from their starting point with a finite effective diffusivity.

The analogy between the CA and hyperscale hydrodynamic descriptions of the fluids goes even further. The equations for the large-scale velocity field derived in Refs. 7-10 are based on the neglect of the higher-order nonlinear terms generated by the scale elimination procedure and thus they are valid when the ratio $v^L \ll v^l$. The same holds for the CA hydrodynamics:[2] the Navier-Stokes equation is an approximation valid only when the Mach number $\text{Ma} = v/v_{th} \ll 1$, where v_{th} is the velocity of the particles on the lattice.

Based on the analogy between hyperscale hydrodynamics and the CA description of the fluids, we argue that lattice gas models are equivalent to the Navier-Stokes equation with an external force having the symmetry of the lattice.

ACKNOWLEDGMENTS

This work was supported by the Air Force Office of Scientific Research under Contract No. F49620-85-C-0026 and by the Office of Naval Research under Contract No. N0014-82-C-0451.

REFERENCES

1. J. Hardy, O. de Pazzis, and Y. Pomeau (1976), *Phys. Rev. A* **13**, 1949.

2. U. Frisch, B. Hasslacher, and Y Pomeau (1986), *Phys Rev. Lett.* **56**, 1505.

3. S. Wolfram, submitted to *Phys. Rev. Lett.*

4. D. d'Humieres, Y. Pomeau, and P. Lallemand, *Mech. Fluids* (in press).

5. S.A. Orszag and V. Yakhot (1986), *Phys. Rev. Lett..* **56**, 1693.

6. S. Wolfram (private communicaton).

7. d'Humieres, Lallemand, and Frisch have recently observed that the 3-D projection of the 4-D Bravais 24-hedral lattice leads to isotropic 3-D fluid dynamics at low Mach numbers.

8. G. Sivashinsky and V. Yakhot (1985), *Phys. Fluids* **28**, 1040.

9. B. Bayly and V. Yakhot, submitted to *Phys. Rev. A.*

10. G. Sivashinsky (1985), *Physica D* **17**, 243.

11. V. Yakhot and G. Sivashinsky, submitted to *Phys. Fluids.*

Guy G. McNamara and Gianluigi Zanetti†

The Research Institutes, The University of Chicago, 5640 South Ellis Avenue, Chicago, IL 60637 and † present address: Program in Applied & Computational Mathematics, Princeton University, Princeton, NJ 08544

Use of the Boltzmann Equation to Simulate Lattice-Gas Automata

This paper originally appeared in *Physical Review Letters*, volume 61, number 20, November 14, 1988, pages 2332 to 2335.

We discuss an alternative technique to the lattice gas automata for the study of hydrodynamic properties; namely, we propose to model the lattice gas with a Boltzmann equation. This approach completely eliminates the statistical noise that plagues the usual lattice gas simulations and therefore permits simulations that demand much less computer time. It is estimated to be more efficient than the lattice gas automata for intermediate to low Reynolds number $R \lesssim 100$.

PACS numbers: 47.10+g

In this Letter we describe a simulation technique that can be used in some situations as an alternative to the lattice gas automaton. The latter has been proposed as a new technique[1] for the numerical study of the Navier-Stokes equation and is based on the simulation of a very simple microscopic system, rather than on the direct integration of partial differential equations. Particles hop between the sites of a regular lattice and may have collisions only on the lattice sites. The collision process is deterministic and is controlled by a set of collision rules chosen so that, for instance, they conserve the number of particles and linear momentum. The transition from the microscopic to the macroscopic description of the lattice gas automata (LGA) is done by defining coarse-grained conserved densities, e.g., momentum density, obtained by averaging their microscopic equivalents on subregions of the lattice. The presence of microscopic conservation laws then reappears in the macroscopic dynamic as hydrodynamic modes and, if the underlying regular lattice has been properly chosen, one can argue that the form of the hydrodynamic equations is very similar to that found for simple fluids.

Many authors[2-4] have tested the validity of the lattice-gas scheme by simulating specific flow configurations that are analytic solutions to the Navier-Stokes equation. All these simulations required a rather massive use of computer resources because the microscopic dynamic of the LGA is intrinsically noisy[5,6] and to obtain reasonably resolved coarse-grained densities, it is necessary to average over a combination of large subregions of the lattice, long times, and numerous initial conditions.

We would like to point out that the hydrodynamic properties of the lattice-gas automata can be determined very efficiently by using an alternative technique. What we suggest is to translate the LGA into a related Boltzmann model. The Boolean site populations of the LGA then become real numbers between 0 and 1 representing their average value and their time evolution controlled by a Boltzmann equation (BE)[7] derived from the lattice-gas model.

It is clear that what we have just described is not equivalent to the LGA since, by factorizing the LGA collision operator on the one-particle distribution functions, we completely neglected all the effects due to the correlations between the particles. Nevertheless, the Boltzmann model shares many features with the LGA. In particular, it has the same hydrodynamic behavior, even though some details, such as the transport coefficients, can be slightly different. This implies that basically all the peculiarities of the LGA, e.g., the lack of Galilean invariance, are inherited by the BE. Also, the Boltzmann gas is, like the LGA, a stable numerical scheme since its time evolution satisfies an H theorem.[8] We note that the BE approximations of the LGA are very common in the literature.[8-10] However, their use was previously restricted to analytical calculations and not to actual numerical simulations.

There is a feature of the LGA that is not shared by the Boltzmann model, namely, the noise. In the latter approach we directly study the time evolution of the mean values of the one-particle distribution functions and we therefore completely bypass the statistical averaging step needed in the LGA simulations. We may then obtain accurate results even when using small lattices (say, 8 × 8 sites in two dimensions).

The Boltzmann scheme is obviously applicable to any lattice-gas automaton and in particular it can be used for the testing of 2-D and 3-D models. For the purpose of the present Letter we constrain ourselves to the case of the original

2-D lattice gas on an hexagonal lattice. In this model, the particles have momenta chosen from the vector

$$\mathbf{C}_a = \left[\cos\left(\pi\frac{a-1}{3} \right), \sin\left(\pi\frac{a-1}{3} \right) \right], \quad a = 1, \ldots, 6.$$

The microscopic densities corresponding to the number of particles and momentum conservation are, respectively, $n(\mathbf{r}, t) = \sum_a f_a(\mathbf{r}, t)$, and $g_l(\mathbf{r}, t) = \sum_a C_{a;l} f_a(\mathbf{r}, t)$, where $f_a(\mathbf{r}, t)$ is the Boolean population field that indicates the presence (1) or absence (0) of a particle moving with momentum $\mathbf{C}_a$ at site $\mathbf{r}$ and time step t, and we use $i, j, k, l, \ldots$ to label Cartesian coordinates. The time evolution of the particle populations can then be written, with the assumption that the particles first hop in the direction of their velocities and then are subject to collisions, as

$$f_a(\mathbf{r}, t+1) = f_a(\mathbf{r} - \mathbf{C}_a, t) + T_a\left(\{f_b(\mathbf{r} - \mathbf{C}_c, t)\} \right),$$

where T_a is the microscopic collision operator,[8] that is the Boolean algebra expression that corresponds to the chosen set of collision rules. In the BE approximation of the previous equation we understand $f_a(\mathbf{r}, t)$ as a continuous variable between 0 and 1 and replace the Boolean operations in T_a with the appropriate arithmetic operations.

The theory of the Boltzmann equation for the lattice gas been extensively treated in Refs. 8-10 and various authors[2,8] have given formulas for the transport coefficients based on the Chapman-Enskog[7] approximation. As a straightforward application of this new tool we computed the values of the bulk and kinematic viscosities of the Boltzmann gas by studying the decay of shear and sound waves

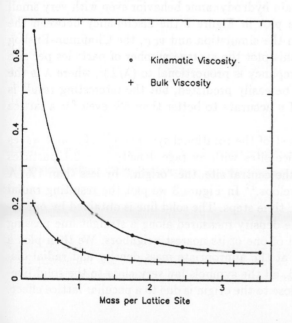

FIGURE 1 Transport coefficients as a function of the density for the Boltzmann gas. The data were obtained by relaxation measurements. The wavelength of the waves used is $L = 64\sqrt{3}$. The set of rules used is the FHP-III defined in Ref. 2. The solid lines are the Chapman-Enskog values.

Kinematic Viscosity
Bulk Viscosity

Mass per Lattice Site

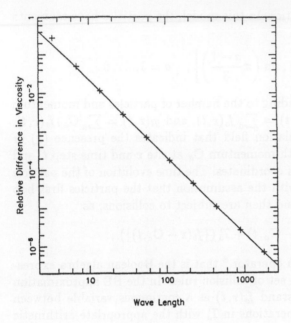

FIGURE 2   Relative discrepancy between the measured kinetic viscosity and the corresponding Chapman-Enskog value as a function of th shear wave wavelength L. The straight line is fitted on all the points except the two leftmost. The slope of the fit is -2.009. It is not exactly -2 because of higher-order corrections for the very small wavelength. These corrections are particularly evident for the first two points.

of wavelength L. The set of collision rules used is FHP-III described in Ref. 2. This set includes all the possible collisions that conserve a number of particles and their total momentum. Because of the simulation geometry we can suppress one of the dimensions and run on a $L \times 1$ lattice. The simulation data are then compared with the Chapman-Enskog predictions indicated as solid lines in Figure 1. The agreement between the two, better than one part in 10^3 for $L > 40$, is not surprising, but it is interesting to note that we can obtain hydrodynamic behavior even with very small lattices. To support this claim we plot in Figure 2 the discrepancy between the kinematic viscosity measured from the simulation and v_{CE}, the Chapman-Enskog prediction, as a function of L. In this plot the average number of particles per site is $\rho = 2.1$. The above-defined discrepancy is proportional to $(\lambda/L)^2$, where λ is the mean free path of the gas, as can be easily predicted, but the interesting result is that the simulation gives values of v accurate to better than 5% even for a lattice as small as 4×1.

The last plot we present is a test of the rotational symmetry[11] of sound waves in a small system of 33×33 lattice sites with average density $\rho = 2.1$ particles per site. We increase the mass of the central site, the "origin," by less than 1%. A "cylindrical" sound wave then develops.[12] In Figure 3 we plot the resulting radial density profile after 16 microscopic time steps. The solid line is obtained by our interpolating with a cubic spline, the density measured along a straight line passing through the origin in the direction of one of its nearest neighbors. We then plot a symbol for each site of the lattice at the appropriate mass density and radial distance from the origin. As expected, all the symbols lay very close to the solid line. The short-wavelength oscillation close to the origin is due to a peculiar lattice effect,

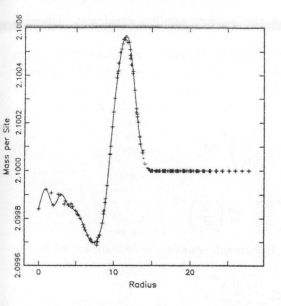

FIGURE 3 Radial density distribution of a cylindrical sound wave 33×33 lattice sites, 16 time steps after initial disturbance, see text. The solid line is obtained by our interpolating the mass density along a line passing through the site initially excited and one of its nearest neighbors. For each site of the lattice we mark a point corresponding to its radial distance from the origin (the locus of the original disturbance) and its mass density.

common to both the LGA and the Boltzmann model, which has very interesting consequences that we will discuss elsewhere.[13,14]

Unfortunately the Boltzmann scheme described here does not seem to be an optimal tool for the study of highly turbulent flows. In fact, let us consider the simulation, using the LGA, of an incompressible, two-dimensional flow of Reynolds number R. We require resolution η in the velocity field and spatial resolution χ. If we introduce L, v, ν, and l as, respectively, a characteristic length, velocity, viscosity, and linear dimension of the coarse-graining subregions, we have $R = vL/\nu$,[15] $\eta > \delta v/v$, and $\chi > l/L$. Since the flow is assumed to be incompressible we also have the constant $L > R\nu/c_s$, where c_s is the speed of sound of the gas. Using standard arguments we can evaluate

$$\eta > \frac{\delta v}{v} \approx \frac{C_0}{vl\left(\frac{\tau}{\tau_s}\right)^{\frac{1}{2}}}, \qquad (1)$$

where $(C_0 \rho)^2 = 3\rho(7 - \rho)/49$ is the variance of the momentum fluctuation per site[6] while τ/τ_s is the number of independent samples obtained by averaging over a time τ. Assuming that the physical process relevant to the destruction of correlations in a region of size l is sound wave propagation, we get $\tau_s \approx l/c_s$. How large can τ be? It is the smallest resolved time scale $\tau = 1/v$, and therefore $\tau/\tau_s \approx (c_s/v)LR^{-1}$. Substituting in Eq. (1) we get another constraint on L:

$$L > \left(\frac{C_0}{\eta\chi(\nu c_s)^{\frac{1}{2}}}\right)^2 R^{-1}. \qquad (2)$$

The previous equation, together with the incompressibility constraint, defines a transition Reynolds number $R_1 = C_0/\eta\chi\nu$ below which the minimum size of the lattice is determined mainly by the requirement of velocity resolution rather than the required Reynolds number. We cannot estimate the current amount of computer work, W_l, needed to perform the simulation. We have $W_l = L^2 T$, where $T = L/v$ is the largest eddy turnover time. Therefore $W_l = L^4/\nu R$ and, for $R > R_1$,

$$W_l > \frac{1}{\nu}\left(\frac{\nu}{c_s}\right)^4 R^3 , \tag{3}$$

while

$$W_l > \frac{1}{\nu}\left(\frac{\nu}{c_s}\right)^4 R_1^3 \left(\frac{R}{R_1}\right)^{-5} \tag{4}$$

otherwise. The work required by the Boltzmann equation is instead, for all R,

$$W_b = \xi_w \frac{1}{\nu}\left(\frac{\nu}{c_s}\right)^4 R^3 , \tag{5}$$

where ξ_w is the work needed to update a site of the Boltzmann lattice measured in units of LGA site updates. From Eqs. (3)–(5) we see that, for $R < R_w$, $R_w = \xi_w^{-1/8} R_1$ and it is more convenient to use the Boltzmann equation. An analogous calculation for memory space consideration gives the upper bound $R_s = \xi_s^{-1/4} R_1$ where ξ is the memory requirement per BE site in units of memory needed for a site of the LGA. At density $\rho = 2.1$ particles per site and for $\eta = \xi = 0.1$ we get $R_1 \approx 340$. In the simulations discussed in this Letter $\xi_w = 50$, $\xi_s = 32$ giving $R_w \approx 200$, $R_s = 142$. See Ref. 15. We have just shown that the BE is, for $R < R_s$, more efficient than the LGA. Also, for asymptotically large Reynolds numbers the standard numerical techniques are more economical[16] than the LGA. However, there are indications[17] that the crossover point is much greater than R_s. Thus there is a window in the Reynolds number in which the LGA schemes are computationally superior.

In conclusion, we have presented a very simple tool for some hydrodynamic simulations and for the testing of the hydrodynamic properties of an LGA. This tool is particularly efficient for low Reynolds number simulations.

We have the pleasure to thank L. Kadanoff, M. Marder, and S. Zaleski for interesting and fruitful discussions.

This work was supported by ONR.

REFERENCES

1. Frisch, U., B. Hasslacher, and Y. Pomeau (1986), *Phys. Rev. Lett.* **56**, 1505.
2. D'Humières, D., and P. Lallemand (1986), *C. R. Acad. Sci.* **302**, 983.
3. Balasubramanian, K., F. Hayot, and W. F. Saam (1987), *Phys. Rev. A* **36**, 2248.
4. Kadanoff, Leo P., Guy R. McNamara, and Gianluigi Zanetti (1987), *Complex Systems* **1**, 791.
5. Dahlburg, J., D. Montgomery, and G. Doolen (1987), *Phys. Rev. A* **36**, 2471.
6. Kadanoff, L. P., G. McNamara, and G. Zanetti, to be published.
7. Cercignani, C. (1975), *Theory and Application of the Boltzmann Equation* (New York: Elsevier).
8. Frisch, U., D. d'Humières, B. Hasslacher, P. Lallemand, Y. Pomeau, and J.-P. Rivet (1987), *Complex Systems* **1** 648, 707.
9. Gatignol, T. (1975), in *Theorie Cinetique des Gaz a Repartition Discrete de Vitesses*, Lecture Notes in Physics (Berlin: Springer-Verlag), vol. 36.
10. Wolfram, S. (1986), *J. Stat. Phys.* **45**, 471.
11. Margolus, N., T. Toffoli, and G. Vichniac (1986), *Phys. Rev. Lett.* **56**, 1694.
12. Landau, L., and E. Lifshitz (1986), *Fluid Mechanics* (New York: Pergamon).
13. McNamara, Guy, and Gianluigi Zanetti, to be published.
14. Zanetti, G., to be published.
15. The Reynolds number defined here is not the relevant one for comparisons between the LGA (or the BE for what it matters) simulation and real simple fluids flows. In fact, because of the non-Galilean invariance of the LGA it should be multiplied by a density-dependent coefficient which is anyhow irrelevant to our comparisons since it afflicts both models.
16. Orszag, S. S., and V. Yakhot (1986), *Phys. Rev. Lett.* **56**, 1691.
17. Zaleski, Stèphane, unpublished.

Dominique d'Humières and Pierre Lallemand
CNRS, Laboratoire de Physique de l'Ecole Normale Supérieure, 24 rue Lhomond, 75231 Paris Cedex 05, France

Numerical Simulations of Hydrodynamics with Lattice Gas Automata in Two Dimensions

This paper originally appeared in *Complex Systems* (1987), Volume 1, pages 599–632.

ERRATA

Equation (2.7) should read

$$\partial_t \rho + \text{div}(\rho \mathbf{u}) = 0,$$
$$\partial_t(\rho u_\alpha) + \partial_\beta\left(g(\rho)\rho u_\alpha u_\beta\right) = -\partial_\alpha P(\rho, u^2) + \partial_\beta\left(\nu\partial_\beta(\rho u_\alpha)\right) + \partial_\alpha\left(\zeta \text{div}(\rho\mathbf{u})\right),$$
$$(2.7)$$

Equations (B.13) and (B.16) should read

$$t_i = a_i \otimes a_{i+1}, \qquad u_i = t_i \cdot \overline{a}_{i+4} \qquad i = \{1, \cdots, 6\}$$
$$v_i = u_i \cdot u_{i+3} \cdot (t_{i+2} \cdot t_{i+5}) \qquad i = \{1, 2, 3\}, \tag{B.13}$$

and

$$\delta = (t_1 \cdot t_4) \cdot (t_2 \cdot t_5) \cdot (t_3 \cdot t_6),$$
$$\epsilon_i = u_i \cdot u_{i+5} \cdot (a_0 \otimes a_{i+1}) \cdot \overline{a}_{i+2} \qquad i = \{1, \cdots, 6\}. \tag{B.16}$$

Complex Systems **1** (1987) 599–632

Numerical Simulations of Hydrodynamics with Lattice Gas Automata in Two Dimensions

Dominique d'Humières
Pierre Lallemand
CNRS, Laboratoire de Physique de l'Ecole Normale Supérieure,
24 rue Lhomond, 75231 Paris Cedex 05, France

Abstract. We present results of numerical simulations of the Frisch, Hasslacher, and Pomeau lattice gas model and of some of its variants. Equilibrium distributions and several linear and nonlinear hydrodynamics flows are presented. We show that interesting phenomena can be studied with this class of models, even for lattices of limited sizes.

1. Introduction

Since Frisch, Hasslacher, and Pomeau [1] have shown that particles moving on a triangular lattice with very simple collisions on the nodes of the lattice obey the Navier-Stokes equation, the use of lattice gas models to study hydrodynamics has received considerable interest during the last two years [2]. However, the Navier-Stokes equation is recovered only in the limit of large systems and for incompressible flows. More theoretical analysis remains to be done to bound the errors of the lattice gas numerical scheme for finite lattice sizes and velocities. At present, some partial answers to this question can be obtained by numerical simulations of the dynamical behavior of the triangular lattice gases and by comparison of these results with classical hydrodynamics.

In section 2, we will describe precisely the models we have studied and recall briefly their theoretical properties obtained with the Boltzmann approximation [2,3]. Section 3 will be devoted to a detailed description of how these models can be simulated on general-purpose computers. In section 4, numerical evidence of the Fermi-Dirac distribution for equilibrium will be presented. The measurement of the transport coefficients will be given in section 5. Finally, we will report in section 6 some examples of nonlinear flows either for stationary or nonstationary situations for moderate Reynolds numbers.

2.   The models

We consider particles moving on a triangular lattice with unit velocity c_i in direction i between a node and one of its six neighbors ($i = 1, \ldots, 6$). At each time step, particles incoming on a node interact together according to collision laws assumed to conserve the number of particles and the total linear momentum on the node. The particles then propagate according to their new velocity. In addition, there is an exclusion principle such that no two particles with the same velocity may occupy the same node at the same time (0 or 1 particle per cell, as defined in reference 2. Thus, each node of the lattice can be described by a six-bit word whose ones represent particles moving with the velocities associated with their bit positions within the word.

We have also used variants of these six-bit models which allow additional particles with zero velocity at each site ($i = 0$ for notational purposes). Thus, we can introduce seven-bit models with at most one "rest particle" per lattice node, or eight-bit models with up to three rest particles per node.[1] In the eight-bit models, two bits are used to code the presence of rest particles, one for mass-one particles and one for a new kind of rest particle of mass two, equivalent to two rest particles of mass one. To handle the case of particles with zero velocity and different masses, the theoretical results of reference 2 need some modifications, given in Appendix A. Here, we will give the general results for the case of b_m moving particles with unit mass and b_r rest particles with mass $m_k = 2^k, k \in \{0, \ldots, b_r - 1\}$. The macroscopic quantities: density ρ and momentum ρu, are related to the local average populations N_{ik} of particles with mass 2^k and velocity c_i, by

$$\rho = \sum_{i,k} 2^k N_{ik} \, ; \rho u = \sum_{i,k} 2^k N_{ik} c_i. \tag{2.1}$$

These populations are given by the following Fermi-Dirac distributions

$$N_{ik} = \frac{1}{1 + \exp(2^k(h + q \cdot c_i))} \tag{2.2}$$

where h and q are nonlinear functions of ρ and u. When $u = 0$, the average density is the same for all the particles with same mass 2^k and will be denoted d_k; taking the mass of the lightest particles as unit mass and $d_0 = d$, one gets

$$d_k = \frac{d^{2^k}}{d^{2^k} + (1-d)^{2^k}} \tag{2.3}$$

Thus, when particles with mass greater than one are added, the density is related by a nonlinear law to the average density of particles of mass one,

---

[1]These particles may be considered to have an internal energy to satisfy energy conservation which is undistinguishable from mass conservation.

which will be called "density per cell" in what follows, as was done for the case where all the particles have the same mass. This nonlinear relation implies that all the expansions around equilibrium cannot take the simple expressions used in reference 2. The density of moving and rest particles, ρ_m and ρ_r respectively, can be defined as

$$\rho_m = b_m d \quad \text{and} \quad \rho_r = \sum_{k=0}^{b_r-1} 2^k d_k, \quad \text{with} \quad \rho = \rho_m + \rho_r; \qquad (2.4)$$

For small u, the expansion of (2.2) up to first order gives

$$N_{i0}^{eq} = d(1 + \frac{2\rho}{\rho_m}c_i \cdot u),$$

$$N_{0k}^{eq} = d_k. \qquad (2.5)$$

The speed of sound is given by

$$c_s^2 = \frac{b_m \, d(1-d)}{2(b_m d(1-d) + \sum_{k=0}^{b_r-1} 4^k d_k(1-d_k))} \qquad (2.6)$$

and, up to second-order terms in velocity and gradients, the lattice gas dynamics is described by[2]

$$\partial_t \rho + \text{div}(\rho u) = 0$$

$$\partial_t(\rho u_\alpha) + \partial_\beta(g(\rho)\rho u_\alpha u_\beta) =$$

$$\partial_\alpha P(\rho, u^2) + \partial_\beta(\nu \partial_\beta(\rho u_\alpha)) + \partial_\alpha(\varsigma \text{div}(\rho u)) \qquad (2.7)$$

with

$$g(\rho) = \frac{\rho}{2\rho_m}\frac{1-2d}{1-d}, P(\rho, u^2) = \frac{\rho_m}{2} - \frac{\rho}{2}g(\rho)(4c_s^2 - 1)u^2,$$

and $\nu = -\frac{1}{4\lambda} - \frac{1}{8}$ \qquad (2.8)

where the kinematic shear viscosity ν is related to the eigenvalue λ of the linearized collision matrix $[A_{IJ}]$ corresponding to the eigenvector $[c_{ix}c_{iy}]$, and the kinematic bulk viscosity ς is related in a complicated way to $[A_{IJ}]$, as shown in Appendix A.[3]

---

[2]Greek and Roman indices refer respectively to components, and velocity labels and the summations over repeated Greek indices are implicit.

[3]Capital Roman indices I and J refer to double lowercase roman indices ik and jl.

2.1 Six-bit model: model I

Table 1 gives all the possible configurations of the six-bit models, listing only the cases such that $j_z \geq j_y \geq 0$ [4]. The first three columns give the number of particles and the total momentum.[4] The fourth column shows the different configurations, the legal collisions exchanging configurations appearing within the same row, and the last column gives the number of different configurations obtained by the application of the symmetry group. To avoid bit representation of the collision probabilities $A(s \to s')$ between states s and s', we have chosen to write this probability $A_I(j \to k)$, where I is an index representing the number of particles, and the total momentum as it is given in the three first columns of table 1 and j and k are the positions of configurations in the rows of this table. The legal collision rules exclude the exchange of configurations belonging to different rows of the configuration table, and in the case of only one configuration, $A_I(1 \to 1) = 1$. With this notation, the semi-detailed balance is written:

$$\sum_j A_I(j \to k) = \sum_k A_I(j \to k) = 1 \quad \forall I$$

The original FHP (model I) model uses only the collision rules exchanging the configurations within rows 3 and 6:

$$A_I(j \to k) = \delta_{jk} \quad \forall I \neq \{200,\ 300\};$$

$$A_{200}(j \to k) = \frac{1 - \delta_{jk}}{2} \quad j,k \in \{1,2,3\};$$

$$A_{300}(j \to k) = 1 - \delta_{jk} \quad j,k \in \{1,2\}.$$

These collision rules: three two-body head-on collisions and two symmetric three-body collisions, changes five configurations among the 64 possible ones. They are the minimal set of rules to prevent spurious conservation laws.

For this model, the transport coefficients are given by

$$\rho = \rho_m = 6d, \quad c_s = \frac{1}{\sqrt{2}},$$

$$g(\rho) = (\rho - 3)/(\rho - 6), \quad P = \frac{\rho}{2}(1 - g(\rho)u^2) \tag{2.9}$$

$$\nu = \frac{1}{12d(1 - d)^3} - \frac{1}{8}, \quad \varsigma = 0 \tag{2.10}$$

n	jx	jy		folds
0	0	0		1
1	2	0		6
2	0	0		1
2	2	0		6
2	3	2		6
3	0	0		1
3	2	0		6
3	4	0		6
4	3	2		6
4	2	0		6
4	0	0		1
5	2	0		6
6	0	0		1

Table 1: List of configurations for the six-bit FHP model. Column 1 gives the number of particles, column 2 gives $2j_x$, column 3 gives $4j_y/\sqrt{3}$, and the last column gives the number of equivalent collisions that can be obtained by successive $\frac{\pi}{3}$ rotations when "fold" $\neq 1$.

n	jx	jy		folds
0	0	0		1
1	0	0		1
1	2	0		6
2	0	0		1
2	2	0		6
2	3	2		6
3	0	0		1
3	2	0		6
3	3	2		6
3	4	0		6

Table 2: List of configurations for the seven-bit models. Configurations with four particles and more are obtained by duality replacing particles by holes and holes by particles.

2.2 Seven-bit models: models II and III

Actually, most of the computer simulations were done with the model that includes rest particles. Table 2 lists all the configurations up to three particles; the configurations for more than three particles are obtained by duality (exchange of particles and holes) from the listed configurations. For the seven-bit models, the "universal" transport coefficients are given by

$$\rho = 7d , \qquad \rho_m = 6d , \qquad c_s = \sqrt{\frac{3}{7}},$$

$$g(\rho) = \frac{7}{12}\frac{7-2\rho}{7-\rho}, \qquad P = \frac{3\rho}{7}\left(1 - \frac{5}{6}g(\rho)u^2\right) \tag{2.11}$$

Model II is defined by the following collision rules:

$$A_I(j \to k) = \delta_{jk} \qquad \forall I \neq \{200, 2\overleftrightarrow{20}, 300, 400\};$$

$$A_{200}(j \to k) = \frac{1-\delta_{jk}}{2} \ j,k \in \{1,2,3\};$$

$$A_{2\overleftrightarrow{20}}(j \to k) = 1 - \delta_{jk} \ j,k \in \{1,2\};$$

$$A_{300}(j \to k) = \frac{1-\delta_{jk}}{2} \quad j,k \in \{1,2,3\};$$

$$A_{300}(j \to k) = 1 - \delta_{jk} \quad j,k \in \{4,5\};$$

$$A_{300}(j \to k) = A_{300}(k \to j) = 0 \quad j \in \{1,2,3\} \quad k \in \{4,5\};$$

$$A_{400}(j \to k) = \delta_{jk} \quad j,k \in \{1,2,3\};$$

$$A_{400}(j \to k) = 1 - \delta_{jk} \quad j,k \in \{4,5\};$$

$$A_{400}(j \to k) = A_{400}(k \to j) = 0 \quad j \in \{1,2,3\} \quad k \in \{4,5\};$$

representing 22 configurations giving active collisions: ten identical to model I with or without rest particles as "spectator" plus twelve two-body collisions changing the number of rest particles. The viscosities are given by

$$\nu = \frac{1}{28d(1-d)^3(1-4d/7)} - \frac{1}{8} , \qquad \varsigma = \frac{1}{98d(1-d)^4} - \frac{1}{28} \tag{2.12}$$

Model III is defined using effective collision rules for all the possible configurations, with the following rules:

---

[4] To use only integers, j_x is multiplied by two and j_y by $4/\sqrt{3}$. Starting from momentum $(2,0)$, successive $\frac{\pi}{3}$ rotations give $(1,2)$, $(-1,2)$, $(-2,0)$, $(-1,-2)$, and $(1,-2)$. In this paper, such sets of configurations with p particles will be denoted $p\overleftrightarrow{20}$.

$$A_I(1 \to 1) = 1 \qquad \forall I \text{ such that } q_I = 1;$$

$$A_I(j \to k) = \frac{1 - \delta_{jk}}{q_I - 1} \forall I \quad \text{such that} \quad 1 < q_I < 4 \text{ , } j, k \in \{1, \cdots, q_I\};$$

$$A_{300}(j \to k) = \frac{1 - \delta_{jk}}{2} j, k \in \{1, 2, 3\}; A_{300}(j \to k) = 1 - \delta_{jk} j, k \in \{4, 5\};$$

$$A_{300}(j \to k) = A_{300}(k \to j) = 0 \; j \in \{1, 2, 3\} \; k \in \{4, 5\};$$

$$A_{400}(j \to k) = \frac{1 - \delta_{jk}}{2} j, k \in \{1, 2, 3\}; A_{400}(j \to k) = 1 - \delta_{jk} j, k \in \{4, 5\};$$

$$A_{400}(j \to k) = A_{400}(k \to j) = 0 \; j \in \{1, 2, 3\} \; k \in \{4, 5\};$$

where q_I is the number of configurations in the Ith row of table 2. With these rules, which are self-dual, 76 configurations give active collisions. The viscosities are given by

$$\nu = \frac{1}{28d(1 - d)(1 - 8d(1 - d)/7)} - \frac{1}{8},$$

$$\varsigma = \frac{1}{98d(1 - d)(1 - 2d(1 - d))} - \frac{1}{28}. \tag{2.13}$$

2.3   Eight-bit model: model IV

We also used a model with rest particles of mass two in addition to the seven particles used in models II and III. Table 2 lists all the configurations up to four particles; the configurations for more than four particles can be obtained taking the dual of the listed configurations. For the eight-bit models, the "universal" transport coefficients are given by

$$\rho = d(7 + \frac{2d}{d^2 + (1 - d)^2}), \quad \rho_m = 6d,$$

$$c_s = \sqrt{\frac{3(d^2 + (1 - d)^2)^2}{7(d^2 + (1 - d)^2)^2 + 4d(1 - d)}},$$

$$g(\rho) = \frac{1}{12}\left(7 + \frac{2d}{d^2 + (1 - d)^2}\right)\frac{1 - 2d}{1 - d},$$

$$P = 3d - \frac{\rho}{2}g(\rho)\left(\frac{5(d^2 + (1 - d)^2)^2 - 4d(1 - d)}{7(d^2 + (1 - d)^2)^2 + 4d(1 - d)}\right)u^2. \tag{2.14}$$

Model IV is defined by the following collision rules:

$$A_I(j \to k) = \delta_{jk} \qquad \forall I \neq \{200, 2\overleftrightarrow{20}, 300, 3\overleftrightarrow{20}, 400, 4\overleftrightarrow{20}, 500, 600\};$$

$$A_{200}(1 \to 1) = 1; \qquad A_{200}(j \to k) = \frac{1 - \delta_{jk}}{2};$$

$$j, k \in \{2, 3, 4\}; \qquad A_{200}(1 \to j) = 0 \quad j \in \{2, 3, 4\};$$

$$A_{220}^{\rightarrow}(j \to k) = 1 - \delta_{jk} \quad j, k \in \{1, 2\};$$

$$A_{300}(1 \to 1) = 1 \; ; \qquad A_{300}(j \to k) = \frac{1 - \delta_{jk}}{2} \quad j, k \in \{2, 3, 4\};$$

$$A_{300}(j \to k) = 1 - \delta_{jk} \quad j, k \in \{5, 6\};$$

$$A_{300}(1 \to j) = 0 \quad j \in \{2, 3, 4, 5, 6\};$$

$$A_{300}(j \to k) = A_{300}(k \to j) = 0 \quad j \in \{2, 3, 4\} \quad k \in \{5, 6\};$$

$$A_{320}^{\rightarrow}(j \to k) = 1 - \delta_{jk} \quad j, k \in \{1, 2\};$$

$$A_{320}^{\rightarrow}(j \to k) = \delta_{jk} \quad jk \in \{3, 4\};$$

$$A_{320}^{\rightarrow}(j \to k) = A_{320}^{\rightarrow}(k \to j) = 0 \quad j \in \{1, 2\} \quad k \in \{3, 4\};$$

$$A_{400}(j \to k) = \frac{1 - \delta_{jk}}{2} \quad j, k \in \{1, 2, 3\};$$

$$A_{400}(j \to k) = 1 - \delta_{jk} \quad j, k \in \{4, 5\};$$

$$A_{400}(j \to k) = \delta_{jk} \quad j, k \in \{6, 7, 8\};$$

$$A_{400}(j \to k) = A_{400}(k \to j) = 0$$

$$j \in \{1, 2, 3\} \quad k \in \{4, 5, 6, 7, 8\};$$

$$A_{400}(j \to k) = A_{400}(k \to j) = 0 \quad j \in \{4, 5\} \quad k \in \{6, 7, 8\};$$

$$A_{420}^{\rightarrow}(j \to k) = 1 - \delta_{jk} \quad j, k \in \{1, 2\};$$

$$A_{420}^{\rightarrow}(j \to k) = \delta_{jk} \quad j, k \in \{3, 4, 5\};$$

$$A_{420}^{\rightarrow}(j \to k) = A_{420}^{\rightarrow}(k \to j) = 0 \quad j \in \{1, 2\} \quad k \in \{3, 4, 5\};$$

$$A_{500}(j \to k) = \delta_{jk} \quad j, k \in \{1, 2, 3\}; \qquad A_{500}(j \to k) = 1 - \delta_{jk} \quad j, k \in \{4, 5\};$$

$$A_{500}(j \to k) = \delta_{jk} \quad j, k \in \{6, 7, 8\}; \qquad A_{500}(j \to k) = A_{500}(k \to j) = 0$$

$$j \in \{1, 2, 3\} \quad k \in \{4, 5, 6, 7, 8\};$$

$$A_{500}(j \to k) = A_{500}(k \to j) = 0 \quad j \in \{4, 5\} \quad k \in \{6, 7, 8\};$$

$$A_{600}(j \to k) = \delta_{jk} \quad j, k \in \{1, 2, 3, 4\};$$

$$A_{600}(j \to k) = 1 - \delta_{jk} \quad j, k \in \{5, 6\};$$

$$A_{600}(j \to k) = A_{600}(k \to j) = 0 \quad j \in \{1, 2, 3, 4\} \quad k \in \{5, 6\};$$

representing 56 configurations giving collisions: 20 identical to model I with or without rest particles as "spectator", 24 two-body collisions changing the number of rest particles of mass one with or without rest particles of mass two as "spectator", and 12 two-body collisions exchanging the number of rest particles of mass one and two. The viscosities are given by

$$\nu = \frac{1}{28d(1-d)^3 \left(1 - \frac{4d}{7}\left(1 - \frac{d(1-d)}{d^2 + (1-d)^2}\right)\right)} - \frac{1}{8}$$

$$\varsigma = \frac{2c_s^4}{d(1-d)^4}\left(\frac{1}{16}\left(\frac{1}{c_s^2} - \frac{5}{3}\right)^2 + \frac{1}{9}\frac{d(1-d)}{(d^2 + (1-d)^2)^3}\right) - \frac{1}{2}\left(\frac{1}{2} - c_s^2\right). \quad (2.15)$$

3. Computer simulations

The present work was done by simulation of the lattice gas models on an FPS-164 using lattices of order 10^6 nodes with a typical speed of 10^6 updates per second [5–8]. The evolution of the lattice gas is computed according to a parallel iteration in five steps.

1. The states of the nodes of obstacles and boundaries are saved in a temporary storage.

2. During this second step, the collision step, the new state of each node is computed as a function of its old state according to the collision rules.

3. The third step is used to determine the new states of the nodes of obstacles and boundaries which are computed as a function of the saved states and the collision rules on obstacles.

4. During the fourth step, the propagation step, the bits of states of each node are propagated toward one of the neighbor nodes according to the physical interpretation of the different bits.

5. This last step is used to set the lattice boundary conditions.

All the programs were written in FORTRAN, using a few tricks to take advantage of the architecture of the computer. Since these tricks are introduced to make the best use of a pipeline architecture, they can be used for most of the vectorized machines and thus will be described in some detail.

3.1 Boundary conditions on the lattice edges and on the obstacles

The basic boundary conditions on the lattice edges can be periodic in the two directions; the particles exiting from one edge are reinjected into the other edge in the same direction. In what follows, we refer to this case as periodic boundary condition.

Another situation, related to a wind-tunnel experiment, consists in providing a flux of fresh particles on one side of the lattice and allowing an output flux on the other side.[5] The exact distribution of input and output particles is derived from equation (2.5) with an appropriate density and velocity. The balance between the input and output fluxes leads to an adjustment of the average speed of the flow inside the "wind tunnel", which depends upon the presence of lateral boundaries or obstacles.

Obstacles are first decomposed into a series of continuous links which approximate its geometrical shape. At nodes which represent an obstacle, particles are either bounced back ($c_i \rightarrow -c_i$) or can be reflected by the boundary of the obstacle. The first case corresponds to a very strict "no-slip" condition, whereas the second case is more closely related to the "slip" condition. One could also diffuse the particles by re-emitting them at random on available links, but that would be more complicated to implement. Momentum transfer between the gas and the obstacle can be computed during step one, leading to forces experienced by the obstacle.

For the present, triangular lattice of the FHP lattice gas, the natural way to label nodes uses nonorthonormal coordinates so that lattices are diamonds. As this is awkward for most situations, we have used lattices whose shape is rectangular. This can be implemented by taking different propagation rules for lines of odd or even parity. On even lines, directions 2 and 6 imply a change of the horizontal coordinate, whereas on odd lines, directions 3 and 5 are associated with a change of the horizontal coordinate. This is displayed in figure 1. This feature complicates only very slightly the computer program corresponding to step 4 of the simulation. However, when a specialized hardware is designed, provision must be made for different propagation rules on odd and even lines as is done in the RAP-1 machine [9]. When working with rectangular physical space, the number of nodes has to be multiplied by $\frac{\sqrt{3}}{2}$ to measure lengths along the vertical axis.

3.2 Collision and propagation steps

During the second step, the post-collision state is computed using either a look up table or the appropriate combination of Boolean operators. In the case of a look-up table, the states are coded with eight bits (one byte)

---

[5]More precisely, in order to keep the density almost constant, particles are also injected on the output side along the links directed toward the inside of the lattice. The physical meaning of these boundary conditions remains to be clarified.

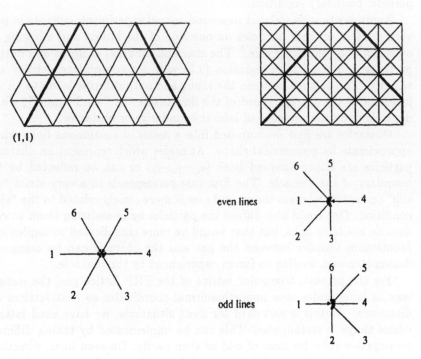

Figure 1: Schematic representation of the six velocity directions and of the lattice used in the FHP lattice gas: Left with 60° angles, Right adapted to orthonormal coordinates.

and eight nodes are packed in each of the 64-bit words of the FPS. During the collision step, the bytes are extracted from the word and used as an address to fetch their new value stored in a look-up table computed once at the beginning of the program. For the six- or seven-bit models, the eighth bit is used as a "random" bit to obtain probabilities 1/2 when it is needed. This bit is initially set at random with an average density 1/2 and is complemented each time a collision occurs. For the eight-bit models, two different look-up tables are used for the odd and even iterations of the whole lattice. It was checked on the seven-bit models that this procedure does not significantly bias the results. During the propagation step, the bits coding the different particles are extracted from the word using masks and moved as a whole, thus saving computation steps.

When Boolean operators are used to code the collision rules, 64 nodes are packed in a word and six to eight words are used to code the different particles. The noninteger probabilities of transition are implemented using a new collision rule at each iteration of the lattice. The minimum number of Boolean operators[6] needed to implement models I to III are given in Appendix B along with the basic tools we used to obtain good results with a reasonable amount of work. The propagation step is obtained by the motion of a full word; thus, during the collision and the propagation steps, 64 nodes are computed simultaneously, allowing high computation speed to be reached. The use of the Boolean rules needs more work than that of the look-up table to obtain efficient codes, but is more suited to pipeline or vector computers; thus, we used the first solution for the mature programs, while the latter was used for preliminary investigations of the various models.

3.3 Initialization and measurements

Initial flows were generated by Monte-Carlo procedure with average population N_{ik} related to the local density and velocity by equation (2.5). The use of the linear expansion restricts the available speeds to $\frac{\ell m}{2\rho}$, greater values giving negative or greater-than-one probabilities which will introduce initial conditions far from equilibrium. Note that we do not take into account the corrections of the equilibrium distributions with the gradients of the density and velocity fields, since they require the knowledge of the viscosities and are considered along with the nonlinear terms as corrections to the leading orders. Macroscopic quantities are obtained by averaging the N_{ik} according to equation (2.1) over rectangular regions with shapes and sizes adapted to the flow under study. For nonlinear flow simulations, we always used the momentum instead of the velocity, since the momentum is

---

[6]We have used only the Boolean operators available on a general purpose computer: and, or, exclusive-or, and complement. More compact rules can certainly be obtained on computers with more Boolean operators as the Connection Machine [10]. In addition, minimum must be taken as the lowest number of operators we found; some operators may probably be saved working harder than we did.

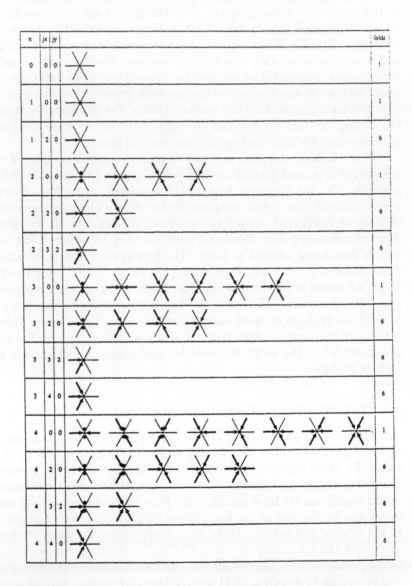

Table 3: List of configurations for the eight-bit model. Configurations with five particles and more are obtained by duality replacing particles by holes and holes by particles.

the relevant variable in the incompressible steady cases [3,11]. Moreover, it naturally comes out of the simulations.

4. Equilibrium distributions

Theoretically, the equilibrium distributions of the local averaged densities N_{ik} in direction i are given by Fermi-Dirac functions depending upon the density ρ and the average velocity u. This property was checked by the simulation of the time evolution of a 512×256 lattice with periodic boundary conditions in both directions. The first 24 time steps were discarded to remove the transient behavior. We checked that this number of time steps was larger than the duration of the transient. We then measured the average populations for each direction averaged on the next eight time steps. These simulations were done on model II for densities per moving cell 0.2 and 0.5 for velocities from $-6/7\sqrt{3}$ to $6/7\sqrt{3}$ and for velocities along the bisector of two c_i. In this case, the average densities of rest particles and particles moving in the directions perpendicular to the velocity are equal. The densities of particles moving on symmetric directions are also equal; thus, there are only three independent unknowns which can be obtained exactly from the solution of a third-degree equation [12]. Figure 2 shows the equilibrium distributions as a function of the velocity for the different populations along with the theoretical curves obtained from the exact solution. These distributions were normalized by the average population at rest. Clearly, the results of the simulations agree very well with the predicted Fermi-Dirac distributions with an error smaller than one percent. Note that for $d = 0.5$, the nonlinearity vanishes due to the Fermi-Dirac distributions and the particular orientation of the velocity with respect to the lattice.

Another test was performed with model IV in which the population of rest particles of mass two depends upon the population of moving particles in a nonlinear way, given by equation (2.3). Figure 3 shows the variation of d_1 as a function of d together with its value obtained by simulations.

5. Linear hydrodynamics

The velocity of sound c, and the kinematic shear and bulk viscosities ν and ς of the lattice gas models described above have been measured using the relaxation of an initial periodic perturbation u(r, 0) of the velocity field [13]: $\mathbf{u}(\mathbf{r}, 0) = (\mathbf{u}_\parallel + \mathbf{u}_\perp)\cos(\mathbf{k} \cdot \mathbf{r})$, where k is the wave vector of the perturbation and $\mathbf{u}_\parallel$ and $\mathbf{u}_\perp$ are the velocity components parallel and perpendicular to the wave vector [14].

The relaxation in time of the velocity u(r, t) and of the density perturbation $\delta\rho(\mathbf{r}, t)$ are given by

$$\mathbf{u}(\mathbf{r}, t) =$$

$$(\mathbf{u}_\parallel \cos(\omega t + \varphi)\exp(-k^2(\nu + \varsigma)t/2) + \mathbf{u}_\perp \exp(-k^2\nu t))\cos(\mathbf{k} \cdot \mathbf{r}), \quad (5.1)$$

Dominique d'Humières and Pierre Lallemand

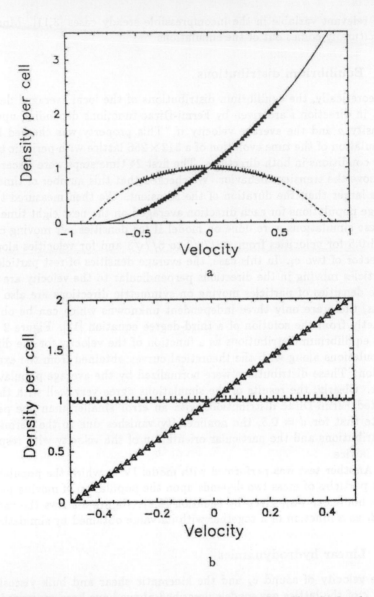

Figure 2: Variation of the density N_i versus velocity. Closed triangles correspond to angle 60° between u and c_i, apex correspond to angle 90° between u and c_i. Solid lines are obtained theoretically (a) at a mean density of 0.2, (b) at a mean density of 0.5.

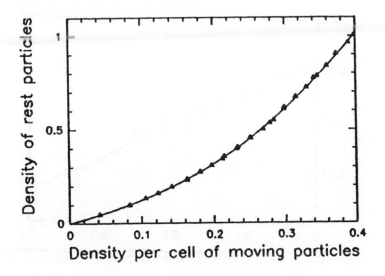

Figure 3: Comparison of theoretical and measured values of the density of rest particles versus density of moving particles for the eight-bit model.

$$\delta\rho(\mathbf{r},t) = (\rho u_\parallel/c)\sin(\omega t)\exp(-k^2(\nu+\varsigma)t/2)\sin(\mathbf{k}\cdot\mathbf{r}), \qquad (5.2)$$

with $\omega = c_s k$, $\tan\varphi = k(\nu+\varsigma)/2c_s$.

Starting from the initial conditions, at each time step, the momentum and the density are averaged along lines perpendicular to the wave vector. The result is Fourier transformed to get the components of the momentum and density corresponding to k. From the relaxation curves, c_s, ν, and ς are measured by least squares fits of equations (5.1) and (5.2) to the time evolution of $u_\perp(k)$, $u_\parallel(k)$, and $\rho(k)$.

The measured sound velocities are isotropic and agree with theoretical values $1/\sqrt{2}$ for model I and $\sqrt{3/7}$ for models II and III. Figure 4 shows the dependence of the speed of sound with the density per cell for the eight-bit model, compared to the theoretical given by equation (2.14). The measured values of the viscosities are summarized in figures 5a to d, along with the theoretical curves computed from equations (2.10), (2.12), (2.13), and (2.15). These measurements were obtained on 256×512 lattices with periodic boundary conditions and for wavelength between 30 and 80 nodes, with no observable effects of the wave numbers on the viscosities over a factor ten on the relaxation times. The size of the symbols corresponds roughly to the error bars. Without rest particles, the experimental values of ν are above the theoretical curve for the measurements corresponding to sound waves, while those corresponding to shear waves are below. Thus, the kinematic bulk viscosity ς is found negative, which is an unphysical result. At present, no convincing explanation has been found for this effect, which may be related to the fact that there are few triple symmetric

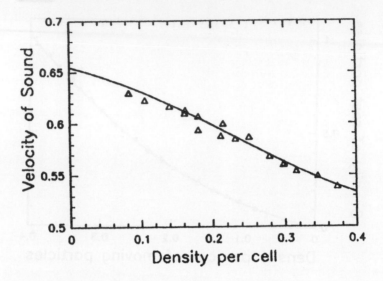

Figure 4: Theoretical and experimental values of the speed of sound for the eight-bit model.

collisions needed to remove a spurious invariant, leading to very long relaxation times of the associated microscopic quantity in comparison with the hydrodynamics time scales. The measured viscosities agree with theoretical predictions for models II, III, and IV. Thus, the presence of rest particles apparently improves the behavior of the lattice gas while decreasing the viscosity significantly, leading to higher Reynolds numbers [15]. Moreover, these results show that the Boltzmann approximation is well verified, even for high densities.

In many simulations, we have considered several disturbances at the same time, taking as initial conditions:

$$\sum_{l=1}^{3}(u_{l\parallel}\cos k_{l}r + u_{l\perp}\cos k_{l}r) \qquad (5.3)$$

and found essentially no coupling between these various waves. Furthermore, we found that the acoustic properties are particularly insensitive to the amplitude of waves.

A uniform motion of the fluid, at speed u, advects sound or shear waves at speed $g(\rho)u$, as discussed in reference 16.

6. Nonlinear flow simulations

We now present a few examples of flows computed by the lattice gas method. These flows were chosen in order to perform quantitative com-

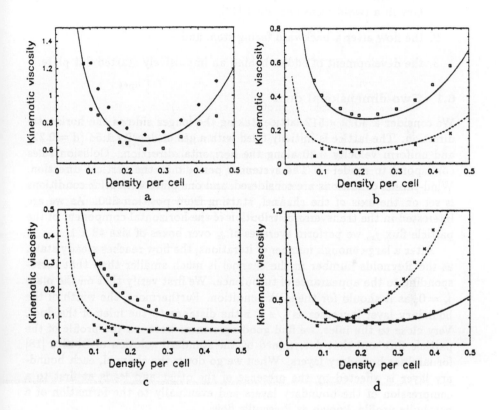

Figure 5: Theoretical shear (solid lines) and bulk (dashed lines) reduced viscosities as a function of the reduced density, compared with the results of numerical simulations for different lattice gas models: (a) original FHP model, (b) model II with rest particles and limited collision rules, (c) model III with rest particles and all possible collisions, and (d) eight-bit model IV.

parisons with results obtained either in real experiments or using standard solutions of the Navier-Stokes equations. Three problems are considered here:

1. the formation of boundary layers and the development of a Poiseuille flow in a two-dimensional duct [17],

2. the flow after a backward facing step, and

3. the development of eddies behind an impulsively started flat plate.

6.1 Two-dimensional duct

We consider a 3072 × 512 lattice, taking the longer side as the horizontal direction. The lattice is initially filled with a gas of density 1.54 ($d = 0.22$) and uniform velocity 0.30 along the horizontal direction. Collision rules correspond to model II. The system is periodic in the vertical direction. Wind-tunnel conditions are considered, and one plate with stick conditions is set on the axis of the channel, starting from position 300. As we are interested in the transverse distribution of the horizontal component of the particle flux j_x, we perform averages of j_x over boxes of size 48 × 1.

After a large enough number of iterations, the flow reaches steady state, as the Reynolds number in the channel is much smaller than that corresponding to the appearance of turbulence. We first verify that on the plate $j_x = 0$, as it should for the stick condition. Furthermore, the width of the boundary layer increases as $\sqrt{x}$ if x the distance to the inlet of the duct. Very close to the inlet, we find good agreement between the profile of the particle flux and that determined by the standard solution of Blasius [18] for laminar boundary layers. When we go down the channel, each boundary layer is affected by the presence of the other and leads at first to a compression of the boundary layers and eventually to the formation of a parabolic profile, known as Poiseuille flow.

The complete description of such an experimental situation has been performed by Slichting [19]. We have followed his solution by calculating the profile of the flow versus distance to the inlet of the duct using an iterative method to solve the Navier-Stokes equations. The velocity of the incoming fluid is measured in the steady-state regime, and the viscosity is measured in a separate experiment performed at the same density; thus, there is no adjustable parameter for the comparison of theory and experimental data obtained with the lattice gas. Figure 6 shows successive profiles of the horizontal component of the particle flux for different values of the relative distance to the inlet of the duct $z = x/w$. Very good agreement between theory and experiment is obtained provided one uses $\nu_{\it eff} = \nu/g(\rho)$.

6.2 Backward facing step

Another well-known flow situation is that of a two-dimensional backward facing step at low Reynolds number. This situation was considered as a

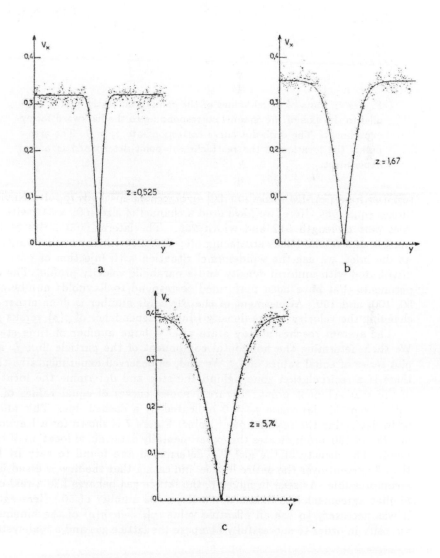

Figure 6: Velocity component parallel to a plate set in the middle of a channel at relative distance from the apex of the plate (a) 0.525, (b) 1.67, and (c) 5.74. The dots are obtained by the lattice gas simulations; the solid lines are calculated using the Slichting method.

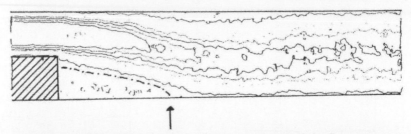

Figure 7: Plots of equal values of the component of the flux j_x parallel to the axis of the channel corresponding to the backward facing experiment. The dash-dot curve corresponds to $j_x = 0$. The arrow points the location of the reattachment point determined in a real experiment.

test case for a GAMM workshop [20] on numerical methods to solve Navier-Stokes equations. Here, we have used a channel of size 4608 × 512 with an inlet part of length 512 and width 256. The lateral boundaries of the channel and of the backward facing step are set with the stick condition. At the inlet, we use the wind-tunnel situation with injection of particles distributed with uniform density and a parabolic velocity profile. The experiments that have been performed correspond to Reynolds numbers of 50, 100, and 150. Adjustment of the Reynolds number is done either by changing the velocity or the density (due to dependence of $g(\rho)$ versus ρ).

The system reaches steady state after a large number of time steps. We then determine the horizontal component of the particle flux j_x and plot series of equal values of j_x. We find, as observed experimentally, that there is a recirculation zone behind the step and determine the location of the reattachment point. Figure 7 shows curves of equal values of j_x, with the particular value $j_x = 0$ indicated by a dashed line. This allows us to determine the reattachment point. Figure 7 is shown for a Reynolds number of 150 and indicates the experimentally determined location of that point. The density of the gas was determined and found to vary by less than 3 percent over the entire lattice, indicating that the flow is essentially incompressible. As seen in figure 7, the lattice gas behaves like a real one. Similar agreement is obtained for a Reynolds number of 50. Here again, it was necessary to use an effective value $\nu_{eff} = \nu/g(\rho)$ of the kinematic viscosity in order to successfully compare the lattice gas and a real system.

6.3 Impulsively started flat plate

This is the simplest case of interaction of a lattice gas flow with an obstacle. Here, we consider a 2816 × 1024 lattice with periodic conditions along y and the wind-tunnel condition at the left and right edges. Initially, the lattice is filled with a gas of uniform density and speed. Here we take $d = 0.30$, $(\rho = 2.1)$, and $v = 0.428$. The collision rules correspond to model III, so that the Mach number is 0.654 and the effective Reynolds number is

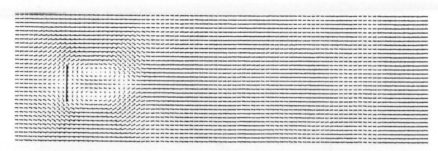

Figure 8: Map of the flux of particles in a 886.8 × 2048 channel 3000 time steps after the introduction of a flat plate of real size 216.5. The distance between the back of the plate and the point where $j_x = 0$ on the axis of the channel is defined as the size of the wake.

approximatively 300. The value of d is close to that of the minimum of the effective kinematic viscosity $\nu_{\text{eff}} = \nu/g(\rho)$, so that small compressibility effects should not be important.

Now, at $t = 0$, we insert a flat plate of real width $w = 216.5$ perpendicular to the input flow. It is set with stick conditions. The presence of the plate first produces shock waves due to the reflection of particles at the surface of the plate, then eddies start to develop symmetrically on either edges of the plate, as shown in figure 8.

Here, we present detailed data concerning the location of the wake, defined as the point where the horizontal component of the particle flux j_x is 0 for the symmetry axis of the problem. If s is the distance from this point to the plate, we consider the relative size of the wake s/w. We then measure s/w as a function of a reduced time $\theta = vt/w$, where v is the incoming velocity and t the real time. This choice of variables corresponds to that used in the analysis of real experiments performed in water by Taneda and Honji [21]. These authors found that

$$s/w = 0.89(vt/w)^{2/3}$$

independently of the Reynolds number Re, when $18 < \text{Re} < 1100$.

We show in figure 9 the "experimental" value of s/w determined for the lattice gas flow versus $(vt/w)^{2/3}$ for a Mach number equal to 0.327 ($v = 0.214$). We find a linear relationship, with a slope of 0.46.

To compare our value of the slope to that of a real experiment, we have again to consider that a lattice gas follows the Navier-Stokes equation, provided the velocity is multiplied by a factor $g(\rho)$. This means we have to compare the slope measured with the lattice gas (0.46) with a renormalized slope $0.89g^{2/3} = 0.423$ in the present case.

Again, we reach almost quantitative agreement between a lattice gas flow and a real experiment, provided we use a properly renormalized value of the velocity.

When time reaches sufficiently large values, it is found that the symmetry of the flow is broken and vortex shedding by the plate occurs. This

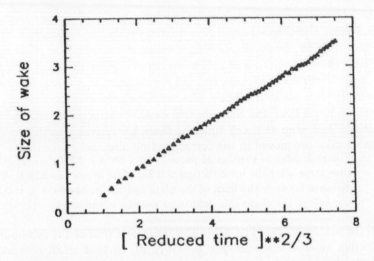

Figure 9: Relative size of the wake behind an impulsively started flat plate s/w as a function of $(vt/w)^{2/3}$.

leads to the formation of a two-dimensional Von Karman street, an example of which is shown in figure 10.

In another experiment, performed at a larger input speed, we have determined the period of the vortex shedding. The value corresponds to a Strouhal number of approximately 0.20, to be compared to a measured value of 0.16 for a truly incompressible flow at roughly the same Reynolds number.

Detailed comparisons between lattice gas flows and real flows like those presented here should be performed at higher velocities to find out which effects are produced by a breakdown of the incompressibility conditions.

7.   Conclusion

Quantitative agreement between theory and simulation has been demonstrated in both the linear and nonlinear regimes for moderate Reynolds numbers, provided a properly renormalized value of the fluid velocity ($v \rightarrow g(\rho)v$) is used in the nonlinear advection term of the Navier-Stokes equation. It should be noted that the introduction of obstacles into the flow is particularly simple and represents, for the cases studied, a computational overhead of a few percent. The present models are limited to Reynolds numbers of order of 10^3 and incompressible flows. Moreover, since local equilibrium is a function of ρ and ρu only, such models cannot simulate thermal phenomena. However, more complicated models derived from the FHP model [22–24] may overcome most of these limitations in a near future. In this case, lattice gas simulations will be a new tool for experimental work

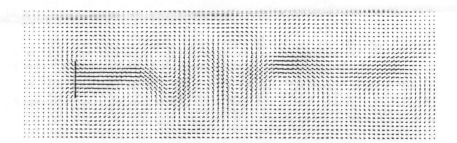

Figure 10: Similar to figure 8, but after 40000 iterations showing the formation of a Karman street. To emphasize the vortices, the mean momentum has been subtracted from the local ones.

in hydrodynamics, with the main advantage of being inherently stable.

Acknowledgments

We thank U. Frisch, B. Hasslacher, M. Hénon, Y. Pomeau, Y. H. Qian, and J. P. Rivet for helpful discussions; A. Noullez for participation in some of the simulations; and GRECO 70 "Expérimentation numérique" for support.

Appendix A.   Boltzmann approximation with rest particles

In this appendix, we will indicate how the theoretical results of reference 2 must be modified to handle the cases with particles with zero velocity and different masses. These modifications will be given for the Boltzmann approximation only.

The macroscopic quantities density, ρ, and momentum, ρu, are related to the local average populations N_{ik} of particles with mass m_k and velocity c_i, by

$$\rho = \sum_{i,k} m_k N_{ik}; \qquad \rho u = \sum_{i,k} m_k N_{ik} c_i. \qquad (A.1)$$

The demonstration used in Appendix C of reference 3 proves that, at equilibrium and for uniform density and velocity, $\{\log(N_{ik}) - \log(1 - N_{ik})\}$ is a collision invariant if the collisions verify the semi-detailed balance. The conclusion is now

$$N_{ik} = \frac{1}{1 + \exp(m_k(h + q \cdot c_i))}, \qquad (A.2)$$

where h and q are nonlinear functions of ρ and u. When $u = 0$, the average density is the same for all the particles with same mass m_k and will be denoted d_k. Taking the mass of the lightest particles as unit mass and $d_0 = d$, one gets

$$d_k = \frac{d^{m_k}}{d^{m_k} + (1-d)^{m_k}} \tag{A.3}$$

The general expansion of the Fermi-Dirac distributions around small u is quite complicated, and we will restrict it to the case of b_m moving particles with unit mass and velocity c and b_r rest particles with mass $m_k = 2^k, k \in \{0, \ldots, b_r - 1\}$. The density of moving and rest particles, ρ_m and ρ_r respectively, can be defined as

$$\rho_m = b_m d \quad \text{and} \quad \rho_r = \sum_{k=0}^{b_r-1} m_k d_k, \quad \text{with} \quad \rho = \rho_m + \rho_r \tag{A.4}$$

Let

$$c_s^2 = \frac{b_m\, d(1-d)c^2}{D(b_m d(1-d) + \sum_{k=0}^{b_r-1} m_k^2 d_k(1-d_k))} \tag{A.5}$$

For small u, the expansion of equation (A.2) up to second order gives in D dimensions:

$$N_{i0}^{eq} = d\Big(1 + \frac{D}{c^2}\frac{\rho}{\rho_m}\mathbf{c}_1 \cdot \mathbf{u} + G(\rho)(Q_{i\alpha\beta} + (\frac{c^2}{D} - c_s^2)\delta_{\alpha\beta})u_\alpha u_\beta\Big)$$

$$N_{0k}^{eq} = d_k\Big(1 - m_k c_s^2 G(\rho)\frac{1-d_k}{1-d}\delta_{\alpha\beta}u_\alpha u_\beta\Big) \tag{A.6}$$

where

$$Q_{i\alpha\beta} = \begin{cases} c_{i\alpha}c_{i\beta} - \frac{c^2}{D}\delta_{\alpha\beta} & \text{if } i \neq 0 \\ 0 & \text{if } i = 0 \end{cases},$$

$$\text{and} \quad G(\rho) = \frac{D^2}{2c^4}\frac{\rho^2}{\rho_m^2}\frac{(1-2d)}{(1-d)}. \tag{A.7}$$

When there are density and velocity gradients, this equilibrium distribution is modified by first-order correction in gradients:

$$N_{i0}^{(1)} = (\psi Q_{i\alpha\beta} - X\delta_{\alpha\beta})\partial_{1\alpha}(\rho u_\beta) = \psi Q_{i\alpha\beta}\partial_{1\alpha}(\rho u_\beta) + X\delta_{\alpha\beta}\partial_{t_1}\rho$$

$$N_{0k}^{(1)} = \chi_k\delta_{\alpha\beta}\partial_{1\alpha}(\rho u_\beta) = -\chi_k\delta_{\alpha\beta}\partial_{t_1}\rho \tag{A.8}$$

where

$$X = \frac{1}{b_m}\sum_{k=0}^{b_r-1} m_k\chi_k \tag{A.9}$$

Using the same Chapman-Enskog expansion as in references 2, 3, and 25 up to second-order terms in velocity and gradients, one gets

$$\partial_t \rho + \text{div}(\rho u) = 0$$

$$\partial_t(\rho u_\alpha) + \partial_\beta\Big(g(\rho)\rho u_\alpha u_\beta\Big) =$$

$$- \partial_\alpha P(\rho, u^2) + \partial_\beta(\nu\partial_\beta(\rho u_\alpha)) + \partial_\alpha((\frac{D-2}{D}\nu + \varsigma)\text{div}(\rho u)), \qquad \text{(A.10)}$$

with

$$g(\rho) = \frac{D}{D+2}\frac{\rho}{\rho_m}\frac{1-2d}{1-d};$$

$$P(\rho, u^2) = \frac{\rho_m}{D}c^2 - \rho g(\rho)\frac{c_s^2}{c^2}\Big(1 + \frac{D}{2} - \frac{c^2}{2c_s^2}\Big)u^2 \qquad \text{(A.11)}$$

and where the kinematic shear and bulk viscosities, ν and ς, are given by

$$\nu = -\frac{b_m c^4}{D(D+2)}\psi - \frac{c^2}{D(D+2)}\; ; \qquad \varsigma = \frac{b_m c^2}{D}X - \frac{1}{2}(\frac{c^2}{D} - c_s^2) \;\; \text{(A.12)}$$

and the speed of sound is c_s, since:

$$\Big(\frac{\partial P}{\partial \rho}\Big)_{u=0} = \frac{b_m\, d(1-d)c^2}{D(b_m d(1-d) + \sum_{k=0}^{b_r-1} m_k^2 d_k(1-d_k))} = c_s^2. \qquad \text{(A.13)}$$

Since the density is no longer linearly related to the density per cell d, the perturbation $N_{ik}^{(1)}$ must be taken as $d_k(1-d_k)N_{ik}^{*(1)}$,

$$\psi = d(1-d)\psi^*,$$

$$\chi_k = d_k(1-d_k)\chi_k^*,$$

$$X = d(1-d)X^* \qquad \text{(A.14)}$$

and, in the Boltzmann approximation, $N_I^{*(1)}$ is related to N_I^{eq} by

$$\partial_{t_1}N_I^{eq} + c_{i\alpha}\partial_{1\alpha}N_I^{eq} = \sum_J \mathcal{A}_{IJ}^* N_J^{*(1)} \qquad \text{(A.15)}$$

where the linearized collision matrix is given by

$$\mathcal{A}_{IJ}^* = -\frac{1}{2}\sum_{ss'}(s_I - s_I')(s_J - s_J')A(s \to s')\prod_K d_K^{s_K}(1-d_K)^{1-s_K} \qquad \text{(A.16)}$$

It is more convenient to rewrite the matrix $[\mathcal{A}_{IJ}^*]$ as

$$[\mathcal{A}_{IJ}^*] = \begin{pmatrix} [R_{kl}] & [C_{kj}] \\ [\widetilde{C}_{il}] & [M_{ij}] \end{pmatrix} \qquad \text{(A.17)}$$

where $[\mathcal{R}_{kl}]$ and $[\mathcal{M}_{ij}]$ are square submatrices with respectively b_r and b_m rows, $[C_{kj}]$ is a submatrix with b_r rows of b_m identical columns $[C_k]$, and $[\widetilde{C}_{il}]$ is the transposed of $[C_{kj}]$. Since $[m_I]$ is an eigenvector of $[A^*_{IJ}]$ with a zero eigenvalue, one must have

$$b_m C_k + \sum_l m_l \mathcal{R}_{kl} = 0 \qquad \forall k$$

$$\sum_l m_l C_l + \sum_j \mathcal{M}_{ij} = 0 \qquad \forall i \tag{A.18}$$

Then, ψ^* and χ^*_k are obtained by the solubility condition of the following linear system:

$$\frac{Dc_s^2}{b_m c^2} m_k \frac{d_k(1-d_k)}{d(1-d)} \partial_{t_1}\rho = \left(b_m C_k X^* - \sum_l \mathcal{R}_{kl}\chi^*_l\right)\partial_{t_1}\rho$$

$$\frac{D}{b_m c^2}\left((c_s^2 - \frac{c^2}{D})\partial_{t_1}\rho + Q_{i\alpha\beta}\partial_{1\alpha}(\rho u_\beta)\right) = -\sum_l (C_l(m_l X^* + \chi^*_i))\partial_{t1}\rho$$
$$+ \sum_j \mathcal{M}_{ij}Q_{j\alpha\beta}\psi^* \partial_{1\alpha}(\rho u_\beta) \tag{A.19}$$

Using equation (A.18), this condition gives

$$\frac{Dc_s^2}{b_m c^2} m_k \frac{d_k(1-d_k)}{d(1-d)} = \left(b_m C_k X^* - \sum_l \mathcal{R}_{kl}\chi^*_k\right), \tag{A.20}$$

$$\frac{D}{b_m c^2}Q_{i\alpha\beta} = \sum_j \mathcal{M}_{ij}Q_{j\alpha\beta}\psi^*. \tag{A.21}$$

Thus, ψ^* is simply related to the eigenvalue λ of the linearized collision matrix A^*_{IJ} corresponding to the eigenvector $[c_{ix}c_{iy}]$:

$$\psi^* = \frac{D}{b_m c^2}\frac{1}{\lambda} \tag{A.22}$$

but the χ^*_k are given by the solution of the linear system of b_r equations given by equation (A.20). We will only sketch how the values of ς were obtained for equations (2.12), (2.13), and (2.15).

The following results are valid when $m_k = 2^k$ and when the collisions are such that they change the total mass of rest particles only by one mass unit. In this case, the submatrix $[\mathcal{R}_{kl}]$ and C_k can be written

$$[\mathcal{R}_{kl}] = b_m \sum_{n=0}^{b_r-1} a_n [\mathcal{R}_{kl}^{(n)}]$$

$$C_k = a_k - \sum_{n=k+1}^{b_r-1} a_n \qquad (\text{A.23})$$

with:

$$[\mathcal{R}_{kl}^{(n)}] = \begin{pmatrix} \overbrace{0 \;\cdots\; 0}^{b_r-n-1 \; times} & 0 & \overbrace{0 \;\cdots\; 0}^{n \; times} \\ \vdots \;\ddots\; \vdots & \vdots & \vdots \;\ddots\; \vdots \\ 0 \;\cdots\; 0 & 0 & 0 \;\cdots\; 0 \\ 0 \;\cdots\; 0 & -1 & 1 \;\cdots\; 1 \\ 0 \;\cdots\; 0 & 1 & -1 \;\cdots\; -1 \\ \vdots \;\ddots\; \vdots & \vdots & \vdots \;\ddots\; \vdots \\ 0 \;\cdots\; 0 & 1 & -1 \;\cdots\; -1 \end{pmatrix} \qquad (\text{A.24})$$

Equation (A.20) becomes

$$\frac{Dc_s^2}{b_m c^2} m_k \frac{d_k(1-d_k)}{d(1-d)} = b_m \left(X^* + \chi_k^* - \sum_{l=0}^{k-1} \chi_l^* \right) a_k$$

$$- b_m \sum_{n=k+1}^{b_r-1} \left(X^* + \chi_n^* - \sum_{l=0}^{n-1} \chi_l^* \right) a_n \qquad (\text{A.25})$$

since the solution of the linear system

$$Y_k - \sum_{l=0}^{k-1} Y_l = A_k$$

is given by

$$Y_k = \frac{1}{2}\left(A_k + \sum_{l=0}^{k} 2^{k-l} A_l \right).$$

Equation (A.25) gives:

$$\chi_k^* - \sum_{l=0}^{k-1} \chi_l^* = \frac{Dc_s^2}{2b_m c^2} \frac{E_k}{m_k a_k} - X^* \qquad (\text{A.26})$$

with

$$E_k = \frac{1}{b_m}\left(m_k^2 \frac{d_k(1-d_k)}{d(1-d)} + \sum_{l=k}^{b_r-1} m_l^2 \frac{d_l(1-d_l)}{d(1-d)} \right), \qquad (\text{A.27})$$

then

$$\chi_k^* = \frac{Dc_s^2}{4b_m c^2}\Big(\frac{E_k}{m_k a_k} + \sum_{l=0}^{k} 2^{k-l}\frac{E_l}{m_l a_l}\Big) - m_k X^* \tag{A.28}$$

$$X^* = \frac{D^2 c_s^4}{4b_m c^4}\sum_{k=0}^{b_r-1}\frac{E_k^2}{a_k m_k^2} \tag{A.29}$$

and, finally,

$$\varsigma = d(1-d)\frac{Dc_s^4}{4c^2}\Big(\frac{1}{a_0}\big(\frac{c^2}{Dc_s^2}-1+\frac{1}{b_m}\big)^2 + \sum_{k=1}^{b_r-1}\frac{E_k^2}{a_k m_k^2}\Big) - \frac{1}{2}\big(\frac{c^2}{D}-c_s^2\big).$$

$$\tag{A.30}$$

Thus, for $D = 2$, $c^2 = 1$, $b_m = 6$, and $b_r = 1$ and 2, equation (A.30) respectively gives

$$\varsigma = \frac{d(1-d)}{98a_0} - \frac{1}{28}, \tag{A.31}$$

and

$$\varsigma = d(1-d)c_s^4\Big(\frac{1}{16a_0}\big(\frac{1}{c_s^2}-\frac{5}{3}\big)^2 + \frac{1}{9a_1}\frac{d_1^2(1-d_1)^2}{d^2(1-d)^2}\Big) - \frac{1}{2}\big(\frac{1}{2}-c_s^2\big). \tag{A.32}$$

Appendix B.   Boolean laws

The basic concept used to look for minimal set of Boolean functions to implement the collision rules is a derivation of the algorithm used by Hardy, de Pazzis, and Pomeau [26] for the four-bit model on a square lattice. A basic collision operator can be written

$$c = a_1 \cdot \bar a_2 \cdot a_3 \cdot \bar a_4 + \bar a_1 \cdot a_2 \cdot \bar a_3 \cdot a_4 = (a_1 \oplus a_2) \cdot (a_2 \oplus a_3) \cdot (a_3 \oplus a_4) \tag{B.1}$$

where $a \cdot b$, $a + b$, $a \oplus b$, and $\bar b$ correspond respectively to the and, or, exclusive-or, and not operators. The new states are then obtained by

$$a_i = c \oplus a_i. \tag{B.2}$$

The computation of the new configuration needs only nine Boolean operators. This basic algorithm can be extended to the triangular case by the definition of several collision operators. Each collision exchanges at least two particles and two holes, when this exchange does not change the total momentum; thus, we can define four-bit collisions operators similar to the one given in equation (B.1).

Appendix B.1   Model I

In this model, the basic collision operators give the possibility of head-on collisions γ_i and of triple symmetric collisions δ:

$$
\begin{aligned}
t_i &= a_i \oplus a_{i+1} \quad i = \{1, \cdots, 6\} \\
u_i &= t_i \cdot t_{i+3} \quad i = \{1, 2, 3\} \\
\gamma_i &= u_i \cdot (a_{i+1} \oplus a_{i+3}) \quad i = \{1, 2, 3\} \\
\delta &= u_1 \cdot u_2 \cdot u_3
\end{aligned} \tag{B.3}
$$

Note that these operators are unchanged by duality, and for two- or four-body collisions, two of them are non-zero. Thus, for a model with all the possible collisions, the collision operators must choose between the different possible collisions using two different sets of collision operators:

$$
c_i = \delta + \gamma_i + \gamma_{i+2} \cdot \bar{\gamma}_{i+1} , \quad i = \{1, 2, 3\} \tag{B.4}
$$

or

$$
c_i = \delta + \gamma_{i+2} + \gamma_i \cdot \bar{\gamma}_{i+1} , \quad i = \{1, 2, 3\} \tag{B.5}
$$

and the new configuration is given by

$$
\begin{aligned}
a_i &= c_i \oplus a_i, \\
a_{i+3} &= c_i \oplus a_{i+3}, \\
i &= \{1, 2, 3\}
\end{aligned} \tag{B.6}
$$

for a total of 35 elementary Boolean operators.

For model I, the collision operators defined in equation (B.4) must be modified to allow collisions only for two-body configurations. This requires checking the absence of particles in the two directions not used to define the γ_i:

$$
c_i = \delta + \gamma_i \cdot \overline{(a_{i+2} + a_{i+5})} , \quad i = \{1, 2, 3\} \tag{B.7}
$$

or

$$
c_i = \delta + \gamma_{i+2} \cdot \overline{(a_{i+1} + a_{i+4})} , \quad i = \{1, 2, 3\} \tag{B.8}
$$

The algorithm to compute the collision step for model I, uses equations (B.3), (B.7) or (B.8) and (B.6), for a total of 35 elementary operators.

Appendix B.2   Model III

The algorithm for the model III can be easily derived for the algorithm for the six-bit model with all the possible collisions. A new step must be added to equation (B.3) to handle the case of rest particles:

$$\epsilon_i = t_i \cdot t_{i+5} \cdot (a_0 \oplus a_{i+1}) \quad i = \{1, \cdots, 6\} \tag{B.9}$$

The choice between two possible collisions is now done by the collision operator c_0 of the rest particles:

$$
\begin{aligned}
c_0 &= \overline{\delta + (\gamma_1 + (\epsilon_1 \oplus \overline{\epsilon}_4)) \cdot (\gamma_2 + (\epsilon_2 \oplus \overline{\epsilon}_5)) \cdot (\gamma_3 + (\epsilon_3 \oplus \overline{\epsilon}_6))}, \\
c_i &= \delta + \overline{c}_0 \cdot (\gamma_i + \gamma_{i+2} \cdot \overline{\gamma}_{i+1}) + c_0 \cdot (\epsilon_{i+5} + \epsilon_i \cdot \overline{\epsilon}_{i+2} + \epsilon_{i+1} \cdot \overline{\epsilon}_{i+3}) \\
&\quad i = \{1, 2, 3\}, \\
c_{i+3} &= \delta + \overline{c}_0 \cdot (\gamma_i + \gamma_{i+2} \cdot \overline{\gamma}_{i+1}) + c_0 \cdot (\epsilon_{i+2} + \epsilon_{i+3} \cdot \overline{\epsilon}_{i+5} + \epsilon_{i+4} \cdot \overline{\epsilon}_i) \\
&\quad i = \{1, 2, 3\}
\end{aligned}
\tag{B.10}
$$

or

$$
\begin{aligned}
c_0 &= \overline{\delta + (\gamma_1 + (\epsilon_2 \oplus \overline{\epsilon}_5)) \cdot (\gamma_2 + (\epsilon_3 \oplus \overline{\epsilon}_6)) \cdot (\gamma_3 + (\epsilon_1 \oplus \overline{\epsilon}_4))}, \\
c_i &= \delta + \overline{c}_0 \cdot (\gamma_{i+2} + \gamma_i \cdot \overline{\gamma}_{i+1}) + c_0 \cdot (\epsilon_{i+1} + \epsilon_i \cdot \overline{\epsilon}_{i+4} + \epsilon_{i+5} \cdot \overline{\epsilon}_{i+3}) \\
&\quad i = \{1, 2, 3\}, \\
c_{i+3} &= \delta + \overline{c}_0 \cdot (\gamma_{i+2} + \gamma_i \cdot \overline{\gamma}_{i+1}) + c_0 \cdot (\epsilon_{i+4} + \epsilon_{i+3} \cdot \overline{\epsilon}_{i+1} + \epsilon_{i+2} \cdot \overline{\epsilon}_i) \\
&\quad i = \{1, 2, 3\}.
\end{aligned}
\tag{B.11}
$$

The new configurations are computed using equation (B.5) and

$$a_0 = c_0 \oplus a_0. \tag{B.12}$$

Thus, the algorithm for the model III uses equations (B.3), (B.9), (B.10), or (B.11), (B.6), and (B.12), for a total of 103 elementary Boolean operators,[7] if $\overline{\epsilon}_i$, $\epsilon_i \cdot \overline{\epsilon}_{i+4}$ and $\delta + \overline{c}_0 \cdot (\gamma_i + \gamma_{i+2} \cdot \overline{\gamma}_{i+1})$ are computed only once.

Appendix B.3   Model II

The algorithm for the model II is slightly different of the previous one, since many possible configurations must be removed.

$$
\begin{aligned}
t_i &= a_i \oplus a_{i+1}, \quad u_i = t_i \cdot \overline{a}_{i+4} \quad i = \{1, \cdots, 6\} \\
v_i &= \quad (t_i \cdot t_{i+3}) \cdot u_{i+1} \cdot u_{i+4} \quad i = \{1, 2, 3\}
\end{aligned}
\tag{B.13}
$$

The choice between head-on collisions is done by

---

[7]Two additional operators can be saved in equation (B.9) using $a_0 \oplus a_2$ for $i = 1, 3$ and $a_0 \oplus a_3$ for $i = 2, 4$.

$$\gamma_i = v_i + v_{i+1}, \qquad i = \{1, 2, 3\} \tag{B.14}$$

or

$$\gamma_i = v_i + v_{i+2}, \qquad i = \{1, 2, 3\} \tag{B.15}$$

and δ is computed using only two and operators (the quantities inside parentheses being computed during (B.13) step) by

$$
\begin{aligned}
\delta &= (t_1 \cdot t_4) \cdot (t_2 \cdot t_5) \cdot (t_3 \cdot t_6) \\
\epsilon_i &= t_i \cdot t_{i+5} \cdot (a_0 \oplus a_{i+1}) \cdot \bar{a}_{i+2} \quad i = \{1, \cdots, 6\}
\end{aligned} \tag{B.16}
$$

The collision operators c_i are given by

$$
\begin{aligned}
c_0 &= \epsilon_1 + \epsilon_2 + \epsilon_3 + \epsilon_4 + \epsilon_5 + \epsilon_6 \\
c_i &= (\delta + \gamma_i) + \epsilon_i + \epsilon_{i+1} + \epsilon_{i+5} \quad i = \{1, 2, 3\} \\
c_{i+3} &= (\delta + \gamma_i) + \epsilon_{i+3} + \epsilon_{i+4} + \epsilon_{i+2} \quad i = \{1, 2, 3\}
\end{aligned} \tag{B.17}
$$

The new configurations are computed using equations (B.6) and (B.12). Thus, the algorithm for the model II uses equations (B.13), (B.14), or (B.15), (B.16), (B.17), (B.6), and (B.12), for a total of 82 elementary Boolean operators, if some or operations are computed only once in equation (B.17).

References

[1] U. Frisch, B. Hasslacher and Y. Pomeau, *Phys. Rev. Lett.*, **56** (1986) 1505.

[2] U. Frisch, D. d'Humières, B. Hasslacher, P. Lallemand, Y. Pomeau and J. P. Rivet, *Complex Systems*, **1** (1987) 646. (This paper lists most of the relevant references in the field of lattice gases.)

[3] U. Frisch and J.P. Rivet, *C. R. Acad. Sci. Paris XII*, **303** (1986) 1065.

[4] M. Hénon, "Calcul de la viscosité dans le réseau triangulaire" and "Viscosité d'un réseau", preprints, Observatoire de Nice, B.P.139, 06003 Nice Cedex, France.

[5] D. d'Humières, Y. Pomeau and P. Lallemand, *C. R. Acad. Sci. Paris XII*, **301** (1985) 1391.

[6] D. d'Humières, P. Lallemand and T. Shimomura, "An experimental study of lattice gas hydrodynamics", *Los Alamos preprint*, LA-UR-85-4051 (1985).

[7] D. d'Humières, Y. Pomeau and P. Lallemand, *Innovative Numerical Methods in Engineering*, (A Computational Mechanics Publication, Springer-Verlag, Berlin, 1986) 241.

[8] D. d'Humières and P. Lallemand, *C. R. Acad. Sci. Paris XII*, **302** (1985) 983.

[9] A. Clouqueur and D. d'Humières, "RAP1, a cellular automaton machine for fluid dynamics", *Complex Systems*, **1** (1987) 584.

[10] S. Hillis, "The Connection machine", (M.I.T. Press, Camdridge, MA, 1986).

[11] D. Levermore, private communication.

[12] G. Doolen, private communication.

[13] R. D. Mountain, *Rev. of Mod. Physics*, **38** (1966) 205.

[14] Other methods can be used to measure ν; see, for example, F. Hayot, "Viscosity in lattice gas automata", *Physica D*, (1987), in press, or L. Kadanoff, G. MacNamara, and G. Zanetti, "Size-dependence of the shear viscosity for a two-dimensional lattice gas", University of Chicago preprint (1987).

[15] D. d'Humières and P. Lallemand, *Physica*, **140A** (1986) 326.

[16] D. d'Humières, P. Lallemand and G. Searby, "Numerical Experiments on Lattice Gases: Mixtures and Galilean Invariance", *Complex Systems*, **1** (1987) 632.

[17] D. d'Humières and P. Lallemand, *C. R. Acad. Sci. Paris XII*, **302** (1985) 983.

[18] H. Slichting, *Boundary layer theory*, (Pergamon Press, London, 1955).

[19] H. Slichting, *Z.A.M.M*, **14** (1934) 368.

[20] GAMM workshop, in *Note on Numerical Fluid Mechanics*, **9**, K. Morgan, J. Périaux and F. Thomasset, eds. (Vieweg and Sons, 1984).

[21] S. Taneda and H. Honji, *J. Phys. Soc. Japan*, **30** (1971) 262.

[22] C. Burges and S. Zaleski, *Complex Systems*, **1** (1987) 31.

[23] P. Clavin, D. d'Humières, P. Lallemand and Y. Pomeau, *C. R. Acad. Sci. Paris*, **303** (1986) 1169.

[24] P. Clavin, P. Lallemand, Y. Pomeau and G. Searby, *J. Fluid Mech.*, in press.

[25] S. Wolfram, *J. of Stat. Phys.*, **45** (1986) 471.

[26] J. Hardy, O. de Pazzis and Y. Pomeau, *Phys. Rev.*, **A13** (1976) 1949.

Dominique d'Humières and Pierre Lallemand
CNRS, Laboratoire de Physique de l'Ecole Normale Supérieure, 24 rue Lhomond, 75231 Paris Cedex 05, France

Numerical Experiments on Lattice Gases: Mixtures and Galilean Invariance

This paper originally appeared in *Complex Systems* (1987), Volume 1, pages 633–647.

Lattice Gas Methods for Partial Differential Equations, SFI SISOC, Eds. Doolen et al., Addison-Wesley Publishing Co., 1990

333

Complex Systems 1 (1987) 633 647

Numerical Experiments on Lattice Gases:
Mixtures and Galilean Invariance

Dominique d'Humières
Pierre Lallemand
Laboratoire de Physique de l'Ecole Normale Supérieure,
24 rue Lhomond, 75231 Paris Cedex 05, France

Geoffrey Searby
Laboratoire de recherche sur la combustion,
Centre de St Jérôme, 13397 Marseille Cedex 13, France

Abstract. In this paper, we first describe an extension of the standard Frisch, Hasslacher, Pomeau hexagonal lattice gas to study reaction-diffusion problems. Some numerical results are presented. We then consider the question of Galilean invariance from an "experimental" point of view, showing cases where the standard model is inadequate. Finally, we introduce a way to cure the Galilean disease and present some results of simulations for a few typical cases.

1. Introduction

The lattice gas technique first introduced by Pomeau et al. [1] and later refined by Frisch, Hasslacher, and Pomeau (FHP) [2] is now considered as an efficient way to simulate viscous flows at moderate Mach numbers in situations involving complex boundaries. However, it is unable to represent thermal or diffusional effects since all particles have the same speed and are of the same nature. In addition, the macroscopic behavior of the FHP (or standard) model is not Galilean invariant in the sense that the nonlinear advection term in the momentum equation involves an the original FHP model to study reaction-diffusion problems and then discuss the implications of $g(d) \neq 1$. Finally, we shall present a way to design a model that is Galilean invariant at least for low Mach numbers.

2. Extension of the FHP model

The original FHP model involves particles whose velocities c_i have the same modulus and point in the six possible directions corresponding to the links between one node of the lattice and its six nearest neighbors. It is useful to add rest particles (0 or 1 in the Boolean model) that are involved in collisions of the type rest particle and n_i, with $i \in \{1, \ldots, 6\}$, represents one of the six possible moving particles. Using the Rivet-Frisch method [3] to calculate the viscosity, or measuring it by relaxation of shear waves, it has been shown [4] that the maximum possible Reynolds number is obtained using the seven-bit model with all possible non-transparent collisions. With this criterion, the above model is thus preferred over the simpler models, and there are 76 non-transparent collisions among the 2^7 possible precollision states.

In a way similar to that of Burges and Zaleski [5], who introduced cellular automata adapted to the description of fluid mixtures, we start now from the original seven-bit FHP model and add seven other bits, one for each bit of the basic model, that may be interpreted as the type (or color) of the particle [6]. The state of the system is thus fully determined by giving at each node of the lattice $\{(n_i, t_i)\}$, $i \in \{0, \ldots, 6\}$ where $t_i = 0$ for $n_i = 0$ and $t_i = 0$ or $t_i = 1$ for $n_i = 1$, ($t_i = 0$ particle of type A, $t_i = 1$ particle of type B). There are now 3^7 possible states.

The bit t_i is attached to and propagates as bit n_i. The collision rules can be of several types according to the type of problem being investigated.

This new model can be analyzed in the same way as the basic model. One may define as macroscopic quantities:

$$\rho = \sum_i n_i \qquad \text{total density}$$
$$\rho \mathbf{u} = \sum_i n_i \mathbf{c}_i \qquad \text{total flux}$$
$$C = \frac{1}{\rho} \sum_i n_i t_i \qquad \text{concentration of B particles.}$$

At the macroscopic level, these quantities satisfy the continuity equation, the Navier-Stokes equation with a factor $g(d)$, and a concentration equation.

$$\partial_t \rho + div \rho \mathbf{u} = 0$$

$$\partial_t \rho u_\alpha + \partial_\beta (\rho g(d) u_\alpha u_\beta) = -\partial_\alpha P + \partial_\beta (\nu(\rho) \partial_\beta \rho u_\alpha)$$

$$\partial_t \rho C + \partial_\alpha \rho C u_\alpha = \partial_\alpha D(\rho) \partial_\alpha \rho C + \varpi(C)$$

where D is a diffusion coefficient and $\varpi(C)$ a source term when type exchange collisions are included.

Here, d is the average density of particles per cell, so that

$$g(d) = \frac{7(1 - 2d)}{12(1 - d)}.$$

2.1 Definition of collisions

The general rules that allow one to determine the output of any precollision situation are the following:

1. The total number of particles is conserved.

2. The total linear momentum is conserved.

Now the collisions can be nonreactive, which means that the numbers of A and B particles are conserved, or they can be reactive, which means that the numbers of A and B particles can vary.

Note that even for nonreactive collisions and distributions of velocities that correspond to no change in the $\{n_i\}$, the values of the $\{t_i\}$ can be redistributed when $\sum t_i$ differs from 0 or $\sum n_i$. In some cases, it is found that there are as many as 35 equivalent distributions of the $\{t_i\}$. This means that a poor model from the point of view of the viscosity (few efficient velocity redistributing collisions) may nevertheless lead to a small value of the diffusion coefficient.

The choice for the redistribution of the $\{t_i\}$ by reactive collisions is obviously huge. Here, we have first considered situations of the type:

$$A + A + B \underset{k_a^-}{\overset{k_a^+}{\rightleftharpoons}} A + A + A$$

$$A + B + B \underset{k_b^-}{\overset{k_b^+}{\rightleftharpoons}} B + B + B$$

where either A or B acts as some sort of catalyzer. In that case, the production rate is given by

$$\varpi(C) = k_a^+ C(1-C)^2 + k_b^- C^3 - k_a^- (1-C)^3 - k_b^+ C^2 (1-C).$$

If we assume symmetry in the reaction rates, $k_a^+ = k_b^+ = k^+$ and $k_a^- = k_b^- = k^-$, then ϖ is an odd function of $C - 0.5$.

Let us define $\chi = k^-/k^+$. If $\chi < 1/3$, ϖ has three roots C_1, $C_2 = 1 - C_1$ and $1/2$. For $\chi = 0$, $C_1 = 0$ and $C_2 = 1$, which means that the system will exhibit segregation into regions which are filled with pure A or pure B.

This result can be extended to situations where all reactions

$$A + nB \underset{0}{\overset{k_+^+}{\rightleftharpoons}} (n+1)B$$

$$nA + B \underset{0}{\overset{k_+^+}{\rightleftharpoons}} (n+1)A$$

with $n > 1$ are included.

Note that this model implies no coupling between the type exchange and the dynamics of the particles and thus cannot represent such effects as surface tension.

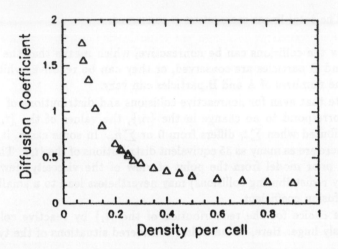

Figure 1: Measured value of the diffusion coefficient D versus density of moving particles per cell.

2.2   Experimental study

The dynamics of lattice gas mixtures has been studied by simulation on a FPS164 computer using the "table method", in which we use the state of each node of the lattice as an address in a collision table to determine the redistribution of the $\{(n_i, t_i)\}$. This method is not very efficient, as it implies reading the memory in a random manner, but it is very simple. The large indetermination in the outcome of some precollisional states (up to 35-fold in this model) is dealt with in the following manner: before each sweep through the lattice, we choose at random one particular outcome among tables containing the equivalent states.

2.2.1   Basic properties

We have determined the diffusion coefficient D by studying the time relaxation of either a concentration step or a sinusoidal modulation of the concentration. In the first case, a very good fit of the data with the standard solution of the diffusion equation is found. The resulting experimental determinations of D are shown in figure 1. Contrary to all results on the standard FHP model (which is dual in the sense that they depend only upon $| 0.5 - d |$) here D is smaller for $d > 0.5$. This comes from the fact that we have taken only $t_i = 0$ for $n_i = 0$. We could have decided to color the holes by taking $t_i = 1$ for $n_i = 1$ and $t_i = 0$ or $t_i = 1$ for $n_i = 0$. The model can be used to analyze how a tracer is dispersed by a flow [7].

We now allow type exchange reactions to take place. In that case, sharp interfaces are observed, the thicknesses of which are on the order of $\sqrt{D/k}$.

Detailed results are described in reference 6.

As mentioned previously, there is no surface tension for the present model when $k^+ = k^-$; thus, the interface between A and B exhibits a diffusive behavior. It can be shown that the dynamics of the interface is governed by $dx/dt = -D/R$ if x is a coordinate normal to the interface and R its radius of curvature. A sinusoidal interface will relax exponentially, and the radius of a circular bubble will decrease as $R^2(t) = R^2(0) - 2Dt$ [8].

The model will be adequate to study hydrodynamics with free boundaries, provided that the relevant time scales are short compared to that of the diffusion of the interfaces. It is therefore preferable to use a density of 0.75 rather than 0.25; the hydrodynamics is the same, but D is smaller.

This model, to which gravity can be added, provides a simple way to study phenomena such as the Rayleigh-Taylor instability [9]. Here we will illustrate its possibilities with some results for the Kelvin-Helmholtz instability.

2.2.2   Kelvin-Helmholtz instability

Two parallel plates of length 1024, separated by 256 lattice sites, are set as boundaries for a channel assumed to be periodic in the direction of the plates. Stick conditions are implemented by imposing velocity reversal of particles colliding with the boundary. The initial conditions for the experiment are

1. half of the channel is filled with particles of type A and velocity $+v$ parallel to the plates, and

2. the other half is filled with B particles and velocity $-v$.

The mean density of moving particles is 0.75 and $v = 0.21$. Maps of the local average of the particle fluxes are computed for various values of time. Note that color-blind maps are identical to those which would be obtained with the standard FHP model with the same microscopic initial distribution of the $\{n_i\}$.

In a first experiment, the collision rules include only nonreactive diffusion of A and B particles. Figure 2 shows isoconcentration curves for A at time $t = 3000$. For this particular run, in one portion of the channel, the interface was destabilized by the Kelvin-Helmholtz instability and A particles were advected by the flow, whereas in another portion of the channel, the interface remained stable and uniform diffusion took place. (This particular result is not typical; usually, the $k = 4$ mode is the most unstable.)

In a second experiment, we use the autocatalytic reactive rules. Figure 3 shows the distribution of the flux of B at time $t = 2000$, and figure 4 shows several successive shapes of the interface that remains sharp for several thousand time steps.

Figure 2: Isoconcentration curves (for $C = 0.10 to .90$) in the situation leading to the Kelvin-Helmholtz instability at time $t = 3000$. Purely diffusive case.

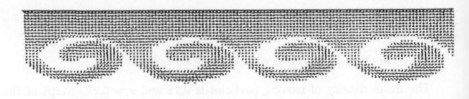

Figure 3: Map of B particle flux at time $t = 2000$ in the situation leading to the Kelvin-Helmholtz instability, using type exchange reactions that allow phase separation.

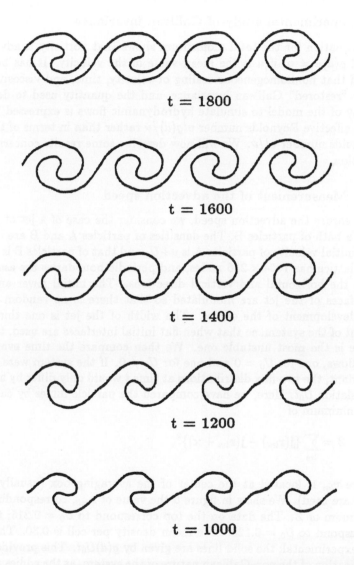

t = 1800

t = 1600

t = 1400

t = 1200

t = 1000

Figure 4: Series of shapes of the interface separating phase A and phase B in the situation leading to the Kelvin-Helmholtz

The results shown here need to be supplemented by quantitative tests. The fact that the advection terms of the total momentum and concentration are different should be analyzed in more detail.

3.   Experimental study of Galilean invariance

The equation for the total momentum shows that vorticity is advected at speed $g(d)u \neq u$, if u is the mean value of the velocity. It has been proposed that nonhomogeneous scaling of velocity, time, and viscosity would allow "restored" Galilean invariance, and the quantity used to define the ability of the model to simulate hydrodynamic flows is expressed in terms of an effective Reynolds number $ulg(d)/\nu$ rather than in terms of the usual Reynolds number ul/ν. We will now describe some results concerning this question.

3.1   Measurement of the advection speed

To measure the advection speed, we consider the case of a jet of particles A in a bath of particles B. The densities of particles A and B are identical. The initial velocity of particles A is $v+U_0$ and that of particles B is $-v+U_0$. The lattice has 1024×256 nodes, and periodic boundaries are assumed in both the horizontal and vertical directions. The initial lower and upper interfaces of the jet are modulated so that there is no random effect in the development of the eddies. The width of the jet is one third of the height of the system, so that when flat initial interfaces are used, the $k = 3$ mode is the most unstable one. We then compare the time evolution of two flows, one for $U_0 = 0$ and one for $U_0 \neq 0$. If the system were Galilean invariant, the two flux distributions at time t would only differ by a uniform translation $U_0 t$. Here, we have compared the particle fluxes by calculating the minimum of

$$Z = \sum_{lm} \{j(r_{lm}) - j(r_{lm} + x)\}^2,$$

where r_{lm} is located at the center of the averaging box (usually 16×16 sites are used). We show in figure 5 the value of x_{min} corresponding to the minimum of Z. The data on the top correspond to $U_0 = 0.215$; the other correspond to $U_0 = 0.128$. The mean density per cell is 0.30. The crosses are experimental; the solid lines are given by $g(d)U_0 t$. This provides a good verification of the non-Galilean nature of the system, as the eddies are found to be advected at speed $g(d)U_0$. Similar results apply up to approximately $U_0 \simeq 0.25$, meaning that the higher order terms in the momentum equation can be neglected in the present situation for Mach numbers up to roughly 0.4.

We also find that the minimum value of Z, the error, remains approximately equal to that obtained when comparing two different microscopic realizations of the same macroscopic conditions.

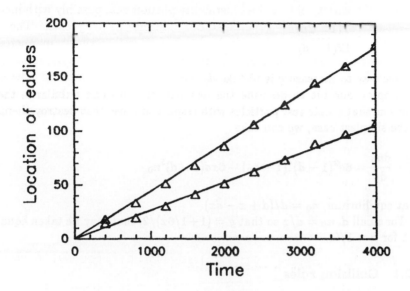

Figure 5: Value of the displacement between two flows. Crosses are determined experimentally; solid curves correspond to the advection speed $U_0 g(d)$.

Another way to study non-Galilean invariant problems would have been to analyze how small eddies are advected by large ones. Significant results would be obtained only for lattices much larger than the one used here in order to avoid viscous relaxation of the small eddies [10].

We have also considered the Kelvin-Helmholtz situation with reactive collisions and found dramatic differences between the results obtained for $U_0 = 0$ and $U_0 \neq 0$. In particular, we find that the roll-up of the interface does not take place when $U_0 \neq 0$.

3.2 New lattice gas with $g(d) = 1$

The absence of Galilean invariance of the original model is due to the use of a finite set of directions for the velocity and to the exclusion principle that leads to the Boolean character of the particles. In general, g is given by

$$g(d) = \frac{D \sum_i d_i \ \sum_i d_i (1 - d_i)(1 - 2d_i) c_i^4}{(D + 2)\{\sum_i d_i (1 - d_i) c_i^2\}^2}$$

if d_i is the density of particles per cell and D the dimensionality. At equilibrium for $u = 0$, the total density is given by $\rho = 6d + d_0$ if d_0 is the density of rest particles. The standard seven-bit model leads to $d_0 = d$ so that

$$g(d) = \frac{7(1-2d)}{12(1-d)}.$$

One way to increase g is to take $d_0 > d$.

Keeping one bit to describe the rest particles, we can unbalance the collisions that create rest particles with respect to those that destroy them. In the simplest case, we can take

$$\frac{dn_0}{dt} = 6d^2(1-d)^4(1-n_0) - 6xd(1-d)^5 n_0$$

so at equilibrium, $n_0 = d/(d + x - dx)$.

For small d, $n_0 = d/x$ so that $g = (1+1/6x)/2$, that may be taken equal to 1 for $x = 1/6$.

3.2.1 Collision rules

Here, we propose to use a two-bit word to represent the population of the center (n_0, m_0), with the following collision rules. All collisions conserving total number and total linear momentum are included [11]. The collisions leading to creation and destruction of rest particles are all included, except a few cases which take place with probability x, y, or z. For details, see table 1, where all collisions considered here are indicated for one of the six possible directions of the velocity vector.

It may be shown that the equilibrium value of n_0 and m_0 are solutions of the following coupled equations:

$$\{(1-d)^4 d^2 + 2(1-d)^3 d^3 + (1-d)^2 d^4 + (1-d)^5\}(1-m_0) = x(1-d)^5 d(1-m_0)n_0$$

$$+ z\{(1-d)^5 d + (1-d)^4 d^2 + 2(1-d)^3 d^3 + (1-d)^2 d^4\}m_0 n_0$$

$$\{(1-d)^4 d^2 + 2(1-d)^3 d^3 + (1-d)^2 d^4 + (1-d)^5\}m_0(1-n_0) = y(1-d)^5 d(1-m_0)n$$

Analyzing the numerical values of n_0, m_0, and the corresponding $g(d)$, it is found that for $x = 0.5, y = z = 0.20$, there exists a value of d for which $g(d) = 1$ and $dg(d)/dd = 0$.

For simplicity of the dynamics, we prefer to use the case $x = 1/2, y = z = 1/6$, which leads to a theoretical value of $g_{max} = 1.072$ for $d = 0.17$ and $g = 1.0$ for $d = 0.21$.

The model can now be analyzed with the Chapman-Enskog method, as was done by Rivet and Frisch for the standard model. This allows us to determine the kinematic shear viscosity

$$\nu = \frac{1}{4(C_{11} - C_{12} - C_{13} + C_{14})} - \frac{1}{8}$$

with

input		output		probability	
moving particles	rest particles	moving particles	rest particles		
(diagram)	1	(diagram)	0	x	
	2		1	y	
	3		2	z	
(diagram)	0, 1 or 2	(diagram)	1, 2 or 3	1	
	3	(diagram) (diagram)	2	z/2	z/2
(diagram)	0, 1, 2 or 3	(diagram) (diagram)	0, 1, 2 or 3	1/2	1/2
(diagram)	0, 1, 2 or 3	(diagram)	0, 1, 2 or 3	1	
(diagram)	0, 1 or 2	(diagram)	1, 2 or 3	1	
	3	(diagram) (diagram)	3 2	1-z	z
(diagram)	0, 1 or 2	(diagram)	1, 2 or 3	1	
	3	(diagram) (diagram)	3 2	1-z	z
(diagram)	0, 1 or 2	(diagram) (diagram)	1, 2 or 3	1/2	1/2
	3	(diagram)	2	z	
(diagram)	0, 1, 2 or 3	(diagram) (diagram)	0, 1, 2 or 3	1/2	1/2
(diagram)	0, 1 or 2	(diagram)	1, 2 or 3	1	

Table 1: Details of the collisions considered in the Galilean invariant lattice gas model.

$$C_{11} - C_{12} - C_{13} + C_{14} =$$

$$5A_{41} + 11A_{32} + 10A_{23} + 6A_{14} + 2A_{05}$$

$$+ \, m_0 n_0 (-2A_{41} + 4A_{32} + 5A_{23} - 3A_{14} - 2A_{05})$$

$$+ \, 2(A_{50} + A_{41})(x m_0 (1 - n_0) + y(1 - m_0)n_0)$$

$$+ \, z m_0 n_0 (2A_{50} + 3A_{41} - 5A_{32} - 4A_{23} + 2A_{14})$$

where $A_{ij} = (1 - d)^i d^j$

We plot in figure 6 the theoretical values of $g(d)$, $g(d)/\nu$, and η which appear in the Reynolds number and characterize the ability of the model to represent flows. It is interesting to note that we have improved the value of the Reynolds number.

3.2.2 Numerical results

The new eight-bit lattice gas has been implemented on a computer. The equilibrium value of $\rho(d)$ is in fair agreement with the theoretical value deduced from the equations for n_0 and m_0.

We have also measured the advection speed. Instead of using the same situation used previously, we have studied the relaxation of a moving shear wave. For this purpose, we consider a 512×512 lattice with periodic boundary conditions and impose as initial conditions

$$v(\mathbf{r}, 0) = v_x + v_y \cos \mathbf{k.r} \text{ with } \mathbf{k} \parallel \text{Ox.}$$

We determine the spatial phase ϕ of the k-Fourier component of $v_y(t)$. It is found to vary linearly with time, and we plot $d\phi/dt$ in figure 7 together with its theoretical value $g(d)v_x k$. Good agreement is obtained, showing that the model is truly Galilean invariant for a particular but adjustable density (here, $d = 0.21$). Moreover, to first order, the invariance is not destroyed by small density fluctuations.

Preliminary measurements of the speed of sound and of the kinematic shear viscosity are in good agreement with the theory.

The new model should be very useful but requires further investigation, especially with regards to the fact that some collision events do not satisfy semi-detailed balance. Adding color to the new model involves no particular difficulty.

4. Conclusion

In this paper, we have presented two extensions of the basic FHP lattice gas model. The first one adds a second bit to each of the seven-bit occupation numbers. It allows the simulation of various diffusion-reaction problems and of hydrodynamic flows with free boundaries. The second allows achievement of true Galilean invariance for a limited range of the density, at the cost of using an extra bit for the rest particles and the lack of semi-detailed balance for a subset of the collisions.

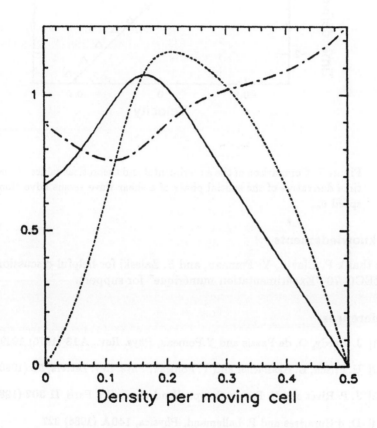

Figure 6: Theoretical values of $g(d)$ (solid line), $0.2 \times g(d)/\nu$ (dotted line), and $25 \times \eta$ (dash-dot line) versus density of moving particles per cell for the new model, when $x = 1/2$ and $y = z = 1/6$.

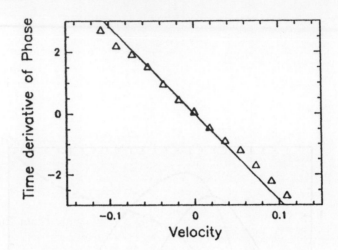

Figure 7: Comparison of the experimental and theoretical values of the time derivative of the spatial phase of a shear wave versus advection speed v_x.

Acknowledgments

We thank P. Clavin, Y. Pomeau, and S. Zaleski for helpful discussions and GRECO 70 "Expérimentation numérique" for support.

References

[1] J. Hardy, O. de Pazzis and Y.Pomeau, *Phys. Rev.*, **A13** (1976) 1949.

[2] U. Frisch, B. Hasslacher and Y. Pomeau, *Phys. Rev. Letters*, **56** (1986) 1505.

[3] J. P. Rivet and U. Frisch, *Compt. Rend. Acad. Sci. Paris*, **II 302** (1986) 267.

[4] D. d'Humières and P. Lallemand, *Physica*, **140A** (1986) 327.

[5] C. Burges and S. Zaleski, *Complex Systems*, **1** (1987) 31.

[6] P. Clavin, P. Lallemand, Y. Pomeau, and G. Searby, *J. Fluid Mech.*, to be published.

[7] J. P. Hulin and C. Baudet, to be published.

[8] D. d'Humières, P. Lallemand and G. Searby, *Proc. of E.P.S. Int. Conf. on Physico-chemical Hydrodynamics* (La Rabida, Spain, July 1986), M. G. Velarde, ed. (Plenum Press 1987).

[9] D. d'Humières, P. Clavin, P. Lallemand, and Y. Pomeau, *Comp. Rend. Acad. Sci. Paris*, **II 303** (1986) 1169.

[10] S. Wolfram, private communication.

[11] D. d'Humières and P. Lallemand, "Numerical Simulations of Hydrodynamics with Lattice Gas Automata in Two Dimensions", *Complex Systems*, **1** (1987) 598.

Tudatsugu Hatori* and David Montgomery†

*Permanent address: Institute of Plasma Physics, Nagoya University, Nagoya 464, Japan, and †Department of Physics and Astronomy, Dartmouth College, Hanover, NH 03755, USA

Transport Coefficients for Magnetohydrodynamic Cellular Automata

This paper originally appeared in *Complex Systems* (1987), Volume 1, pages 735–752.

Complex Systems 1 (1987) 735–752

Transport Coefficients for
Magnetohydrodynamic Cellular Automata

Tudatsugu Hatori*
David Montgomery
Department of Physics and Astronomy,
Dartmouth College, Hanover, NH 03755, USA

Abstract. A Chapman-Enskog development has been used to infer theoretical expressions for coefficients of kinematic viscosity and magnetic diffusivity for a two-dimensional magnetohydrodynamic cellular automaton.

1. Introduction

This article will present some approximate theoretical calculations of transport coefficients for magnetohydrodynamic (MHD) cellular automata (CA). Throughout, the attention will be limited to two-dimensional (2D) models.

CA, as possible fluid simulation tools, have recently become an active research area [1–4], following several years of basic background investigations on lattice gases [5–8]. A cellular automaton for simulating 2D MHD has recently been proposed [9]. As in the fluid case, 2D MHD macroscopic continuum equations can be derived with some degree of conviction from the dynamics represented by the microscopic computer game which the cellular automaton is. In interpreting future results from MHD CA computations, it will be desirable to have theoretical estimates of the kinematic viscosity and magnetic diffusivity derived from a microscopic kinetic theory. The principal purpose of this article is to present such a derivation.

Section 2 reviews the 2D MHD CA previously put forward [9], and relates it to its ancestor, the hexagonal lattice gas [1]. Section 3 presents a Markovian stochastic model of its dynamics. Section 4 is devoted to the calculation of the kinematic viscosity ν and the magnetic diffusivity η, which are expected to characterize the magnetofluid which the CA is simulating. Section 5 briefly discusses the results and remarks upon possible generalizations. In the development, some familiarity with the Chapman-Enskog procedure for deriving fluid equations and transport coefficients from kinetic equations will be assumed; the books of Chapman and Cowling

---

*Permanent address: Institute of Plasma Physics, Nagoya University, Nagoya 464, Japan.

[10], or Ferziger and Kaper [11] are valuable expositions of the Chapman-Enskog method. The systematic derivation of kinetic equations from a more basic description such as Liouville's equation [12,13] is not as far advanced as it is in classical continuum kinetic theory. The kinetic description used is of a more provisional character [2,4].

2. 2D MHD CA

Many properties of the 2D MHD CA [9] are appropriated from the hexagonal lattice gas CA of Frisch, Hasslacher, and Pomeau [1]. Our presentation draws heavily on a treatment of that model due to Wolfram [2]. Both models have particles which reside at the centers of adjacent hexagons in one of six discrete velocity states, $\hat{e}_a = (\cos(2\pi a/6), \sin(2\pi a/6))$, $a = 1, 2, \ldots 6$. The particles all move to the centers of the adjacent hexagons toward which they are directed, at discretized integer time steps. Between time steps, a set of "scattering rules" permit the particles to make transitions among the different $\hat{e}_a$ within a given hexagon. These scattering rules conserve momentum, and (trivially) conserve particle number and kinetic energy. Fermi-Dirac statistics apply, so that no more than one particle per single-particle state per hexagon is ever permitted. Macroscopic fluid variables such as density or fluid velocity are interpreted as averages of these quantities over the particles inside "supercells" consisting of many adjacent hexagons. Existing pilot codes typically have ten million or more hexagons and a thousand or more supercells.

All the properties in the previous paragraph are shared by the 2D MHD and two-dimensional Navier-Stokes models, although the 2D MHD model [9] adds two features to the two-dimensional Navier-Stokes model. First, the particles carry an additional integer index σ corresponding to the usual [14] 2D MHD one-component magnetic vector potential $\mathbf{A} = A_z(x, y, t)\hat{e}_z$. σ is allowed to take on the values ± 1 or 0; thus, there become 18 allowed particle states per hexagon instead of 6. The geometry is such that the magnetic field $\mathbf{B} = \nabla \times \mathbf{A} = (B_x, B_y, 0)$, the fluid velocity $\mathbf{u} = (u_x, u_y, 0)$, and all variables are independent of z. The σ label has variously been alluded to as a "photon" index, a "spin", or a "color"; we will use the term photon index as corresponding most closely to the physics being described. Second, a Lorentz force $(\nabla \times \mathbf{B}) \times \mathbf{B} = -(\nabla A_z)\nabla^2 A_z$ is introduced into the equation of motion in the same way that external forces such as gravity are presently introduced in the hydrodynamic case. This involves stepping outside the pure CA framework to impart the requisite momentum per unit volume by randomly flipping the microscopic distribution over the $\hat{e}_a$ proportionally to $-(\nabla A_z)\nabla^2 A_z$. (For purposes of computing this Lorentz force, standard finite-difference approximations are applied to the supercell averages of σ, which are interpreted as A_z.)

The just-described model has been shown [9] to lead to the following standard set of two-dimensional magnetofluid equations for low Mach numbers ($u^2 \ll 1$):

$$n\left(\frac{\partial \mathbf{u}}{\partial t} + \mathbf{u} \cdot \nabla \mathbf{u}\right) = -\nabla \cdot \mathbf{p} + \mathbf{j} \times \mathbf{B} + n\nu \nabla^2 \mathbf{u}, \tag{2.1}$$

$$\frac{\partial A_z}{\partial t} + \mathbf{u} \cdot \nabla A_z = \eta \nabla^2 A_z, \tag{2.2}$$

$$\nabla \cdot \mathbf{u} = -\frac{1}{n}\left(\frac{\partial}{\partial t} + \mathbf{u} \cdot \nabla\right) n \cong 0. \tag{2.3}$$

The symbols in equations (2.1) through (2.3) mean the following. The field variables are understood as supercell averages over several (typically, $(32)^2$) hexagons adjacent to each other. n is the particle number density. n has a maximum of 18 per hexagon ($\cong 20.8$ per unit area) and has been assumed in equations (2.1) through (2.3) to be $\ll 9$ per hexagon. $\mathbf{u}$ is the fluid velocity and has been assumed to be small in magnitude compared to unity. $\mathbf{j} = j_z \hat{e}_z = \nabla \times \mathbf{B} = -(\nabla^2 A_z)\hat{e}_z$ is the (supercell-averaged) electric current density. $\mathbf{j} \times \mathbf{B} = -(\nabla A_z)\nabla^2 A_z$ is the Lorentz volume force. The pressure dyadic is $\mathbf{p} \cong (n/2)\mathbf{1}(1-u^2/2) - n\mathbf{u}\mathbf{u}/2$. (Here, as elsewhere, more complicated coefficients result unless n is $\ll 9$.)

Finally, ν and η are the coefficients of kinematic viscosity and magnetic diffusivity respectively. This paper is devoted to an approximate theory of their evaluation for the case in which the effect of the Lorentz force on the distribution function is small compared to the effects of particle collisions. Once a kinetic equation is written down, such as a Boltzmann equation, Chapman-Enskog methods can be used to extract transport coefficients [2,10]. This program is carried out in sections 3 and 4.

3.   Markovian stochastic model

The exact one-particle distribution function is discrete. For each hexagon, it consists of 18 numbers, each either 0 or 1, depending upon whether there is or is not present a particle with velocity $\hat{e}_a$ and photon label σ. For fluid calculations, we require a smooth, differentiable distribution function, and we get it by ensemble averaging the exact one. The realizations of the ensemble remain discrete in velocity $\hat{e}_a$ and photon label σ, and differ from each other only by spatial translations, and by temporal translations of the instant which is called $t = 0$. The ensemble-averaged distribution function, $f_{a,\sigma}(\mathbf{x}, t)$, may be thought of as an 18-component vector field ($a = 1, 2, \ldots 6$, and $\sigma = -1,\ 0,\ +1$) which is a continuous function of space coordinates $\mathbf{x}$ and time t, now themselves both continuous. Always, $0 \le f_{a,\sigma} \le 1$. In terms of $f_{a,\sigma}$, the various fluid variables are defined as moments, according to the definitions'

$$lln = \sum_{a,\sigma} f_{a,\sigma} \qquad \text{(number density)} \tag{3.1}$$

$$nu = \sum_{a,\sigma} \hat{e}_a f_{a,\sigma} \quad \text{(fluid velocity} = u) \tag{3.2}$$

$$nA_z = \sum_{a,\sigma} \sigma f_{a,\sigma} \quad \text{vector potential} = A_z) \tag{3.3}$$

$$p = \sum_{a,\sigma} (\hat{e}_a - u)(\hat{e}_a - u) f_{a,\sigma} \quad \text{(pressure tensor)} \tag{3.4}$$

$$\pi = \sum_{a,\sigma} \hat{e}_a \hat{e}_a f_{a,\sigma} \quad \text{(momentum flux tensor)}. \tag{3.5}$$

$$\phi = \sum_{a,\sigma} \hat{e}_a \sigma f_{a,\sigma} \quad A_z\text{—flux vector).} \tag{3.6}$$

In equations (3.1) through (3.6), the $\sum_a$ always runs over $a = 1$ to 6, and the $\sum_\sigma$ always runs over $-1, 0, +1$.

Though it is not a deduction, it is a reasonable assumption that $f_{a,\sigma}$ may be advanced from its value at a previous time step according to the Markovian recipe:

$$f_{a,\sigma}(\mathbf{x}, t) =$$

$$\sum_{b,\lambda} \int d\mathbf{x}' P(a, \sigma, \mathbf{x}, t | b, \lambda, \mathbf{x}', t-1) f_{b,\lambda}(\mathbf{x}', t-1). \tag{3.7}$$

P is an assumed probability for a particle with coordinates $\hat{e}_b$, λ, $\mathbf{x}'$ at time $t - 1$ to find itself with coordinates $\hat{e}_a$, σ, $\mathbf{x}$ at time t. Until an explicit expression for P is given, of course, the content of equation (3.7) is only formal.

Spatial translations from one hexagon to another are always regarded as sharp, so that $\mathbf{x}' = \mathbf{x} - \hat{e}_b$. In the instant before time t, collisions are understood to occur which may scatter the particle from $(\hat{e}_b, \lambda)$ to $(\hat{e}_a, \sigma)$. All possibilities $\hat{e}_b, \lambda$ are understood to be summed over. In the spirit of Boltzmann, the transition probability P may be considered to depend upon the distribution function f itself, which will make the right-hand side of equation (3.7) nonlinear in f. (We will frequently omit subscripts and arguments when they are obvious from the context.)

Under the above assumptions, equation (3.7) reduces to

$$f_{a,\sigma}(\mathbf{x}, t) = \sum_{b,\lambda} P_{a\sigma;b\lambda}(\mathbf{x}, t) f_{b,\lambda}(\mathbf{x} - \hat{e}_b, t-1). \tag{3.8}$$

In the absence of collisions, $P_{a\sigma;b\lambda} = \delta_{ab}\delta_{\sigma\lambda}$. Subtracting the collisionless version of equation (3.8) from both sides gives

$$f_{a,\sigma}(\mathbf{x}, t) - f_{a,\sigma}(\mathbf{x} - \hat{e}_a, t-1) =$$

$$\sum_{b,\lambda} (P_{a\sigma;b\lambda}(\mathbf{x}, t) - \delta_{ab}\delta_{\sigma\lambda}) f_{b\lambda}(\mathbf{x} - \hat{e}_b, t-1). \tag{3.9}$$

For particle-conserving transition probabilities, which is what we are dealing with here, it is useful to rewrite P as

$$P_{a\sigma;b\lambda}(\mathbf{x},\ t) = \delta_{ab}\delta_{\sigma\lambda}\left[1 - \sum_{c,\mu}W_{c\mu;b\lambda}(\mathbf{x},\ t)\right] + W_{\omega\omega;b\lambda}(\mathbf{x},\ t) \qquad (3.10)$$

in terms of a non-negative transition probability W.

We insert equation (3.10) into equation (3.9), and further make the assumption that $f_{a,\sigma}(\mathbf{x},\ t)$ varies over characteristic lengths and times large compared to unity. This latter assumption permits us to Taylor expand the second term on the left-hand side of equation (3.9) with respect to $\mathbf{x}$ and t. The zeroth-order terms cancel, and we are left with

$$\left[\left(\frac{\partial}{\partial t} + \hat{e}_a \cdot \nabla\right) - \frac{1}{2}\left(\frac{\partial}{\partial t} + \hat{e}_a \cdot \nabla\right)^2 + \ldots\right] f_{a,\sigma}(\mathbf{x},\ t)$$

$$= \sum_{b,\lambda}[-W_{b\lambda;a\sigma}(\mathbf{x},\ t)f_{a,\sigma}(\mathbf{x} - \hat{e}_a,\ t - 1)$$

$$+ W_{a\sigma;b\lambda}(\mathbf{x},\ t)f_{b,\lambda}(\mathbf{x} - \hat{e}_b,\ t - 1)]. \qquad (3.11)$$

In terms of the conventional usage applied to the Boltzmann equation, the terms on the right-hand side of equation (3.11) correspond to the effects of "direct" and "inverse" collisions, respectively.

If we further Taylor-expand the f's on the right-hand side of equation (3.11) and equate the leading terms on both sides to each other, we get

$$\left(\frac{\partial}{\partial t} + \hat{e}_a \cdot \nabla\right)f_{a,\sigma}(\mathbf{x},\ t) = \sum_{b,\lambda}[-W_{b\lambda;a\sigma}(\mathbf{x},\ t)f_{a,\sigma}(\mathbf{x},\ t)$$

$$+ W_{a\sigma;b\lambda}(\mathbf{x},\ t)f_{b,\lambda}(\mathbf{x},\ t)] \equiv \Omega_{a\sigma}(f). \qquad (3.12)$$

Equation (3.12) is the starting point for our Chapman-Enskog development. The right-hand side determines $\Omega_{a\sigma}(f)$, which we call the collision integral. Its content is still formal until we adopt explicit expressions for the W's. Also absent from our present considerations is the effect of the higher-order terms in the Taylor expansions which are dropped in getting from equations (3.11) to (3.12); it may be that modifications to the transport coefficients will result from them [2,15]. This should be regarded as an important modification to be considered in the future.

The W's must always be chosen to satisfy the three non-trivial conservation laws:

$$\sum_{a,\sigma}\Omega_{a\sigma} = 0, \qquad (3.13)$$

$$\sum_{a,\sigma}\hat{e}_a\Omega_{a\sigma} = 0, \qquad (3.14)$$

$$\sum_{a,\sigma}\sigma\Omega_{a\sigma} = 0, \qquad (3.15)$$

for all $f_{a,\sigma}$, in addition to the trivial conservation law for kinetic energy. We defer until section 4 writing down explicit forms for equation (3.12) which will satisfy equations (3.13) through (3.15).

It will be assumed that the net effect of $\Omega_{a\sigma}$ in equation (3.12) is to drive $f_{a,\sigma}$ to a local thermal equilibrium over length and time scales short compared to those over which n, $\mathbf{u}$, p, A_z, etc., vary. This makes possible an iterative Chapman-Enskog approach to equation (3.12):

$$f_{a,\sigma}(\mathbf{x},\ t) = f_{a,\sigma}^{(0)}(\mathbf{x},\ t) + f_{a,\sigma}^{(1)}(\mathbf{x},\ t) + \dots, \tag{3.16}$$

where $f_{a,\sigma}^{(0)}(\mathbf{x},\ t)$ is the local Fermi-Dirac distribution, which depends on $\mathbf{x}$ and t only functionally through dependence on $n(\mathbf{x},\ t)$, $\mathbf{u}(\mathbf{x},\ t)$, and $A_z(\mathbf{x},\ t)$. $f_{a,\sigma}^{(1)}$ is the first-order deviation of $f_{a,\sigma}$ from $f_{a,\sigma}^{(0)}$, where the parameter of smallness is the usual one: the ratio of the collision mean-free path to the characteristic length scale for the variation of the moments, or the mean-free time to the time scale for the variation of the moments.

Since $\Omega_{a\sigma}(f^{(0)}) = 0$, the first-order terms in the smallness parameter are $(\partial/\partial t + \hat{e}_a \cdot \nabla)f_{a,\sigma}^{(0)}$ on the left-hand side and the $f^{(1)}$-proportional part of the right-hand side, which we abbreviate as $\Omega_{a\sigma}^{(1)}$:

$$\left(\frac{\partial}{\partial t} + \hat{e}_a \cdot \nabla\right) f_{a,\sigma}^{(0)}(\mathbf{x},\ t) = \Omega_{a\sigma}^{(1)} = \sum_{b,\lambda} C_{a\sigma,b\lambda}^{(0)} f_{b,\lambda}^{(1)}(\mathbf{x},\ t). \tag{3.17}$$

The right-hand side of equation (3.17) stands symbolically for the result of linearizing the right-hand side of equation (3.12) about $f_{a,\sigma}^{(0)}$ in powers of $f_{a,\sigma}^{(1)}$. Explicit forms for $\Omega_{a\sigma}^{(1)}$ and $C_{a\sigma,b\lambda}^{(0)}$ will be given in section 4 and the appendices. $C_{a\sigma,b\lambda}^{(0)}$ is in effect an 18×18 collision matrix. Some difficulty is involved in inverting it, but that is what is required in order to express $f_{a,\sigma}^{(1)}$ in terms of $(\partial/\partial t + \hat{e}_a \cdot \nabla)f_{a,\sigma}^{(0)}$. Once we have $f_{a,\sigma}^{(1)}$, we may express such first-order moments as

$$\pi^{(1)} = \sum_{a,\sigma} \hat{e}_a\hat{e}_a f_{a,\sigma}^{(1)}(\mathbf{x},\ t), \tag{3.18}$$

and

$$\phi^{(1)} = \sum_{a,\sigma} \hat{e}_a\sigma f_{a,\sigma}^{(1)}(x,\ t), \tag{3.19}$$

in terms of the (unexpanded) moments n, $\mathbf{u}$, A_z.

The three exact conservation laws which follow from equations (3.12) and (3.13) through (3.15) are:

$$\frac{\partial n}{\partial t} + \nabla \cdot (n\mathbf{u}) = 0, \tag{3.20}$$

$$\frac{\partial(n\mathbf{u})}{\partial t} + \nabla \cdot \pi = 0, \tag{3.21}$$

$$\frac{\partial}{\partial t}(nA_z) + \nabla \cdot \phi = 0. \tag{3.22}$$

If the moments in equations (3.20) through (3.22) are evaluated using only $f_{a,\sigma}^{(0)}$, the ideal Euler equations result. The expressions (3.18) and (3.19) add dissipative corrections to (3.21) and (3.22) which, in the usual way, lead to equations (2.1) through (2.3). The reader is reminded that the Lorentz force, $\mathbf{j} \times \mathbf{B}$, in equation (2.1) has to be put in as an "external" force in the way already indicated.

4. Evaluation of Chapman-Enskog transport coefficients

In this section, we first specialize $\Omega_{a,\sigma}$ to a Boltzmann-like collision term which permits two- and three-body collisions. In the notation of Wolfram [2], the scattering events allowed are the $2L$, $2R$ and $3S$ collisions. In addition to these mechanical collisions, simultaneous σ-conserving transitions of the photon index are allowed and may change the numbers of the $+1, 0, or -1$ σ-values in some collisions [9]. Second, we linearize this collision integral about a local thermodynamic equilibrium distribution and calculate the first Chapman-Enskog correction to the distribution function. Finally, we use this calculated distribution function, substituting it into the exact differential form of the conservation laws, to infer coefficients of kinematic viscosity and magnetic diffusivity. The calculation is algebraically lengthy and tedious and the details are relegated to four appendices, to which the reader who is primarily interested in the results may wish to skip.

The shorthand notation $\tilde{f}_{a,\sigma} = f_{a,\sigma}/(1-f_{a,\sigma})$ is adopted, and the collision term is of the generic form [2]:

$$\left(\frac{\partial}{\partial t} + \hat{e}_a \cdot \nabla\right) f_{a,\sigma} = \{\Pi_{b,\lambda} (1 - f_{b,\lambda})\} \left\{ \sum_{\mu\xi} S_{\mu\xi}^{2L}(\sigma) \tilde{f}_{a+1,\mu} \tilde{f}_{a+4,\xi} \right.$$

$$+ \sum_{\mu\xi} S_{\mu\xi}^{2R}(\sigma) \tilde{f}_{a+2,\mu} \tilde{f}_{a+5,\xi} - \sum_{\mu} S_{\mu}^{2}(\sigma) \tilde{f}_{a,\sigma} \tilde{f}_{a+3,\mu}$$

$$\left. + \sum_{\mu\xi\eta} S_{\mu\xi\eta}^{3}(\sigma) \tilde{f}_{a+1,\mu} \tilde{f}_{a+3,\xi} \tilde{f}_{a+5,\eta} - \sum_{\mu\xi} S_{\mu\xi}^{3}(\sigma) \tilde{f}_{a,\sigma} \tilde{f}_{a+2,\mu} \tilde{f}_{a+4,\xi} \right\}. \tag{4.1}$$

The coefficients $S_{\mu\xi}^{2L}(\sigma)$, $S_{\mu\xi}^{2R}$, $S_{\mu}^{2}(\sigma)$, ... in front of the distribution-function products in equation (4.1) are numbers which play the role of the differential scattering cross sections in the usual continuous Boltzmann equation. They are to some extent arbitrary and depend upon how the CA are set up. They are chosen to vanish for any collision which does not satisfy the conservation laws. Their numerical values are constrained by detailed balance considerations and by symmetry requirements. The terms with positive signs in equation (4.1) add particles to the states a, σ, and the terms with negative signs remove particles from a, σ. H-theorems have been proved for simpler but similar collision integrals [8], and it is reasonable to assume that an H-theorem also holds for equation (4.1). It may be readily verified that the form for $\Omega_{a,\sigma}$ given in detail in Appendix A vanishes when we set $f_{a,\sigma} = f_{a,\sigma}^{(0)} \equiv [1 + \exp(\alpha + \beta \mathbf{u} \cdot \hat{e}_a + \gamma \sigma A_z)]^{-1}$, where the coefficients in $f_{a,\sigma}^{(0)}$ are arbitrary. These Lagrange multipliers α, β, γ are determined locally in terms of n, $\mathbf{u}$ and A_z by equations (3.1) to (3.3), each moment being permitted a slow $\mathbf{x}$ and t dependence. $\Omega_{a\sigma}(f)$ is separated into four terms, $\Omega_{a\sigma}(f) = \Omega_{a\sigma}^{I} + \Omega_{a\sigma}^{II} + \Omega_{a\sigma}^{III} + \Omega_{a\sigma}^{IV}$, given in Appendix A, with all coefficients enumerated.

We seek the second term of the Chapman-Enskog solution by solving (3.17) for $f_{a,\sigma}^{(1)}$. This solution is then fed, in turn, into equations (3.18) through (3.19) to give first-order dissipative contributions to equations (3.21) and (3.22).

The linearized collision term from equation (3.17) is identified in detail in Appendix B. It is found that the collision matrix in equation (3.17) whose element is $C_{a\sigma,b\lambda}^{(0)}$ can be written as the sum of four direct products of matrices which act in the $\hat{e}_a$ space and the σ space separately. To represent this conveniently, we write the matrix as

$$\vec{\vec{C}}^{(0)} = \sum_{j=1}^{4} \vec{\vec{\omega}}^{(j)} \otimes \tau^{(j)}. \tag{4.2}$$

Arrows over vectors and dyads remind the reader that they refer to the six-dimensional $\hat{e}_a$-space, and boldface vectors and dyads without arrows refer to the three-dimensional σ-space. In component notation, we have that

$$\left(\vec{\vec{\omega}}^{(j)} \otimes \tau^{(j)} \right)_{a\sigma,b\lambda} = \omega_{ab}^{(j)} \tau_{\sigma\lambda}^{(j)}, \tag{4.3}$$

(a and b run from 1 to 6, and σ and λ run from -1 to +1). All the $\vec{\vec{\omega}}^{(j)}$, $j = 1$, 2, 3, 4, are 6×6 circulant matrices [2,16], and represent scattering processes for velocity, while all $\tau^{(j)}$, $j = 1, 2, 3, 4$, represent 3×3 matrices describing scattering events for the photon index σ, but are not all circulant. The linearization of $\Omega_{a\sigma}$ is tedious, and leads to expressions for the $\vec{\vec{\omega}}^{(j)}$ and $\tau^{(j)}$ which are enumerated in Appendix B. They are given there for the limit in which both $\mathbf{u}$ and A_z are small compared to unity.

The inversion of the collision matrix $\vec{\vec{C}}^{(0)}$ is made possible by the fact that all circulant matrices of a given dimension M (here, $M = 6$) have the same set of right eigenvectors [2,16] (see Appendix C). Use of this fact will reduce the inversion of the (18×18) matrix $\vec{\vec{C}}^{(0)}$ to that of inverting a new matrix which is only 3×3.

The circulant matrices $\vec{\vec{\omega}}^{(i)}$ have the same set of eigenvectors $\vec{v}^c$, and eigenvalues $\lambda^{c(i)}$ (see Appendix C):

$$\vec{\vec{\omega}}^{(i)} \cdot \vec{v}^c = \lambda^{c(i)} \vec{v}^c, \tag{4.4}$$

where $i = 1, 2, 3, 4$ and c runs from 1 to 6. We may expand the $f_{b,\lambda}^{(1)}$ in equation (3.17) in terms of the $\vec{v}^c$, so that

$$f_{b,\lambda}^{(1)} = \sum_c \psi_\lambda^c v_b^c, \tag{4.5}$$

so that finding the expansion coefficients, ψ_λ^c, is equivalent to finding $f_{b,\lambda}^{(1)}$.

Using equation (4.5) in $\Omega_{a\sigma}$, we have, because of equations (4.2) and (4.3),

$$\sum_{b,\lambda} C_{a\sigma,b\lambda}^{(0)} f_{b,\lambda}^{(1)} = \sum_\lambda \sum_c \tau_{\sigma\lambda}^c \psi_\lambda^c v_a^c, \tag{4.6}$$

in terms of a new matrix τ^c whose elements are defined by

$$\tau_{\sigma\lambda}^c = \sum_{j=1}^4 \lambda^{c(j)} \tau_{\sigma\lambda}^{(j)}. \tag{4.7}$$

We need to find eigenvectors $w_r^{c,\nu}$ ($\nu = 1, 2, 3$) and corresponding eigenvalues $\xi^{c,\nu}$ for τ^c

$$\sum_{\lambda=1}^3 \tau_{\sigma\lambda}^c w_\lambda^{c,\nu} = \xi^{c,\nu} w_\sigma^{c,\nu}. \tag{4.8}$$

This is done in Appendix D.

The three-component vector ψ^c is represented in terms of the eigenvectors $\mathbf{w}^{c,\nu}$ ($\nu = 1, 2, 3$) as

$$\psi_\lambda^c = \sum_{\nu=1}^3 \rho^{c,\nu} w_\lambda^{c,\nu}, \tag{4.9}$$

so that

$$\sum_{b,\lambda} C_{a\sigma,b\lambda}^{(0)} f_{b,\lambda}^{(1)} = \sum_{c=1}^6 \sum_{\nu=1}^3 v_a^c w_\sigma^{c,\nu} \xi^{c,\nu} \rho^{c,\nu}. \tag{4.10}$$

The solution of equation (3.17) is then

$$f_{a,\sigma}^{(1)} = \sum_{c=1}^6 \sum_{\nu=1}^3 v_a^c w_\sigma^{c,\nu} \rho^{c,\nu}, \tag{4.11}$$

with the coefficient $\rho^{c,\nu}$ given by

$$\rho^{c,\nu} = \frac{1}{\xi^{c,\nu}} \sum_{a=1}^6 \sum_{\sigma=-1}^{+1} (v_a^c w_\sigma^{c,\nu})^* \left(\frac{\partial}{\partial t} + \hat{e}_a \cdot \nabla \right) f_{a,\sigma}^{(0)}. \tag{4.12}$$

4.1　Evaluation of the viscosity

Using the Euler equations [9] obtained from putting $f_{a,\sigma}^{(0)}$ into equations (3.20) through (3.22), we have, for the left-hand side of equation (3.17),

$$\left(\frac{\partial}{\partial t} + \hat{e}_a \cdot \nabla\right) f_{a,\sigma}^{(0)} = \frac{(2\hat{e}_a\hat{e}_a - 1)}{18} : n\nabla\mathbf{u} + \frac{\sigma\hat{e}_a}{12} n \cdot \nabla A_z. \tag{4.13}$$

It is the first term on the right-hand side of (4.13) which makes a non-vanishing contribution to the viscosity; it leads to a contribution to $\rho^{c,\nu}$ of

$$\rho^{c,\nu}(\text{visc.}) = \frac{1}{\xi^{c,\nu}} \sum_{a,\sigma} (v_a^c w_\sigma^{c,\nu})^* \left(\frac{2\hat{e}_a\hat{e}_a - 1}{18}\right) : n\nabla\mathbf{u}. \tag{4.14}$$

It can be shown using the material in Appendix C that

$$\sum_{a=1}^6 (v_a^c)^* (2\hat{e}_a\hat{e}_a - 1) = 0, \quad \text{for } c = 1, 2, 4, 6, \tag{4.15}$$

so that coefficients $\rho^{c,\nu}(\text{visc.})$ survive only for $c = 3$ and 5. Moreover, the eigenvectors $w^{c,\nu}$ (Appendix D) have symmetries such that

$$\sum_\sigma w_\sigma^{3,\nu} = \sum_\sigma w_\sigma^{5,\nu} = 0 \quad \text{for } \nu = 1, 2. \tag{4.16}$$

It follows that only $\rho^{3,3}(\text{visc.})$ and $\rho^{5,3}(\text{visc.})$ remain finite.

Inserting the solution (4.14) into equation (3.18) and using the results just quoted yields

$$\pi^{(1)} = \sum_{a=1}^6 \sum_{\sigma=1}^3 \hat{e}_a\hat{e}_a \left(v_a^3 w_\sigma^{3,3}\rho^{3,3} + v_a^5 w_\sigma^{5,3}\rho^{5,3}\right)$$

$$= \left\{\left(\sum_a \hat{e}_a\hat{e}_a v_a^3\right)\left(\sum_{a'} \hat{e}_{a'}\hat{e}_{a'} v_{a'}^3\right)^*\right.$$

$$\left. + c.c.\right\} \left[\left(\sum_\sigma w_\sigma^{3,3}\right)^2 \frac{1}{3\xi^{3,3}}\right] : \frac{n\nabla\mathbf{u}}{3}, \tag{4.17}$$

from which the kinematic viscosity ν can be inferred. The expression in equation (4.17) is quite similar to the corresponding momentum tensor expression for the pure two-dimensional Navier-Stokes case, viz.

$$\pi_{NS}^{(1)} = \left\{\left(\sum_a \hat{e}_a\hat{e}_a v_a^3\right)\left(\sum_{a'} \hat{e}_{a'}\hat{e}_{a'} v_{a'}^3\right)^* + c.c.\right\} \left[\frac{1}{\lambda_3}\right] : \frac{n\nabla\mathbf{u}}{3}, \tag{4.18}$$

where λ_3 is the eigenvalue of $\bar{v}^3$ for the Navier-Stokes case. λ_3 is given by

$$\frac{1}{\lambda_3(NS)} = -\frac{2}{n(1 - n/6)^3}. \tag{4.19}$$

The factor in equation (4.17) which corresponds to the factor $1/\lambda_3$ in equation (4.18) is, from Appendix D,

$$\left(\sum_\sigma w_\sigma^{3,3}\right)^2 \frac{1}{3\xi^{3,3}} = \frac{-2}{n\,(1-n/18)^{15}},\tag{4.20}$$

Comparing equations (4.20) and (4.19), we see that in the limit of zero density $(n \to 0)$, the viscosities for the 2D MHD CA are the same as those for the two-dimensional Navier-Stokes one, and differ by the factor $(1-n/18)^{15}/(1-n/6)^3$. In the important case of very low density,

$$\nu = \frac{1}{2n}, \ (n \to 0),\tag{4.21}$$

in agreement with the result of Wolfram [2].

4.2 Evaluation of the magnetic diffusivity

The coefficient $\rho^{c,\nu}$ relevant to the magnetic diffusivity is the second term on the right-hand side of equation (4.13):

$$\rho^{c,\nu}(\text{mag}) = \frac{1}{\xi^{c,\nu}} \sum_a \sum_\sigma (v_a^c w_\sigma^{c,\nu})^* \frac{\hat{e}_a \sigma}{12} n \cdot \nabla A_z.\tag{4.22}$$

Since

$$\sum_a v_a^{c*} \hat{e}_a = 0 \quad \text{for } c = 1,3,4,5,\tag{4.23}$$

and

$$\sum_\sigma \sigma w_\sigma^{2,\nu} = \sum_\sigma \sigma w_\sigma^{6,\nu} = 0 \quad \text{for } \nu = 2,3,\tag{4.24}$$

only $\rho^{2,1}(\text{mag})$ and $\rho^{6,1}(\text{mag})$ have finite values. The first-order flux $\phi^{(1)}$ is

$$\phi^{(1)} = \sum_a \sum_\sigma \hat{e}_a \sigma \left(v_a^2 w_\sigma^{2,1} \rho^{2,1}(\text{mag}) + v_a^6 w_\sigma^{6,1} \rho^{6,1}(\text{mag})\right)$$

$$\left\{\left(\sum_a \hat{e}_a v_a^2\right)\left(\sum_{a'} \hat{e}_{a'} v_{a'}^2\right)^* + c.c.\right\} \cdot \left(\sum_\sigma \sigma w_\sigma^{2,1}\right)^2 \frac{n}{2\xi^{2,1}} \nabla A_z.\tag{4.25}$$

Using the formulas for $w^{2,1}$ and $\xi^{2,1}$ from Appendix D and the identity

$$\sum_{a=1}^{3} \hat{e}_a \hat{e}_a = \frac{3}{2} \mathbf{1},$$

we finally have, from equation (4.25),

$$\phi^{(1)} = \frac{1}{2} \frac{n\nabla A_z}{\left(3\tilde{f} + 9\tilde{f}^2\right)(1-n/18)^{16}},\tag{4.26}$$

where $\tilde{f} \equiv n/(18-n)$. As $n \to 0$, the magnetic diffusivity tends toward

$$\eta = \frac{3}{n}, \quad (n \to 0), \tag{4.27}$$

which is to be compared with equation (4.21).

In summary, we have for the magnetic diffusivity η,

$$\eta = \frac{1}{2} \frac{1}{\left(3\tilde{f} + 9\tilde{f}^2\right)\left(1 - n/18\right)^{16}}, \tag{4.28}$$

where $\tilde{f} = n/(18-n)$. For the kinematic viscosity, we have

$$\nu = \frac{1}{2n} \frac{1}{\left(1 - n/18\right)^{15}}. \tag{4.29}$$

5. Discussion and concluding remarks

Equations (4.28) and (4.29) are the principal results of this paper. Their computational verification or disproof awaits the results of 2D MHD CA codes now in preparation. It is also worth a reminder that additional contributions to ν and η connected with finite discrete lattice size may contribute additive additional terms to equations (4.28) and (4.29) and remain to be evaluated.

It should also be noted that a definite magnetic Prandtl number $\nu/n \equiv P_{mag}$ ($\cong 1/6$, at low densities) is implied by equations (4.28) and (4.29). This parameter is one which, in computations, is one that it would be desirable to vary. It is clear that some variation of it should be possible by varying the fraction of $2R$, $2L$, and $3S$ collisions which are permitted to exchange σ-values among the particles. Just how wide the range in P_{mag} that this will permit is uncertain, and probably must be determined by computational practice.

The least-satisfying features of the 2D MHD CA remains the need for the microscopic velocity-flipping routine for the inclusion of the Lorentz force in equation (2.1). The code inevitably will be slowed down by its operation, and how serious a limitation is also something that can be decided only by computational practice.

Acknowledgments

One author (T. H.) would like to express his gratitude to Dartmouth College for hospitality during a visit there during the autumn of 1986.

This work was supported in part by the U. S. Department of Energy grant number DE-FG02-85ER53194 and in part by NASA grant number NAG-W-710.

Appendix A.   Enumeration of the terms in equation (4.1)

The terms on the right-hand side of equation (4.1), written out in detail, are $\Omega_{a,\sigma}^{I} + \Omega_{a,\sigma}^{II} + \Omega_{a,\sigma}^{III} + \Omega_{a,\sigma}^{IV}$, where

$$\Omega_{a,\pm1}^{I}/\left(\Pi_{a,\sigma}(1-f_{a,\sigma})\right) =$$

$$\frac{1}{2}\left[\tilde{f}_{a+1,\pm1}\tilde{f}_{a+4,\pm1} + \tilde{f}_{a+2,\pm1}\tilde{f}_{a+5,\pm1}\right.$$

$$+ \frac{1}{2}\left(\tilde{f}_{a+1,\pm1}\tilde{f}_{a+4,0} + \tilde{f}_{a+1,0}\tilde{f}_{a+4,\pm1} + \tilde{f}_{a+2,\pm1}\tilde{f}_{a+5,0} + \tilde{f}_{a+2,0}\tilde{f}_{a+5,\pm1}\right)$$

$$+ \frac{1}{3}\left(\tilde{f}_{a+1,1}\tilde{f}_{a+4,-1} + \tilde{f}_{a+1,0}\tilde{f}_{a+4,0} + \tilde{f}_{a+1,-1}\tilde{f}_{a+4,1}\right.$$

$$\left.\left. + \tilde{f}_{a+2,1}\tilde{f}_{a+5,-1} + \tilde{f}_{a+2,0}\tilde{f}_{a+5,0} + \tilde{f}_{a+2,-1}\tilde{f}_{a+5,1}\right)\right]. \tag{A.1}$$

$$\Omega_{a,0}^{I}/\left(\Pi_{a,\sigma}(1-f_{a,\sigma})\right) =$$

$$\frac{1}{2}\left[\frac{1}{2}\left(\tilde{f}_{a+1,1}\tilde{f}_{a+4,0} + \tilde{f}_{a+1,0}\tilde{f}_{a+4,1} + \tilde{f}_{a+2,1}\tilde{f}_{a+5,0} + \tilde{f}_{a+2,0}\tilde{f}_{a+5,1}\right)\right.$$

$$+ \frac{1}{3}\left(\tilde{f}_{a+1,1}\tilde{f}_{a+4,-1} + \tilde{f}_{a+1,0}\tilde{f}_{a+4,0} + \tilde{f}_{a+1,-1}\tilde{f}_{a+4,1} + \tilde{f}_{a+2,1}\tilde{f}_{a+5,-1}\right.$$

$$\left. + \tilde{f}_{a+2,0}\tilde{f}_{a+5,0} + \tilde{f}_{a+2,-1}\tilde{f}_{a+5,1}\right)$$

$$\left. + \frac{1}{2}\left(\tilde{f}_{a+1,0}\tilde{f}_{a+4,-1} + \tilde{f}_{a+1,-1}\tilde{f}_{a+4,0} + \tilde{f}_{a+2,0}\tilde{f}_{a+5,-1} + \tilde{f}_{a+2,-1}\tilde{f}_{a+5,0}\right)\right].$$

$$\tag{A.2}$$

$$\Omega_{a,\sigma}^{II}/\left(\Pi_{a,\sigma}(1-f_{a,\sigma})\right) = -\tilde{f}_{a,\sigma}\left(\tilde{f}_{a+3,1} + \tilde{f}_{a+3,0} + \tilde{f}_{a+3,-1}\right). \tag{A.3}$$

$$\Omega_{a,\pm1}^{III}/\left(\Pi_{a,\sigma}(1-f_{a,\sigma})\right) = \tilde{f}_{a+1,\pm1}\tilde{f}_{a+3,\pm1}\tilde{f}_{a+5,\pm1}$$

$$+ \frac{2}{3}\left(\tilde{f}_{a+1,\pm1}\tilde{f}_{a+3,\pm1}\tilde{f}_{a+5,0} + \tilde{f}_{a+1,\pm1}\tilde{f}_{a+3,0}\tilde{f}_{a+5,\pm1} + \tilde{f}_{a+1,0}\tilde{f}_{a+3,\pm1}\tilde{f}_{a+5,\pm1}\right)$$

$$+ \frac{2}{3}\left(\tilde{f}_{a+1,\pm1}\tilde{f}_{a+3,\pm1}\tilde{f}_{a+5,\mp1} + \tilde{f}_{a+1,\pm1}\tilde{f}_{a+3,\mp1}\tilde{f}_{a+5,\pm1} + \tilde{f}_{a+1,\mp1}\tilde{f}_{a+3,\pm1}\tilde{f}_{a+5,\pm1}\right)$$

$$+ \frac{1}{3}\left(\tilde{f}_{a+1,\pm1}\tilde{f}_{a+3,0}\tilde{f}_{a+5,0} + \tilde{f}_{a+1,0}\tilde{f}_{a+3,\pm1}\tilde{f}_{a+5,0} + \tilde{f}_{a+1,0}\tilde{f}_{a+3,0}\tilde{f}_{a+5,\pm1}\right)$$

$$+ \frac{1}{3}\left(\tilde{f}_{a+1,\pm1}\tilde{f}_{a+3,0}\tilde{f}_{a+5,\mp1} + \tilde{f}_{a+1,\pm1}\tilde{f}_{a+3,\mp1}\tilde{f}_{a+5,0}\right.$$

$$+ \tilde{f}_{a+1,0}\tilde{f}_{a+3,\pm1}\tilde{f}_{a+5,\mp1} + \tilde{f}_{a+1,0}\tilde{f}_{a+3,\mp1}\tilde{f}_{a+5,\pm1}$$

$$+ \tilde{f}_{a+1,\mp1}\tilde{f}_{a+3,0}\tilde{f}_{a+5,\pm1} + \tilde{f}_{a+1,\mp1}\tilde{f}_{a+3,\pm1}\tilde{f}_{a+5,0}\Big)$$

$$+ \frac{1}{3}\left(\tilde{f}_{a+1,\mp1}\tilde{f}_{a+3,\mp1}\tilde{f}_{a+5,\pm1} + \tilde{f}_{a+1,\mp1}\tilde{f}_{a+3,\pm1}\tilde{f}_{a+5,\mp1}\right.$$

$$\left. + \tilde{f}_{a+1,\pm1}\tilde{f}_{a+3,\mp1}\tilde{f}_{a+5,\mp1}\right). \tag{A.4}$$

$$\Omega^{III}_{a,0}/\left(\Pi_{a,\sigma}(1 - f_{a,\sigma})\right)$$

$$= \frac{1}{3}\left(\tilde{f}_{a+1,1}\tilde{f}_{a+3,1}\tilde{f}_{a+5,0} + \tilde{f}_{a+1,1}\tilde{f}_{a+3,0}\tilde{f}_{a+5,1} + \tilde{f}_{a+1,0}\tilde{f}_{a+3,1}\tilde{f}_{a+5,1}\right.$$

$$+ \frac{2}{3}\left(\tilde{f}_{a+1,1}\tilde{f}_{a+3,0}\tilde{f}_{a+5,0} + \tilde{f}_{a+1,0}\tilde{f}_{a+3,1}\tilde{f}_{a+5,0}\right.$$

$$\left. + \tilde{f}_{a+1,0}\tilde{f}_{a+3,0}\tilde{f}_{a+5,1}\right)$$

$$+ \frac{1}{3}\left(\tilde{f}_{a+1,1}\tilde{f}_{a+3,0}\tilde{f}_{a+5,-1} + \tilde{f}_{a+1,1}\tilde{f}_{a+3,-1}\tilde{f}_{a+5,0} + \tilde{f}_{a+1,0}\tilde{f}_{a+3,1}\tilde{f}_{a+5,-1}\right.$$

$$+ \tilde{f}_{a+1,0}\tilde{f}_{a+3,-1}\tilde{f}_{a+5,1}$$

$$\left. + \tilde{f}_{a+1,-1}\tilde{f}_{a+3,0}\tilde{f}_{a+5,1} + \tilde{f}_{a+1,-1}\tilde{f}_{a+3,1}\tilde{f}_{a+5,0}\right) + \tilde{f}_{a+1,0}\tilde{f}_{a+3,0}\tilde{f}_{a+5,0}$$

$$+ \frac{2}{3}\left(\tilde{f}_{a+1,-1}\tilde{f}_{a+3,0}\tilde{f}_{a+5,0} + \tilde{f}_{a+1,0}\tilde{f}_{a+3,-1}\tilde{f}_{a+5,0} + \tilde{f}_{a+1,0}\tilde{f}_{a+3,0}\tilde{f}_{a+5,-1}\right)$$

$$+ \frac{1}{3}\left(\tilde{f}_{a+1,-1}\tilde{f}_{a+3,-1}\tilde{f}_{a+5,0} + \tilde{f}_{a+1,-1}\tilde{f}_{a+3,0}\tilde{f}_{a+5,-1} + \tilde{f}_{a+1,0}\tilde{f}_{a+3,-1}\tilde{f}_{a+5,-1}\right). $$

$$\tag{A.5}$$

$$\Omega^{IV}_{a,\sigma}/\left(\Pi_{a,\sigma}(1 - f_{a,\sigma})\right) = -\tilde{f}_{a,\sigma}\left(\tilde{f}_{a+2,1}\tilde{f}_{a+4,1} + \tilde{f}_{a+2,1}\tilde{f}_{a+4,0} + \tilde{f}_{a+2,0}\tilde{f}_{a+4,1}\right.$$

$$+ \tilde{f}_{a+2,0}\tilde{f}_{a+4,0} + \tilde{f}_{a+2,1}\tilde{f}_{a+4,-1} + \tilde{f}_{a+2,-1}\tilde{f}_{a+4,1}$$

$$\left. + \tilde{f}_{a+2,-1}\tilde{f}_{a+4,0} + \tilde{f}_{a+2,0}\tilde{f}_{a+4,-1} + \tilde{f}_{a+2,-1}\tilde{f}_{a+4,-1}\right). \tag{A.6}$$

Appendix B. Explicit form of the matrices $\vec{\bar{\omega}}^{(j)}$ and $\tau^{(j)}$ of equation (4.2)

The fluid velocity u and the magnetic vector potential A_z are both considered to be small compared to unity. For purposes of inverting $\vec{\bar{C}}^{(0)}$, they may be dropped, leaving $f_{a,\sigma}^{(0)} = n/18 + $ (higher-order terms). Calling $n/18 \equiv f$, and $\tilde{f} = f/(1-f)$, the explicit forms are:

$$\vec{\bar{\omega}}^{(1)} = (1-f)^{16} \text{ circ } [0, 1, 1, 0, 1, 1](\tilde{f}/2), \tag{B.1}$$

(where "circ" means a 6×6 circulant matrix),

$$\tau^{(1)} = \begin{pmatrix} 11/6 & 5/6 & 2/6 \\ 5/6 & 8/6 & 5/6 \\ 2/6 & 5/6 & 11/6 \end{pmatrix}, \tag{B.2}$$

$$\vec{\bar{\omega}}^{(2)} = (1-f)^{16} \text{ circ } [0, 1, 0, 1, 0, 1]\tilde{f}^2 \tag{B.3}$$

$$\tau^{(2)} = \begin{pmatrix} 5 & 2 & 2 \\ 2 & 5 & 2 \\ 2 & 2 & 5 \end{pmatrix}, \tag{B.4}$$

$$\vec{\bar{\omega}}^{(3)} = -(1-f)^{16} \text{ circ } [9\tilde{f}^2 + 3\tilde{f}, 0, 0, 0, 0, 0], \tag{B.5}$$

$$\tau^{(3)} = \begin{pmatrix} 1 & 0 & 0 \\ 0 & 1 & 0 \\ 0 & 0 & 1 \end{pmatrix}, \tag{B.6}$$

$$\vec{\bar{\omega}}^{(4)} = -(1-f)^{16} \text{ circ } [0, 0, 3\tilde{f}^2, \tilde{f}, 3\tilde{f}^2, 0], \tag{B.7}$$

$$\tau^{(4)} = \begin{pmatrix} 1 & 1 & 1 \\ 1 & 1 & 1 \\ 1 & 1 & 1 \end{pmatrix}. \tag{B.8}$$

Note that only $\tau^{(1)}$, among the τ's, is anything other than a circulant matrix.

Appendix C. Eigenvectors $\vec{v}^c$ and eigenvalues $\lambda^{c(i)}$ for the $\vec{\bar{\omega}}^{(i)}$ matrices, equation (4.4)

The six eigenvectors and corresponding eigenvalues are [2,16]:

$$\vec{v}^1 = (1, 1, 1, 1, 1, 1)/\sqrt{6}$$

$$\vec{v}^2 = (1, \sigma_0, -\sigma_0^*, -1, -\sigma_0, \sigma_0^*)/\sqrt{6}$$

$$\vec{v}^3 = (1, -\sigma_0*, -\sigma_0, 1, -\sigma_0^*, -\sigma_0)/\sqrt{6}$$

$$\vec{v}^4 = (1, \ -1 \ , \ 1, \ -1, \ 1, \ -1)/\sqrt{6}$$

$$\vec{v}^5 = (1, \ -\sigma_0, \ -\sigma_0^*, \ 1, \ -\sigma_0, \ -\sigma_0^*)/\sqrt{6}$$

$$\vec{v}^6 = (1, \ \sigma_0^*, \ -\sigma_0, \ -1, \ , -\sigma_0^*, \ \sigma_0)/\sqrt{6}, \tag{C.1}$$

where $\sigma_0 = (1 + i\sqrt{3})/2$.

The corresponding eigenvalues are

$$\lambda^{1(1)} = 2\tilde{f}(1-f)^{16}$$

$$\lambda^{2(1)} = 0$$

$$\lambda^{3(1)} = -\tilde{f}(1-f)^{16}$$

$$\lambda^{4(1)} = 0$$

$$\lambda^{5(1)} = -\tilde{f}(1-f)^{16}$$

$$\lambda^{6(1)} = 0, \tag{C.2}$$

$$\lambda^{1(2)} = 3\tilde{f}^2(1-f)^{16}$$

$$\lambda^{2(2)} = 0$$

$$\lambda^{3(2)} = 0$$

$$\lambda^{4(2)} = -3\tilde{f}^2(1-f)^{16}$$

$$\lambda^{5(2)} = 0$$

$$\lambda^{6(2)} = 0, \tag{C.3}$$

$$\lambda^{c(3)} = -(3\tilde{f} + 9\tilde{f}^2)(1-f)^{16}, \text{ for arbitrary c.}$$

$$\lambda^{1(4)} = -(\tilde{f} + 6\tilde{f}^2)(1-f)^{16}$$

$$\lambda^{2(4)} = (\tilde{f} + 3\tilde{f}^2)(1-f)^{16}$$

$$\lambda^{3(4)} = -(\tilde{f} - 3\tilde{f}^2)(1-f)^{16}$$

$$\lambda^{4(4)} = (\tilde{f} - 6\tilde{f}^2)(1-f)^{16}$$

$$\lambda^{5(4)} = -(\tilde{f} - 3\tilde{f}^2)(1-f)^{16}$$

$$\lambda^{6(4)} = (\tilde{f} + 3\tilde{f}^2)(1-f)^{16}. \tag{C.4}$$

As for f and $\tilde{f}$, see Appendix B; again, $f = n/18$, $\tilde{f} = f/(1-f)$, for purposes of inverting $\overset{\Rightarrow(0)}{C}$.

Appendix D.   The matrix τ^c of equation (4.7) and its eigenvectors $w^{c,\nu}$ and eigenvalues $\xi^{c,\nu}$

Explicitly, τ^c is a symmetric matrix:

$$\tau^c = \begin{pmatrix} a^c & d^c & e^c \\ d^c & b^c & d^c \\ e^c & d^c & a^c \end{pmatrix}, \tag{D.1}$$

where

$$a^c = \frac{11}{6}\lambda^{c(1)} + 5\lambda^{c(2)} + \lambda^{c(3)} + \lambda^{c(4)}$$

$$b^c = \frac{4}{3}\lambda^{c(1)} + 5\lambda^{c(2)} + \lambda^{c(3)} + \lambda^{c(4)}$$

$$d^c = \frac{5}{6}\lambda^{c(1)} + 2\lambda^{c(2)} + \lambda^{c(4)}$$

$$e^c = \frac{1}{3}\lambda^{c(1)} + 2\lambda^{c(2)} + \lambda^{c(4)}. \tag{D.2}$$

The $\lambda^{c(i)}$ are given in Appendix C.

The first eigenvector and eigenvalue are

$$w^{c,1} = \frac{1}{\sqrt{2}}(1,\ 0,\ -1)$$

$$\xi^{c,1} = \frac{3}{2}\lambda^{c(1)} + 3\lambda^{c(2)} + \lambda^{c(3)}, \tag{D.3}$$

and in particular,

$$\xi^{2,1} = -(3\tilde{f} + 9\tilde{f}^2)(1 - f)^{16}. \tag{D.4}$$

The second and third eigenvectors are not simple; but those for $c = 3$ and $c = 5$, which are identical and are required to evaluate the viscosity, are

$$w^{3,2} = w^{5,2} = \frac{1}{\sqrt{6}}(1,\ -2,\ 1)$$

$$w^{3,3} = w^{5,3} = \frac{1}{\sqrt{3}}(1,\ 1,\ 1), \tag{D.5}$$

and the eigenvalues are

$$\xi^{3,2} = \xi^{5,3} = b^c - d^c = \left(-\frac{7}{2}\tilde{f} - 9\tilde{f}^2\right)(1 - f)^{16}$$

$$\xi^{3,3} = \xi^{5,3} = b^c + 2d^c = -9\tilde{f}(1 - f)^{16}. \tag{D.6}$$

Again, $f = n/18$, and $\tilde{f} = f/(1 - f)$ (see Appendix B).

References

[1] U. Frisch, B. Hasslacher, and Y. Pomeau, *Phys. Rev. Lett.*, **56** (1986) 1505.

[2] S. Wolfram, *J. Stat. Phys.*, **45** (1986) 471.

[3] D. d'Humiéres, Y. Pomeau, and P. Lallemand, *C.R. Acad. Sci. Paris, Ser. I*, **302** (1985) 1391.

[4] J. P. Rivet and U. Frisch, *C.R. Acad. Sci. Paris, Ser. II*, **302** (1986) 267.

[5] S. Harris, *Phys. Fluids*, **9** (1966) 1328.

[6] J. Hardy, O. de Pazzis, and Y. Pomeau, *J. Math Phys.*, **14** (1973) 1746.

[7] J. Hardy and Y. Pomeau, *J. Math Phys.*, **13** (1972) 1042.

[8] R. Gatignol, "Théorie Cinétique des Gaz á Répartition Discréte des Vitesses", *Lecture Notes in Physics*, **36** (Springer-Verlag, Berlin, 1975).

[9] D. Montgomery and G. D. Doolen, "Magnetohydrodynamic Cellular Automata," *Physics Letters A*, **120** (1987) 229.

[10] S. Chapman and T. G. Cowling, *The Mathematical Theory of Non-Uniform Gases*, (Cambridge University Press, Cambridge, 1970) 3rd edition.

[11] J. H. Ferziger and H. G. Kaper, *Mathematical Theory of Transport Processes in Gases*, (North-Holland, Amsterdam, 1972).

[12] N. N. Bogolyubov, *Problems of a Dynamical Theory in Statistical Physics*, (Moscow: State Technical Press, 1946). English translation by E. K. Gora in *Studies in Statistical Mechanics*, Vol. I., J. deBoer and G. E. Uhlenbeck, eds. (North-Holland, Amsterdam, 1962).

[13] W. Brittin, ed., *Lectures in Theoretical Physics, Vol. IXC: Kinetic Theory*, (Gordon & Breach, New York, 1967).

[14] D. Fyfe, D. Montgomery, and G. Joyce, *J. Plasma Phys.*, **17** (1977) 369.

[15] D. Boghosian and D. Levermore, "A Cellular Automaton for Burgers' Equation", *Complex Systems*, **1** (1987) 17.

[16] P. J. Davis, *Circulant Matrices*, (Wiley, New York, 1979).

Fernand Hayot

Department of Physics, Ohio State University, Columbus, OH 43210, USA; On leave from Service de Physique Theorique, C.E.N. Saclay, 91191 Gif-sur-Yvette, France

The Effect of Galilean Non-Invariance in Lattice Gas Automaton One-Dimensional Flow

This paper originally appeared in *Complex Systems* (1987), Volume 1, pages 753–761.

Complex Systems 1 (1987) 753 761

The Effect of Galilean Non-Invariance in Lattice Gas Automaton One-Dimensional Flow

Fernand Hayot*
Department of Physics, Ohio State University,
Columbus, OH 43210, USA

Abstract. In the simple case of one-dimensional flow between plates, we show the effect of Galilean non-invariance of the usual hexagonal lattice gas mode. This effect leads to a distorted velocity profile when the velocity exceeds a value of 0.4. Higher-order corrections to the Navier-Stokes equations are considered in a discussion of the numerical importance of the distortion.

1. Introduction

It was argued by Frisch, Hasslacher, and Pomeau [1] that a (two-dimensional) hexagonal lattice gas model reproduces—upon coarse-graining—the fluid behavior described ordinarily by the Navier-Stokes equation. The significance of this proposal is that it could lead to new ways to simulate fluid flow, based upon simple binary arithmetic rather than high-precision floating-point calculations.

A characteristic feature of lattice gas automata is the appearance of higher-order corrections to the Navier-Stokes equation, once coarse-graining is performed [2]. These corrections are due to the discrete nature of both coordinate and velocity spaces. Since they are manifestations of the discrete lattice dynamics, they break Galilean invariance and show up as soon as the fluid velocity is no longer negligible compared to the microscopic velocity, or, equivalently, as soon as the Mach number (the ratio of fluid to sound velocity) approaches one. This transonic regime is easily obtained in lattice gases because here, the macroscopic fluid velocity is bounded by one, and the sound velocity itself takes the value $v_s = 1/\sqrt{2} \simeq 0.717$. The transonic regime can occur already at small Reynolds numbers, which is not normally the case in compressible fluids.

It is our aim here to study how large the Mach number can actually be (in the case studied) before effects due to the breaking of Galilean invariance set in. In particular, in as simple a fluid flow as one-dimensional flow

---

*On leave from Service de Physique Theorique, C.E.N. Saclay, 91191 Gif-sur-Yvette, France.

between two plates, effects due to those higher-order lattice gas corrections to Navier-Stokes show up at high enough velocity. This was pointed out in the discussion of unsteady, one-dimensional flow in reference 3, and the demonstration and analysis of these effects is carried further in this work.

It is not that we believe it is very useful to study these corrections quantitatively. However, the case of one-dimensional fluid flow between plates is one of the simplest possible where Galilean invariance breaking will simply manifest itself as a distortion of the usual linear velocity profile given in the beginning of all textbooks on fluid mechanics. Moreover, while the linear profile is *independent* of fluid viscosity, the distorted one will depend on it. Also, while there is no pressure or density gradient in the "textbook" flow, a density gradient perpendicular to the direction of flow appears in the lattice gas. This one-dimensional flow thus provides a very clear case where some physical consequences of Galilean non-invariance and of higher-order corrections can be exhibited and discussed. Therefore, we consider the demonstration of the presence of these corrections and their consequences to be the main point of this work, not the crude analytical study which is only semi-quantitative.

In section 2, after briefly reviewing the physical system and the lattice gas model, we discuss the distortion of the velocity profile, the appearance of a density gradient, and the form of higher-order corrections to the Navier-Stokes equation for the flow considered. We summarize our results in section 3.

2.   Numerical data and analysis

The lattice gas model is the usual one [1], with particles permanently moving on a two-dimensional hexagonal grid with coordination number equal to six. These particles moreover undergo two- and four-particle and so-called three-particle symmetric collisions, all of which conserve energy and momentum. The initial random distribution of particles corresponds to the same average number of particles in each of six possible directions at each site, and therefore leads to zero macroscopic velocity. The particles are enclosed [3] in a rectangle of dimensions $L = 84$ and $L' = 240$, where L is the width and L' the length. The direction of length corresponds to the y-axis, and the perpendicular one is taken to be the x-axis. Periodic boundary conditions are imposed at top $(y = L')$ and bottom $(y = 0)$ of the system. The boundary condition at $x = L$ corresponds to no-slip, where particles bounce back along the incoming direction after hitting the wall. At the left wall $(x = 0)$, boundary conditions are taken to be specular. As in reference 3, a tangential instability is introduced at $x = 0$ which creates a flow in the positive y direction by permitting some particles that normally would reflect on the wall going from top to bottom to bounce back into the direction of increasing y. We have checked that our results are insensitive to how the instability is created, another way being to start from no-slip boundary conditions at $x = 0$ and introduce a bias toward increasing y by allowing some particles to reflect specularly. In reference 3, the main object of study was how the instability propagates into the system with time, and it was shown that in the obtained steady state the expected linear velocity profile $u_y(x)$ is obtained, except for distortions at small x and high velocity. This steady state is the starting point of the present investigation.

In figure 1, the velocity profile $u_y(x)$ is shown as a function of the fractional distance in the direction perpendicular to the flow. The macroscopic average of density and velocity is done over cells of width 6 (in the lattice units) and length 200, after checking that the flow is y independent and the velocity component in the x direction is negligible. The density is $1/3$. The profile is linear for large x, the linear part extrapolating at $x = 0$ to a maximum allowed velocity of $\sqrt{3/2} = 0.866$. For small x, the velocity profile is rounded off with an effective intercept of 0.53 at $x = 0$. This distortion, moreover, extends over half of the width.

Further study shows on one hand that the round-off disappears at small velocity, and on the other that it extends over larger absolute distances when the width of the system is increased, scaling approximately with the system width (cf. figure 2). It cannot therefore be a result of gas slip velocity at the left boundary ($x = 0$) of the rectangular box in which the fluid flows.

In figure 2, the data are shown for two different widths $L = 84$ and $L = 156$, everything else being equal. As mentioned, the rounding-off of the velocity profile occurs in both cases, from $x = 0$ to $x/L \simeq 0.5$, the linear part extrapolating to $u_y(0) = 0.866$. (The data for the system $L = 156$ do not fall exactly on top of those for $L = 84$; however, one must not forget that the larger system is less one-dimensional than the smaller one, the length of the system ($L' = 240$) being the same in both cases). The fact that the velocity profile depends on the ratio x/L is a characteristic of the linear profile (see below). The data of figure 2 show that this remains true through the region of rounding off.

In figure 3, data are shown for the system of width $L = 84$, which correspond to different maximum velocities at $x = 0$. The data demonstrate that when this velocity drops below 0.4, the rounding-off mostly disappears and the full linear profile is recovered. A small effect, presumably due to gas slip velocity, remains close to $x = 0$.

The distortion cannot be due to gas slip velocity, and we conclude that what is seen goes beyond the usual fluid description of the Navier-Stokes equation. The distortion shows up as soon as the velocity reaches a value of 0.4, which corresponds to a Mach number of $0.4\sqrt{2} = 0.56$. However, it disappears for small enough velocities. Clearly, Galilean invariance can only be restored if the fluid velocity is sufficiently small compared to the unit microscopic velocity. This was already pointed out for the discrete velocity model in reference 4.

In the textbook case where one of the plates, say the left plate, moves with velocity U, the linear profile (with respect to the choice of axes of

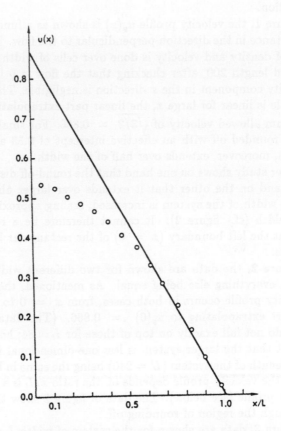

Figure 1: Velocity profile (open dots) $u(x) \equiv u_y(x)$ of the flow between two plates corresponding to the maximum instability on the left plate, as a function of x/L, where $L = 84$ is the width of the system. (The length of the system is $L' = 240$.) The straight line, which is a fit for $x/L > 0.5$, intersects the velocity axis at $\sqrt{3/2} = 0.866$, the maximum velocity in the y direction.

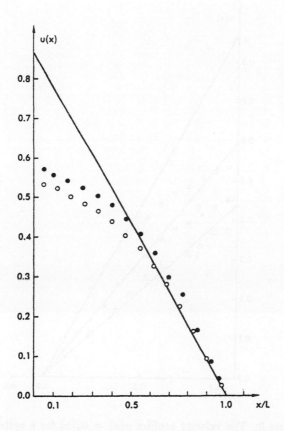

Figure 2: Velocity profile $u(x) \equiv u_y(x)$ as a function of x/L, where the open dots correspond to a system of width $L = 84$ (the same as in figure 1), and the full dots to a system of width $L = 156$. The length $L' = 240$ is the same in both cases and the straight line is the same as in figure 1.

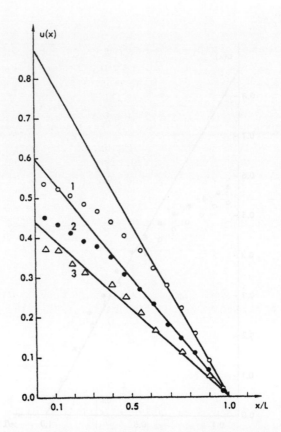

Figure 3: The velocity profiles $u(x) = u_y(x)$ for a system of width $L = 84$ and length $L' = 240$, of decreasing maximum velocity. Curve 1 is the same as in figure 1, with its straight line of intercept 0.866; curve 2 has lower maximum velocity, with a corresponding straight line of intercept 0.6. For curve 3, where the maximum velocity is about 0.4, the straight line has approximately the same intercept. The respective straight lines fit the profile for x/L closer and closer to zero as the maximum velocity is lowered.

figure 1) is given by

$$u(x) = U\left(1 - \frac{x}{L}\right), \tag{2.1}$$

where u is the component of velocity in the y direction. This is a solution of the Navier-Stokes equation:

$$\nu \frac{d^2u}{dx^2} = 0 \tag{2.2}$$

with the appropriate boundary conditions (ν denotes the kinematic viscosity). Notice that the velocity profile, expression (2.1), is independent of ν. The linear shape of the profile is the same as long as the relative velocity of the two plates is U. However, from the distorted profile of figure 2, one can tell at which plate tangential flow occurs; this is a sign of Galilean invariance breaking.

Let us now assume that the distortion of the velocity profile can be represented by corrections to the usual Navier-Stokes equations. The form of higher-order corrections, both in the convective and viscous part of the Navier-Stokes equations, is given in reference 2. Because of the symmetry of one-dimensional flow, only corrections to the viscous term appear in our case. To lowest order in the velocity, they are of the form

$$u^2\frac{d^2u}{dx^2} \quad \text{and} \quad u\left(\frac{du}{dx}\right)^2.$$

Adding these to the right-hand side of equation (2.2) preserves the $u \to -u$ symmetry of the equation, which remains a property of the flow in figure 2. The term in $u^2(d^2u/dx^2)$ leads to a velocity-dependent kinematic viscosity. Replace this by an effective ν_{eff} averaged over velocity and consider the influence of term $u(du/dx)^2$, which plays a more important role. The extended Navier-Stokes equation reads:

$$\frac{d^2u}{dx^2} = \frac{\alpha}{\nu_{\text{eff}}}\, u\, \left(\frac{du}{dx}\right)^2, \tag{2.3}$$

where α measures the strength of the extra term. (The profile now depends on viscosity.) This can be written as (in obvious notation)

$$\frac{u''}{u'} = \frac{\alpha}{\nu_{\text{eff}}}\, u\, u' = \frac{\alpha}{2\nu_{\text{eff}}}(u^2)'.$$

The solution of this equation is related for $\alpha < 0$ to the error function $w(u)$ [5]. An expansion to the lowest orders in velocity gives the solution

$$u + \frac{|\alpha|}{6\nu_{\text{eff}}}u^3 = U\left(1 - \frac{x}{L}\right), \tag{2.4}$$

where we have written $\alpha = -|\alpha|$. Obviously, to even approach a description of the round-off in figure 3, α must be taken negative, the u^3 term compensating for the difference between the measured u and the extrapolated velocity profile. An estimate of $|\alpha|/\delta\nu_{eff}$ from the curves in figure 3 leads to a value between 1 and 2 for curve 1 (intercept $U = 0.866$) and between 1 and 1.5 for curve 2 (intercept $U = 0.6$). This is not really satisfactory. However, the error on the coefficient is large since it is related to that of $1/u^3$. Moreover, if the effective viscosity increases with u, the spread in values is reduced. (The same result is achieved if fifth-order terms in the velocity add to the third-order ones.) The point is, however, not a detailed numerical study of these higher-order terms, but the realization that for high velocities (Mach number greater than 0.56) they are very important, because $|\alpha|$ turns out to be of order $6\nu_{eff}$.

Our last remark concerns density. To the same order (order u^2) in which there are corrections to the Navier-Stokes equation, there appears a correction to the pressure [2]. The pressure becomes

$$p = 3\rho\left(1 - \frac{c^{(2)}}{4}u^2\right),$$

where ρ is the density and $c^{(2)}$ is a density dependent coefficient, equal to 1 when the average density is $1/3^{(2)}$. Thus, although there is no contribution from pressure to equation (2.3), one now obtains a transverse density gradient from $dp/dx = 0$, namely (with $c^{(2)} = 1$)

$$\frac{d\rho}{dx}\left(1 - \frac{u^2}{4}\right) = \frac{\rho}{4}\frac{d}{dx}u^2, \tag{2.5}$$

where $u = u_y(x)$. As long as u_x is small compared to u_y, and it is, the equation of continuity continues to be satisfied. This density gradient is seen in the data. Excluding the two points closest to the plates where the density is always smaller, the density drops continuously from small x, where it is 0.3541, to high x, where it is 0.3243. A fit to the data with formula (2.5) works well until one gets close to $x = 0$, where the data rises faster than the numbers given by the fit, indicating as in the previous discussion of the velocity profile the need for even higher-order corrections in velocity.

3. Summary

We have shown how at large fluid velocity (Mach number above 0.56), the lattice gas automaton fails to reproduce the linear velocity profile of one-dimensional flow between two plates. We interpret this as a sign of Galilean invariance breaking due to the discrete nature in coordinate and velocity space of the hexagonal lattice gas model of fluid mechanics. Higher-order corrections to the Navier-Stokes equations are introduced and their numerical importance estimated by comparison with the observed effects. The appearance of a density gradient in the direction transverse to the flow is discussed.

Acknowledgment

I am grateful to W. Saam for a reading of the manuscript and his very helpful comments.

References

[1] U. Frisch, B. Hasslacher, and Y. Pomeau, *Phys. Review Letters*, **56** (1986) 1505.

[2] S. Wolfram, *Journal of Statistical Physics*, **45** (1986) 471.

[3] F. Hayot, OSU preprint, *Physical Review A*, in press.

[4] J. E. Broadwell, *Journal of Fluid Mechanics*, **19** (1964) 401.

[5] M. Abramowitz and I. A. Stegun, eds., *Handbook of Mathematical Functions*, (NBS, 1970).

Leo P. Kadanoff, Guy R. McNamara and Gianluigi Zanetti
The Enrico Fermi and James Franck Institutes, The University of Chicago, 5640 South Ellis Avenue, Chicago, IL 60637, USA

A Poiseuille Viscometer for Lattice Gas Automata

This paper originally appeared in *Complex Systems* (1987), Volume 1, pages 791–803.

Complex Systems **1** (1987) 791–803

A Poiseuille Viscometer for Lattice Gas Automata

Leo P. Kadanoff
Guy R. McNamara
Gianluigi Zanetti
*The Enrico Fermi and James Franck Institutes, The University of Chicago,
5640 South Ellis Avenue, Chicago, IL 60637, USA*

Abstract. Lattice gas automata have been recently proposed as a new technique for the numerical integration of the two-dimensional Navier-Stokes equation. We have accurately tested a straightforward variant of the original model, due to Frisch, Hasslacher, and Pomeau, in a simple geometry equivalent to two-dimensional Poiseuille (Channel) flow driven by a uniform body force.

The momentum density profile produced by this simulation agrees well with the parabolic profile predicted by the macroscopic description of the gas given by Frisch et al. We have used the simulated flow to compute the shear viscosity of the lattice gas and have found agreement with the results obtained by d'Humiéres et al. [10] using shear wave relaxation measurements, and, in the low density limit, with theoretical predictions obtained from the Boltzmann description of the gas [17].

1. Introduction

In a now classic paper, Frisch, Hasslacher, and Pomeau [1] proposed a new technique for solving the two-dimensional Navier-Stokes equation based on the implementation of a lattice gas automaton. Their original idea has recently been extended to two-dimensional binary fluids, two-dimensional magnetohydrodynamics, three-dimensional Navier-Stokes, and other interesting problems [4].

Two-dimensional lattice gas automata have been described in great detail in reference 3. We will therefore give only a very short description of the model in order to define the nomenclature used.

Lattice gas automata are based on the construction of an idealized microscopic world of particles living on a lattice. The particles can move on the lattice by "hopping" from site to site. In the specific examples considered in this paper, we allow only hops from a site to its nearest neighbors (a particle may also remain stationary at its current site) and we indicate these motions with the vectors $\vec{C}^\alpha$. The $\vec{C}^\alpha$ are traditionally interpreted as the momenta of the particles. (We are using the lattice spacing, the "mass" of a particle, and the simulation time step as fundamental units.) To simplify even further, we suppose that there cannot be more than one particle with a given momentum at a given site. The population at each site can then be represented by an $l + 1$ element binary vector, $\{f^\alpha(\vec{x})\}$, where l is the number of nearest neighbors and $\vec{x}$ is the label of a lattice site. We can now define the microscopic number density

$$\hat{\rho}(\vec{x}) = \sum_\alpha f^\alpha(\vec{x}) \tag{1.1}$$

and the microscopic momentum density

$$\vec{\hat{g}}(\vec{x}) = \sum_\alpha \vec{C}^\alpha f^\alpha(\vec{x}). \tag{1.2}$$

The time evolution of the gas is produced by the effect of two alternating steps: the "hopping" phase we described above and a collision phase. In the latter, the $\{f^\alpha\}$ of each site are transformed according to a set of collision rules. The rules can change from site to site or from time step to time step, but in any case they will conserve the microscopic densities $\hat{\rho}$ and $\vec{\hat{g}}$ on each site.

It is possible to construct macroscopic densities from $\hat{\rho}$ and $\vec{\hat{g}}$ by averaging in space and time over appropriate regions. The time evolution of the macroscopic number and momentum densities, ρ and $\vec{g}$, can be expressed, in the appropriate limit, in terms of the conservation laws

$$\partial_t \rho + \partial_k g_k = 0, \tag{1.3}$$
$$\partial_t g_i + \partial_j T_{ij} = 0, \tag{1.4}$$

(where Latin indices now denote Cartesian coordinates). It should be noted that we express the above densities in units of mass and momentum per unit area rather than units of mass and momentum per lattice site, as used by other authors (for instance, [9,10]). We completely ignore all the mathematical difficulties implied in the derivation of equations (1.3) and (1.4) [3], but we note that ρ and $\vec{g}$ in equations (1.3) and (1.4) are intended to be small perturbations from the equilibrium state, $\vec{g} = 0$ and $\rho =$ constant.

The structure of the stress tensor T_{ij} reflects the symmetries of the underlying lattice. Frisch et al. have shown that a hexagonal lattice possesses sufficient symmetry to obtain the right structure for T_{ij}. By this we mean that up to higher derivatives and $O(g^4)$, it is possible to write

$$T_{ij} =$$

$$\lambda(n)g_i g_j + p(n, g^2)\delta_{ij} - \nu(n)(\partial_j g_i + \partial_i g_j - \delta_{ij}\partial_k g_k) - \xi(n)\delta_{ij}\partial_k g_k, \tag{1.5}$$

where the quantities ν and ξ can be interpreted as transport coefficients while λ (which equals 1 for standard Navier-Stokes) arises from the absence of Galilean invariance for the lattice gas [3,13]. In the limit of incompressible flow, equations (1.3) and (1.4) together with the constitutive relation

(1.5), can be rescaled to the incompressible Navier-Stokes equation [3]. Thus, we can interpret this lattice gas as an analog computer capable of solving the two-dimensional incompressible Navier-Stokes equation.

Note that nowhere is there an attempt to simulate the microscopic behavior of a real fluid. Lattice gas automata are quite distinct from molecular dynamical simulations [20]. While both kinds of simulations seem to produce the expected macroscopic behavior for the fluid (in the sense of giving the expected constitutive relations for the macroscopic currents), they represent two completely different approaches to the problem. Molecular dynamical simulations attempt to faithfully model the microscopic behavior of a real fluid, while lattice gas automata extract only the minimal microscopic properties required to obtain the desired macroscopic behavior [5–7].

This suggests two interesting paths of research. The first, more technically oriented, concerns how well the results obtained from this new technique agree with real fluids, while the second concerns the more profound question of the connection between the microscopic and macroscopic aspects of many body systems [21]. In this paper, we principally address technical questions: the quantitative accuracy of the constitutive relation, equation (1.5), in a particular simple example, and the comparison of the effective kinematic viscosity measured in our steady non-equilibrium simulation with the values obtained by shear wave relaxation methods.

2. The simulation model

The object of our simulation is a steady forced flow between two walls with no-slip boundary conditions. We are simulating a steady flow because it allows us to obtain good accuracy in the measurements of ρ and $\vec{g}$ by extensive time averaging. We are simulating a channel with null velocity at the walls because for weak forcing (low Reynolds number), the g profile is expected to be a parabola and there is a simple relation between the maximum g, the forcing level, and kinematic viscosity ν. The actual simulation setup described below is conceptually very different from a direct implementation of a no-slip boundary channel flow but, as we will show, gives the same parabolic momentum profile.

The simulation system we have employed is a model of forced two-dimensional Poiseuille flow [13,22–24]. The system is a hexagonal lattice with an equal number of rows and columns (figure 1). Note that the system width, W, is $\sqrt{3}/2$ times the length, L, due to the unequal row and column spacings. The flow is forced by adding momentum in the positive x direction to the system at a constant rate: After each time step, we randomly select a lattice site and, if possible, apply one of the microscopic forcing rules described in figure 2. Each successful application of a forcing rule adds one unit of momentum to the system. The forcing process is repeated until the desired amount of momentum has been transferred to the gas; fractional amounts of momentum to be added to the system are accumulated across

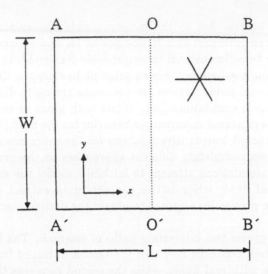

Figure 1: The simulation model. The walls AA′ and BB′ are joined by periodic boundary conditions while "Möbius strip" boundary conditions (see text) are used to connect the AB and A′B′ walls. The representative lattice site shown in the upper right-hand corner illustrates the orientation of the underlying hexagonal lattice.

time steps until they sum to an amount greater than 1, at which time one additional unit of momentum is added to the gas. The result of this process is a constant body force applied to the gas uniformly across the width and length of the channel.[1]

The forcing level employed in the present work varies from 0.3 to 2.8 units of momentum per time step. Within this range, the resulting flow is steady when averaged over a period of the order of a few diffusion times, L^2/ν. For a steady flow, the equations for the forced flow [22,3] become

$$0 = \partial_k g_k \tag{2.1}$$
$$\partial_k(\lambda g_k g_l) = -\partial_l p + \partial_k(\nu \partial_k g_l) + f_l \tag{2.2}$$

where $\vec{f} = (f, 0)$ is the average force per unit area.

The two walls perpendicular to the flow, AA′ and BB′ in figure 1, are mapped onto each other by periodic boundary conditions. The walls parallel to the flow, AB and A′B′, are mapped onto each other by "Möbius strip" boundary conditions. This boundary condition can be described as a two-step process whereby particles crossing the boundary have their position

---

[1]The actual forcing scheme is slightly more complicated since it must compensate for inhomogeneity in the momentum and number densities due to the macroscopic flow (see [11]). The forcing algorithm randomly selects a lattice row and column and then searches along the row until it finds a site where a forcing rule may be successfully applied. The program terminates if no forcing operation can be performed on a selected row. This guarantees that forcing operations will be uniformly distributed across the width of the channel, despite variations in the mass and momentum densities.

Before   After

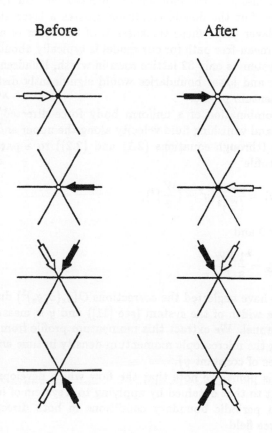

Figure 2: Forcing rules. The four pairs of diagrams represent the
microscopic forcing rules used in the simulation. The black symbols
indicate occupied states while the outlined symbols indicate vacant
states. States not indicated in a diagram may be either filled or vacant. Each forcing operation adds one unit of momentum in the x
direction.

and velocity reflected with respect to the line OO' and then standard periodic boundary conditions are applied. The alternative to the Möbius strip boundary would be the use of no-slip boundary conditions, for instance, random scattering of the particle impinging on the walls. Both boundary conditions dissipate the momentum injected into the gas by the uniform body force, but the no-slip condition creates a layer at the boundary (a Knudsen layer [14]) whose thickness is of the order of a mean-free path.[2] Since the mean-free path for our model is typically about 5 lattice spacings and the system is only 32 lattice rows in width, Knudsen layers along both the upper and lower boundaries would significantly distort the Poiseuille flow momentum profile.

The combination of a uniform body force directed in the positive x direction and vanishing fluid velocity along the upper and lower boundaries gives rise (through equations (2.1) and (2.2)) to a parabolic momentum density profile

$$g_x(y) = \frac{g_{max}}{(\frac{W}{2})^2}(y^2 - (\frac{W}{2})^2),$$ (2.3)

with $g_y = 0$ and

$$g_{max} = \frac{1}{8}\frac{FW}{L\nu}$$ (2.4)

where we have neglected the corrections $O(g_x(\partial_y g_x)^2)$ due to variation of ρ across the width of the system (see [11]) and y is measured from the axis of the channel. We extract this momentum profile from the simulation by averaging the microscopic momentum density in time and along the lattice rows (lines of constant y).

At this point, we note that the flow which develops in the channel is equivalent to that obtained by applying to a system of length L and width $2W$, with periodic boundary conditions in both directions, the "square wave" force field

$$\vec{f}^*(x,y) = (f,0) \quad for\ 0 \leq y < W,$$ (2.5)

$$\vec{f}^*(x,y) = (-f,0) \quad for\ W \leq y < 2W.$$ (2.6)

We have verified this correspondence by comparing simulation results from runs employing Möbius strip boundary conditions with runs using square wave forcing (see figure 3). Both types of flow exhibit long wavelength instabilities related to the existence of inflection points in the momentum profile at $y = 0$ and $y = W$. The critical Reynolds number given by linear stability analysis for infinitely long channels (Kolmogorov flow [15,16]) is quite small, $Re_{cr} \approx 1.11$, but a finite length-to-width ratio increases Re_{cr}. The particular length-to-width ratio used in our simulation, $1/\sqrt{3}$,

---

[2]The Knudsen layer is caused by the matching between the artificial particle distribution imposed by this kind of boundary condition and the non-equilibrium particle distributions imposed by the macroscopic flow in the bulk.

appears to be stable even for the largest Reynolds number obtainable in our simulation (≈ 50), as was suggested by the linear stability analysis of the problem.

3. Results

In this section, we will occasionally refer to mass density in number of particles per site $n, n = (\sqrt{3}/2)\rho$, instead of number of particles per unit area, ρ; this is done for notational convenience. Figure 4 shows a typical momentum profile obtained from our simulation. The average number of particles per site in this run is $n = 2.1$, the system dimensions are $W = 16\sqrt{3}$ and $L = 32$ (corresponding to a 32×32 lattice), and we have used the model-II collision rules described in reference 3. The profile was obtained by averaging the microscopic momentum density $\vec{g}$ in the direction parallel to the flow and on one million iterations. The g_y component appears to be due entirely to statistical noise; it is small: $\max |g_y(y)/g_{max}| < 0.01$, where g_{max} is g_x at the center of the channel.

The solid line represent a parabola fitted to the simulation results which are shown as symbols. The fit is very good; if we define

$$e(y) = \frac{|g_x(y) - h(y)|}{g_x(y)}$$

where $h(y)$ is the fitted parabola, then $\max |e(y)| < 0.01$ over the central region of the parabola (roughly the 26 centermost rows).

The result quoted above can be improved by increasing the number of time steps on which the simulation in figure 4 is averaged. However, improvements obtained by averaging are limited by systematic deviations from a parabolic profile which can be reduced only by decreasing the amplitude of $\vec{g}$. These systematic deviations are due to higher-order terms, $O(g^4)$, neglected in equation (1.5), and to the presence of a term proportional to g^2 in the expression, derived in [1,2,3]; for the pressure p, see [11].

We can use equation (2.4) to relate the maximum measured velocity to the applied force. This permits us to define an effective kinematic viscosity

$$\nu = \frac{1}{8} \frac{FW}{Lg_{max}}. \tag{3.1}$$

In figure 5, we plot the measured ν as a function of the reduced density for a set of simulations using the model II collision rules of reference 3. The system used for the measurement was 32×32, and the forcing level was very weak, so that the typical Mach number, defined as the ratio between the speed of sound and the hydrodynamical velocity, $\vec{g}/\rho$, is approximately ≈ 0.1.

In the same figure, we compare our viscosity measurements with data obtained by relaxation measurements [10]. The errors bars for the latter set of data are set at about 3 percent of the actual measured viscosity [12], while the errors on our data set are about 1 percent and are not indicated.

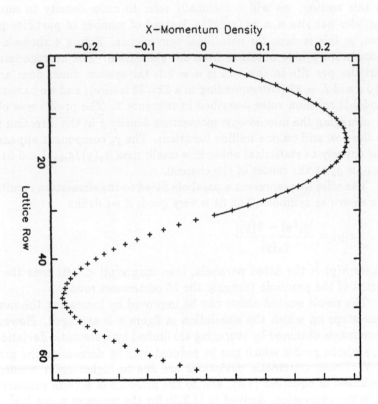

Figure 3: Möbius strip boundary conditions versus square wave forcing. The symbols represent the x momentum density profile for square wave forcing (see text) with periodic boundary conditions in both directions. The solid line is a parabolic fit to the momentum density profile obtained using Möbius strip boundary conditions on a system half as wide. Both simulations were run at a density of 2.1 particles per site and the profiles were averaged over one million time steps.

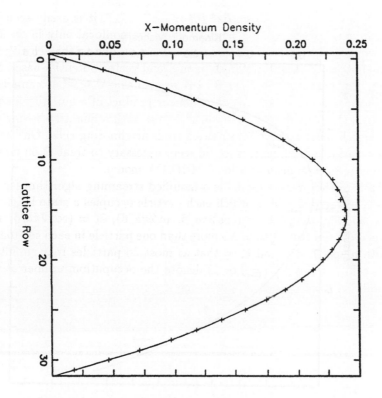

Figure 4: Typical momentum profile. The x momentum density profile for a 32×32 system run at $n = 2.1$ particles per site and a forcing level $F = 0.76$ momentum units per time step. The profile was averaged over one million time steps. The solid line is a parabolic fit to the simulation data points (symbols).

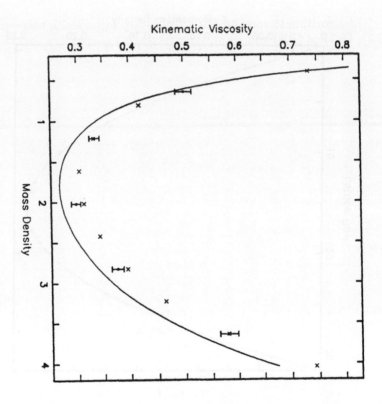

Figure 5: Kinematic viscosity versus mass density. Viscosity values derived from the present work are shown as crosses. Symbols with error bars are the results obtained by d'Humières et al. The solid line is the theoretical value obtained using Chapman-Enskog techniques.

In both cases, the errors are only rough estimates since they were not computed from first principles but were estimated by comparing similar runs with different initial conditions.

The two sets of viscosity data appear to be consistent, except in the range $2 < \rho < 3$, where our data are consistently greater than the results of reference 10. It would be tempting to relate the discrepancy between the two viscosities to viscosity renormalization effects [18], but we presently do not have any conclusive evidence.

The solid line is the shear viscosity as calculated by the technique of Michel Hénon [25] and by other authors using Chapman-Enskog techniques [17]. Both methods are based on an approximate theoretical description of the gas in which the correlations between particles are completely neglected.

In the low density limit, there is a very good agreement with the theory, as is expected since the relative importance of particle correlation becomes negligible in that limit. Note that we do not quote results for $\rho < 0.4$ because for these densities the mean-free path (see [22]) of the particles becames comparable to the width of the channel.

4. Conclusion

We have given some precise simulation evidence that LGAs are correctly represented by the constitutive relation, equation (1.5). We have also shown that the simulation of channel flow gives the expected parabolic profile to a good degree of accuracy and that the effective kinematic viscosity obtained by these steady non-equilibrium flows is in reasonable agreement with that obtained by d'Humières et al. using shear waves relaxation measurements.

The technique used for these simulations is capable of providing reasonably accurate measurement of viscosity; the particular kind of boundary conditions used allow a wide range of lattice size width and makes feasible the study of the dependence of the kinematic viscosity on the width of the simulation channel. For two-dimensional fluids, there are rather precise predictions, based on renormalization group arguments and other techniques, for this dependence [18]. In first-order perturbation theory, the kinematic viscosity diverges as the logarithm of the box size. We have some preliminary results, to be published elsewhere, which indicate the presence of this effect even within the range of channel widths accessible by our method of simulation.

Acknowledgments

We would like to thank B. Hasslacher and U. Frisch for interesting and helpful discussions, and G. Doolen and T. Shimomura for the original version of the FORTRAN program we have used. This work was supported by ONR and NSF-MRL, and we gratefully acknowledge the hospitality of the Los Alamos National Laboratory, where a portion of this work was done.

References

[1] U. Frisch, B. Hasslacher, and Y. Pomeau, *Phys. Rev. Lett.*, **56** (1986) 1505.

[2] S. Wolfram, *J. Stat. Phys.*, **45** (1986) 471.

[3] U. Frisch, D. d'Humières, B. Hasslacher, P. Lallemand, Y. Pomeau, and J. Rivet, "Lattice Gas Hydrodynamics in Two and Three Dimensions", *Complex Systems*, **1** (1987) 648.

[4] See the bibliography of [3].

[5] L. P. Kadanoff and J. Swift, *Phys. Rev.*, **165** (1968) 310.

[6] J. Hardy and Y. Pomeau, *J. of Math. Phys.*, **13** (1972) 1042.

[7] J. Hardy, O. de Pazzis, and Y. Pomeau, *J. of Math. Phys.*, **14** (1973) 1746; *Phys. Rev.*, **A13** (1976) 1949.

[8] D. d'Humières, P. Lallemand, and U. Frisch, *Europhys. Lett.*, **2** (1986) 297.

[9] D. d'Humières, Y. Pomeau, and P. Lallemand, *C. R. Acad. Sci. Paris II*, **301** (1985) 1391.

[10] D. d'Humières, P. Lallemand, and T. Shimomura, "Computer simulations of lattice gas hydrodynamics", LANL preprint.

[11] J. Dahlburg, D. Montgomery, and G. Doolen, "Noise and Compressibility in Lattice-Gas Fluids", *Physical Review*, in press.

[12] D. d'Humières, personal communication.

[13] D. d'Humières and P. Lallemand, *C. R. Acad. Sci. Paris II*, **302** (1985) 983.

[14] C. Cercignani, *Theory and Application of the Boltzmann Equation*, (New York, Elsevier, 1975).

[15] L. D. Meshalkin and Ya. G. Sinai, *J. Appl. Math. (PMM)*, **25** (1979) 1700.

[16] Z. S. She, "Large scale Dynamics and Transition to Turbulence in the Two-dimensional Kolmogorov Flow", *Proceed. Fifth Intern*, Beer-Sheva Seminar on MHD Flows and Turbulence, Israel, March 2–6, 1987.

[17] J. P. Rivet and U. Frisch, *C. R. Acad. Sci. Paris II*, **302** (1986) 267.

[18] Y. Pomeau and P. Résibois, *Phys. Rep.*, **19** (1975) 63.
K. Kawasaki and J. D. Gunton, *Phys. Rev.*, **A 8** (1973) 2048;
D. Forster, D. Nelson, and M. Stephen, *Phys. Rev.*, **A 16** (1977) 732;
T. Yamada and K. Kawasaki, *Prog. of Theor. Phys.*, **53** (1975) 111; Y. Pomeau and P. Résibois, *Phys. Rep.*, **19** (1975) 63.

[19] B. Alder and T. Wainwright, *Phys. Rev. Lett.*, **18** (1970) 968.

[20] D. J. Evans and G. P. Morriss, *Phys. Rev. Lett*, **51** (1983) 1776; D. H. Heyes, G. P. Morriss, and D. J. Evans, *J. of Chem. Phys.*, **83** (1985) 4760.

[21] L. P. Kadanoff, *Physics Today*, 39 (1986) 7.

[22] C. Burges and S. Zaleski, "Buoyant Mixtures of cellular Automaton Gases", *Complex Systems*, 1 (1987) 31.

[23] K. Balasubramanian, F. Hayot, and W. F. Saam, "Darcy's law for lattice gas hydrodynamics", Ohio State University preprint (1987).

[24] D. H. Rothman, MIT preprint (1987), submitted to *Geophysics*.

[25] M. Hénon, *Complex Systems*, 1 (1987) 762.

Jean-Pierre Rivet
Observatoire de Nice, BP 139, 06003 Nice Cedex, France and Ecole Normale Supérieure, 45 rue d'Ulm, 75230 Paris Cedex 05, France

Green-Kubo Formalism for Lattice Gas Hydrodynamics and Monte-Carlo Evaluation of Shear Viscosities

This paper originally appeared in *Complex Systems* (1987), Volume 1, pages 839–851.

Lattice Gas Methods for Partial Differential Equations, SFI SISOC, Eds. Doolen et al., Addison-Wesley Publishing Co., 1990

399

Complex Systems **1** (1987) 839–851

Green-Kubo Formalism for Lattice Gas Hydrodynamics and Monte-Carlo Evaluation of Shear Viscosities

Jean-Pierre Rivet
Observatoire de Nice,
BP 139, 06003 Nice Cedex, France
and
Ecole Normale Supérieure,
45 rue d'Ulm, 75230 Paris Cedex 05, France

Abstract. A Green-Kubo formula, relating the shear viscosity to discrete time correlation functions, is derived via a Liouville equation formalism for a class of non-deterministic lattice gas models. This allows a Monte-Carlo calculation of the viscosity. Preliminary results are presented for the Frisch-Hasslacher-Pomeau two-dimensional lattice gas model.

1. Introduction

When a physical system at thermodynamical equilibrium is subject to a weak large-scale perturbation (say a temperature gradient), a flux of a conjugated quantity (say a heat flux) results, which is linear in the gradient. In an isotropic newtonian fluid, a gradient of velocity creates a *momentum flux*, related by a linear relation involving a fourth-order tensor. Isotropy implies that this tensor is expressible in terms of two scalars, the shear and bulk viscosities. Fluctuation-dissipation theory relates such *transport coefficients* to time-integrated correlation functions. The earliest results in that line was obtained by Einstein in the study of Brownian motion [1]. In the fifties, systematic fluctuation-dissipation relations were developed for classical and quantum mechanical systems by Green [2,3] and Kubo [4].

Cellular automata with discrete state variables attached to a lattice and suitable conservation relations (lattice gases) present thermodynamic equilibria, as continuous systems do, and they can display large-scale hydrodynamic behavior [5,6]. Fluctuation-dissipation relations for lattice gases have been considered in references 6 through 9. Due to discreteness, there are novel features in the theory of transport coefficients, such as "propagation viscosities" [10]. Typically, there are two possible approaches. One is based on "noisy hydrodynamics" [6]. The other one, in the spirit of Green

[3], uses a Liouville equation approach and is developed here for a quite general class of D-dimensional, non-deterministic, one-speed models. For a more restricted class of deterministic two-dimensional models, results were already announced by Frisch and Rivet [11].

In section 2, we formulate the problem; we will use the same notation as in reference [6]; however, in order to make the paper reasonably self-contained, we will reintroduce briefly some of the basic concepts. In section 3, we perturbatively solve the Liouville equation around an equilibrium state and find the discrete Green-Kubo formula for the shear viscosity. In section 4, we show how to use the discrete Green-Kubo formula for a Monte-Carlo calculation of the shear viscosity; numerical results are given only for the simplest FHP model [5,6]. Comparisons are made with theoretical values obtained from the lattice Boltzmann approximation [6,10] and with results of numerical experiments based on relaxation of large scale shear waves [12,13].

2. The class of models and the formalism

In order to avoid heavy notation as much as possible, we limit the following study to the class of non-deterministic, one-speed models whose complete definition is given in reference 6. This includes several two-dimensional and three-dimensional (pseudo-four-dimensional) models known as HPP, FHP-I, and FCHC. We will also give the final results for FHP models with rest-particles, which do not belong to this class. We recall briefly the main features of the one-speed models: unit mass particles are moving with speed c along links of a regular D-dimensional Bravais lattice, where each node is connected to its b nearest neighbors by a set of b vectors c_i, $i = 1, \ldots, b$ of equal modulus c. This set is supposed to verify some further geometric conditions given in reference 6. The fact that two particles with the same velocity vector are not allowed to be at the same node at the same time (exclusion principle) enables us to describe the state of one node at any integer time by a b-bit binary word: $s = \{s_i, i = 1, \ldots, b\}$ where $s_i = 1$ if a particle is present at the node, in the cell corresponding to the velocity vector c_i, and $s_i = 0$ otherwise.

If initial conditions (time $t = 0$) are taken such that all particles are located at the nodes, the free propagation along links ensures that at any integer time[1] t_*, all particles are at the nodes. At any node, incoming particles can perform *local* collisions according to a non-deterministic rule; that is, an input state described by the binary word s will be changed into an output state s' with the transition probability $A(s \rightarrow s')$. These transition probabilities are taken node-independent. $A(s \rightarrow s')$ is zero if input and output states have different total mass $(\sum_i s_i)$ or momentum $(\sum_i s_i c_i)$.

The state of the whole lattice $\mathcal{L}$ at integer time t_* may be described by the so-called Boolean field:

---

[1] As in [6], an index "star" denotes the discrete independent variables.

$$n(t_*) = \left\{ n_i(t_*, \mathbf{r}_*), \; i = 1, .., b; \; \mathbf{r}_* \in \mathcal{L} \right\}. \tag{2.1}$$

The time evolution of this Boolean field is governed by the *microdynamical equation*, (see [6], section 3.1) which can be formally written using streaming, collision, and evolution operators S, C, and $\mathcal{E}$:

$$n(t_* + 1) = S \circ C n(t_*) = \mathcal{E} n(t_*). \tag{2.2}$$

For non-deterministic collision rules, the operator $\mathcal{E}$ is itself non-deterministic. The conservation laws induce two exact relations for the Boolean field:

$$\sum_i n_i(t_* + 1, \mathbf{r}_* + \mathbf{c}_i) = \sum_i n_i(t_*, \mathbf{r}_*), \tag{2.3}$$

$$\sum_i \mathbf{c}_i n_i(t_* + 1, \mathbf{r}_* + \mathbf{c}_i) = \sum_i \mathbf{c}_i n_i(t_*, \mathbf{r}_*). \tag{2.4}$$

The lattice gas may be described statistically by a probability distribution $P(s(.))$ that gives the probability of occurrence of a configuration $s(.) = \{s(\mathbf{r}_*), \mathbf{r}_* \in \mathcal{L}\}$. The time-evolution of this probability distribution is given by a Liouville equation (see [6], section 3.3)

$$P\Big(t_* + 1, S s'(.)\Big) =$$

$$\sum_{s(.) \in \Gamma} \prod_{\mathbf{r}_* \in \mathcal{L}} A[s(\mathbf{r}_*) \to s'(\mathbf{r}_*)] \, P\Big(t_*, s(.)\Big), \quad \forall s'(.) \in \Gamma, \tag{2.5}$$

where Γ denotes the set of all possible configurations of the lattice $\mathcal{L}$. For further use, it is convenient to introduce a global transition probability $A^{(g)}(s \to s')$ which is $\prod_{\mathbf{r}_* \in \mathcal{L}} A[s(\mathbf{r}_*) \to s'(\mathbf{r}_*)]$ and to write equation (2.5) in the more compact form

$$P\Big(t_* + 1, S s'(.)\Big) = \sum_{s(.) \in \Gamma} A^{(g)}(s \to s') \, P\Big(t_*, s(.)\Big). \tag{2.6}$$

The following mean[2] quantities will be useful in the sequel:

mean population: $N_i(t_*, \mathbf{r}_*) = \sum_{s. \in \Gamma} s_i(\mathbf{r}_*) P\Big(t_*, s(.)\Big),$

density: $\rho(t_*, \mathbf{r}_*) = \sum_i N_i(t_*, \mathbf{r}_*),$

mass current (momentum): $\mathbf{j}(t_*, \mathbf{r}_*) = \sum_i \mathbf{c}_i N_i(t_*, \mathbf{r}_*),$

mean velocity: $\mathbf{u}(t_*, \mathbf{r}_*) = \mathbf{J}(t_*, \mathbf{r}_*)/\rho(t_*, \mathbf{r}_*). \tag{2.7}$

---

[2] Averaged over the probability distribution $P\Big(s(.)\Big)$

The Liouville equation admits a family of homogeneous factorized equilibrium *solutions* of the form [6]:

$$P^{(eq)}\left(s(.)\right) = \prod_{r_* \in \mathcal{L}} \prod_i N_i^{(eq)\,s_i}(1 - N_i^{(eq)})^{(1-s_i)}. \tag{2.8}$$

For low-speed equilibria, $N_i^{(eq)}$ may be expressed in terms of the density and mean velocity:

$$N_i^{(eq)}(\rho, \mathbf{u}) = \frac{\rho}{b} + \frac{\rho D}{bc^2} c_{i\alpha} u_\alpha + O(u^2). \tag{2.9}$$

Averages over the the equilibrium distribution with $\mathbf{u} = 0$ will be denoted by angular brackets $\langle \rangle$. *Local equilibria* having the above form but with slowly varying parameters ρ and $\mathbf{u}$ will be the zero-order terms of an expansion in powers of the scale separation ϵ between the lattice constant and the smallest excited scale. As has been shown in reference 6, sections 5 through 7, when the lattice has sufficient isotropy, as we will assume here,[3] hydrodynamical equations are obtained for the density and momentum. The momentum equation involves a kinematic shear viscosity[4]

$$\nu(\rho) = -\frac{bc^4}{D(D+2)}\psi(\rho) - \frac{c^2}{2(D+2)}. \tag{2.10}$$

The coefficient $\psi(\rho)$ relates the first-order perturbation $\epsilon N_i^{(1)}$ of the mean population to the gradient of the mass current $\mathbf{j} = \rho \mathbf{u}$ through [6]

$$\epsilon N_i^{(1)} = \psi(\rho) Q_{i\alpha\beta} \partial_\alpha(j_\beta),$$

$$\text{where} \quad Q_{i\alpha\beta} = c_{i\alpha} c_{i\beta} - \frac{c^2}{D}\delta_{\alpha\beta} \tag{2.11}$$

is the (traceless) microscopic stress-tensor. To determine the viscosity, we must find the shear-induced perturbation of the mean population. As the mean population does not satisfy a closed set of equations,[5] we must revert to the full probability distribution satisfying the Liouville equation.

3.   Perturbative resolution of the Liouville equation

Global homogeneous equilibrium distributions are exact steady solutions of the Liouville equation, but local equilibria are not.[6] We will look for a solution of the Liouville equation (2.5) in the form

$$P\left(t_*, s(.)\right) = P^{(0)}\left(t_*, s(.)\right) + \epsilon P^{(1)}\left(t_*, s(.)\right) + O(\epsilon^2), \tag{3.1}$$

---

[3]The formalism presented here is easily extended to anisotropic cases in terms of an anisotropic viscosity tensor.

[4]The bulk viscosity is zero for one-speed models [6].

[5]Except in the lattice Boltzmann approximation which we are not using here.

[6]Local equilibria have the same dependence in ρ and $\mathbf{u}$ as a global equilibria but with ρ and $\mathbf{u}$ allowed to be space-dependent.

where $P^{(0)}\left(t_*, s(.)\right)$ is a low-speed local equilibrium distribution whose parameters have slow variations on spatial scale ϵ^{-1}. We assume that the perturbation $P^{(1)}$ vanishes initially. Using equation (2.5), we find that the perturbation $P^{(1)}$ satisfies the inhomogeneous Liouville equation

$$P^{(1)}\left(t_* + 1, S s'(.)\right) - \sum_{s(.)\in\Gamma} A^{(g)}(s \to s') P^{(1)}\left(t_*, s(.)\right) =$$

$$-\epsilon^{-1}\left[P^{(0)}\left(t_* + 1, S s'(.)\right) - \sum_{s(.)\in\Gamma} A^{(g)}(s \to s') P^{(0)}\left(t_*, s(.)\right)\right]. \quad (3.2)$$

We have now to make some straightforward manipulations of the r.h.s of the above equation.

Using the fact that s_i' is 0 or 1 and performing for each value of i the spatial shift $\mathbf{r}_* \to \mathbf{r}_* + \mathbf{c}_i$, we can write $P^{(0)}\left(t_* + 1, S s'(.)\right)$ in the form

$$P^{(0)}\left(t_* + 1, S s'(.)\right)$$

$$= \prod_{j,\rho_*}\left[N_j^{(0)}(t_*+1, \rho_*+\mathbf{c}_i) s'(\rho_*) + \left(1 - N_j^{(0)}(t_*+1, \rho_*+\mathbf{c}_i)\right)\left(1 - s'(\rho_*)\right)\right]. (3.3)$$

As the mean populations $N_i^{(0)}$ are supposed to have slow space and time variations, we Taylor-expand all finite differences up to the first order in the gradients, and make the rescalings (see [6], section 5)

$$\partial_t \to \epsilon\partial_{t_1} \quad \text{and} \quad \partial_r \to \epsilon\partial_{r_1}. \quad (3.4)$$

We thereby obtain

$$P^{(0)}\left(t_* + 1, S s'(.)\right) = P^{(0)}\left(t_*, s'(.)\right)$$

$$+\epsilon\sum_{j,\rho_*}(\partial_{t_1} + c_{i\alpha}\partial_{1\alpha}) N_j^{(0)}(t_*,\rho_*)\left(P_{j,\rho_*}^{(+eq)}(s'(.)) - P_{j,\rho_*}^{(-eq)}(s'(.))\right)$$

$$+ O(\epsilon^2). \quad (3.5)$$

Here, we have introduced

$$P_{j,\rho_*}^{(+eq)}(s(.))$$

$$= s_j(\rho_*) \prod_{(i',\mathbf{r}_*')\neq(j,\rho_*)}\left[N_{i'}^{(eq)} s_{i'}(\mathbf{r}_*') + (1 - N_{i'}^{(eq)})(1 - s_{i'}(\mathbf{r}_*'))\right],$$

$$P_{j,\rho_*}^{(-eq)}(s(.))$$

$$= (1 - s_j(\rho_*)) \prod_{(i',\mathbf{r}_*')\neq(j,\rho_*)}\left[N_{i'}^{(eq)} s_{i'}(\mathbf{r}_*') + (1 - N_{i'}^{(eq)})(1 - s_{i'}(\mathbf{r}_*'))\right], (3.6)$$

which have an interesting interpretation. $P_{j,\rho_*}^{(+eq)}(s(.))$ (respectively $P_{j,\rho_*}^{(-eq)}(s(.))$) is the probability distribution corresponding to a state where all nodes and all cells are occupied with the zero-speed global equilibrium probability, except the ith cell of the node ρ_* which is occupied with probability 1 (respectively 0). These states are referred to as "SBSE" for Single Bit Set Equilibrium (respectively, "SBCE" for Single Bit Cleared Equilibrium).

Note that $P^{(0)}(s'(.))$ is the same as $\sum_{s(.)\in\Gamma} A^{(g)}(s \to s')P^{(0)}(s(.))$, because $P^{(0)}$ has *locally* the same analytic form as $P^{(eq)}$. This allows us to rewrite equation (3.2) for the perturbation $P^{(1)}$ as

$$P^{(1)}\Big(t_* + 1, Ss'(.)\Big) - \sum_{s(.)\in\Gamma} A^{(g)}(s \to s')P^{(1)}\Big(t_*, s(.)\Big) =$$

$$-\sum_{j,\rho_*}(\partial_{t_1} + c_{i\alpha}\partial_{1\alpha})N_i^{(0)}\left[P_{j,\rho_*}^{(+eq)}(s'(.)) - P_{j,\rho_*}^{(-eq)}(s'(.))\right],$$

$$\forall s'(.) \in \Gamma. \tag{3.7}$$

We can now re-express the time derivatives in terms of space derivatives by using the macrodynamical Euler equations ([6], section 5) and the low speed equilibrium form (2.9); the expression $(\partial_{t_1} + c_{i\alpha}\partial_{1\alpha})N_i^{(0)}$ becomes then $\frac{D}{bc^2}Q_{i\alpha\beta}\partial_{1\alpha}(\rho u_\beta)$ and equation (3.7) becomes

$$P^{(1)}\Big(t_* + 1, Ss'(.)\Big) - \sum_{s(.)\in\Gamma} A^{(g)}(s \to s')P^{(1)}\Big(t_*, s(.)\Big) =$$

$$-\frac{D}{bc^2}\sum_{j,\rho_*}\Big(P_{j,\rho_*}^{(+eq)}(s'(.)) - P_{j,\rho_*}^{(-eq)}(s'(.))\Big)Q_{j\alpha\beta}\partial_{1\alpha}(\rho u_\beta),$$

$$\forall s'(.) \in \Gamma. \tag{3.8}$$

In order to solve equation (3.8), we consider the probability distributions $P_{j,\rho_*}^{(+eq)}(t_*, s(.))$ and $P_{j,\rho_*}^{(+eq)}(t_*, s(.))$ obtained after t_* evolution steps, starting from an SBSE and an SBCE. The solution of (3.8) with vanishing initial condition is

$$P^{(1)}\Big(t_*, s'(.)\Big) =$$

$$-\frac{D}{bc^2}\sum_{\tau_*=0}^{t_*-1}\sum_{j,\rho_*}\Big[P_{j,\rho_*}^{(+eq)}(\tau_*, S^{-1}s'(.))$$

$$-P_{j,\rho_*}^{(-eq)}(\tau_*, S^{-1}s'(.))\Big]Q_{j\alpha\beta}\partial_{1\alpha}(\rho u_\beta). \tag{3.9}$$

For further use, we introduce the following notation:

$$N^{(+eq)}(\tau_*; i, \mathbf{r}_* | j, \rho_*) = \sum_{s(.) \in \Gamma} s_i(\mathbf{r}_*) P_{j,\rho_*}^{(+eq)}(\tau_*, s'(.)),$$

$$N^{(-eq)}(\tau_*; i, \mathbf{r}_* | j, \rho_*) = \sum_{s(.) \in \Gamma} s_i(\mathbf{r}_*) P_{j,\rho_*}^{(-eq)}(\tau_*, s'(.)). \tag{3.10}$$

$N^{(+eq)}(\tau_*; i, \mathbf{r}_* | j, \rho_*)$ (respectively, $N^{(-eq)}(\tau_*; i, \mathbf{r}_* | j, \rho_*)$) is the conditional probability to find at time τ_*, a particle at node $\mathbf{r}_*$ in cell i, knowing that at time 0 there was one (respectively none) at node ρ_* in cell j. We can now express the perturbation of mean population $N_i^{(1)}(t_*, \mathbf{r}_*) = \sum_{s'(.) \in \Gamma} s_i'(\mathbf{r}_*) P^{(1)}(t_*, s'(.))$ in the form

$$N_i^{(1)}(t_*, \mathbf{r}_*) =$$

$$-\frac{D}{bc^2} \sum_{\tau_*=0}^{t_*-1} \sum_{j,\rho_*} \left(N^{(+eq)}(\tau_*; i, \mathbf{r}_* | j, \rho_*) \right.$$

$$\left. -N^{(-eq)}(\tau_*; i, \mathbf{r}_* | j, \rho_*) \right) Q_{j\alpha\beta} \partial_{1\alpha}(\rho u_\beta). \tag{3.11}$$

We now use the two following identities:

$$N_i^{(eq)} = \frac{\rho}{b} = N^{(+eq)}(\tau_*; i, \mathbf{r}_* | j, \rho_*) \frac{\rho}{b} + N^{(-eq)}(\tau_*; i, \mathbf{r}_* | j, \rho_*)(1 - \frac{\rho}{b}), \tag{3.12}$$

$$\langle n_i(\tau_*, \mathbf{r}_*) \, n_j(0, \rho_*) \rangle = N^{(+eq)}(\tau_*; i, \mathbf{r}_* | j, \rho_*) N_j^{(eq)}(0, \rho_*). \tag{3.13}$$

Equation (3.12) expresses that the equilibrium populations can be recovered from transition probabilities. Equation (3.13) expresses the two-point equilibrium probability in terms of the transition probability and the single-point probability. Using (3.12) and (3.13), we can rewrite (3.11) as

$$N_i^{(1)}(t_*, \mathbf{r}_*) = -\frac{Db}{c^2\rho(b-\rho)} \sum_{\tau_*=0}^{t_*-1} \sum_{j,\rho_*} \langle \tilde{n}_i(\tau_*, \mathbf{r}_*) \tilde{n}_j(0, \rho_*) \rangle Q_{j\alpha\beta} \partial_{1\alpha}(\rho u_\beta),$$

$$\tilde{n}_i = n_i - \langle n_i \rangle = n_i - \frac{\rho}{b}. \tag{3.14}$$

The average is over the zero-speed global equilibrium. From the isotropy of fourth-order tensors, it follows that

$$\sum_{i\alpha\beta} Q_{i\alpha\beta} Q_{i\alpha\beta} = \frac{bc^4(D-1)}{D}. \tag{3.15}$$

Using equations (2.10), (2.11), (3.15), reversal and translation invariances, we finally obtain

$$\nu(\rho) = \nu^{(prop)} + \sum_{\tau_*=0}^{t_*-1} \Gamma(\tau_*), \qquad \nu^{(prop)} = -\frac{c^2}{2(D+2)}, \tag{3.16}$$

where

$$\Gamma(\tau_*) =$$

$$\frac{bD}{c^2(D-1)(D+2)} \frac{1}{\rho(b-\rho)} \sum_{\rho_*} \sum_{i,j} \sum_{\alpha\beta} Q_{i\alpha\beta} \langle \tilde{n}_i(\tau_*,\rho_*)\tilde{n}_j(0,0) \rangle Q_{j\alpha\beta} \quad (3.17)$$

is (within a numerical factor) the correlation function of the microscopic stress-tensor. If $\Gamma(\tau_*)$ falls off sufficiently fast as $\tau_* \to \infty$, the summation over τ_* can be extended to infinity. This is the case in three dimensions but not in two, where the viscosity may at best display a quasi-steady plateau as τ_* increases (see section 4), so we will keep a finite upper bound for the time-summation.

Let us now specialize the results for the FHP-I model. We just substitute 6 for b, 2 for D, and 1 for c in the above formulæ and get

$$\nu(\rho) = -\frac{1}{8} + \sum_{\tau_*=0}^{t_*-1} \Gamma(\tau_*)$$

$$\Gamma(\tau_*) = \frac{3}{\rho(6-\rho)} \sum_{\rho_*} \sum_{i,j} \sum_{\alpha\beta} Q_{i\alpha\beta} \langle \tilde{n}_i(\tau_*,\rho_*)\tilde{n}_j(0,0) \rangle Q_{j\alpha\beta}. \quad (3.18)$$

For variants of FHP-I called FHP-II and FHP-III [6,14], which include rest particles, a simple generalization leads to

$$\nu(\rho) = -\frac{1}{8} + \sum_{\tau_*=0}^{t_*-1} \Gamma(\tau_*),$$

$$\Gamma(\tau_*) = \frac{49}{6\rho(7-\rho)} \sum_{\rho_*} \sum_{I,J} \sum_{\alpha\beta} (1 - \frac{1}{2}\delta_{\alpha\beta}) Q_{I\alpha\beta} \langle \tilde{n}_I(\tau_*,\rho_*)n_J(0,0) \rangle Q_{J\alpha\beta},$$

$$Q_{I\alpha\beta} = c_{I\alpha}c_{I\beta} - \frac{3}{7}\delta_{\alpha\beta}. \quad (3.19)$$

The capital indices I and J which take the values $(\star, 1, 2, \ldots, 6)$ refer to the seven possible velocities, namely, the velocity zero $(I = \star)$ and the six non-zero velocities of FHP-I.

4. Monte-Carlo calculation of the shear viscosity

In lattice gases as in (continuous) molecular dynamics (MD), there are broadly two strategies for calculating transport coefficients: *macroscopic* strategies involving large-scale gradients and *microscopic* strategies based on Green-Kubo relations. Their relative merits in MD have been recently discussed in reference 15. For lattice gases, macroscopic calculations of the shear viscosity have been performed by d'Humières and Lallemand [13] and by Zanetti [9]. The existence of a discrete analog for lattice gases of the MD Green-Kubo formula provides us with an alternative numerical procedure. It is our purpose here to describe this path for a simple case, the two-dimensional model FHP-I, which has binary head-on and triple collisions.

Since lattice gases are governed by *local* cellular automata rules, there are some interesting simplifications not present in MD. These are consequences of the following properties concerning equilibrium finite-time correlations:

P1 Finite $L \times L$ periodically wrapped-around lattices and infinite lattices admit the same factorized single-time equilibrium solutions. Multi-time distributions are identical when the maximum time-separation between the arguments is less than $, L/(2c)$, where c is the particle speed (unity for FHP-I model).

P2 The binary equilibrium correlation function $\langle \tilde{n}_i(\tau_*, \rho_*)\tilde{n}(0,0)\rangle$ vanishes if $c|\tau_*| < |\rho_*| < L/2$. (Actually, for the FHP-I model, it vanishes outside of a hexagonal domain of influence $D(\tau_*)$ inscribed in the circle of radius $c|\tau_*|$.)

The first part of **P1** has been established in reference 6 (section 4.1). For the second part, we observe that multiple-time distributions are expressible in terms of single-time distributions and iterated evolution operators. The evolution operator propagates information at the maximum speed c_o; thus, for time-separation less than $L/(2c)$ no finite-size effects are felt. **P2** is a consequence of **P1** and of the absence of single-time correlations between different nodes.

A consequence of **P1** is that finite-time correlations appearing in the Green-Kubo formula can be evaluated on a sufficiently large finite lattice by ensemble and space averaging. A consequence of **P2** is that the spatial summation over ρ_* in the Green-Kubo formula 3.18 can be restricted to the domain of influence $D(\tau_*)$; otherwise, as we will see, trouble arises. We have found time averaging to be unreliable even for correlation functions involving only small (or vanishing) time-separations. This may be due to a finite-size effect and/or a bias in the total momentum of the randomly generated initial configuration.

After these preliminaries, we describe a simulation strategy. We combine two kinds of averaging:

spatial averaging over all lattice nodes. All nodes of the lattice are successively shifted to the origin $r_* = 0$; the results of these elementary experiments are cumulated and divided by the number of nodes L^2.

ensemble averaging in which N elementary experiments are performed. Each elementary experiment involves an independent random generation of the Boolean variables $n_i(0, r_*)$ with mean $N_i = \rho/6$, the zero-velocity equilibrium value. A small number of samples (typically, N is about 40) ensures an adequately low level of random noise, provided the spatial summation over ρ_* is restricted to the domain of influence $D(\tau_*)$. If we simply use (3.18), ignoring that the spatial summation can be restricted to the domain of influence, the results will be far too noisy for realistic values of N. Consider, for example, the $\tau_* = 0$ term of the time-summation; its relative Monte-Carlo noise (the inverse of the signal to noise ratio) is $O(\sqrt{L^2/N})$; this tends to zero when $N \to \infty$ but far too slowly: it is still about 30 for a 64×64 lattice filled with a density $\rho = 1.2$, averaged over $N = 50$ shots. On the other hand, if the summation is restricted to the domain of influence, namely $\rho_* = 0$ for the $\tau_* = 0$ term, the relative Monte-Carlo noise becomes $O(1/\sqrt{N\,L^2})$ which is far smaller. We see how important it is not to sum over terms which are known by **P2** to be zero. For $\tau_* \neq 0$, this is also true, although to a lesser extent since the size of the domain of influence grows like $|\tau_*|$.

Numerical results

The characteristics of the numerical experiments were:

    lattice size: 64×64 nodes

    averaging over $N = 40$ independent realizations

    number of time steps $\tau_{maz} = 15$.

Three independent experiments (with different seeds for the pseudo-random generator) have been done in order to have an experimental estimate of the Monte-Carlo noise. The function $\Gamma(\tau_*)$ in the expression 3.18 of the kinematic shear viscosity will be referred to as the *correlation function*. The cumulated correlation function up to time $t_* - 1$ plus the propagation viscosity $(-1/8)$ will be referred to as the *viscosity*. Figures 1a and b present for a density per node $\rho = 1$, the correlation function and the viscosity for τ_* between 0 and 15. The black circles are averages over the three experiments. The absolute Monte-Carlo noise grows in a way consistent with our theoretical estimate. Clearly, beyond $\tau_* = 10$, the results are too noisy to be significant, but for our purpose there is no need to go beyond $\tau_* = 10$. The viscosity exhibits beyond $\tau_* = 8$ a plateau defined with an accuracy of about 3 percent at the value $\nu = 0.7 \pm 0.02$. Similarly, evaluated viscosities for various values of the density are represented as black circles (with error bars) on figure 2. The stars (also with error bars) are the viscosities observed by d'Humières and Lallemand [13] from macroscopic simulations with relaxation of shear waves. Whenever both data are available, the error bars overlap except at very low densities. This probably reflects a pathology of the FHP-I model: at low densities, triple collisions are so rare that the dynamics are affected by a spurious invariant that would be present in the absence of triple collisions.

The continuous curve of figure 2 is the viscosity calculated from the lattice Boltzmann approximation [6,10]

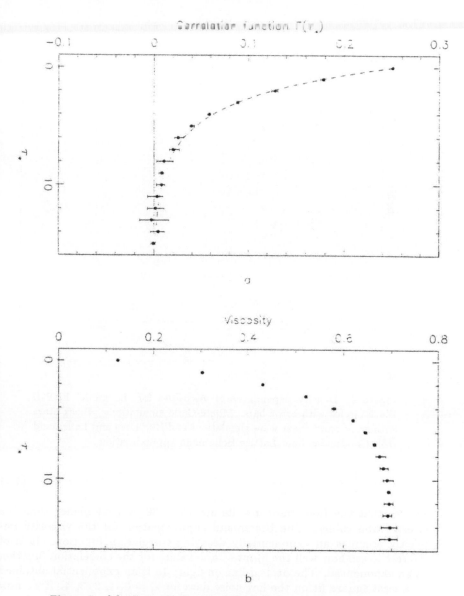

Figure 1: (a) Monte-Carlo simulation of the correlation function of the microscopic stress-tensor for the FHP-I model at density $\rho = 1$. Black circles are data points with error bars. The dashed line is a least square fit of an exponential to the first five points. (b) Viscosity with error bars in the same conditions as figure 1a. The plateau gives the effective kinematic shear viscosity.

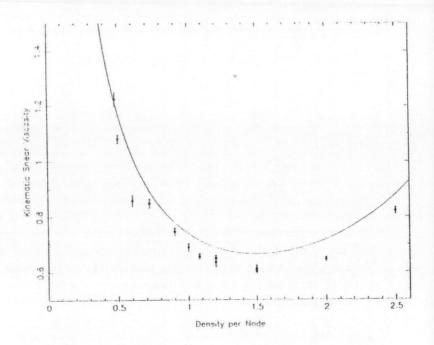

Figure 2: Density dependence of viscosities for the model FHP-I. Black circles with error bars: Monte-Carlo simulations. Black stars with error bars: shear wave simulations of d'Humières and Lallemand [13]. Continuous line: Lattice Boltzmann approximation.

$$\nu = -\frac{1}{8} + \frac{1}{2\rho(6-\rho)^3}. \tag{4.1}$$

We see that the Boltzmann results are 10 to 30 percent higher than the Green-Kubo values. The Boltzmann approximation for the viscosity implicitly assumes an exponentially decaying correlation function. It is of interest to see how well the Monte-Carlo values for the correlation function fit an exponential. The dashed line on figure 1a is an exponential obtained by a least square fit on the low noise data for $\tau_* = 0, 1, 2, 3, 4$. If we now assume that the exponential behavior holds all the way to infinity and sum the geometric series, we obtain a value of the viscosity which lies within a few percent of the Boltzmann value.

The correlation function $\Gamma(\tau_*)$ cannot be exponential all the way to infinity; theoretical arguments on "long time tails" predict that the viscosity for very long times and very large systems is logarithmically divergent [6,7,16,17]. There have been various attempts to find long time tails in lattice gas correlation functions [7,18]. To unambiguously reveal long time tails in simulations of the FHP model is definitely beyond the scope of the present work.

Acknowledgments

I am grateful to U. Frisch and M. Hénon for useful discussions. Computations were done on an IBM 3081. This work was supported by an EC grant ST2J-0190-3F.

References

[1] A. Einstein, *Ann. Phys.*, **17** (1905) 549, **19** (1906) 371.

[2] H. S. Green, *J. Chem. Phys.*, **20** (1952) 1281, **22** (1954) 398.

[3] H. S. Green, *J. Math. Phys.*, **2** (1961) 344.

[4] R. Kubo, *J. Phys. Soc. Jpn.*, **12** (1957) 570.

[5] U. Frisch, B. Hasslacher, and Y. Pomeau, *Phys. Rev. Lett.*, **56** (1986) 1505.

[6] U. Frisch, B. Hasslacher, D. d'Humières, P. Lallemand, Y. Pomeau, and J. P. Rivet, "Lattice gas hydrodynamics in two and three dimensions", *Complex Systems*, **1** (1987) 648.

[7] J. Hardy, Y. Pomeau, and O. de Pazzis, *Phys. Rev.*, **A13** (1976) 1949.

[8] F. Hayot, "Viscosity in lattice gas automata", *Physica D*, in press.

[9] L. Kadanoff, G. McNamara, and G. Zanetti, "Size-dependence of the shear viscosity for a two-dimensional lattice gas", preprint Univ. Chicago (1987).

[10] M. Hénon, "Viscosity of a lattice gas", *Complex Systems*, **1** (1987) 762.

[11] U. Frisch and J. P. Rivet, *Comptes Rendus Acad. Sci. Paris*, **303** (1986) 1065.

[12] D. d'Humières, P. Lallemand, and T. Shimomura, "An experimental study of lattice gas hydrodynamics", Los Alamos Report LA-UR-85-4051 (1985).

[13] D. d'Humières and P. Lallemand, *Physica*, **140A** (1986) 337.

[14] D. d'Humières and P. Lallemand, "Numerical simulations of hydrodynamics with lattice gas automata in two dimensions", *Complex Systems*, **1** (1987) 598.

[15] J. J. Erpenbeck, *Phys. Rev.*, **A35** (1986) 218.

[16] B. J. Alder and T. E. Wainwright, *Phys. Rev.*, **A1** (1970) 18.

[17] Y. Pomeau and P. Résibois, *Phys. Rep.*, **19C** (1975) 64.

[18] N. Margolus, T. Toffoli, and G. Vichniac, *Phys. Rev. Lett.*, **56** (1986) 1694.

P. Clavin,† D. d'Humières,‡ P. Lallemand,‡ and Y. Pomeau‡

†Combustion Research Laboratory, St. Jerome Center, University of Provence, 13397 Marseille, Cedex 13, France and ‡Physics Laboratory, École Normale Supérieure, 75231 Paris, Cedex 05, France

Cellular Automata for Hydrodynamics with Free Boundaries in Two and Three Dimensions

This paper originally appeared in *Comptes Rendus de l'Académie des Sciences de Paris* 303, II (1986), p. 1169; it was translated by Marie Hasslacher.

Cellular automata are used to simulate two-dimensional hydrodynamic flows with free boundaries as found in the Rayleigh-Taylor instability. We propose an extension of these rules for three-dimensional flows with free boundaries.

INTRODUCTION

It was recently shown[1] that the microscopic description of fluids, currently called *molecular dynamics*, could be radically simplified by considering a *lattice gas*. This is an ensemble of point particles moving at a constant speed along the links of a regular lattice with collision rules satisfying microscopic conservation laws.

Allowing an extremely fast numerical treatment that is well adapted to the *cellular automaton* technique, this method provides a simple way of calculating viscous

2-D flows for moderate Reynolds numbers.[2,3,4] It allows one to obtain nonstationary flow fields comparable to those obtained through the most refined traditional numerical methods. Since the first attempts, a notable gain in Reynolds numbers has been obtained by decreasing the viscosity.[5] However, since "particles" essentially behave like hard spheres with zero volume, the equation of state of the models considered was too simple to get the analog of a phase separation, and consequently a flow with a free surface. Here, we propose a method of creating stable interfaces between different thermodynamical phases, a method that will be used to simulate the Rayleigh-Taylor instability in 2-D. We will finally show how to generalize this method to 3-D by starting from d'Humières et al.'s model .[6]

FREE BOUNDARIES

The automaton equation of state proposed by Pomeau et al. [1] is: $P = \alpha\rho$, where α is 1/2 for the 6-bit automata (simple hexagonal lattice) or 3/7 for the 7-bit automaton (center hexagonal lattice), where P is the pressure, and ρ the average density. This equation of state can be extended for example, by adding two bits to code the number of zero speed particles.

$$\rho = \frac{7P}{3} + \frac{2P^2}{9 - 6P + 2P^2} \ .$$

The gas compressibility, however, always remains positive, and it doesn't seem that variations could lead to a first-order phase transition, or to the existence of free surfaces.

We propose using two types of particles, A and B, moving at the same speed on a same-hexagonal lattice (planar case) with centers. This will lead to a 14-bit automaton, considering the Boolean character of the particles used in the lattice gas models.

The collision rules are derived from the following principles:

Collisions A–A or B–B are similar to those of the automaton proposed by Pomeau et al.[1] For collisions A–B, we apply either elastic collisions where the number of each type of particle is conserved, or reactive collisions where only the total number of particles is conserved (with conservation of total momentum).

If we only consider elastic collisions, the ensemble of both fluids satisfies, at the macroscopic level, equations of the Navier-Stokes type with an isotropic term for molecular diffusion $D \, \Delta C$, where C is the relative concentration of types of particles, and D the diffusion coefficient. This can easily be checked experimentally as shown in Clavin et al.[7]

The boundaries arise spontaneously through the introduction of autocatalytic reactive collisions on the lattice sites:

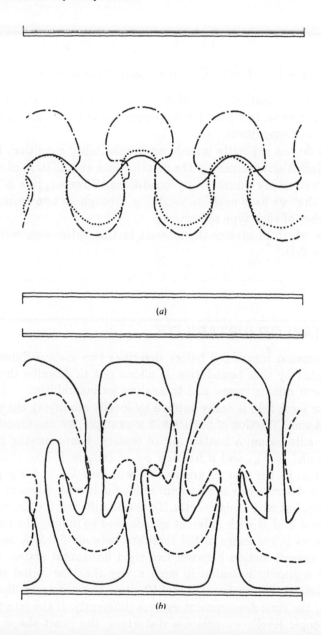

FIGURE 1   Series of shapes of the interface between two fluids A and B in the situation of Rayleigh-Taylor instability. Gravity is applied to fluid A that is initially at the top. The initial interface was taken as a sinusoid of wavelength 1/3 the width of the system with periodic boundary conditions in the horizontal direction. (a) at times t=200 (——), t=800 (----) and t=1,200 (–.–.). (b) at times t=1,600 (----) and t=2,400 (——).

$$A + A + B \xrightarrow{k_1} 3A \text{ with a reaction constant}, k_1 .$$

$$A + B + B \xrightarrow{k_2} 3B \text{ with a reaction constant}, k_2 .$$

When $k_1 = k_2 = k$, the stationary interface thickness is proportional to $\sqrt{D/k}$, i.e., of the order of the mean free-path for frequencies of reactive and elastic collisions of the same order of magnitude.

These rules do not explicitly include an immiscibility condition, but we could imagine in a detailed arrangement of the creation and annihilation of particles near the interfaces that satisfy immiscibility conditions. However, this is not essential since we notice that we have near-immiscibility through an approximate conservation of the number of each type of particle.

An example which illustrates the interest in this automaton will be analyzed in more detail in 2-D.[7]

RAYLEIGH-TAYLOR INSTABILITY

The 14-bit automaton mentioned before describes two viscous fluids, almost immiscible, separated by thin boundaries. It allows one to describe the evolution of drops, to represent mixing layers, and to analyze jet instabilities.

Any exterior force field is easily applied by locally modifying the orientation of the velocity of a small fraction of particles. For example, we can transform, at each iteration of the automaton, a fraction ρ_A of centers A into moving particles with velocity c_A parallel to F_A, and a fraction ρ_B of centers B into moving particles c_B parallel to F_B. In the case of gravity, we will take F_A and F_B parallel to g. This allows us to describe two almost immiscible fluids with different densities and subject to gravity, but with equal inertia. To obtain different inertia, we would need to take particles A and B with different speeds, and to modify the collision rules.

Figure 1 shows a few forms of the interface between fluids A and B, initially set up without overall motion between two rigid horizontal plates. At the initial time, we apply a gravity potential in such a way that the initial state becomes unstable. The lattice we use has 512×512 sites. Depending on the initial form of the interface, the time development evolves differently. If the interface is planar (at the macroscopic level), we notice a distortion, the exact shape of which depends on the initial microscopic conditions. On the other hand, if we start with a sinusoidal interface, with a wavelength 512/K (we assume periodic boundary conditions in the horizontal direction), the distortion amplitude increases exponentially, with a growth rate proportional to $\sqrt{gK}$, for low values of K. After some time, the distortions are no longer harmonic, and we get the typical forms observed in experiments.[8] Figure 2 shows velocity fields of A or B particle flows at the same time.

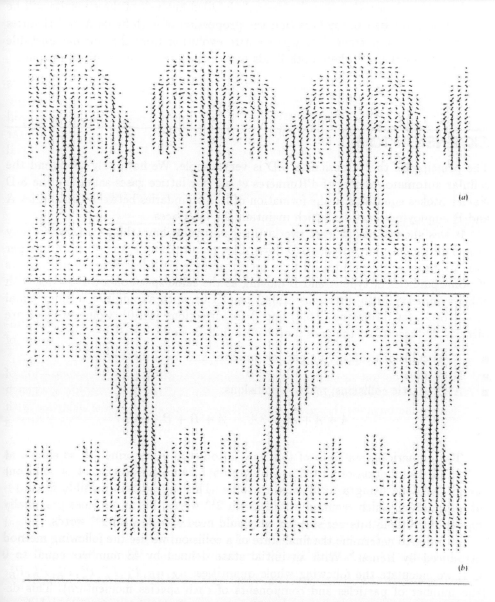

FIGURE 2 Maps of particle fluxes at time t=1,600. (a) case of fluid A; (b) case of fluid B.

These few results show that this automaton allows us to calculate flows with a free boundary, in a rather complex case. A detailed analysis of these results will be shown later, but we can say here that the proportion of both fluids A and B varies at most by a few percent during the entire evolution from the original unstable state to the final state where both fluids have exchanged their positions in space.

EXTENSION TO 3-D

The principle of the extension to 3-D is very simple. We have only to extend the cellular automaton model of d'Humières et al.[6] for lattice gases satisfying the 3-D Navier-Stokes equations to the formation of free boundaries between two species A and B employing reactions which maintain the interfaces.

It was shown that at least two lattice gas models have the necessary isotropy properties to calculate viscous 3-D flows. One require 19 bits, the other 24 bits per site. We propose to consider a 2×24 bit automaton that describes the movements of point particles of type A and B moving at the same speed on a 3-D lattice with cubic symmetry, with directions $(1,0,0)$ or $(1,1,0)$, with an additional index equal to $+1$ or -1 for directions $(1,0,0)$ and 0 for directions $(1,1,0)$. Collision rules are directly deduced from the ones used before in 2-D, i.e.;

- A–A: collisions conserving particle number and total momentum.
- B–B: same.
- A–B: elastic collisions; reactive collisions:

$$A + A + B \to 3A, \quad A + B + B \to 3B .$$

The numerical treatment of this new automaton is, in principle, as simple as in 2-D for the motion of particles. On the other hand, the treatment of collisions cannot be done using a fixed look-up table. This indeed was possible for a 14-bit automaton, which required a table with 2^{14} words, but it becomes practically impossible with 48 bits per site, for we would need a table with 2^{48} words.

In order to determine the final state of a collision, we use the following method introduced by Hénon.[9] With an initial state defined by 48 numbers equal to 0 or 1, we calculate the following whole quantities: $n_A, n_B, P_A^x, P_A^y, P_A^z, P_B^x, P_B^y, P_B^z$ (the number of particles and components of each species momentum). This defines the class of the initial states. Then, we only need another state belonging either to the same class, or to another class defined by the same values of $n_A, n_B, P_A^x + P_B^x, P_A^y + P_B^y, P_A^z + P_B^z$ (the case of elastic collisions), or of $n_A + n_B, P_A^x + P_B^x, P_A^y + P_B^y, P_A^z + P_B^z$ (the case of reactive collisions).

This does not require a complicated algorithm. However, we must make sure that on average, there is no bias in the selection of the various space directions, as it would be the same as applying an external field. For some values of quantities $n_A, n_B, P_A^x, P_A^y, \ldots$, there exist several equivalent configurations among which we can arbitrarily choose. Using all of the symmetries present in the problem will probably allow us to solve this difficulty.

CONCLUSION

We described a way to simulate viscous flows of two fluids practically immiscible and separated by thin free boundaries. The utility of this model was qualitatively illustrated by presenting a few simulations of the time development of the Rayleigh-Taylor instability in 2-D. We described the extension of the model to 3-D and showed what problems would arise for numerical implementation of the model to calculate 3-D viscous flows with free boundaries. The inherent problems seem minor compared to the advantages of the technique of flow calculation by the lattice gas method, if we consider applying an external field and how easily we can include arbitrarily shaped obstacles.

REFERENCES

1. Frisch, U., B. Hasslacher, and Y. Pomeau. *Phys. Rev. Lett.* **56** (1986):1505–1508.
2. d'Humières, D., Y. Pomeau, and P. Lallemand. *Comptes Rendus* **301 II** (1985):1391–1396.
3. Salem, P., and S. Wolfram. In *Theory and Applications of Cellular Automata*. World Scientific, 1986.
4. Doolen, G., and U. Frisch, private communication.
5. d'Humières, D., and P. Lallemand. *Physica* **140 A** (1986):326–335.
6. d'Humières, D., P. Lallemand, and U. Frisch. *Europhysics Letters* **2** (1986): 291–297.
7. Clavin, P., P. Lallemand, Y. Pomeau, and G. Searby. *J. Fluid Mech.* **188** (1986):437–469.
8. Lord Rayleigh. *Scientific Papers* **II** (1900):200–207; G. I. Taylor. *Proc. Roy. Soc. London* **A 201** (1950):192–196.
9. Hénon, M. *Complex Systems* **1** (1987):763–789.

D. d'Humières, Y. Pomeau, and P. Lallemand
Physics Laboratory, École Normale Supérieure, 75231 Paris Cedex 05, France

Simulation of 2-D Von Karman Streets using a Lattice Gas

This paper originally appeared in *Comptes Rendus de l'Académie des Sciences de Paris* 301, II (1985), p. 1391; it was translated by Marie Hasslacher.

A lattice gas model is introduced in which particles travel along the links of a planar triangular lattice, with constant speed and synchronized to collide at the vertices of the lattice. The macroscopic hydrodynamic behavior of the gas is demonstrated for channel flow around a plate perpendicular to the mean flow velocity, leading to the creation of nonstationary eddies.

INTRODUCTION

A common simplification in the study of disordered systems[1,4] is the restriction of the position of the system's particles to the sites of a regular lattice with adequate symmetry. This quantification simplifies the description of microscopic systems while conserving system properties on a large scale. This method is widely used to calculate the statistical properties of disordered systems such as spin glass phases,[1] structure sizes in metallic alloys,[2] percolation in a porous medium,[3] or the size of "clusters" in aggregation phenomena.[4]

In the same way, the quantization of space directions has also been used in the kinetic theory of gases and the Boltzmann equation.[5] It has been shown that a gas whose particles move with discrete speeds, will satisfy the Navier-Stokes equation on a large scale and at long times, when all allowed collisions obey momentum and energy conservation.

An early attempt to quantize both space coordinates and velocity directions simultaneously was made by Hardy, de Pazzis and Pomeau,[6] by using a gas of particles moving on a square lattice. Only four velocities of same modulus and oriented along the axes of the square lattice are allowed in this model, and there is at most one particle per site in each allowed direction. The movement of all particles is simultaneous so they periodically meet at the lattice nodes where allowed collisions occur.

Each time step can be further split into two stages: a "collision" stage where particles modify their speed according to rules defined by a number of reversible and deterministic collision rules conserving both particle number and momentum (and therefore energy since all speeds have the same modulus), and a "propagation" stage in which particles travel toward the nearest site in the direction defined by their new velocity. Due to the symmetries of the square lattice, the hydrodynamical equations of this model show qualitative differences from Navier-Stokes equations. Also, since there are only two collision rules, spurious conservation properties lead to pathological unphysical behavior.

THE PRINCIPLE OF THE METHOD

We introduce results obtained with a similar model, whose underlying lattice is composed of equilateral triangles, a lattice free of the problems mentioned above.[7] On this lattice, six directions with indexes, $i \in \{1, \ldots, 6\}$, converge toward each node, i.e., six possible incident particles represented by the corresponding index i modulo 6. Also, we introduce a population of rest particles represented at the given node by index, k. The properties of this gas are completely defined by the following collision rules between particles: $(i, i+3) \rightarrow (i+1, i+4)$ or $(i-1, i+2)$ head-on collisions, $(i, k) \rightarrow (i+2, i-4)$ collisions on a rest particle, $(i+2, i+4) \rightarrow (i, k)$ collisions at 120°, and $(i, i+2, i+4) \rightarrow (i+1, i+3, i+5)$ collisions between 3 particles. A theoretical analysis of the system shows that the average quantities ρ and $\mathbf{j} = \rho\mathbf{v}$ satisfy the continuity and Navier-Stokes equations (2-D in this case) in their usual form.[7]

Theoretically, the presence of a fixed lattice breaks Galilean invariance, which appears as a distortion of the Navier-Stokes equation. However, these distortions remain small when the average speeds are low. (We note that this problem is common to all numerical methods used to solve the Navier-Stokes equation that use discretization of time and space.) Furthermore, since local equilibrium distribution is determined by only two parameters ρ and $\rho \mathbf{v}$, there is no equivalent of temperature which can be written so: $\gamma = C_p/C_v = 1$. Linear hydrodynamical properties were studied elsewhere by numerical simulation.[8] These simulations confirmed the theoretical sound speed, $(\sqrt{3/7})$, and its isotropy, and measured shear and bulk viscosities as a function of density.

IMPLEMENTATION

Here we describe results obtained for flow around a plate. We used two rules for particle-wall interactions: the particles underwent either specular reflections or speed reversal. The simulations show no qualitative difference between either choice.

The simulations described here were done on an FPS 164 array processor, whose 64-bit words were used to store 8 sites coded with 8 bits (6 directions, 1 center and 1 bit to select one of the two types of direct collisions). During the "collision" stage, the eight state bits of a site are used as an address to look up their new values in a 256-entry table containing the collision rules. The "propagation" stage is then implemented using an appropriate shifting of the 6 state bits corresponding to the six possible directions. Since these simulations are done on a finite size lattice (M × N sites, M= 1024, N= 512), we must define the boundary conditions on the edge of the lattice. The results described below were obtained with periodic boundary conditions in the vertical direction, while particles were injected from the left edge with a constant average density ρ_0 and a constant average speed $\mathbf{v}_0$. Particles reaching the right edge were absorbed. We initialize a gas flow with average speed $\mathbf{v}_0$. We must then allow time for the flow to settle down. Then separation of the flow occurs in front of the plate and eddies appear at the right and left edges of the plate. When the speed $\mathbf{v}_0$ is high enough (for the densities used, the Reynolds number is about 144 $\| \mathbf{v}_0 \|$), we get alternating detachment of eddies created by the plate, which then flow downstream.

RESULTS AND CONCLUSION

In order to reduce statistical fluctuations inherent in the small number of particles per site, we average over several sites, restricting the finest scales we can study. To visualize the flow, the lattice is subdivided into cells of 16 × 16 sites from which we calculate the corresponding average flux $\langle \mathbf{j} \rangle$; this local average flux is represented by a proportional vector for each of the cells covering the lattice. Figures 1 and 2 show five eddies for a velocity, $\mathbf{v}_0 = 0.5$ (Re $\simeq$ 70) at times $t_1 = 5000$ and $t_2 = 5500$. The comparison of both velocity fields shows a change in the sense of rotation of the large eddy located behind the plate, clearly illustrating the unstable character of the flow. (The corresponding Strouhal number is approximately 0.2.)

The qualitative results shown here will be complemented in future work by quantitative measurements of flow properties, calculated according to the local populations of the seven degrees of freedom mentioned earlier.

The study of the dynamics of the hexagonal lattice gas allows one to calculate flows for Reynolds numbers around 10^2 (increasing linearly with the size of the lattice), with a resolution of about 10^{-2} times the size of the system. The method we briefly described here is interesting for two reasons:

1. the extremely simple character of the operations implemented at each site of the lattice (looking up a table and exchanging bits between nearest neighbors) allows one to consider, in the very near future, building a specialized computer with a very high degree of parallelism (as high as one processor per site), allowing very high computation speeds for the solution of Navier-Stokes equations, and

2. the microscopic nature of collisions with the walls allows one to place plates of any shape into the flow without much difficulty (contrary to the case of finite element methods and, *a fortiori* spectral methods).

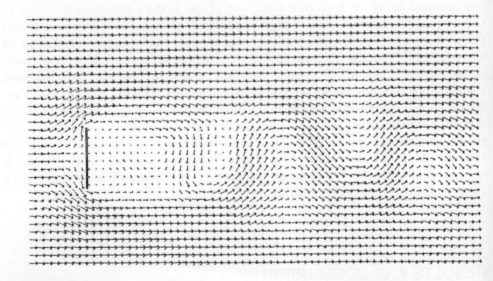

FIGURE 1   Two-dimensional flow around a flat plate or a Reynolds number of order 70 after 5,000 time steps. The direction and modulus of arrows are proportional to the mass flux on a 64 × 32 grid.

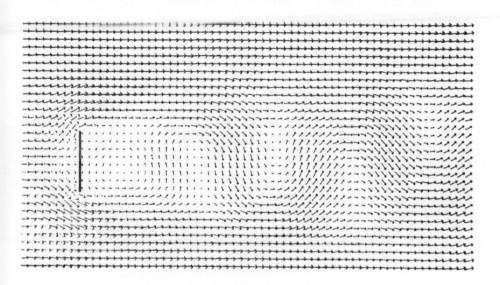

FIGURE 2   Flow under same conditions as in Figure 1 1,500 time steps later.

REFERENCES

1. van Hemmen, J. L., and I. Morgenstern, eds. "Heidelberg Colloquium on Spin Glasses." *Lecture Notes in Physics*. Berlin: Springer-Verlag, 1983.
2. Mott, N. F., and H. Jones. *The Theory of the Properties of Metals and Alloys*. New York: Dover, 1958, 30.
3. Deutcher, G., R. Zallen, and J. Adler, eds. *Percolation Structures and Processes*. Bristol: Hilger, 1983.
4. Meakin, P. *Phys. Rev.* **A 27** (1983):1495–1507.
5. Gatignol, R. "Théorie Cinétique des Gaz à Répartition Discrète de Vitesses." *Lecture Notes in Physics*. Berlin: Springer-Verlag, 1975, and references.
6. Hardy, J., O. de Pazzis, and Y. Pomeau. *Phys. Rev.* **A 13** (1976):1949–1961.
7. Frisch, U., B. Hasslacher, and Y. Pomeau. *Phys. Rev. Lett.* **56** (1986):1505–1508 .
8. d'Humières, D., P. Lallemand, and T. Shimomura. *Los Alamos report LA-UR-85-4051*, 1985.

D. d'Humières and P. Lallemand
Physics Laboratory, École Normale Supérieure, 75231 Paris Cedex 05, France

Flow of a Lattice Gas between Two Parallel Plates: Development of the Poiseuille Profile

This paper originally appeared in *Comptes Rendus de l'Académie des Sciences de Paris* **302**, II (1986), p. 983; it has been translated by Marie Hasslacher.

After recalling briefly the use of a lattice gas to simulate aerodynamical flows, we describe an experimental situation corresponding to a 2-D flow between two parallel plates. A series of velocity distributions shows the progressive deformation from a Blasius profile for a boundary layer on a flat plate to a parabolic Poiseuille distribution. The comparison of the present results to those computed by Slichting shows the excellent quality of the predictions that can be obtained with the new simulation method proposed by Frisch, Hasslacher, and Pomeau.

INTRODUCTION

It was recently shown[1] that particles moving at a constant speed on a planar triangular lattice may exhibit general behavior similar to that of a viscous gas. For this purpose, the rules defining collisions between these particles must be chosen correctly. This gas was theoretically analyzed by Rivet and Frisch[2] using the Chapman-Enskog expansion adapted to the case of discrete speeds[3,4]; they were able to calculate viscosity as a function of model details and density. The dynamics of this gas can be calculated by computer simulation, with extremely simple programs, allowing the treatment of a very large number of particles and sites of the triangular lattice on which they move.

In a previous note,[5] we described an initial qualitative study of a 2-D flow around a linear plate placed normally to the incident flow. We showed that when the Reynolds number is high enough, a nonstationary regime appears, forming eddies. However, the compressible character of the flow makes it difficult to compare quantitatively with the results obtained by more classical methods.

The initial purpose of the work described here was to use the lattice gas method for a well-known flow: the development of a laminar boundary layer along a flat plate parallel to the direction of a uniform incident flow. In fact, and as we show later, the limited size of the lattices used prevented us from treating this problem exactly. Instead we simulated the case of a uniform incident flow along the axis of a flat channel bounded by two parallel faces. We compare our results with the theoretical profiles calculated by Slichting[7] for the development of a Poiseuille profile from the channel entrance. We did not simulate the Blasius profiles.

SUCCINCT DESCRIPTION OF THE CALCULATION

Starting from a lattice formed of N × M sites linked between nearest neighbors according to a regular lattice with hexagonal symmetry, we initially place a maximum of seven particles per site: one particle per direction and one stopped per site. The moving particles have constant speed and move toward their six nearest neighbors where they can undergo collisions. One stopped particle, whose speed is zero, is also allowed at each site. The particles undergo local collisions that conserve the mass and momentum. However, the two boundaries, parallel to the bisectrix of two adjacent sides of the basic triangle, have special rules allowing one to describe the bouncing of particles from planar walls. These two boundaries pass through the middle of the lattice, but because of the periodic boundary conditions in the direction perpendicular to the walls, it is equivalent to work with a channel. Furthermore, the selection of bounce conditions insures that the average speed at each point of the channel walls is zero.

As an initial condition for the simulation, we take a random particle distribution such that the average density is uniform, and the average speed is constant and parallel to the walls. At each iteration, we inject at one end of the lattice particles with the same statistical distribution, and at the other end particles with a Poiseuille distribution. We then let the system evolve for a large number of iterations (of the order of 10,000 for a 512×3072 site lattice), then measure the density and the speed in the entrance zone together with the speed component parallel to the channel walls, by taking the spatial average of the particle flow on a 1×48 site rectangle, located parallel to the plates. In order to decrease statistical noise due to the rather low number of particles located in the averaging rectangle, we took the time average of 48 samples obtained from successive intervals of 20 iterations (no analysis was done to check the statistical independence of the successive samples).

The data we are about to introduce were obtained for a lattice 512 lattice units wide and $3072 \times \sqrt{3}/2$ lattice units long. The walls begin at abscissa $301 \times \sqrt{3}/2$. The density and the speed imposed at the start are respectively 1.4 particle per site and 3/7th of the microscopic speed. Actually, the presence of the walls slows the gas, and the stationary values measured are respectively 1.54 and 0.30, this last value corresponding to Mach number 0.454.

The qualitative behavior of these "experimental" results are *a priori* satisfying: we first check that the speed component parallel to the walls is zero at the surface. Then, we observe a monotonous increase of the tangential speed between the walls and the edges of the lattice—the latter are actually the middle of the channel. As we get farther away from the channel entrance, the speed gradients decrease. At a great distance, we get a parabolic profile. As expected, speed along the channel axis increases to an asymptotic value. This shows that experimental data should not be compared with the classic profiles described by the Blasius theory for a boundary layer in a semi-infinite medium.

COMPARISON WITH SLICHTING THEORY

Slichting[7,8] calculated the stationary distribution of the speed of an incident fluid of constant speed between two flat plates parallel to the incident direction. The calculation is done differently according to whether we are near the entrance, or in the region where the Poiseuille profile applies. We used the development of the flow function used by Slichting.

Near the entrance, the small parameter in the problem is:

$$\varepsilon = \sqrt{\nu_{eff} x / a^2 U_0}$$

where ν_{eff} is the kinematic viscosity divided by a factor $g(\rho)$ due to the Boolean character of the particles used.[1,2] x is the distance to the entrance of the channel, a is half the distance between the walls and U_0 is the speed of the incident flow. We calculated numerically the values of the three first terms in the development of flow function.

Far away from the entrance, the small expansion parameter is:

$$w = \exp(-\lambda_1 \varepsilon^2).$$

where λ_1 is the eigenvalue of the differential equation satisfied by the first correction term of the flow function with a Poiseuille form, an eigenvalue introduced to satisfy the boundary conditions of the problem. We calculated the two first correction terms with Poiseuille formula. The connection between both types of solutions was done for $\varepsilon = 0.016$, which is not particularly good since the parameter w is quite large for this value of ε.

In the figures, we show both the theoretical curves and simulation results. Since the incident speed in the entrance zone is determined by a local measurement of ρ and ρv, the only unknown of the problem left is the kinematic viscosity measured independently by studying shear wave relaxation.[6] Also the factor $g(\rho)$ is 0.419 for our experiment, so that there is no adjustable parameter. For each section perpendicular to the plates, we measured the speed in the central area, and we used this value to establish the scale of the theoretical curve.

Figure 1 shows the speed in a region near the entrance ($x = 134$, i.e., a relative distance $z = x/a$ of 0.525 and $\varepsilon = 0.0674$). Figure 2 shows the speed for $x = 425$($z = 1.67$ and $\varepsilon = 0.120$), at the limit of validity in the expansion in ε. Figure 3 shows the speed for $x = 633$($z = 2.48$ and $\varepsilon = 0.146$) compared to the expansion in w. Figure 4 corresponds to $x = 1464$($z = 5.74$ and $\varepsilon = 0.223$), i.e., to a region where Slichting expansion works.

We notice that the agreement between experimental results and theoretical calculations is very good. Although the entrance Mach number is not negligible, effects of gas compressibility do not matter. The modifications of boundary layers of real gases in fast flows are essentially due to thermal effects that don't exist in the current lattice gas model.

CONCLUSION

This note describes the first example of a quantitative comparison of results of flows simulated by the lattice gas method with predictions obtained by classic methods in fluid mechanics. We studied the case of Poiseuille flow. This case involves both the nonlinear terms and the viscous term of Navier-Stokes equation and is therefore adequate to show possible problems in the model. The present work experimentally confirms the validity of the lattice gas method proposed by Frisch, Hasslacher, and Pomeau.

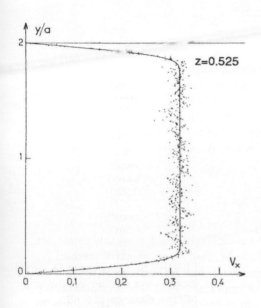

FIGURE 1 Component of the velocity parallel to a flat plate vs. distance in a section of channel located at a relative distance of 0.525 from the entrance. The dots are obtained by the lattice gas method, the solid line is calculated following Slichting.[7]

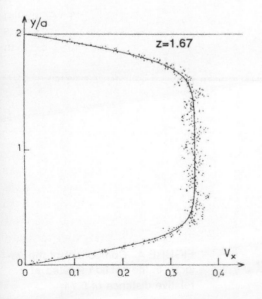

FIGURE 2 Component of the velocity in the section located at a relative distance of 1.67.

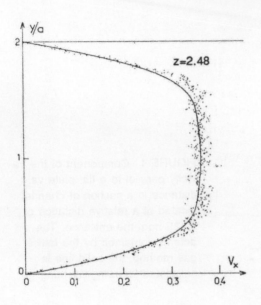

FIGURE 3 Component of the velocity in the section located at a relative distance of 2.48.

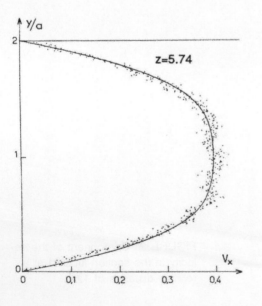

FIGURE 4 Component of the velocity in the section located at a relative distance of 5.74.

REFERENCES

1. Frisch, U., D. Hasslacher, and Y. Pomeau. *Phys. Rev. Lett.* **56** (1986):1505–1508.
2. Rivet, J. P., and U. Frisch *Comptes Rendus* 302 **II** (1986):267–272.
3. Gatignol, R. *Zeitschrift für Flugwissenschaften* **18** (1970):93–97.
4. Gatignol, R. "Théorie Cinétique des Gaz à Répartition Discrète de Vitesses." *Lecture Notes in Physics*, vol. 36. Berlin: Springer Verlag, 1975.
5. d'Humières, D., Y. Pomeau, and P. Lallemand. *Comptes Rendus* **301 II** (1985):1391–1394.
6. d'Humières, D., P. Lallemand, and T. Shimomura. *Los Alamos report LA-UR-85-4051*, 1985.
7. Slichting, H. *Z.A.M.M* **14** (1934):368–373.
8. Slichting, H. *Boundary Layer Theory.* London: Pergamon Press, 1955.

U. Frisch† and J. P. Rivet‡

†CNRS, Observatoire de Nice BP 139, 06003 Nice Cedex, France and ‡École Normale Supérieure, 45 rue d'Ulm, 75230 Paris Cedex 05, France

Lattice Gas Hydrodynamics, Green-Kubo Formula

This paper originally appeared in *Comptes Rendus* **303**, II (1986), p. 1065 and was translated by Maria Hasslacher.

The Green-Kubo formalism, allowing the evaluation of transport coefficients of macroscopic systems in terms of equilibrium correlation functions, is extended to lattice gas hydrodynamics.

INTRODUCTION

In a previous note,[1] we used Boltzmann's formalism with discrete speeds to calculate the shear and bulk viscosities of the lattice hydrodynamics models introduced by Frisch, Hasslacher, and Pomeau (FHP).[2] That approach ignored on the one hand the space and time discretization effects, and on the other, correlations between sites. It was therefore limited in principle to low densities.

In this paper, we show how to include all these effects. Basically, we develop a Green-Kubo-type formalism, adapted to a gas discrete in time, position, and speed (cellular automaton). The discrete-speed case with particles that keep their individuality when colliding, had been studied by Hardy and Pomeau.[4] Here, we will develop this theory for the case of the simplest FHP model (triangular lattice, double and triple collisions, no rest particles).

THE MICRODYNAMICAL EQUATION

We start from an equation for the Boolean field $n_i(t_*, \mathbf{r}_*)$ that takes values 1 (particle present) and 0 (particle absent). Index i, defined modulo six, corresponds to the six speed directions: t_* and $\mathbf{r}_*$ correspond to the lattice discrete time and discrete position vectors. The exact dynamics of the automaton lattice is defined by the Boolean rules of automaton updating, given in ref. 2. By recoding this Boolean logic in an arithmetic form, we reach the microdynamical equation (a similar equation for the so-called HPP model can be found in ref. 5, see also ref. 6).

$$
\begin{aligned}
n_i(t_* + 1, \mathbf{r}_* + \mathbf{c}_i) = & n_i + \xi_{\mathbf{r}*} n_{i+1} n_{i+4} \bar{n}_i \bar{n}_{i+2} \bar{n}_{i+3} \bar{n}_{i+5} \\
& + \bar{\xi}_{\mathbf{r}*} n_{i+2} n_{i+5} \bar{n}_i \bar{n}_{i+1} \bar{n}_{i+3} \bar{n}_{i+4} - n_i n_{i+3} \bar{n}_{i+1} \bar{n}_{i+2} \bar{n}_{i+4} \bar{n}_{i+5} \\
& + n_{i+1} n_{i+3} n_{1+5} \bar{n}_i \bar{n}_{i+2} \bar{n}_{i+4} - n_i n_{i+2} n_{i+4} \bar{n}_{i+1} \bar{n}_{i+3} \bar{n}_{i+5}
\end{aligned}
$$

$$(1)$$

Here, the RHS is evaluated at time t_* and at node $\mathbf{r}_*$; $\bar{n} = 1 - n$ is the negation of n and $\mathbf{c}_i$ is the (unit) speed vector in direction i. In the LHS of Eq. (1), the spatial difference corresponds to the free propagation of particles. The Boolean variable $\xi_{\mathbf{r}}$ selects one of two possible output channels (rotation $\pm\pi/3$) depending, for example, on the parity of the line index of the site, with weights that are chosen equal for simplicity. Later on, we will use a condensed notation: $n.(t_*)$ to describe the Boolean field at time t_*. We will also define $\mathcal{E}$ the inverse evolution operator that gives $n.(t_*)$ when operating on $n.(t_* + 1)$. The microdynamical equation then reads:

$$
\mathcal{E}(n.(t_* + 1)) = n.(t_*) \ .
$$

$$(2)$$

An important aspect of the lattice gas models is that the evolution operator globally conserves the particle number $\sum_i n_i$ and the momentum $\sum_i \mathbf{c}_i n_i$. The collision operator defined by the RHS in (1) conserves number and momentum at each site.

THE LIOUVILLE EQUATION AND GLOBAL EQUILIBRIUM SOLUTIONS

In order to do the statistical mechanics of the automaton lattice, we assume an ensemble Γ of Boolean fields. We write $P(t_*, m.)$ for the probability of having at time t_* a *configuration* $m.$ of the Boolean field $n.(t_*)$. This satisfies a Liouville equation, expressing probability conservation:

$$P(t_* + 1, m.) = P(t_*, \mathcal{E}(m.)) \ , \qquad \forall m. \in \Gamma \ . \tag{3}$$

We verify that Eq. (3) admits *global equilibrium* stationary solutions, for which P is a product on all sites and directions of the $N_i^{(eq)}(\rho, \mathbf{u}) m + (1 - N_i^{(eq)}(\rho, \mathbf{u})) \overline{m}$ form factors. The $N_i^{(eq)}$ equilibrium populations are given by a Fermi-Dirac distribution,[2] of which we will use only a first-order expansion with respect to $\mathbf{u}$.

Starting from P, we get populations N_i (ensemble averages of the n_i Boolean variables) given by:

$$N_i(t_*, \mathbf{r}_*) = \sum_{m.} m_i(\mathbf{r}_*) P(t, m.) \ . \tag{4}$$

The conservation of the particle number and of the momentum by the collision operator implies *exact* relations for populations:

$$\sum_i N_i(t_* + 1, \mathbf{r}_* + \mathbf{c}_i) = \sum_i N_i(t_*, \mathbf{r}_*)$$

and

$$\sum_i \mathbf{c}_i N_i(t_* + 1, \mathbf{r}_* + \mathbf{c}_i) = \sum_i \mathbf{c}_i N_i(t_*, \mathbf{r}_*) \ . \tag{5}$$

that will help us deduce the hydrodynamical equations.

Starting from the N_i, we define local hydrodynamical quantities (at point $\mathbf{r}_*$ at time t_*), density, $\rho = \sum_i N_i$, and momentum, $\rho \mathbf{u} = \sum_i \mathbf{c}_i N_i$; $\mathbf{u}$ is the local speed.

MULTISCALE DEVELOPMENT FOR HYDRODYNAMICS

Lattice gas hydrodynamics describes the re-gluing on very large lattices of *local* equilibria with parameters ρ and $\mathbf{u}$ slowly varying in space and time. For the derivation of transport coefficients, we will only use low speed regimes with negligible nonlinearities. This derivation uses a multiscale method: ρ and $\mathbf{u}$ depend on the slow spatial variable $\mathbf{r}_1 = \varepsilon \mathbf{r}_*$; this dependence is assumed to be regular enough to allow interpolations between lattice points and therefore Taylor expansions. Physically, we expect: (1) a relaxation toward an equilibrium on the ε^0 time scale; (2) sound waves propagating on time scales ε^{-1}, and (3) momentum diffusion on time scales ε^{-2}. We therefore use a formalism with three times: t_* (discrete), $t_1 = \varepsilon t_*$ and $t_2 = \varepsilon^2 t_*$. We can now apply to the lattice gas case the method developed by Green[3] for the Liouville equation of a system of interacting classical particles. The idea is to expand the populations and the probability P in powers of ε (dependences on the various time and/or space variables are assumed).

$$N_i = N_i^{(0)} + \varepsilon N_i^{(1)} + \varepsilon^2 N_i^{(2)} + O(\varepsilon^3) \ , \tag{6}$$

$$P(m.) = P^{(0)}(m.) + \varepsilon P^{(1)}(m.) + \varepsilon^2 P^{(2)}(m.) + O(\varepsilon^3) \ . \tag{7}$$

$N^{(0)}$ is the equilibrium distribution,[2] based on the *local* values of ρ and $\mathbf{u}$ and, similarly, $P^{(0)}(m.)$ is the local equilibrium distribution. We now outline the strategy of the perturbation calculation. Starting from conservation relations (5), we use form (6) for the N_i's and we expand in a Taylor series to second order in ε (with $\partial_t \rightarrow \varepsilon\partial_{t_1} + \varepsilon^2\partial_{t_2}$ and $\partial_{\mathbf{r}} \rightarrow \varepsilon\partial_{\mathbf{r}_1}$). The partial derivative with respect to component α of $\mathbf{r}_1$ is written c_α. We get at order ε:

$$\partial_{t_1}\rho + \partial_\beta(\rho u_\beta) = 0, \qquad \partial_{t_1}(\rho u_\alpha) + \partial_\alpha\left(\frac{\rho}{2}\right) = O(u^2) \ , \tag{8}$$

and at order ε^2:

$$\partial_{t_2}\rho = 0 \ ,$$

$$\partial(\rho u_\alpha) = -\partial_\beta \sum_i c_{i\alpha} c_{i\beta} N_i^{(1)} + \left(\frac{1}{4}\delta_{\beta\delta}\delta_{\alpha\gamma} - \frac{1}{6}\sum_i c_{i\alpha} c_{i\beta} c_{i\gamma} c_{i\delta}\right)\partial_\beta\partial_\gamma(\rho u_\delta). \tag{9}$$

The calculation of $P^{(1)}(m.)$ (which fixes $N_i^{(1)}$) is done as in [3], by noticing that $P^{(1)}(m.)$ satisfies a Liouville equation with a second term (with a few changes due to discrete variables; see ref. 7). Eq. (9) for momentum diffusion then reads:

$$\partial_{t_2}(\rho u_\alpha) = \left(S_{\alpha\beta\gamma\delta}^{coll} + S_{\alpha\beta\gamma\delta}^{prop}\right)\partial_\beta\partial_\gamma(\rho u_\delta) \ . \tag{10}$$

The tensor, $S_{\alpha\beta\gamma\delta}^{prop}$, comes only from free particle propagation: it has a negligible effect at low density that comes from the discrete character of space in this model. This effect is not seen in real gases and doesn't exist in models where only speeds are discrete. The tensor $S_{\alpha\beta\gamma\delta}^{coll}$ represents the collision contribution and is expressed through the correlation functions "at equilibrium."[7] These fourth-order tensors are obviously symmetrical in $\alpha\beta$ and $\gamma\delta$; since the lattice and the collision rules are invariant under the (hexagonal) group of multiple angle rotations of $\pi/3$. These tensors are isotropic. Eq. (10) then takes the simpler vector form:

$$\partial_{t_2}(\rho\mathbf{u}) = \left(\nu^{coll} + \nu^{prop}\right)\Delta(\rho\mathbf{u}) + \left(\mu^{coll} + \mu^{prop}\right)\nabla(\nabla \cdot \rho\mathbf{u}) \tag{11}$$

with

$$\nu^{coll} = \frac{3}{\rho(6-\rho)}\sum_{\alpha,\beta}\sum_{\tau=0}^{t_*-1}\sum_{\rho_*}\sum_{i,j} Q_{i\alpha\beta}\left\langle\left(n_j(\tau,\rho_*) - \frac{\rho}{6}\right)\left(n_j(0,0) - \frac{\rho}{6}\right)\right\rangle Q_{j\alpha\beta}$$

and

$$\nu^{prop} = -\frac{1}{8} \quad \text{and} \quad \mu^{coll} = \mu^{prop} = 0 \ .$$

The brackets $\langle\rangle$ mean an average over the *global* equilibrium of density ρ and speed zero. The $Q_{i\alpha\beta}$ *microscopic constraint tensor* is given by $Q_{i\alpha\beta} = c_{i\alpha}c_{i\beta} - (1/2)\delta_{\alpha\beta}$ (FHP expression). We note that, in a similar 3-D formalism, the summation on τ could be extended to infinity. In 2-D, there is a divergence that appears only for very long times and infinite lattices (as in the HPP model, see ref. [8]), because two is a crossover dimension. These problems, which are not peculiar to lattice gases, will not be studied further here.

That the bulk collision viscosity is zero is a consequence of the zero trace of the microscopic constraint tensor.

The formalism can be considerably simplified if we only consider the *lattice Boltzmann* approximation, and ignore the correlations of particles participating in collisions. This does not change the propagation speeds, and for the dynamical collision viscosities it leads to explicit expressions that were already derived in our previous note.[1]

We note that the propagation viscosity ν^{prop} is *negative*; the free propagation of particles at local equilibrium does tend to accentuate homogeneities. The presence of this negative contribution was noticed first by Henon[9] who calculated the shear viscosity of the FHP model by looking for static solutions of the lattice Boltzmann equation.

Our Green-Kubo formalism extends easily to other lattice gases. If we add in the FHP model binary collisions with zero speed particles (center hexagonal model), we find $v^{prop} = -1/8$ and $\mu^{prop} = -1/28$. For the 3-D model called "pseudo-4-D,"[10] we find $v^{prop} = -1/6$. The inclusion of propagation terms considerably reduces the difference at finite density between viscosities measured by simulations[11] and calculations in the Boltzmann approximation.

REFERENCES

1. Rivet, J. P., and U. Frisch. "Automates sur Gaz de Réseau dans l'Approximation de Boltzmann." *Comptes Rendus* **302(II)** (1986):267–272.
2. Frisch, U., B. Hasslacher, and Y. Pomeau. "Lattice Gas Automata for the Navier Stokes Equation." *Phys. Rev. Lett.* **56** (1986):1505–1508.
3. Green, H. S. "Theories of Transport in Fluids." *J. Math. Phys.* **2** (1961):344–349.
4. Hardy, J., and Y. Pomeau. "Thermodynamics and Hydrodynamics for a Modeled Fluid." *J. Math. Phys.* **13** (1972):1042–1051.
5. Hardy, J., Y. Pomeau, and O. de Pazzis. "Time Evolution of a Two-Dimensional Model System. I. Invariant States and Time Correlation Functions." *J. Math. Phys.* **14** (1973):1746–1759.
6. Pomeau, Y. "Invariant in Cellular Automata." *J. Phys.* **A 17** (1984):L415–L418.
7. Rivet, J. P. "Hydrodynamique sur Réseau d'Automates." *Nice Observatory report*, 1986.
8. Hardy, J., O. de Pazzis, and Y. Pomeau. "Molecular Dynamics of a Classical Lattice Gas: Transport Properties and Time Correlation Functions." *Phys. Rev.* **A 13** (1986):1949–1961.
9. Hénon, M. "Calcul de la Viscosité dans le Réseau Triangulaire." *Nice Observatory internal report*, 1986.
10. d'Humières, D., P. Lallemand, and U. Frisch. "Lattice Gas Models for 3-D Hydrodynamics." *Europhys. Lett.* **2**, 291–297.
11. D. d'Humières, P. Lallemand, and T. Shimomura (1986), "An Experimental Study of Lattice Gas Hydrodynamics." *Los Alamos report LA-UR-85-4051*, 1986.

J. P. Rivet† and U. Frisch‡

†École Normale Supérieure, 45 rue d'Ulm, 75230 Paris Cedex 05, France and ‡Centre National de la Recherche Scientifique, Observatoire de Nice, 06003 Nice Cedex, France

Lattice Gas Automata in the Boltzmann Approximation

This paper originally appeared in *Comptes Rendus* **302**, II (1986), p. 267, and was translated by Maria Hasslacher.

Shear and bulk viscosities are determined for two lattice gas automata simulating the two-dimensional Navier-Stokes equations.

INTRODUCTION

Lattice gases introduced by Hardy, de Pazzis, and Pomeau,[1,2] are attracting interest again for their applications to numerical fluid mechanics, and for the possibility of creating massively parallel, specialized machines.[3,4,5]

We recall that the currently known lattice gas models that lead to the 2-D Navier-Stokes equations are formed of "Boolean molecules," with speed zero or unity, moving on a plane with a triangular grid. Lattice gases differ from kinetic theory models with discrete speeds (see monograph[6]) on two points: the discrete character of the positions and an exclusion principle (no more than one particle per site and per direction). The *simple hexagonal model* includes six possible speed vectors with a unit modulus, and the collision rules shown in Figure 1a. In this model, triple collisions are essential in order to avoid a spurious conservation law. The *center hexagonal* model also allows "rest" particles on the lattice sites with additional collision rules shown in Figure 1b. In this case, triple collisions are not mandatory. Numerical experiments[4] show that Reynolds numbers measured on the *center hexagonal* model are about twice as high as those measured on the *simple hexagonal* model. Here, we will compute the bulk and shear viscosities of these lattice gases in the Boltzmann approximation. This approximation has validity only at low density, but we may skip this restriction.

BOLTZMANN'S EQUATION

We derive this method only for the *simple hexagonal* case, since it is easily generalized to the *center hexagonal* case. We write $C_i, (i = 1, \ldots, 6)$ for the six possible unit vectors and $N_i(\mathbf{r}, t), i = 1, \ldots, 6$ for the respective populations in

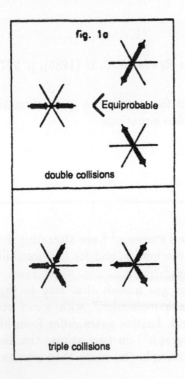

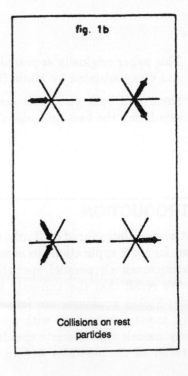

FIGURE 1

these six directions at a lattice site $\mathbf{r}$ at time t. By ignoring correlations between sites (Boltzmann's approximation), we obtain a system of six equations for a microscopic regime governing the evolution of these populations, i.e.,

$$\partial_t N_i + \mathbf{C}_i \cdot \nabla N_i + T_i \ (N_1, \ldots, N_6) = \tilde{T}_i(\tilde{N}_1, \ldots, \tilde{N}_6) \times \prod_{j=1}^{6} (1 - N_j),$$

$$i = 1, \ldots, 6 \ . \tag{1}$$

The six quantities, $\tilde{N}_i = N_i/(1 - N_i)$, are introduced because of the exclusion principle. In these conditions, $\tilde{T}_i$, expressed in terms of $\tilde{N}_j$, have the same form as without the exclusion principle, i.e.,

$$\begin{aligned}
\tilde{T}_i(\tilde{N}) = &+ \frac{1}{2}\tilde{N}_{i+1}\tilde{N}_{i+4} + \frac{1}{2}\tilde{N}_{i-1}\tilde{N}_{i-4} \\
&- \tilde{N}_i\tilde{N}_{i+3} + \tilde{N}_{i+1}\tilde{N}_{i+3}\tilde{N}_{i+5} - \tilde{N}_i\tilde{N}_{i+2}\tilde{N}_{i+4} \ .
\end{aligned} \tag{2}$$

Index i is defined modulo 6. The three first terms of the right-hand side of Eq. (2) describe the double collisions (with equal weights on both outgoing channels in Figure 1a). The two last terms describe triple collisions. For the *center hexagonal* case, we add the index "*" for quantities relative to rest particles, and add terms to Eqs. (1) and (2) describing collisions involving rest particles (see Figure 1b).

The Boltzmann equation (1) with an exclusion principle has, like the usual Boltzmann equations, an "H theorem"[6] that predicts a decreasing monotone evolution of the quantity: $H = \sum_{i=1}^{6} N_i log N_i + (1 - N_i)log(1 - N_i)$ toward its thermodynamical equilibrium value. We show that, as in ref. (6), at equilibrium, the quantities $log(\tilde{N}_i)$ are linear combinations of collision invariants, i.e., density:

$$\rho = \sum_{i=1}^{6} N_i \tag{3}$$

and momentum:

$$\rho\mathbf{u} = \sum_{i=1}^{6} N_i \mathbf{C}_i \ . \tag{4}$$

In this model (as in its center variant), the energy invariant is not independent of the density invariant; therefore, thermal phenomena cannot be described by the simple model, but this limitation can be overcome.

PASSAGE TO HYDRODYNAMICS

We know that from a microscopic point of view, hydrodynamics can be seen as the re-gluing of local thermodynamical equilibria with parameters (here, ρ and u) that are slowly varying in position and time. Frisch, Hasslacher, and Pomeau[3] showed that hexagonal symmetry models lead within reasonable limits to the 2-D Navier-Stokes equations:

$$\partial_t \rho + \nabla \cdot (\rho \mathbf{U}) = 0 \ , \tag{5}$$

$$\partial_t(\rho u_\alpha) + \sum_\beta \partial_{x_\beta}(\rho g(\rho) u_\alpha u_\beta) = -\partial_{x_\alpha} p + \eta_1 \nabla^2 u_\alpha + \eta_2 \partial_{x_\alpha} \nabla \cdot \mathbf{u} \ . \tag{6}$$

The form of these equations can be obtained by simple invariance arguments. Evaluating coefficients requires an asymptotic expansion of the Chapman-Enskog type on Boltzmann's equation. In order to obtain the $g(\rho)$ coefficient of the nonlinear term and the equation-of-state relating pressure p and density, we must expand to second order in u, but only to first order in gradients (ref. (3)). We obtain:

$$g(\rho) = \frac{\rho - 3}{\rho - 6} \ , \quad p(\rho) = \frac{\rho}{2} \quad \text{for simple hexagonal}$$

$$\text{and} \ \ g(\rho) = \frac{7}{6} \cdot \frac{\rho - (7/2)}{\rho - 7}, \quad p(\rho) = \frac{3}{7}\rho \quad \text{for center hexagonal} \ . \tag{7}$$

The calculation of shear viscosity, η_1, and bulk viscosity, η_2, requires an expansion to second order in gradients. We restrict ourselves to low Mach numbers. This restriction allows us to work with the Boltzmann equation linearized in the neighborhood of the equilibrium solution at zero speed where $N_i = \rho/6$. We can first look only at sound waves, i.e., at perturbations of the form $n_i(t)exp[i\varepsilon\mathbf{k} \cdot r]$, where ε is the spatial expansion parameter. The vector, $\mathbf{n}$ formed out of the n_i, satisfies the equation:

$$\partial_t \mathbf{n} = (A + \varepsilon B)\mathbf{n} \ . \tag{8}$$

A, the linearized collision operator, is a real symmetrical matrix, six by six with three zero eigenvalues (related to collisional invariants) and B is the $\text{diag}(-jC_i \cdot k)$ antihermitian matrix. Under the perturbation, εB, the Goldstone modes (eigenvectors with a zero eigenvalue) of A become hydrodynamical modes. Since $A + \varepsilon B$ is hermitian for ε pure imaginary, the dynamics is dispersive on times scales $0(1/\varepsilon)$ and dissipative on times scales $0(1/\varepsilon^2)$. We can then use a multi-scale technique with two times:

$$t_1 = \varepsilon t \ , \quad t_2 = \varepsilon^2 t \ . \tag{9}$$

The population is expanded in powers of ε:

$$\mathbf{n}(t) = \mathbf{n}^{(0)}(t_1,t_2) + \varepsilon\mathbf{n}^{(1)}(t_1,t_2) + \varepsilon^2\mathbf{n}^{(2)}(t_1,t_2) + 0(\varepsilon^3) \ . \tag{10}$$

By substituting in Eq. (8) and identifying terms to the order ϵ^2, we obtain:

$$A\mathbf{n}^{(0)} = 0 \ , \quad A\mathbf{n}^{(1)} = \partial_{t_1}\mathbf{n}^{(0)} - B\mathbf{n}^{(0)} \ ,$$
$$A\mathbf{n}^{(2)} = \partial_{t_2}\mathbf{N}^{(0)} + \partial_{t_1}\mathbf{n}^{(1)} - B\mathbf{n}^{(1)} \ . \tag{11}$$

This system is studied by standard techniques[6]; linearized hydrodynamic equations are obtained as solubility conditions (orthogonality at the kernel of adjoint of A). Operator A can be explicitly diagonalized, which simplifies the calculation of its pseudo-inverse, involved in the solution of Eq. (11).

TRANSPORT COEFFICIENTS

The result is (c is the sound wave speed):
simple hexagonal:

$$c = \sqrt{\frac{1}{2}} \ , \quad \eta_1 = \frac{1}{2}(1-d)^{-3} \ , \quad \eta_2 = 0 \ . \tag{12}$$

center hexagonal:

$$c = \sqrt{\frac{3}{7}} \ , \quad \eta_1 = \frac{1}{4}(1-d)^{-4} \ , \quad \eta_2 = \frac{1}{14}(1-d)^{-4} \ . \tag{13}$$

d is the *reduced density*, i.e., $\rho/6$ in the simple hexagonal case, and $\rho/7$ in the center hexagonal case. We note that the above results are not changed by introducing a weight coefficient different from unity in front of the triple collision terms. Figure 2 shows these theoretical results compared with the numerical simulation results obtained in ref. (4). In this reference, the definition of viscosities is different; the necessary corrections were made. We note that the agreement is good at low density. In the simple hexagonal case, the bulk viscosity is zero in the Boltzmann approximation, but it is *negative* in simulations. We note that the usual derivation of the positivity of the bulk viscosity[7] doesn't apply, due to the absence of a thermodynamic entropy variable.

An interesting application of the above results is the density value maximizing the Reynolds number for the lattice gas simulation of an incompressible flow. Usually, the Reynolds number is defined as the product of a scale and a speed characteristic of the flow, divided by the kinematic viscosity η_1/ρ. Here, the nonlinear term is renormalized by factor $g(\rho)$ [see Eq. (6)] that must be included in the Reynolds number. The maximum occurs for values ρ_* and d_* of density and reduced density, given here:

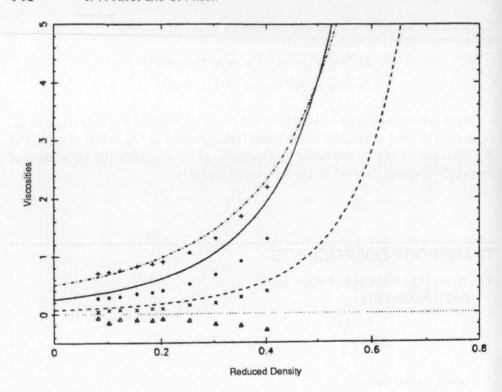

FIGURE 2

simple hexagonal:

$$\rho_* = 1.074 \text{ and } d_* = 0.179 , \tag{14}$$

center hexagonal:

$$\rho_* = 1.085 \text{ and } d_* = 0.155 . \tag{15}$$

In simulations with the center model, d_* is about 0.2 which is slightly higher.

Operating near the optimal density (and also at low density), the Reynolds number for the case with centers is about twice as high as the Reynolds number for the simple hexagonal case.

REFERENCES

1. Hardy, J., Y. Pomeau, and O. de Pazzis. *J. Math. Phys.* **14** (1973):1746.
2. Hardy, J., O. de Pazzis, and Y. Pomeau. *Phys. Rev.* **A 13** (1976):1949.
3. Frisch, U., B. Hasslacher, and Y. Pomeau. "Lattice Gas Automata for the Navier-Stokes Equation." *Phys. Rev. Lett.* **56** (1986):1505–1508.
4. d'Humières, D., P. Lallemand, and T. Shimomura. "An Experimental Study of Lattice Gas Hydrodynamics." *Los Alamos report LA-UR-85-4051*, 1985.
5. d'Humières, D., Y. Pomeau, and P. Lallemand. "Simulation d'Allées de von Karman à l'Aide d'un Gaz Réseau." *Comptes Rendus* **II 301** (1985):1391–1394.
6. Gatignol, R. "Théorie Cinétique des Gaz à Répartition Discrète des Vitesses." *Lecture Notes in Phys.*, vol. 36. New York: Springer-Verlag, 1975.
7. Landau, L., and E. Lifchitz. *Mécanique des Fluides*. Moscow, 1971.

More Partial Differential Equations

More Partial Differential Equations

Hudong Chen†‡ and William H. Matthaeus†

†Bartol Research Institute, University of Delaware, Newark, Delaware 19716 and ‡present address: Center for Nonlinear Studies, Los Alamos National Laboratory, Los Alamos, NM 87545

New Cellular Automaton Model for Magnetohydrodynamics

This paper originally appeared in *Physical Review Letters*, Volume 58, Number 18, May 4, 1987, pp. 1845–1848.

A new type of two-dimensional cellular automation method is introduced for computation of magnetohydrodynamic fluid systems. Particle population is described by a 36-component tensor referred to a hexagonal lattice. By appropriate choice of the coefficients that control the modified streaming algorithm and the definition of the macroscopic fields, it is possible to compute both Lorentz-force and magnetic-induction effects. The method is local in the microscopic space and therefore suited to massively parallel computations.

PACS numbers: 52.30.–q, 47.10+g, 47.65.+a, 52.65+z

The recent development of a hexagonal lattice-gas (HLG) model[1] for two-dimensional hydrodynamics has led to a considerable level of interest in the use of cellular automata[2] (CA) for the study of fluid and fluid-like physical systems.[3] CA fluid models may offer significant computational advantages[1] and provide insights into the relationship between macroscopic physics and the nature of the microphysical world.[1-3] Recently, Montgomery and Doolen[4] introduced a two-dimensional magnetohydrodynamics (MHD) model that makes use of both microscopic cellular-automata and macroscopic finite-difference methods. Their model departs from the usual notion[2] of cellular automata by a nonlocal computation of the Lorentz force, involving spatial differences of the coarse-grained magnetic potential. It has been suggested[5] that nonlocal features of CA models of MHD and other plasma systems may be inescapable in view of the nonlocal physics that pervades the approximations leading to MHD. In the context of our own investigations[6] of a passive scalar CA model and its generalization to MHD, we concluded, in accordance with Ref. 5, that nonlocal features are inevitable for this type of MHD model, but we attribute the nonlocality to the formulation in terms of the vector potential. In this Letter we present an alternative formulation of MHD cellular automaton in which the microscopic dynamical rules are completely local in both time and space. The new method would appear to be well suited to large-scale parallel computation.

The system of two-dimensional incompressible MHD equations[7,8] for which we develop a CA model may be written as

$$\rho(\partial \mathbf{v}/\partial t + \mathbf{v} \cdot \nabla \mathbf{v}) = -\nabla p + (\nabla \times \mathbf{B}) \times \mathbf{B} + \nu \nabla^2 \mathbf{v}, \tag{1}$$

$$\partial \mathbf{B}/\partial t + \mathbf{v} \cdot \nabla \mathbf{B} = \mathbf{B} \cdot \nabla \mathbf{v} + \mu \nabla^2 \mathbf{B}, \tag{2}$$

where $\mathbf{B}$, $\mathbf{v}$, p, ρ, μ, and ν are the magnetic field, velocity field, pressure, mass density, resistivity, and viscosity, respectively. For incompressible flow $\nabla \cdot \mathbf{v} = 0$ and $\rho = $ const, while the pressure is determined from the Poisson equation that results from computation of the divergence of Eq. (1). In the relevant two-dimensional (x, y) geometry, $\mathbf{v}$ and $\mathbf{B}$ lie in the x-y plane and depend only on those coordinates and time. The magnetic potential A_z is related to $\mathbf{B}$ by $\mathbf{B} = \nabla \times A_z \hat{\mathbf{e}}_z$ where $\hat{\mathbf{z}} = \hat{\mathbf{x}} \times \hat{\mathbf{y}}$.

The possibility that all the nonlinearities in Eqs. (1) and (2) might be handled locally by a CA model may be easily motivated by consideration of the structure of Eqs. (1) and (2), with neglect of pressure and dissipation, in terms of the Elsässer[9] variables $\mathbf{z}^{\pm} = \mathbf{v} \pm \mathbf{B}/\rho^{1/2}$. From $\partial \mathbf{z}^{\pm}/\partial t \approx -\mathbf{z}^{\pm} \cdot \nabla \mathbf{z}^{\pm}$, it is easily seen that the relevant nonlinearities, including Lorentz force, are nonlocal only in an appropriately generalized advective sense. Ordinary advection due to the velocity field can be adequately treated in both hydrodynamic[1] and vector-potential-based MHD[4-6] CA models. The Lorentz force $(\nabla \times \mathbf{B}) \times \mathbf{B}$ written as $-\nabla A_z \nabla^2 A_z$ cannot be dealt with in this way because it involves a nonlinear product of a quantity having components of the gradient that are not parallel to $\mathbf{v}$, with a quantity having a second derivative which must involve information from neighboring cells. On the basis of the Elsässer-variable argument, it would seem necessary to treat $\mathbf{B}$ on more nearly equal footing with $\mathbf{v}$ to achieve a local MHD CA model.

The basis of the present model is a modified streaming algorithm for particles moving on a hexagonal grid in which each particle occupies a state labeled by two vectors, $\hat{\mathbf{e}}_a$ and $\hat{\mathbf{e}}_b$, where $\hat{\mathbf{e}}_a = (\cos 2\pi a/6, \sin 2\pi a/6)$, $\hat{\mathbf{e}}_b = (\cos 2\pi b/6, \sin 2\pi b/6)$ and both a and b run from 1 to 6. No more than one particle in each cell may occupy a state with a specified a and b, so that at most 36 particles may simultaneously reside in a cell. Letting N_b^a ($= 0$ or 1) denote the occupation number at a certain location, we define $f_a^b \equiv \langle N_a^b \rangle$ to be the ensemble-averaged particle distribution. At each CA time level, streaming, by which we mean the noncollisional component of particle motion, consists of motion to the adjacent cell in the direction $\hat{e}_a$ with probability $1 - |P_{ab}|$. Alternatively, with probability $|P_{ab}|$ the particle moves to the adjacent cell in the direction $\hat{e}_b P_{ab}|P_{ab}|$. This leads to a kinetic equation[3] for the tensor particle distribution f_a^b,

$$\partial f_a^b(\mathbf{x}, t)/\partial t = -\{(1 - |P_{ab}|)\hat{e}_a + P_{ab}\hat{e}_b\} \cdot \nabla f_a^b(\mathbf{x}, t) + \Omega_{ab}, \tag{3}$$

where Ω_{ab} represents the effect of all collisions that modify f_a^b.

The macroscopic number density, fluid velocity field, and magnetic field will be designated as n, $\mathbf{v}$, and $\mathbf{B}$, respectively, and will be related to the microstate by the relations

$$n = \sum_{a,b} f_a^b, \tag{4}$$

$$n\mathbf{v} = \sum_{a,b} \{(1 - |P_{ab}|)\hat{e}_a + P_{ab}\hat{e}_b\} f_a^b, \tag{5}$$

$$n\mathbf{B} = \sum_{a,b} \{Q_{ab}\hat{e}_b + R_{ab}\hat{e}_a\} f_a^b, \tag{6}$$

where the 6×6 matrices $\mathbf{P}$, $\mathbf{Q}$, and $\mathbf{R}$ are as yet undetermined constants that must be selected on theoretical grounds to give the desired physical behavior of MHD. By requiring that the behavior of the system be locally invariant under both proper and improper rotations, we conclude that $\mathbf{P}$, $\mathbf{Q}$, and $\mathbf{R}$ must be circulant[3] matrices, and that P_{ab}, Q_{ab}, and R_{ab} depend only on $|a - b|$. There remain twelve coefficients in these matrices that must be selected to specify the streaming behavior of the model, from which all the nonlinearities of MHD will emerge.

Further simplification is obtained by consideration of symmetries of the Lorentz force that are implied by the structure of Eqs. (1) and (2). Notice that the Lorentz force is unchanged by the transformation $\mathbf{B} \to -\mathbf{B}$. We wish to have this macroscopic transformation correspond to the microscopic transformation $\hat{e}_b \to -\hat{e}_b$, so that $\hat{e}_b$ will act as the magnetic quantum in the same way that $\hat{e}_a$ controls undeflected momentum transport. To achieve this, we require that $\mathbf{B} \to -\mathbf{B}$ everywhere whenever $\hat{e}_b \to -\hat{e}_b$, but that neither $\mathbf{v}$ nor $\hat{e}_a$ is affected. Upon consideration of Eqs. (5) and (6), one can see that this requires $P_{ab} \equiv -P_{ab+3}$ to ensure that the velocity is unchanged, while $Q_{ab} \equiv Q_{ab+3}$ and $R_{ab} \equiv -R_{ab+3}$. Combining this with the rotational and reflection symmetries, we are left with just six independent coefficients, chosen to be P_{aa}, P_{aa+1}, Q_{aa}, Q_{aa+1}, R_{aa}, and R_{aa+1}. With these choices the transformation $\hat{e}_b \to -\hat{e}_b$ will cause $\mathbf{B} \to -\mathbf{B}$ without changing $\mathbf{v}$ while the transformation $\hat{e}_a \to -\hat{e}_a$ will lead to $\mathbf{v} \to -\mathbf{v}$ without modifying $\mathbf{B}$. This choice of symmetry also accounts for physically correct behavior of fluid elements in simple macroscopic field geometries. Bidirectional streaming allows the average trajectory of a particle to vary relative to $\hat{e}_a$ and $\hat{e}_b$. Consider for the moment the most probable $\hat{e}_b$ to be a good estimate of the local macroscopic $\mathbf{B}$. Particles streaming across a simple sheared **B** can easily be understood to experience a deflecting force that is qualitatively consistent with the Lorentz force since deflection will be towards the most probable $\hat{e}_b$ when the angle $\theta = \cos^{-1}(\hat{e}_a \cdot \hat{e}_b)$ is acute or towards $-\hat{e}_b$ when the angle is obtuse. Moreover, the effect of magnetic pressure is also correctly accounted for by acceleration of particles towards the weaker magnetic field region where the most probable $\hat{e}_b$ will be encountered less frequently.

Following previous CA fluid-model developments,[1,3,4] collision rules are adopted to randomize the microscopic state while preserving macroscopic quantities that are found to be necessary to give correct ensemble averaged behavior. Essential to the approach are inequalities between collisional and macroscopic time and length scales that allow the local microstate to be treated as near to equilibrium. The same approach is adopted here, with the requirement that the collision term, Ω_{ab} in Eq. (3), satisfies $\sum_{a,b} \Omega_{ab} = 0$,

$$\sum_{a,b} \{(1 - |P_{ab}|)\hat{e}_a + P_{ab}\hat{e}_b\} \Omega_{ab} = 0,$$

and $\sum_{a,b} \{Q_{ab}\hat{e}_b + R_{ab}\hat{e}_a\} \Omega_{ab} = 0$. These represent conservation of particle number, the momentum $n\mathbf{v}$, and the density-weighted magnetic field $n\mathbf{B}$. A large number of possible collision rules satisfy these requirements, many of these straightforward extensions of hydrodynamic CA collisions.[1,3]

Without further approximation, and with no assumptions about the form of the distribution function, the equation of particle transport is obtained from the kinetic equation (3) by summation over a and b, and then use of conservation of particles by collisions and Eqs. (4) and (5). This leads to

$$\partial n/\partial t + \nabla \cdot (n\mathbf{v}) = 0. \tag{7}$$

This is the ordinary fluid-continuity equation.

In order to deduce transport equations for $\mathbf{v}$ and $\mathbf{B}$, which will be cast in the form of Eqs. (1) and (2), it is necessary to rely on collisions to produce a local equilibrium state. In the lowest-order Chapman-Enskog expansion,[1,3,4] the collisions lead to a local Fermi-Dirac equilibrium distribution.

$$f_a^b(\text{equil}) = \frac{1}{1 + \exp(\alpha + \beta \cdot \hat{e}_a + \eta \cdot \hat{e}_b)}, \tag{8}$$

where α, β, and η are obtained by the definitions of n, $\mathbf{v}$, and $\mathbf{B}$ in Eqs. (4)–(6).

The momentum equation is obtained by our multiplying both sides of Eq. (3) by $\{(1-|P_{ab}|)\hat{e}_a + P_{ab}\hat{e}_b\}$, summing over a and b, and using Eqs. (3)–(5). To proceed we consider the expansion of Eq. (8) in the limit $|\mathbf{v}| \ll 1$ and $|\mathbf{B}| \ll 1$. After some tedious algebra, which is facilitated by use of the symmetries adopted above, we arrive at

$$\partial(n\mathbf{v})/\partial t = - C_1 \nabla n/6 - C_2 \nabla \cdot [nG(n)\mathbf{v}\mathbf{v}] + C_3 \nabla \cdot [nG(n)\mathbf{B}\mathbf{B}] +$$
$$C_4 \nabla [nG(n)\mathbf{v}^2] + C_5 \nabla [nG(n)\mathbf{B}^2]. \tag{9}$$

In Eq. (9), $G(n) \equiv (18 - n)/(36 - n)$ and C_1, C_2, C_3, C_4, and C_5 are rational functions of the six independent streaming coefficients in the matrices $\mathbf{P}$, $\mathbf{Q}$, and $\mathbf{R}$. The above relation contains terms of the same general form as the momentum equation obtained in other fluid CA models,[1,4] except that correctly structured terms involving the magnetic field also appear.

Similarly, an equation of the same general form as the induction equation (2) is obtained after multiplication of Eq. (3) by $Q_{ab}\hat{e}_b + R_{ab}\hat{e}_a$ and summation over all a and b. The result is

$$\partial(n\mathbf{B})/\partial t = -D_1\nabla \cdot [nG(n)\mathbf{vB}] + D_2\nabla \cdot [nG(n)\mathbf{Bv}] + D_3\nabla[n\mathbf{v} \cdot \mathbf{B}], \qquad (10)$$

where D_1, D_2, and D_3 depend only on the streaming coefficients.

To arrive at a CA model for MHD a number of restrictions must be placed on the coefficients in Eqs. (9) and (10). For example, the last term in Eq. (10) must be eliminated, since it does not appear in Eq. (2) and will generate nonsolenoidal magnetic fields; thus $D_3 = 0$ must be enforced. For nonnegative pressure, $C_1 > 0$ is required. Furthermore, for consistency with Eqs. (1) and (2), we must choose $C_2 = C_3 = D_1 = D_2 > 0$. Along with the constraints that $|P_{ab}| < 1$ (to allow a probabilistic interpretation of the modified streaming) and the vanishing of D_3, we arrive at four constraint equations and four inequalities that restrict allowed values of streaming coefficients P_{aa}, P_{aa+1}, Q_{aa}, Q_{aa+1}, R_{aa}, and R_{aa+1}. We have found numerical solutions to the constraints that indicate the existence of continuous ranges of allowed parameters, all of which have $P_{aa} < 0$. One solution is $P_{aa} = -1/3$, $P_{aa+1} = +2/9$, $Q_{aa} = 1/2$, $Q_{aa+1} \approx 0.065$, $R_{aa} = 0$, $R_{aa+1} \approx -0.232$.

Having solved the constraint equations, the final result of the zeroth-order macroscopic behavior of the MHD CA model is

$$\partial(n\mathbf{v})/\partial t = -C_2\nabla \cdot [nG(n)(\mathbf{vv} - \mathbf{BB})] - \nabla[C_1n/6 - nG(n)(C_4\mathbf{v}^2 + C_5\mathbf{B}^2)],$$
$$(11)$$

$$\partial(n\mathbf{B})/\partial t = -C_2\nabla \cdot [nG(n)(\mathbf{vB} - \mathbf{Bv})]. \qquad (12)$$

For the above solution, we have found $C_1 = 1.77$, $C_2 = 1.09$, $C_4 = -C_5 = C_2/2$. For $|\mathbf{v}| \ll a$ and $|\mathbf{B}| \ll 1$, corresponding to the low-Mach-number flow limit, the equation of state gives a lowest-order relation between pressure and density of the form $p = C_1n/6$, with the additional anisotropic effects of order $\mathbf{v}^2$ and $\mathbf{B}^2$. In the same limit, the density will exhibit only small fluctuations about a uniform constant value[1,3] so that the factors of n in various terms cancel and the factor $C_2G(n)$ may be used to rescale the relationship between microscopic and macroscopic time.[1] This leads to a set of dynamical equations almost identical to incompressible MHD. All numerical solutions to the constraints that we have found have the property that $C_4 = -C_5 = C_2/2$, leading to an exact representation of the Lorentz force in Eq. (11).

There are a number of additional issues important to the development of the MHD CA model that warrant brief mention here; a detailed description of the model will be forthcoming.[10] First, the allowed collisions are closely related to those in the six-state hexagonal lattice gas (HLG),[1,3] and always involve sets of particles with zero net $\mathbf{v}$ and $\mathbf{B}$. At very low densities $n \ll 1$ collisions will be infrequent and the collisional mean free path may be unacceptably large. Fortunately there appears to be no restriction on running the model at higher density, except that $n < 18$ [so that $G(n) > 0$]. However, certain manipulations in the lattice kinetic theory may be difficult to justify[11] for high densities. The key restriction on interpreting Eqs. (11) and (12) as a model of MHD is that the density be very nearly uniform and the flow therefore incompressible. This approximation is favored both by substantial inequality between the characteristic microscopic and macroscopic length and by small amplitudes of the macroscopic fields, equivalent to the low-Mach-number[1] condition for the HLG. This latter conditions should be no more restrictive here than for the HLG since the sound speed for the present model is $c_s \approx (C_1/6)^{1/2}$. Moreover, small departures from incompressibility should behave properly as sound waves in the nearly quiescent state by the same reasoning used in the HLG case.[3] We are currently investigating [10] the behavior of MHD Alfvén and magnetosonic[7] waves. Another issue of importance is the requirement that $\nabla \cdot \mathbf{B} = 0$. The choice of streaming coefficients leading to $D_3 \equiv 0$ eliminated the most seriously offending term in Eq. (10), but it is not possible to eliminate $\nabla \cdot \mathbf{B}$ exactly. On the other hand, from the divergence of Eq. (12), we find that $\partial \nabla \cdot \mathbf{B}/\partial t$ is at most of the order of the density inhomogeneities, presumably Mach number squared. This is no more restrictive than the low-Mach-number requirement for incompressiblity. Moreover, diffusion decreases $\nabla \cdot \mathbf{B}$. It will be important, however, to initialize the model with a magnetic field that is as nearly divergence-free as possible. The viscosity[1,3,4] and magnetic diffusivity[6] in this model have not yet been computed in view of the number of degrees of freedom involved. Nevertheless, we have found that the first-order Chapman-Enskog expansion gives the correct structure of the diffusion terms, i.e., they are proportional to $\nabla^2 \mathbf{v}$ in the momentum equation and $\nabla^2 \mathbf{B}$ in the induction equation.

In summary we find that field-line stretching and Lorentz forces can be incorporated into a local CA model by the introduction of a microscopic bidirectional streaming procedure. The allowed particle states on the hexagonal lattice are labeled by two vectors and the direction of particle motion is selected according to prescribed streaming coefficients, leading to a modified kinetic equation involving a tensor particle distribution. The definition of the macroscopic velocity and magnetic fields also depend on the streaming coefficients. With proper choice of the coefficient matrices $\mathbf{P}$, $\mathbf{Q}$, and $\mathbf{R}$, MHD is recovered for low fluid speed and low magnetic field strength. The success of the model is associated with the choice of symmetries for the streaming coefficients, which supports the notion[2] that simplified microscopic models may exhibit physically meaningful macroscopic behavior when the microscopic conservation laws and symmetries are correctly constructed. It is also likely that systems of equations other than MHD might be modeled by CA methods in a similar fashion. The idea of multidirectional streaming has some precedent in the recent CA model of Bohosian and Levermore[12] for the one-dimensional Burgers equation. The present model is well suited for parallel computation on machines such as the massively parallel processor because the microscopic behavior is independent of the macroscopic state. Numerical experimentation and comparison with standard computational methods will be needed to assess the potential utility[1] and possible limitations[13] of this CA MHD model.

Useful discussions with Professor David Montgomery, Dr. John Dorband, and Larry Klein are gratefully acknowledged. This research supported by the National Science Foundation under Grant No. ATM-8609740, and by NASA through the Director's Office and the Solar Terrestrial Theory Program at Goddard Space Flight Center.

REFERENCES

1. Frisch, U., B. Hasslacher, and Y. Pomeau., *Phys. Rev. Lett.* **56** (1986):1505.
2. Wolfram, S. *Rev. Mod. Phys.* **55** (1983):601.
3. Wolfram, S. *J. Stat. Phys.* **45** (1986):471.
4. Montgomery, D., and G. Doolen. *Phys. Lett.* **120A** (1987):229.
5. Montgomery, D., and G. Doolen. *Los Alamos Report No. LA-UR-86-3649*, 1986; to be published in *Proceedings of the Conference on Modern Approaches to Large Scale Physical Nonlinear Systems, Santa Fe, New Mexico, October 27–29, 1986.*
6. Chen, H., and W. Matthaeus. *Phys. Fluids.* To be published.
7. Cowling, T. G. *Magnetohydrodynamics.* Bristol, England: Adam Hilger Ltd., 1976.
8. Fyfe, D., D. Montgomery, and G. Joyce. *J. Plasma Phys.* **17** (1977):369.
9. Elsässer, W. *Phys. Rev.* **79** (1950):183.
10. Chen, H., W. H. Matthaeus, and L. Klein. *Phys. Fluids* **31** (1988):1439.
11. Montgomery, D. Private communication.
12. Boghosian, B., and D. Levermore. *Complex Sys.* **1** (1987):17.
13. Orszag, S., and V. Yakhot. *Phys. Rev. Lett.* **56** (1986):1691.

David Montgomery∗ **and Gary D. Doolen**†
∗Department of Physics and Astronomy, Dartmouth College, Hanover, NH 03755, USA
†Los Alamos National Laboratory, Los Alamos, NM 87545, USA

Two Cellular Automata for Plasma Computations

This paper originally appeared in *Complex Systems* (1987), Volume 1, pages 831–838.

Complex Systems **1** (1987) 831–838

Two Cellular Automata for Plasma Computations

David Montgomery
Department of Physics and Astronomy,
Dartmouth College, Hanover, NH 03755, USA

Gary D. Doolen
Los Alamos National Laboratory,
Los Alamos, NM 87545, USA

Abstract. Plasma applications of computational techniques based on cellular automata are inhibited by the long-range nature of electromagnetic forces. One of the most promising features of cellular automata methods has been the parallelism that becomes possible because of the local nature of the interactions, leading (for example) to the absence of Poisson equations to be solved in fluid simulations. Because it is in the nature of a plasma that volume forces originate with distant charges and currents, finding plasma cellular automata becomes largely a search for tricks to circumvent this nonlocality of the forces. We describe automata for two situations where this appears possible: two-dimensional magnetohydrodynamics (2D MHD) and the one-dimensional electrostatic Vlasov-Poisson system. Insufficient computational experience has accumulated for either system to argue that it is a serious alternative to existing methods.

1. Two-dimensional magnetohydrodynamics (2D MHD)

The basic equations of two-dimensional incompressible magnetohydrodynamics are a relatively straightforward generalization of those of two-dimensional fluid mechanics, and for our purposes can be written as [1–3]:

$$\frac{\partial(\rho u)}{\partial t} + \nabla \cdot (\rho u u) = -\nabla p + j \times B + \rho \nu \nabla^2 u, \tag{1.1}$$

$$\frac{\partial}{\partial t}(\rho A_z) + \nabla \cdot (\rho u A_z) = \rho \eta \nabla^2 A_z, \tag{1.2}$$

$$\nabla \cdot u = 0. \tag{1.3}$$

Here, $\mathbf{u} = (u_x, u_y, 0)$ is the velocity field and $\mathbf{B} = (B_x, B_y, 0)$ is the magnetic field. The mass density, assumed uniform and constant, is ρ, and the pressure is p. For all variables, $\partial/\partial z = 0$. The magnetic field $\mathbf{B}$ is obtained from a one-component magnetic vector potential $\mathbf{A} = (0, 0, A_z)$ as $\mathbf{B} = \nabla \times \mathbf{A}$. The electric current density is $\mathbf{j} = \nabla \times \mathbf{B} = (0, 0, -\nabla^2 A_z)$ in this geometry. The pressure p is obtained from solving the Poisson equation that results from taking the divergence of equation (1.1) and using $\nabla \cdot \partial(\rho\mathbf{u})/\partial t = \rho\partial(\nabla \cdot \mathbf{u})/\partial t = 0$.

The dissipation coefficients ν and η are the kinematic viscosity and magnetic diffusivity respectively. In the natural dimensionless units of the problem, they may be thought of as the reciprocals of either (a) the Reynolds number ($\nu \to R^{-1}$) and magnetic Reynolds number ($\eta \to R_m^{-1}$) or (b) the Lundquist number ($\eta \to S^{-1}$) and the "viscous" Lundquist number ($\nu \to M^{-1}$). The choice (a) is preferable when comparable magnetic and fluid kinetic energies are expected, and the choice (b) when the fluid kinetic energy is expected to be small or zero. Idiomatically, these are the "astrophysical" and "controlled fusion" regimes, respectively.

The generalization of fluid-dynamic cellular automata methods to equations (1.1) through (1.3) present essentially two challenges, over and above those associated with two-dimensional Navier-Stokes fluids. First, equation (1.1) differs from the two-dimensional Navier-Stokes equation only by the presence of the extra volume force $\mathbf{j} \times \mathbf{B}$ on the right-hand side. For the two-dimensional geometry, $\mathbf{j} \times \mathbf{B} = -(\nabla A_z)\nabla^2 A_z$, so all magnetic quantities are determined *locally* by A_z. Given A_z, the inclusion of $\mathbf{j} \times \mathbf{B}$ presents neither more nor fewer significant complications than the inclusion of external forces such as gravity now present in operating two-dimensional Navier-Stokes codes; this will be discussed presently.

The second new feature is the advancement of A_z, and it can be dealt with by a straightforward extension of methods introduced for the hexagonal lattice gas model [4–7]. The cells and molecules are identical to those of the hexagonal lattice gas, but in addition, each molecule carries with it a "photon label," or quantum of A_z, associated with an index $\sigma = +1, -1$, or 0. Each molecule can execute the same two-dimensional, momentum-conserving, Fermi-Dirac collisions that it executes in fluid simulations, carrying its value of σ with it. Each hexagon now contains eighteen possible single-particle states instead of six. A particle's σ can change only in collisions for which $\Sigma_\sigma \sigma = 0$. The field A_z is interpreted as an average of σ over super-cells containing many adjacent hexagons, in the same way that the fluid velocity $\mathbf{u}$ is interpreted as the average of the particle velocity $\hat{e}_a$ ($a = 1, 2, \ldots 6$) over super-cells containing many adjacent hexagons.

It appears to be possible and desirable to restrict the σ-exchanging collisions to those two-body collisions for which the initial pairs of σ-values are $(+1, -1), (-1, +1)$, and $(0, 0)$. For each of these three initial possibilities, it is convenient to choose as the outcome one of the same three possible pairs of values, randomly, with each one of the three possibilities as equally likely (probability 1/3). In hexagons with two or more particles per $\hat{e}_a$, no collision occurs—the final state is the same as the initial one. Though it is not an essential feature of the physics, the σ variable may be thought of as the z-component of canonical momentum for fluid particles bearing (both signs of) charge for motions confined to a plane. However, the particle motions underlying the true microphysics are far more complicated than this argument would indicate, and they are many approximations removed from 2D MHD. Their inclusion, in a way that reflected the microphysics as faithfully as the mechanical scattering rules represent the molecular collisions, would require particles of two mass species, z-velocities, z-accelerations, and so on.

The inclusion of the $j \times B = -(\nabla A_z)\nabla^2 A_z$ force in equation (1.1) apparently cannot be dealt with within the pure cellular automaton framework without the necessity of replacing the configuration-space cells by phase space cells (see section 2). In this respect, it is apparently a matter similar to the inclusion of gravity in the two-dimensional Navier-Stokes case and can be dealt with in the same way that gravity is now included in the Los Alamos code. Define $\langle A_z \rangle$ as the macroscopic average of σ over, say, $(64)^2$ hexagons. $\nabla \langle A_z \rangle$ and $\nabla^2 \langle A_z \rangle$ can then be obtained from local finite-difference approximations. Inside each super-cell, hexagons may be chosen randomly in proportion to the components of $-(\nabla \langle A_z \rangle)\nabla^2 \langle A_z \rangle$, and particles are flipped to admissible unoccupied values of $\hat{e}_a$ to provide the requisite momentum per unit time per unit volume, as indicated by $-(\nabla \langle A_z \rangle \nabla^2 \langle A_z \rangle)$.

A Chapman-Enskog development can be given for any model kinetic equation which contains a collision term which conserves x and y momenta in particle collisions, conserves $\Sigma_\sigma \sigma$, and obeys an H-theorem [5,8–10]. The Fermi-Dirac statistics require an eighteen-valued distribution function f_a^σ at each hexagon. For theoretical purposes, f_a^σ can be interpreted as a smooth, spatially-differentiable ensemble average. The mass, momentum, and vector potential densities, also assumed smooth, are defined by $\rho = \Sigma_{a,\sigma} f_a^\sigma$, $\rho u = \Sigma_{a,\sigma} \hat{e}_a f_a^\sigma$, and $\rho A_z = \Sigma_{a,\sigma} \sigma f_a^\sigma (\cos(2\pi a/6),\ \sin(2\pi a/6))$, $a = 1, 2, \ldots 6$, and $\sigma = 1, -1, 0$.

The differential conservation laws are:

$$\frac{\partial \rho}{\partial t} + \nabla \cdot (\rho u) = 0 \tag{1.4}$$

$$\frac{\partial}{\partial t}(\rho u) + \nabla \cdot \Pi = 0 \tag{1.5}$$

$$\frac{\partial (\rho A_z)}{\partial t} + \nabla \cdot \phi = 0 \tag{1.6}$$

where $\Pi \equiv \Sigma_{a,\sigma} \hat{e}_a \hat{e}_a f_a^\sigma$ and $\phi = \Sigma_{a,b} \hat{e}_a \sigma f_a^\sigma$. Π and ϕ are the momentum flux tensor and vector potential flux vector respectively.

The local thermodynamic equilibrium distribution is [10]

$$f_a^\sigma(eq.) = [1 + \exp(\alpha + \beta u \cdot \hat{e}_a + \gamma \sigma A_z)]^{-1} \tag{1.7}$$

where the Lagrange multipliers α, β, γ are obtained from requiring that equation (1.7) lead to ρ, $\mathbf{u}$, and A_z. Explicit calculation of α, β, γ requires that, as in the Navier-Stokes case, we assume $u^2 \ll 1$ (small Mach numbers). Solving for α, β, γ, $f_a^\sigma (eq.)$ becomes, up to terms of $0(u^3)$,

$$f_a^\sigma (eq.) = (\rho/18)\Big\{ 1 + 2\hat{e}_a \cdot \mathbf{u} + 3\sigma A_z/2$$

$$+ \left(\frac{9-\rho}{18-\rho}\right) \left[2(2\hat{e}_a\hat{e}_a : \mathbf{uu} - u^2) \right.$$

$$+ \frac{3}{2}\left(\frac{3\sigma^2 A_z^2}{2} - A_z^2\right) + 6\sigma\hat{e}_a \cdot \mathbf{u}A_z\Big]\Big\} \tag{1.8}$$

Inserting $f_a^\sigma(eq.)$ in equations (1.4) through (1.6) gives the Euler equations:

$$\frac{\partial \rho}{\partial t} + \nabla \cdot (\rho \mathbf{u}) = 0 \tag{1.9}$$

$$\frac{\partial}{\partial t}(\rho \mathbf{u}) + \nabla \cdot (\rho \mathbf{uu}) = -\nabla \cdot \mathbf{P} \tag{1.10}$$

$$\frac{\partial}{\partial t}(\rho A_z) + \nabla \cdot \left[\rho \mathbf{u}\frac{2(9-\rho)A_z}{(18-\rho)}\right] = 0. \tag{1.11}$$

In equation (1.10), $\mathbf{P}$ is the pressure tensor is

$$\mathbf{P} = (\rho/2)\left[1 - \left(\frac{9-\rho}{18-\rho}\right)u^2\right]\mathbf{1} - \left(\frac{9}{18-\rho}\right)\rho\mathbf{uu}. \tag{1.12}$$

At low densities ($\rho \ll 9$), the combination $2(9-\rho)/(18-\rho) \to 1$ so that equation (1.11) becomes the nondissipative version of equation (1.2), and $\mathbf{P} \to (\rho/2)(1 - u^2/2)\mathbf{1} - \rho\mathbf{uu}/2$ as in the fluid case [4,5,7].

The imperfections in the model are two-fold: the $\mathbf{u}$-dependent $\mathbf{P}$, which has the same form in the Navier-Stokes case, and the absence of the $\mathbf{j} \times \mathbf{B}$ term on the right-hand side of equation (1.10), which, as already discussed, involves going outside the model. (At this level, the transport of A_z is still that of a "passive scalar.")

The Chapman-Enskog procedure [10] may be carried to first order in the ratio of mean-free path to macroscopic length scale, and leads to ($u^2 \ll 1, n \ll 9$):

$$\left(\frac{\partial}{\partial t} + \hat{e}_z \cdot \nabla\right)f_a^\sigma(eq.) = \left(\frac{2\hat{e}_a\hat{e}_a - 1}{18}\right) : \rho\nabla\mathbf{u} + \frac{\sigma\hat{e}_a}{12} \cdot \rho\nabla A_z = \Omega_{a,\sigma}^{(1)}(f)$$

$$\tag{1.13}$$

The right-hand side of equation (1.13) stands for the linearized (about $f_a^\sigma(eq.)$) collision term which might contain, say, the 2R, 2L, and 3S collisions (Wolfram's notation [5]) appropriately generalized for the σ-exchange. $\Omega_{a,\sigma}^{(i)}(f)$ is, in general, an 18×18 matrix acting on an 18-component column vector $f_{a,\sigma}^{(1)}$, and is difficult to invert, even without allowing for other potential complications such as finite lattice size effects [11]. The solution for $f_{a,\sigma}^{(1)}$ has not been carried out, but from the form of the middle part of equation (1.13), it will be seen to contain terms involving $\rho \nabla u$ and $\rho \nabla A_z$. These, in turn, seem certain to lead to terms in equations (1.5) and (1.6) proportional to $\nabla \cdot (\rho \nabla u)$ and $\nabla \cdot (\rho \nabla A_z)$ respectively. In the incompressible limit, these are the standard viscous and resistive dissipative terms of equations (1.1) and (1.2). The coefficients ν and η have been calculated by Hatori and Montgomery [15].

The inclusion of "stopped" particles will make these coefficients even more difficult to calculate, and as in the case of real substances, "measurement" may provide more reliable values than theoretical calculation.

2.   One-dimensional Vlasov cellular automata

The one-dimensional, electrostatic, Vlasov-Poisson system advances particle distribution functions $f_j(x, v, t)$ and an electric field $E(x, t)$, which accelerates particles in the x direction. In the continuous two-dimensional phase space (x, v), f_j obeys

$$\frac{Df_j}{Dt} \equiv \frac{\partial f_j}{\partial t} + v \frac{\partial f_j}{\partial x} + \frac{e_j E}{m_j} \frac{\partial f_j}{\partial v} = 0 \tag{2.1}$$

in dimensionless units [12]. The symbol f_j represents two different distributions: $j \rightarrow i$ for positive ions and $j \rightarrow e$ for electrons. For the electrons, $e_e = -1$, and for the ions, $e_i = +1$. For the electrons, $M_e = 1$, and for the ions, $M_i = M$, an arbitrary integer > 1. (The cellular automaton is designed to eliminate all floating-point operations.) The velocity space is infinite, $-\infty < v < \infty$, and a variety of boundary conditions may be assumed in x. For present purposes, we will assume ideal reflecting plates at $x = 0$ and $x = L$, no external electric fields, and zero net charge between $x = 0$ and $x = L$. This results in $f_j(0, v, t) = f_j(0, -v, t)$, $f_j(L, v, t) = f_j(L, -v, t)$, $E(0, t) = E(L, t) = 0$, and can be implemented by choosing periodic initial conditions of period $2L$ subject to the initial symmetry $f_j(x, -v, 0) = f_j(-x, v, 0)$, which will be preserved in time. L is a (large) integer.

The electric field $E(x, t)$ obeys Poisson's equation

$$\frac{\partial E}{\partial x} = \int_{-\infty}^{\infty} (f_i - f_e) dv \tag{2.2}$$

and in one dimension may be obtained from Gauss's law [13,14]. For the present assumptions about the boundary conditions, it may be written as

$$E(x,t) = \int_0^x dx' \int_{-\infty}^{\infty} dv(f_i(x',v,t) - f_e(x',v,t)) \qquad (2.3)$$

in $0 < x < L$.

In the cellular automaton version, the phase space is divided into square cells of integer dimension for both species, so that $x \to n$, $n = 1, 2, 3, \ldots L$; $v \to s$, $s = -\infty, \ldots - 1, 0, 1, \ldots \infty$, although in practice, all v-cells above some large but finite $|s|$ will always be empty. Time is discretized into integer steps $\tau = 0, 1, 2, 3, \ldots$. Inside each phase space cell, the discretized distribution functions $f_j(n, s, \tau)$ are always either 1 or 0. Fermi-Dirac statistics are assumed, as initial conditions, and are preserved by the automaton now to be described. The discretized electric field is $E(n, \tau)$, which will also be an integer.

The updating of the $f_j(n, s, \tau)$ and $E(s, \tau)$ takes place in three steps. At $\tau = 1, 4, 7, \ldots$, $f_j(n, s, \tau)$ is updated in the x direction according to

$$f_j(n, s, \tau) = f_j(n - s, s, \tau - 1),$$

$$(\tau = 1, 4, 7, \ldots). \qquad (2.4)$$

At $\tau = 2, 5, 8, \ldots$, $f_j(n, s, \tau)$ is updated in the v direction according to

$$f_i(n, s, \tau) = f_i(n, s - E(n, \tau - 1), \tau - 1),$$

$$f_e(n, s, \tau) = f_e(n, s + ME(n, \tau - 1), \tau - 1),$$

$$(\tau = 2, 5, 8, \ldots). \qquad (2.5)$$

(Equations (2.5) and (2.6) reflect the accelerations of ion and electrons.) Finally, at $\tau = 0, 3, 6, 9, \ldots$, the electric field is updated according to the discretized version of Gauss's law which is

$$E(n, \tau) = \sum_{n'=0}^{n} \sum_{s=-\infty}^{\infty} [f_i(n', s, \tau - 1) - f_e(n', s, \tau - 1)],$$

$$(\tau = 0, 3, 6, 9, \ldots) \qquad (2.6)$$

Equation (2.7) is not a local determination of $E(n, \tau)$ and cannot be made so. It can, however, be updated in a somewhat more local way. Namely, establish $E(n, 0)$ by equation (2.7), then determine $E(n, \tau)$ for $\tau \geq 3$ by counting the net numbers of charges which pass into or out of the region between $x = 0$ and $x = n$ by counting those which go past the nth cell in both directions. Since equation (2.7) simply gives $E(n, \tau)$ as the total number of positive charges minus the total number of negative charges which lie between the first and n^{th} cells, this number, once established, can be updated by the local operation of counting flow past a point (this is the discretized version of the one-dimensional Maxwell equation $\partial E / \partial t + 4\Pi j = 0$).

At this point, so little analysis of the automaton given by equations (2.4) through (2.7) has been done that it is pointless to speculate about how effective a competitor it can be made to established methods of solving the one-dimensional Vlasov equation such as particle-in-cell or Fourier-Hermite spectral techniques. Some points clearly need to be analyzed, such as the manifest nonconservation of energy associated with updating the electric field at different time steps from those at which the velocity space is updated (equations (2.4) to (2.7), though the net nonconservation of energy over a three-unit cycle τ may be quite small). That the recipe is in some sense a convergent algorithm for the Vlasov equations seems almost obvious, with the convergence enhanced by spreading the velocity-space distribution, for a given number of particles per fixed phase-space volume, over more and more cells in s. That no floating-point operations are involved is also manifest. There is nothing sacred about the order in which the three steps (2.5), (2.6), and (2.7) are carried out, and it may be determined in practice that another order is superior.

Equations (2.4) through (2.7) are perhaps the first example of a *phase-space* cellular automaton, as contrasted with a *configuration-space* one. It seems inevitable, in circumstances where in nature the system is such that local fluid variables do not suffice to determine the distribution function, that phase space considerations must arise. The Vlasov equation is perhaps the first and simplest example. Once again, however, a local conservation law has played a crucial role: equation (2.1) is a statement of the local conservation of f_j itself [14].

Acknowledgments

This work was supported in part under the auspices of the U. S. Department of Energy at Los Alamos and in part by NASA Grant NAG-W-710 and U. S. Department of Energy Grant FG02-85ER53194 at Dartmouth.

References

[1] D. Fyfe, D. Montgomery, and G. Joyce, *J. Plasma Phys.*, **17** (1977) 369.

[2] A. Pouquet, *J. Fluid Mech.*, **88** (1978) 1.

[3] S. A. Orszag and C. M. Tang, *J. Fluid Mech.*, **90** (1979) 129.

[4] U. Frisch, B. Hasslacher, and Y. Pomeau, *Phys. Rev. Lett.*, **56** (1986) 1505.

[5] S. Wolfram, *J. Stat. Phys.*, **45** (1986) 471.

[6] D. d'Humières, Y. Pomeau, and P. Lallemand, *C.R. Acad. Sci. Paris, Ser. II*, **301** (1985) 1391.

[7] J. P. Rivet and U. Frisch, *C.R. Acad. Sci Paris, Ser. II*, **302** (1986) 267.

[8] R. Gatignol, "Théorie Cinetique des Gaz á Répartition Discréte des Vitesses", *Lecture Notes in Physics*, **36** (Springer-Verlag, Berlin, 1975).

[9] J. H. Ferziger and H. G. Kaper, *Mathematical Theory of Transport Processes in Gases* (North-Holland, Amsterdam, 1972).

[10] D. Montgomery and G. D. Doolen, *Physics Letters A*, **120** (1986) 229.

[11] B. Boghosian and D. Levermore, "A Cellular Automaton for the Burgers' Equation", *Complex Systems*, **1** (1987) 31.

[12] B. Adler, S. Fernbach, and M. Rotenberg, eds. *Methods in Computational Physics*, **9**: *Plasma Physics* (Academic Press, New York, 1970).

[13] J. M. Dawson, "The Electrostatic Sheet Model for a Plasma and its Modification to Finite-Size Particles", in *Methods in Computational Physics*, B. Adler, S. Fernbach, and M. Rotenberg, eds. (Academic Press, New York, 1970) 1–28.

[14] D. C. dePackh, *J. Electron Contr.*, **10**, (1962) 13a. See also: K. Symon, D. Marshall, and K. W. Li, in *Proc. 4th Conf. on Numerical Simulation of Plasmas*, J. P. Boris and R. A. Shanny, eds. (Washington, D.C.: Office of Naval Res., 1971) 68–125.

[15] T. Hatori and D. Montgomery, "Transport Coefficients for Magnetohydrodynamic Cellular Automata", *Complex Systems*, **1** (1987) 734.

P.-M. Binder
Center for Nonlinear Studies, Los Alamos National Laboratory, Los Alamos, NM 87545
and Applied Physics Section, Dunham Laboratory, Yale University, New Haven, CT 06520

Numerical Experiments with Lattice Lorentz Gases

A new method to simulate point-scatterer systems in a lattice is described.
Simulations are performed to illustrate the effect of static correlations in
the diffusion coefficient and velocity correlation function, and to study the
ergodicity of deterministic models.

INTRODUCTION

The Lorentz[1] and wind-tree models[2] have been crucial in the understanding of the
behavior of fluid transport coefficients.[3,4] These models consist of a classical point
particle moving amidst fixed, randomly placed scatterers which cause the particle
to collide elastically. Because of the linearity of the Boltzmann equation for these
models, much more analytical work has been possible than with a regular fluid.
This has also turned out to be the case with lattice gases; a number of unsus-
pected high-density effects have been found in particle scatterer models. In particular, although in some models numerical measurements agree with low density
theories,[5,6,7] the inclusion of reflective collisions causes the diffusion coefficient to

Lattice Gas Methods for Partial Differential Equations, SFI SISOC,
Eds. Doolen et al., Addison-Wesley Publishing Co., 1990

decrease or vanish.[8,9,10] Under these conditions, the Boltzmann assumption of uncorrelated collisions is invalid.

The lattice Lorentz models obey the following description: a particle moves from one node to a nearest neighbor at integer time steps (streaming or propagation step). It will move in a straight line until it encounters a scatterer, in which case its velocity will change according to one of several possible rules (collision step). The particle could, upon encountering a scatterer, always deflect to the left,[11] deflect left or right according to the parity of the time step,[5,12] deflect left or right according to the direction from which it encounters the scatterer,[7] or have a probabilistic outcome[6,8] which typically could be left-right or isotropic. The most successful method of solution at the Boltzmann level has been that of generating functions,[6,12] although for reflecting collisions a high-density effective medium approximation technique[8,9] is required.

The simulations in Binder,[5] Ruijgrok and Cohen,[7] and van Velzen and Ernst[9] have followed this method: at the nodes of a large (square or hexagonal) $L \times L$ lattice one places fixed scatterers with probability c. Then, in a smaller 1×1 sublattice situated at the center of the larger one, one places one particle per node and assigns it a random velocity among the allowed ones. The particles are independent, and by allowing the system to evolve one can make measurements of velocities or displacements. We will call this the "frozen configuration" method. The main problem with this method is that several particles can follow each other's trajectories for long times, causing artificial correlations. This is especially true at low scatterer densities.

We propose a second method which, although slower, avoids these correlations. This method in addition allows for simulations exactly at the Boltzmann level (that is, uncorrelated collisions). We will present results to illustrate the difference between correlated and uncorrelated models below.

This method is as follows:

One follows one particle at a time.

1. Apply the streaming (propagation) operator to the particle.

2. Check if the site of arrival has been visited. If not,

3. Generate a scatterer with a probability c.

4. Change the velocity if a scatterer is present.

5. Perform the necessary measurements.

6. Return to step 1 until the required time has been reached.

7. Repeat steps 1–6 for each particle.

A listing of a Fortran code for this method is given in appendix 1. This method has the disadvantage that one cannot check the overall concentration of scatterers prior to running the simulation. An interesting feature is that one can simulate exactly a Boltzmann-level gas by omitting step 2. This produces uncorrelated scatterers with probability c at each time step.

Note that the standard way to simulate the Boltzmann equation is to assign initial probabilities to each node, consistent with the equilibrium solution, and to redistribute these probabilities at each node and time step in a way also consistent with the equilibrium distribution.

In the case of the Lorentz gas, the equilibrium solution is $p(n, i, t) = \frac{1}{4}$. We cannot simulate a Boltzmann-level gas by the method described above for a simple reason: the diffusion coefficient for any particle-scatterer collision rule at the Boltzmann level is[6]

$$D = (2\lambda c)^{-1} - (4)^{-1} ,$$

where λ is a relevant eigenvalue of the collision operator. If for the particular model under study $\lambda \neq \frac{1}{4}$, the Boltzmann-level simulation in the previous paragraph leads to the wrong diffusion coefficient.

The existence of a frozen pattern of scatterers with isotropic collision rules is known to have correlations that diminish the Boltzmann level results by about one-third.[8,9] Therefore, one cannot use the "frozen configuration" method to simulate a Boltzmann Lorentz gas, unless one re-generates the entire pattern of scatterers at each time step, which is very slow.

We now report simulations of a model in a square lattice in which a particle-scatterer collision produces a probabilistic outcome of one-quarter in each direction. The diffusion coefficient, obtained by fitting measurements of the mean-squared displacement of Einstein's equation for times $200 < t < 400$ will be compared to the Boltzmann-level result,

$$D = (2c)^{-1} - (4)^{-1} .$$

and an effective medium approximation theory.[9] The averages, obtained by the method described above, are taken over 225000 independent configurations.

Table 1 illustrates the breakdown of the Boltzmann approximation of uncorrelated scatterers; we have performed simulations of both models, which show especially at low densities that the correlated model has a lower diffusion coefficient. This is caused by retracing trajectories, in which the particle returns to a previously visited scatterer via a quasi-one-dimensional path.

The breakdown of the Boltzmann description can also be seen in measurements of another quantity, the velocity correlation function (VCF)

$$\phi(t) = \langle v(0)v(t) \rangle ,$$

where the brackets indicate an average over configurations. The quantity is related to D by the Green-Kubo formula,

$$D = \sum_{t=0}^{\infty} \phi(t) - \frac{1}{4} .$$

TABLE 1 Density, theoretical and simulation diffusion coefficient at the Boltzmann-level, and theoretical and simulation diffusion coefficient with correlated collisions for isotropic scattering

Density	D(Boltz)	D(sim)	D(EMA)	D(sim)
0.1	4.75	4.80 ± 0.2	3.62	3.95 ± 0.16
0.2	2.25	2.22 ± 0.1	1.75	1.77 ± 0.07
0.4	1.00	1.00 ± 0.04	0.81	0.81 ± 0.03
0.6	0.58	0.58 ± 0.02	0.50	0.50 ± 0.02
0.8	0.38	0.375 ± 0.015	0.342	0.345 ± 0.015
1.0	0.25	0.25 ± 0.01	0.25	0.248 ± 0.01

The Boltzmann level prediction is

$$\phi(t) = (1 - c)^t .$$

Figures 1 and 2 are plots of the velocity correlation function for probabilistic outcome models: Figure 1 corresponds to uncorrelated isotropic scattering. It is identical to the VCF for left-right scattering, uncorrelated or not. Figure 2 corresponds

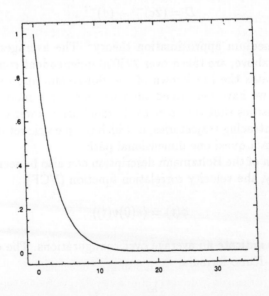

FIGURE 1 Velocity correlation function for a deflection-only stochastic Lorentz gas or uncorrelated isotropic scattering: $c = 0.3, 2.5$ million configurations.

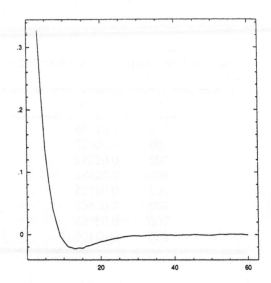

FIGURE 2   Velocity correlation function for an stochastic Lorentz gas with correlated isotropic scattering: $c = 0.3, 2.5$ million configurations. Note the cage effect.

to isotropic scattering with fixed scatterers. The density of scatterers in both cases is $c = 0.3$. While the former decays exponentially and fits the theory quite well, in the latter we see a strong negative part which reduces D, in accord with the results of the table. The negative tail has been observed in the continuum models.[13] Each plot is an average of 2.5×10^6 configurations, which yields error bars of 1/1000. To this accuracy, we observe no algebraic long-time tails.

In calculating diffusion coefficients, the existence of an equilibrium distribution is assumed. This assumption can be questioned in the case of deterministic models. The reason is the following: if a particle re-visits a scatterer with the same arrival velocity, it will be locked in a cyclic trajectory that will force it to revisit the same scatterer an infinite number of times. This is, the system is not ergodic. The relevant question is: how important are these trajectories? If they happen to be very likely, they will not contribute to the mean-squared displacement, and can cause the diffusion coefficient to decrease or even vanish, as observed in random walks on sites below the percolation threshold.[14] This has been predicted for the left-turning model at high densities[11]: for a density of one (i.e., all sites occupied by scatterers), all particles get locked in trajectories of length four, and the diffusion coefficient vanishes. For the odd times-left, right times-even model,[5,12] we have measured with the method proposed in this paper the probability of limit cycle locking. This was done with 100000 independent particles at a concentration of scatterers $c = 0.2$. The results are given in Table 2.

TABLE 2 Fraction of particles locked in a limit cycle after t time steps. Average over 100,000 particles at $c = 0.2$

Time	P(cycle)
100	0.01536
200	0.02037
300	0.02342
400	0.02532
500	0.02705
600	0.02861
700	0.02983
800	0.03106

Therefore, most trajectories are diffusive, and the assumption of the existence of an equilibrium distribution appears to be correct. Such a small percentage of orbits has an effect on the diffusion coefficient which is smaller than the uncertainty in the numerical measurements that we can perform so far. This cumulative fraction of particles travelling in orbiting trajectories fits fairly well a $t^{-1/3}$ curve, indicating that the probability of new particles returning to the origin in a cyclic trajectory decays as $t^{-4/3}$. This indicates that the integral converges, and furthermore, that in this case the cumulative fraction of non-diffusive orbits will never be larger than 4 percent or so.

We have proposed a new method of simulation which avoids correlation effects, and which in addition permits exact Boltzmann-level simulations. With this method we observe the breakdown of the Boltzmann result for isotropic scattering and cage effects in the velocity correlation function. We also report measurements of limit cycle locking in a deterministic lattice Lorentz model.

We thank M. H. Ernst, G. van Velzen and M. Colvin for useful discussions, and the U.S. Department of Energy and AFOSR for financial support.

APPENDIX 1: COMPUTER CODE

This is an example of a Fortran code that generates scatterers (correlated or un-correlaed) along the particle's trajectory.

```
C THIS PROGRAM CALCULATES THE MEAN-SQUARED
C DISPLACEMENT FOR PROBABILISTIC MODEL EXACTLY AT
C THE BOLTZMANN LEVEL
C A,B,C=ALPHA, BETA, GAMMA
C CON=DENSITY OF SCATTERERS
C NREP=NUMBER OF INDEPENDENT SAMPLES
C
      DIMENSION NDX2(50)
C
C INCLUDE THE NEXT TWO LINES FOR (A) CORRELATED
C COLLISIONS AND (B) VCF MEASUREMENTS, RESPECTIVELY
C
      DIMENSION NS(360,360), NVIS(360,360)
      DIMENSION NV(60)
      CALL INTRDM
C
      READ*, CON,NREP
      READ*, A,B,C
      AR=A+B
      AL=A
      AU=A+B+C
C LOOP OVER CONFIGURATIONS
C
      DO 100 NPT=1,NREP
C INITIAL CONDITIONS
C
      NX=400
      NY=400
      NVEL=0
      NSTEP=1
C STREAMING TERM
C GENERATE NEW POSITION
C
  200 IF (NVEL.EQ.0) THEN
      NX=NX+1
      ENDIF
      IF (NVEL.EQ.1)THEN
      NY=NY+1
      ENDIF
      IF (NVEL.EQ.2) THEN
      NX=NX-1
      ENDIF
      IF(NVEL.EQ.3) THEN
      NY=NY-1
      ENDIF
```

```
C COLLISION STEP
C GENERATE SCATTERER AT NEW POSITION
C (REGARDLESS OF PREVIOUS HISTORY)
C THE COMMENTED STATEMENTS BELOW ARE FOR
C CORRELATED COLLISIONS
C IF (NVIS(NX,NY).EQ.0) THEN
C X=RANF(0)
C IF (X.LT.CON) THEN
C NS(NX,NY)=1
C ENDIF
C ENDIF
C IF (NS(NS,NY).EQ.1) THEN
C
      X=RANF(0)
      IF (X.LT.CON) THEN
C GENERATE NEW VELOCITY ACCORDING TO PROBABILITIES
C ALPHA, BETA, GAMMA
C
      X=RANF(0)
      IF (X.LT.AL) THEN
      NVEL=MOD(NVEL+2,4)
      ENDIF
      IF (X.GE.AR.AND.X.LT.AU) THEN
      NVEL=MOD(NVEL+1,4)
      ENDIF
      IF (X.GE.QU) THEN
      NVEL=MOD(NVEL+3,4)
      ENDIF
      EIDIF
C CALCULATE MSD EVERY 20 TIME STEPS UP TO 400
      MTH=NSTEP/20
      MODT=MOD(NSTEP+1)
      IF (MODT.EQ.0) THEN
      NDX(MTH)=NDX2(MTH)+(NX-400)**2+(NY-400)**2
      ENDIF
C THE COMMENTED STATEMENTS BELOW ARE FOR VCF
C MEASUREMENTS FOR DEFLECTIVE MODEL ONLY
C IF (NVEL.EQ.0) THEN
C NV(NSTEP)=NV(NSTEP)+1
C ENDIF
C IF (NVEL.EQ.2) THEN
C NV(NSTEP)=NV(NSTEP)-1
C ENDIF
      IF (NSTEP.LE.400) THEN
      GO TO 200
      ENDIF
  100 CONTINUE
      DO 450 JC=1,20
      PO=FLOAT(NDX2(JC))/(FLOAT(NREP)*4.0)
      PRINT*,JC*20, PO,PO/FLOAT(JC*20)
  450 CONTINUE
      STOP
      END
```

REFERENCES

1. Lorentz, H. A. (1905), *Proceedings of the Royal Academy of Amsterdam* **7**, 438, 585, 684.
2. Ehrenfest, P., and T. Ehrenfest (1959), *The Conceptual Foundations to the Statistical Approach in Mechanics* (Ithaca: Cornell University Press).
3. van Leeuwen, J. M. J., and A. Weyland (1968), *Physica* **36**, 457; A. Weyland and J. M. J. van Leeuwen (1968), *Physica* **38**, 35.
4. Hauge, E. H., and E. G. D. Cohen (1969), *J. Math. Phys.* **10**, 397.
5. Binder, P. M. (1987), *Complex Systems* **1**, 559.
6. Ernst, M. H., and P. M. Binder (1988), *J. Stat. Phys.* **51**, 981.
7. Ruijgrok, Th. W., and E. G. D. Cohen (1988), *Phys. Lett.* **133A**, 415.
8. Ernst, M. H., G. A. van Velzen and P. M. Binder (1989), *Phys. Rev. A*, to appear.
9. van Velzen, G. A., and M. H. Ernst, in *Discrete Kinetic Theory, Lattice Gas Dynamics and Foundations of Hydrodynamics*, Ed. R. Monaco (Singapore: World Scientific), to appear.
10. Binder, P. M. (1989), *Complex Systems* **3**, 1.
11. Gates, D. J. (1972), *J. Math. Phys.* **13**, 1315.
12. Ernst, M. H., G. A. van Velzen, and P. M. Binder, unpublished.
13. Alder, B. J., and W. E. Alley (1978), *J. Stat. Phys.* **19**, 341.
14. Stauffer, D. (1985), *An Introduction to Percolation Theory* (London: Taylor and Francis).

Bruce M. Boghosian† and C. David Levermore‡

†Thinking Machines Corporation, 245 First Street, Cambridge, MA 02142-1214, USA and
‡Lawrence Livermore Laboratory, University of California, Livermore, CA 94550, USA

A Cellular Automaton for Burger's Equation

This paper originally appeared in *Complex Systems* (1987), Volume 1, pages 17–30.

Complex Systems **1** (1987) 17–30

A Cellular Automaton for Burgers' Equation

Bruce M. Boghosian
Thinking Machines Corporation,
245 First Street, Cambridge, MA 02142-1214, USA

C. David Levermore
Lawrence Livermore National Laboratory,
University of California
Livermore, CA 94550, USA

Abstract. We study the approximation of solutions to the Burgers' equation,

$$\frac{\partial n}{\partial t} + c\frac{\partial}{\partial x}\left(n - \frac{n^2}{2}\right) = \nu\frac{\partial^2 n}{\partial x^2} \qquad (1)$$

by spatially averaging a probabilistic cellular automaton motivated by random walks on a line. The automaton consists of moving "particles" on a one-dimensional periodic lattice with speed one and in a random direction subject to the exclusion principle that at most one particle may move in a given direction from a given lattice site, at a given time. The exclusion principle gives rise to the nonlinearity in Eq. (1) and introduces correlations between the particles which must be estimated to obtain statistical bounds on the error. These bounds are obtained in two steps. The first is showing that the ensemble average of the automaton is a stable explicit finite differencing scheme of Eq. (1) over the lattice with a second order convergence in the lattice spacing. The numerical diffusion of this scheme plays an important role in relating the automaton rules to Eq. (1). The next step is showing that the spatial averaging of a single evolution of the automaton converges to the spatial averaging of the ensemble as $1/\sqrt{M}$ where M is the number of lattice sites averaged. Simulations are presented and discussed.

1. Introduction

Recently it has been proposed to use cellular automata on large lattices for obtaining solutions to partial differential equations, in particular the incompressible Navier-Stokes equations [1]. Such automata have rules with locally conserved (or nearly conserved) quantities which, when averaged over microscopic configurations, give macroscopic behavior which is hope-fully described by the PDE's. The analyses justifying such hopes have been largely formal.

In this paper we study as a model an automaton for solving the Burg-ers' equation which is simple enough to analyze. We are not proposing this automaton as an effective method for computing solutions to Burg-ers' equation, but it does allow us to study in a relatively simple context some of the same issues that arise in the application of cellular automaton techniques to solving the incompressible Navier-Stokes equations [1].

Our probabilistic cellular automaton is motivated by considering ran-dom walks on a one dimensional lattice (in this presentation, we shall re-strict the discussion to one spatial dimension, although all that we do has obvious higher dimensional analogs). All "particles" on the lattice move exactly one lattice site to either the right or the left in one time step. Be-tween two neighboring lattice sites we associate a physical distance Δx, and between two successive steps of the random walk we associate a physical time Δt. It is well-known [2] that the density of such a system of parti-cles executing an uncorrelated, unbiased random walk obeys the diffusion equation,

$$\frac{\partial n}{\partial t} = \nu \frac{\partial^2 n}{\partial x^2} \qquad (1.1)$$

where $\nu = (\Delta x)^2/2\Delta t$ is the diffusion coefficient.

Next consider an uncorrelated random walk that is biased so that the probability of a step to the right is $(1 + \overline{\alpha})/2$, and the probability of a step to the left is $(1 - \overline{\alpha})/2$. It is also well-known [2] that such biasing leads to linear advection in the direction of the bias. That is, the density of the system obeys

$$\frac{\partial n}{\partial t} + c \frac{\partial n}{\partial x} = \nu \frac{\partial^2 n}{\partial x^2} \qquad (1.2)$$

where $c = \overline{\alpha}\Delta x/\Delta t$ is the linear advection coefficient, and $\nu = (\Delta x)^2/2\Delta t$ is the diffusion coefficient.

2. The Cellular Automaton

To model systems of the sort described in the Introduction by a cellular automaton, it is most convenient to impose the Fermi exclusion rule that no two particles occupying the same site may be moving in the same di-rection. That way, the state of each site is uniquely specified by two bits of information: the right bit, which is one if there is a rightward moving particle present and zero otherwise, and the left bit, which is one if there is a leftward moving particle present and zero otherwise. Thus each site has four possible states labelled by the four binary numbers from 00 to 11. Each step of the automaton has two substeps: in the first, *the collision substep*, the particles change their direction randomly at the present lattice site (either with or without bias) subject to the exclusion principle; in the second, *the advection substep*, the particles move to the neighboring lattice site in their new direction.

This exclusion rule induces a correlation between particles that are two lattice sites apart. To see this, suppose that three successive lattice sites have states 01, 00, and 10, respectively, after the collision substep of a time step. Then, after the collision substep of the next time step, it follows that the middle lattice site *must* have state 11. Thus the evolution of one particle is affected by that of another particle two sites away, so they are correlated. We shall now show that this correlation naturally gives rise to the nonlinear term in Eq. (1).

Following the nomenclature developed in the introduction, we denote the right bit at lattice site k and time step l by $b_0(k,l)$, and the left bit at lattice site k and time step l by $b_1(k,l)$. After the collision substep of a time step, we denote the new states by $b_0'(k,l)$ and $b_1'(k,l)$. These are given by the truth table,

$b_1(k,l)$	$b_0(k,l)$	$b_1'(k,l)$	$b_0'(k,l)$
0	0	0	0
0	1	$(1-\alpha(k,l))/2$	$(1+\alpha(k,l))/2$
1	0	$(1-\alpha(k,l))/2$	$(1+\alpha(k,l))/2$
1	1	1	1

where $\alpha(k,l)$ is either 1 or -1 with mean $\overline{\alpha}$. The rule in the above table may be written in the form

$$b_0'(k,l) = \frac{1+\alpha(k,l)}{2}b_0(k,l) \vee b_1(k,l) + \frac{1-\alpha(k,l)}{2}b_0(k,l) \wedge b_1(k,l) \quad (2.1)$$

$$b_1'(k,l) = \frac{1-\alpha(k,l)}{2}b_0(k,l) \vee b_1(k,l) + \frac{1+\alpha(k,l)}{2}b_0(k,l) \wedge b_1(k,l). \quad (2.2)$$

Here, $\vee$ denotes the *inclusive or* operation, and $\wedge$ denotes the *and* operation on a pair of bits.

In the advection substep, the particles move to the neighboring lattice site in their new direction. The rule for this is easily seen to be

$$b_0(k+1,l+1) = b_0'(k,l) \quad (2.3)$$

$$b_1(k-1,l+1) = b_1'(k,l). \quad (2.4)$$

By composing the rules for the above two substeps, we arrive at the rule for one full time step of the cellular automaton

$$b_0(k+1,l+1) = \frac{1+\alpha(k,l)}{2}b_0(k,l) \vee b_1(k,l) + \frac{1-\alpha(k,l)}{2}b_0(k,l) \wedge b_1(k,l)$$
$$(2.5)$$

$$b_1(k-1, l+1) = \frac{1-\alpha(k,l)}{2} b_0(k,l) \vee b_1(k,l) + \frac{1+\alpha(k,l)}{2} b_0(k,l) \wedge b_1(k,l).$$

$$(2.6)$$

Now if b and b' are bits, there is a well-known algebraic representation for the $\vee$ and $\wedge$ operations: $b \vee b' = b + b' - bb'$ and $b \wedge b' = bb'$. Using this, the cellular automaton rule can be written in the algebraic form

$$b_0(k+1, l+1) = \frac{1+\alpha(k,l)}{2} (b_0(k,l) + b_1(k,l)) - \alpha(k,l)b_0(k,l)b_1(k,l) \quad (2.7)$$

$$b_1(k-1, l+1) = \frac{1-\alpha(k,l)}{2} (b_0(k,l) + b_1(k,l)) + \alpha(k,l)b_0(k,l)b_1(k,l). \quad (2.8)$$

Note that the nonlinear terms in Eqs. (7) and (8) owe their origin to the exclusion principle.

3. The Ensemble Average

We now turn our attention to *ensemble averages* of the automaton described in the last section; that is, we envision applying the above-described cellular automaton rule to a large set of systems, with possibly different initial conditions. For example, we might perform a large number of simulations of the automaton on a computer, using a grid of fixed size, with the initial conditions $b_i(k, 0)$ chosen randomly from some known distribution. Then, $\overline{b}_i(k, l)$ denotes the value of the ith bit at position k and time step l *averaged over all the simulations*. Henceforth, we shall consistently use overbars to denote ensemble averages.

A word should be said about the random numbers, $\alpha(k, l)$. Throughout this work, we shall assume that they are generated by a "perfect" random number generator. That is, we assume that

$$\overline{\alpha(k,l)} = \overline{\alpha} \tag{3.1}$$

and

$$\overline{\alpha(k,l)\alpha(k',l')} = \delta_{kk'}\delta_{ll'} + (1 - \delta_{kk'}\delta_{ll'})\overline{\alpha}^2. \tag{3.2}$$

Then, since $b_i(k, l)$ depends on past random numbers, $\alpha(k', l')$ with $l' < l$, we can do things like

$$\overline{\alpha(k,l)b_i(k,l)} = \overline{\alpha} \cdot \overline{b_i(k,l)}, \tag{3.3}$$

etc. A study of exactly how "perfect" a random number generator has to be in order to validate our results would be interesting, but is beyond the scope of the present paper. Note that one can regard the cellular automaton defined by Eqs. (7) and (8) as a *stochastic cellular automaton* thanks to the inclusion of the random $\alpha(k, l)$'s; or, if one prefers, one can regard the random number generator as part of the rule, in which case it is a perfectly *deterministic cellular automaton*.

In our simulations, we produced the random bits using a simple cellular automaton due to Wolfram [3] that generates bits with a high degree of randomness. To get random bits from Wolfram's automaton, one applies it to a finite string of bits (we used 59 bits) with periodic boundary conditions, and samples the values at one site as a function of time. To bias the mean of the random bits one can generate more than one unbiased random bit per site and then apply logical operations to them; for example, when two strings of unbiased random bits are combined using the "and" operation the result is a string of random bits with mean 0.25, and when they are combined using the "inclusive or" operation the result is a string of random bits with mean 0.75.

We now take the ensemble average of Eqs. (7) and (8) to get

$$\bar{b}_0(k+1, l+1) = \frac{1+\bar{\alpha}}{2}\left(\bar{b}_0(k,l) + \bar{b}_1(k,l)\right) - \bar{\alpha} \cdot \overline{b_0(k,l)b_1(k,l)} \qquad (3.4)$$

$$\bar{b}_1(k-1, l+1) = \frac{1-\bar{\alpha}}{2}\left(\bar{b}_0(k,l) + \bar{b}_1(k,l)\right) + \bar{\alpha} \cdot \overline{b_0(k,l)b_1(k,l)}. \qquad (3.5)$$

This may be written

$$\bar{b}_0(k+1, l+1) = \frac{1+\bar{\alpha}}{2}\left(\bar{b}_0(k,l) + \bar{b}_1(k,l)\right)$$
$$- \bar{\alpha} \cdot \left(\bar{b}_0(k,l)\,\bar{b}_1(k,l) + C_{01}(k,l;k,l)\right) \qquad (3.6)$$

$$\bar{b}_1(k-1, l+1) = \frac{1-\bar{\alpha}}{2}\left(\bar{b}_0(k,l) + \bar{b}_1(k,l)\right)$$
$$+ \bar{\alpha} \cdot \left(\bar{b}_0(k,l)\,\bar{b}_1(k,l) + C_{01}(k,l;k,l)\right), \qquad (3.7)$$

where we have defined the *covariance*

$$C_{ij}(k,l;k',l') \equiv \overline{(b_i(k,l) - \bar{b}_i(k,l))(b_j(k',l') - \bar{b}_j(k',l'))}. \qquad (3.8)$$

We now begin to establish the relationship between ensemble-averaged quantities and solutions of the Burgers' equation, Eq. (1), by introducing new quantities which are more directly related to those solutions. Define

$$\bar{b} \equiv \bar{b}_0 + \bar{b}_1 \qquad (3.9)$$

$$\bar{v} \equiv \left(\bar{b}_0 - \bar{b}_1\right)/\Delta x, \qquad (3.10)$$

where Δx is the lattice spacing. Thus

$$\bar{b}_0 = \frac{1}{2}\left(\bar{b} + \bar{v}\Delta x\right) \qquad (3.11)$$

$$\bar{b}_1 = \frac{1}{2}\left(\bar{b} - \bar{v}\Delta x\right). \qquad (3.12)$$

Substituting, we can find the update rules for $\bar{b}$ and $\bar{v}$,

$$
\begin{aligned}
\bar{b}(k, l+1) &= \frac{1+\overline{\alpha}}{2}\bar{b}(k-1, l) + \frac{1-\overline{\alpha}}{2}\bar{b}(k+1, l) \\
&\quad + \frac{\overline{\alpha}}{4}[\bar{b}^2(k+1, l) - \bar{b}^2(k-1, l) \\
&\quad - (\Delta x)^2\left(\bar{v}^2(k+1, l) - \bar{v}^2(k-1, l)\right)] \\
&\quad + \overline{\alpha}[C_{01}(k+1, l; k+1, l) - C_{01}(k-1, l; k-1, l)]
\end{aligned} \tag{3.13}
$$

$$
\begin{aligned}
\bar{v}(k, l+1) &= -\frac{\bar{b}(k+1, l) - \bar{b}(k-1, l)}{2\Delta x} + \frac{\overline{\alpha}}{\Delta x}\frac{\bar{b}(k-1, l) + \bar{b}(k+1, l)}{2} \\
&\quad + \frac{\overline{\alpha}}{4\Delta x}[-\bar{b}^2(k+1, l) - \bar{b}^2(k-1, l) \\
&\quad + (\Delta x)^2\left(\bar{v}^2(k+1, l) + \bar{v}^2(k-1, l)\right)] \\
&\quad + \frac{\overline{\alpha}}{\Delta x}[C_{01}(k+1, l; k+1, l) + C_{01}(k-1, l; k-1, l)]
\end{aligned} \tag{3.14}
$$

Now suppose that $n(x, t)$ is the *exact* solution of Eq. (1), and define

$$
w(x, t) \equiv \frac{c}{2\nu}\left(n(x, t) - \frac{1}{2}n^2(x, t)\right) - \frac{\partial}{\partial x}n(x, t). \tag{3.15}
$$

Discretize these by defining

$$
\hat{n}(k, l) \equiv n(k\Delta x, l\Delta t) \tag{3.16}
$$

$$
\hat{w}(k, l) \equiv w(k\Delta x, l\Delta t). \tag{3.17}
$$

Then by Taylor expanding and using the fact that $n(x, t)$ solves Eq. (1), we find

$$
\begin{aligned}
\hat{n}(k, l+1) &= \frac{1+\overline{\alpha}}{2}\hat{n}(k-1, l) + \frac{1-\overline{\alpha}}{2}\hat{n}(k+1, l) \\
&\quad + \frac{\overline{\alpha}}{4}[\hat{n}^2(k+1, l) - \hat{n}^2(k-1, l) \\
&\quad - (\Delta x)^2\left(\hat{w}^2(k+1, l) - \hat{w}^2(k-1, l)\right)] \\
&\quad + O\left((\Delta x)^4\right)
\end{aligned} \tag{3.18}
$$

$$
\begin{aligned}
\hat{w}(k, l+1) &= -\frac{\hat{n}(k+1, l) - \hat{n}(k-1, l)}{2\Delta x} + \frac{\overline{\alpha}}{\Delta x}\frac{\hat{n}(k-1, l) + \hat{n}(k+1, l)}{2} \\
&\quad + \frac{\overline{\alpha}}{4\Delta x}[-\hat{n}^2(k+1, l) - \hat{n}^2(k-1, l) \\
&\quad + (\Delta x)^2\left(\hat{w}^2(k+1, l) + \hat{w}^2(k-1, l)\right)] \\
&\quad + O\left((\Delta x)^2\right).
\end{aligned} \tag{3.19}
$$

Note that these are very similar in form to the update equations for $\bar{b}$ and $\bar{v}$, Eqs. (13) and (14). The covariances, C_{01}, that appeared in the former equations have been replaced by truncation errors from the Taylor expansion in the latter equations.

Thus we can define the *errors*,

$$e(k,l) \equiv \bar{b}(k,l) - \hat{n}(k,l) \tag{3.20}$$

$$f(k,l) \equiv \bar{v}(k,l) - \hat{w}(k,l). \tag{3.21}$$

The update rule for the errors is then

$$
\begin{aligned}
e(k,l+1) &= \frac{1+\bar{\alpha}}{2}e(k-1,l) + \frac{1-\bar{\alpha}}{2}e(k+1,l) \\
&+ \frac{\bar{\alpha}}{4}[\left(\bar{b}(k+1,l) + \hat{n}(k+1,l)\right)e(k+1,l) \\
&- \left(\bar{b}(k-1,l) + \hat{n}(k-1,l)\right)e(k-1,l)] \\
&+ \bar{\alpha}\left[C_{01}(k+1,l;k+1,l) - C_{01}(k-1,l;k-1,l)\right] \\
&+ O\left((\Delta x)^4\right)
\end{aligned}
\tag{3.22}
$$

$$
\begin{aligned}
f(k,l+1) &= -\frac{e(k+1,l) - e(k-1,l)}{2\Delta x} + \frac{\bar{\alpha}}{\Delta x}\frac{e(k-1,l) + e(k+1,l)}{2} \\
&+ \frac{\bar{\alpha}}{4\Delta x}[-\left(\bar{b}(k+1,l) + \hat{n}(k+1,l)\right)e(k+1,l) \\
&- \left(\bar{b}(k-1,l) + \hat{n}(k-1,l)\right)e(k-1,l)] \\
&+ \frac{\bar{\alpha}}{\Delta x}\left[C_{01}(k+1,l;k+1,l) + C_{01}(k-1,l;k-1,l)\right] \\
&+ O\left((\Delta x)^2\right).
\end{aligned}
\tag{3.23}
$$

Note that the evolution equation for $e(k,l)$ has decoupled from that of $f(k,l)$, so that it suffices to consider Eq. (22) alone. If we ignore the correlation (but *not* the truncation) terms, then we have the matrix equation

$$e(j,l+1) = \sum_k L(j,k;l)e(k,l) + O\left((\Delta x)^4\right) \tag{3.24}$$

where $L(j,k;l)$ has positive elements, and columns that sum to unity. If we use the $\mathcal{L}_1$ norm,

$$\|e(l)\| \equiv \frac{\Delta x}{L}\sum_j |e(j,l)|, \tag{3.25}$$

then we may write

$$\|e(l+1)\| = \frac{\Delta x}{L}\sum_j |e(j,l+1)|$$

$$\leq \frac{\Delta x}{L}\sum_j \left|\sum_k L(j,k;l)e(k,l)\right| + \mathcal{O}\left((\Delta x)^4\right)$$

$$\leq \frac{\Delta x}{L}\sum_{jk} L(j,k;l)|e(k,l)| + K(\Delta x)^4$$

$$= \frac{\Delta x}{L}\sum_k |e(k,l)| + K(\Delta x)^4$$

$$= \|e(l)\| + K(\Delta x)^4 \qquad (3.26)$$

where K is some constant. Here we have used the triangle inequality, the positivity of the $L(j,k;l)$, and the fact that the columns of $L(j,k;l)$ sum to unity. Then, supposing $e(0)=0$, we see that iteration for $\mathcal{O}\left((\Delta x)^{-2}\right)$ generations (times of order unity) will still yield $\|e(l)\| = \mathcal{O}\left((\Delta x)^2\right)$.

Now fix a subinterval (x_1, x_2) of the spatial domain of $n(x,t)$, and fix a time t_0. Let $\Delta x \to 0$ such that $x_1 = k\Delta x$, $x_2 = (k+M)\Delta x$, and $t_0 = l\Delta t$. Then by basic quadrature estimates

$$\frac{1}{M}\sum_{i=0}^{M-1} \hat{n}(k+i,l) = \frac{1}{x_2-x_1}\int_{x_1}^{x_2} n(x,t_0)dx + \mathcal{O}\left((\Delta x)^2\right), \qquad (3.27)$$

while by our basic $\mathcal{L}_1$ error estimate we have

$$\frac{1}{M}\sum_{i=0}^{M-1} \bar{b}(k+i,l) = \frac{1}{M}\sum_{i=0}^{M-1} \hat{n}(k+i,l) + \mathcal{O}\left((\Delta x)^2\right). \qquad (3.28)$$

Here we have used the fact that $M\Delta x/L = (x_2-x_1)/L$ is fixed. Combining these results gives

$$\frac{1}{M}\sum_{i=0}^{M-1} \bar{b}(k+i,l) = \frac{1}{x_2-x_1}\int_{x_1}^{x_2} n(x,t_0)dx + \mathcal{O}\left((\Delta x)^2\right), \qquad (3.29)$$

which is our final convergence result for the ensemble average.

Thus, we have shown that the ensemble average of the cellular automaton simulates a stable, second-order accurate, fully-explicit differencing scheme for the Burgers' equation. Note that the proof of this is very similar in form to demonstrations of stability and accuracy for finite difference approximations. The neglect of the correlations is the weakest link in the chain of reasoning, and it will be discussed further in future work.

4. Bounding the Covariance for the Diffusion Equation

For the diffusion equation ($\bar{\alpha}=0$), we can obtain an upper bound on the covariance, C_{ij}. To get dynamical equations for the covariance, we write

$$C_{ij}(k,l+1;k',l+1) \equiv \overline{(b_i(k,l+1) - \bar{b}_i(k,l+1))(b_j(k',l+1) - \bar{b}_j(k',l+1))} \tag{4.1}$$

and use Eqs. (7) and (8) to express the right- hand side in terms of quantities at time step l. For $k \neq k'$ or $i = j$, we get

$$
\begin{aligned}
C_{00}\,(k+1,l+1;k'+1,l+1) &= \tfrac{1}{4}(C_{00}\,(k,l;k',l) + C_{01}\,(k,l;k',l) \\
&\quad + C_{10}\,(k,l;k',l) + C_{11}\,(k,l;k',l)) \\
C_{01}\,(k+1,l+1;k'-1,l+1) &= \tfrac{1}{4}(C_{00}\,(k,l;k',l) + C_{01}\,(k,l;k',l) \\
&\quad + C_{10}\,(k,l;k',l) + C_{11}\,(k,l;k',l)) \\
C_{10}\,(k-1,l+1;k'+1,l+1) &= \tfrac{1}{4}(C_{00}\,(k,l;k',l) + C_{01}\,(k,l;k',l) \\
&\quad + C_{10}\,(k,l;k',l) + C_{11}\,(k,l;k',l)) \\
C_{11}\,(k-1,l+1;k'-1,l+1) &= \tfrac{1}{4}(C_{00}\,(k,l;k',l) + C_{01}\,(k,l;k',l) \\
&\quad + C_{10}\,(k,l;k',l) + C_{11}\,(k,l;k',l))
\end{aligned}
\tag{4.2}
$$

When $k = k'$ and $i \neq j$, however, we get

$$
\begin{aligned}
C_{01}\,(k+1,l+1;k-1,l+1) &= C_{01}(k,l;k,l) - \tfrac{1}{4}\left(\bar{b}_1(k,l) - \bar{b}_0(k,l)\right)^2 \\
C_{10}\,(k-1,l+1;k+1,l+1) &= C_{10}(k,l;k,l) - \tfrac{1}{4}\left(\bar{b}_0(k,l) - \bar{b}_1(k,l)\right)^2 .
\end{aligned}
\tag{4.3}
$$

Note that Eqs. (2) are homogeneous in the covariances, while Eqs. (3) contain forcing terms on the right hand side. Because these forcing terms are negative definite, we can use induction on l to conclude that

$$
C_{ij}(k,l;k',l) \leq 0
\tag{4.4}
$$

if $k \neq k'$ or $i \neq j$.

Suppose that we use spatial averaging to estimate the density n at a given gridpoint k. For example, we could average over M gridpoints to get the density

$$
n_s \equiv \frac{1}{M} \sum_{i=0}^{M-1} [b_0(k+i,l) + b_1(k+i,l)].
\tag{4.5}
$$

We would like to compare this with the ensemble-averaged version of the same thing,

$$
n_e \equiv \frac{1}{M} \sum_{i=0}^{M-1} [\bar{b}_0(k+i,l) + \bar{b}_1(k+i,l)],
\tag{4.6}
$$

at the same gridpoint, k. We find

$$
\overline{(n_s - n_e)^2} = \frac{1}{M^2} \sum_{i,j=0}^{M-1} (C_{00}(k+i,l;k+j,l) + C_{01}(k+i,l;k+j,l)
$$
$$
+ C_{10}(k+i,l;k+j,l) + C_{11}(k+i,l;k+j,l))
$$

$$\leq \frac{1}{M^2} \sum_{i=0}^{M-1} (C_{00}(k+i,l;k+i,l) + C_{11}(k+i,l;k+i,l))$$

$$= \frac{1}{M^2} \sum_{i=0}^{M-1} [\bar{b}_0(k+i,l) - (\bar{b}_0(k+i,l))^2$$

$$+ \bar{b}_1(k+i,l) - (\bar{b}_1(k+i,l))^2]$$

$$\leq \frac{1}{M} n_e \left(1 - \frac{n_e}{2}\right)$$

$$\leq \frac{1}{2M} = \frac{\Delta x}{2(x_2 - x_1)} \tag{4.7}$$

where we have used the Schwarz inequality in the final step. Thus, the spatial averaging of a single evolution of the automaton converges to the spatial averaging of the ensemble as $1/\sqrt{M}$. The simulations presented in the next section were carried out in precisely this fashion; the displayed results are spatial averages for a single evolution of the automaton.

5. Simulations

The equation simulated by the above-described automaton is

$$\frac{\partial n}{\partial t} + c\frac{\partial}{\partial x}\left(n - \frac{n^2}{2}\right) = \nu\frac{\partial^2 n}{\partial x^2}, \tag{5.1}$$

where $0 < n < 2$. To maximize the signal-to-noise ratio, it is best to operate with $n \sim 1$. Then, the transformation

$$u = c(n - 1) \tag{5.2}$$

may be applied to the result, so that u obeys the Burgers' equation in standard form,

$$\frac{\partial u}{\partial t} - u\frac{\partial u}{\partial x} = \nu\frac{\partial^2 u}{\partial x^2}. \tag{5.3}$$

Note that $-c < u < c$, so the parameter c should be chosen greater than $\sup|u|$ to insure that $0 < n < 2$.

We have used the automaton on a Connection Machine [4] computer to simulate the solution to Burgers' equation with periodic boundary conditions on a spatial domain of unit length, and with initial condition

$$n(x,0) = n_a + n_b \cos(2\pi x). \tag{5.4}$$

The exact solution to this problem may be found by application of the Cole-Hopf transformation. It is

$$n = n_a + \frac{2\nu}{c\psi}\frac{\partial\psi}{\partial x}, \tag{5.5}$$

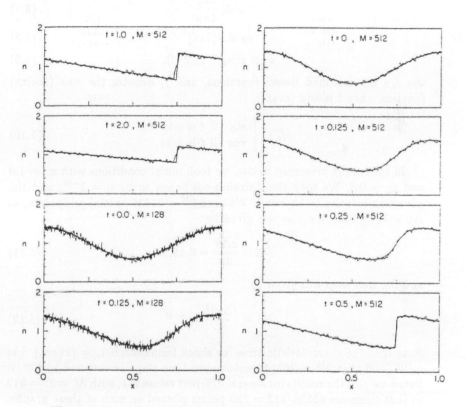

Figure 1: Simulations of the cellular automaton model as a function of time t. The solid curve gives the exact solution to Burgers' equation. The development of a shock is evident. Averages were made over M lattice sites.

where

$$\psi \equiv I_0(z) + 2\sum_{\ell=1}^{\infty}(-1)^{[\ell/2]}I_\ell(z)f_\ell(2\pi\ell x + \nu_\ell t)\exp(-\mu_\ell t), \qquad (5.6)$$

where in turn

$$z \equiv \frac{cn_b}{4\pi\nu} \qquad (5.7)$$

$$\mu_\ell \equiv \nu(2\pi\ell)^2 \qquad (5.8)$$

$$\nu_\ell \equiv c(n_a - 1)(2\pi\ell), \qquad (5.9)$$

the I_ℓ's are modified Bessel functions, and f_ℓ denotes the sine (cosine) function when ℓ is odd (even):

$$f_\ell \equiv \begin{cases} \sin & \text{if } \ell \text{ is odd} \\ \cos & \text{if } \ell \text{ is even.} \end{cases} \qquad (5.10)$$

In the results presented below, we took initial conditions with $n_a = 1.0$ and $n_b = 0.4$. We took the diffusion coefficient to be $\nu = 2^{-15}$, and the advection velocity to be $c = 1$. We used $2^{16} = 65536$ spatial gridpoints, so $\Delta x = 2^{-16}$. Then the bias was given by

$$\overline{\alpha} = \frac{c\Delta x}{2\nu} = 0.25, \qquad (5.11)$$

and the time step was given by

$$\Delta t = \frac{(\Delta x)^2}{2\nu} = 2^{-18}. \qquad (5.12)$$

Note that the characteristic time for shock formation is $t_s = (2\pi cn_b)^{-1} \approx 0.398$, and that $2^{18} = 262144$ automaton time steps correspond to $t = 1$. Below we plot the results for several different values of t, with $M = 2^9 = 512$ so that there are $65536/512 = 128$ points plotted on each of these graphs. The last two plots, however, were made with $M = 128$; note that the amplitude of the noise in these plots is roughly twice that in the other plots, as expected from Eq. (7). The smooth curves are the exact answer, as given by Eq. (5).

6.   Conclusion

We have motivated, developed, and analyzed a cellular automaton for the simulation of Burgers' equation. As stated at the outset of this paper, we are not proposing that this technique be used as an effective method for computing solutions to Burgers' equation, but rather that it be used to study in a relatively simple context important issues about stability and accuracy that arise in the application of cellular automaton techniques to solving the incompressible Navier-Stokes equations [1]. The argument used to show that the cellular automaton does indeed yield an approximation to the solution of the partial differential equation is the same for both cases; including the neglect of the correlations. The general procedure used to get the solution from the cellular automaton is the same in both cases: the spatial average of the cellular automaton is used to approximate the spatial average of the ensemble average of the automaton.

Acknowledgements

The work of BMB was performed in part at Lawrence Livermore National Laboratory under DOE contract number W-7405-ENG-48. The work of CDL was performed in part at Lawrence Livermore National Laboratory under DOE contract number W-7405-ENG-48, and in part at Brown University.

References

[1] See for example, U. Frisch, B. Hasslacher, Y. Pomeau, "Lattice Gas Automata for the Navier-Stokes Equation," *Physical Review Letters*, **56** (1986) 1505; and S. Wolfram, "Cellular Automaton Fluids 1: Basic Theory," *Journal of Statistical Physics*, **45** (1986) 471-526, and references contained therein.

[2] W. Feller, *An Introduction to Probability Theory and its Applications*, Volume I, Chapter 14, Section 6 (J. Wiley, 1970).

[3] S. Wolfram, "Random Sequence Generation by Cellular Automata," *Advances in Applied Mathematics*, **7** (1986) 123. See, in particular, eqn. (3.1a).

[4] D. Hillis, *The Connection Machine*, (M.I.T. Press, 1985).

Hudong Chen, Shiyi Chen, Gary Doolen and Y. C. Lee
Center for Nonlinear Studies, Los Alamos National Laboratory, Los Alamos, NM 87545, USA

Simple Lattice Gas Models for Waves

This paper originally appeared in *Complex Systems* (1988), Volume 2, pages 259–267.

Complex Systems **2** (1988) 259–267

Simple Lattice Gas Models for Waves

Hudong Chen
Shiyi Chen
Gary Doolen
Y. C. Lee
Center for Nonlinear Studies, Los Alamos National Laboratory,
Los Alamos, NM 87545, USA

Abstract. A simple lattice gas model for solving the linear wave equation is presented. In this model a photon representation is used. Energy and momentum are shown to be conserved.

1. Introduction

The rapidly developing cellular automata (CA) theory, known also as the lattice gas method, has recently generated wide interest for modeling many different physical processes described generally by partial differential equations [1–5]. A lattice gas system consists of "particles" moving on a lattice satisfying certain symmetry requirements. The updating of the system is realized by designing microscopic rules for the moving and scattering of the particles. The solution of the partial differential equation of interest is usually approximated in terms of the averaged behavior of a set of microscopic quantities.

We can list three significant aspects of using the lattice gas method to study physical systems. First, since all particle interactions are local, this method provides a way to utilize concurrent architectures [6]. The advent of computers with 10^6 to 10^9 processors would allow computations to be done at 10^4 to 10^7 times present speeds with slower and therefore more reliable processors. Consequently, it is expected that the CA techniques will become a very important computational tool for numerical modeling. Second, because the CA operates with only integers and Boolean algebra, it requires less computer storage so that the spatial and time resolution can be much higher than other methods. For example 8×10^9 particles can be followed on existing Cray Solid State Disks containing 512 megawords. Many realistic physical processes can be simulated more accurately without roundoff or instabilities. Third, this method may provide new insight in understanding the relationship between the microscopic machanisms and the macroscopic behavior for some many-body physical systems.

The CA theory has been successful in modeling fluids [1,2,4], magnetofluids [3,5] and other systems [7]. It also appears in principle that most physical systems with diffusive macroscopic dynamics can be approximated by CA models. The macroscopic behavior of some diffusive systems can be described by the following parabolic equation:

$$\frac{\partial f(\mathbf{x}, t)}{\partial t} = F(t, \mathbf{x}; f) + D\nabla^2 f(\mathbf{x}, t), \tag{1.1}$$

where the macroscopic quantity $f(\mathbf{x}, t)$ represents, for example, the averaged lattice particle density. The parameter D is the diffusivity. $F(t, \mathbf{x}; f)$ is usually a nonlinear function of f.

It is of equal importance to investigate whether or not physical processes described by the hyperbolic differential equations can be modeled within CA framework [8]. An important process of this kind is wave propagation. Wave propagation possesses several features which differ from the diffusion process. In this report, we present a simple CA model which simulates wave propagation and conserves energy and momentum.

2.   The wave equation model

Wave propagation processes are described by many different types of equations. However, the basic mechanics contains a common feature which is described by the simplest linear wave propagation process governed by the following linear hypobolic equation:

$$\frac{\partial^2 u(\mathbf{x}, t)}{\partial t^2} = C^2 \nabla^2 u(\mathbf{x}, t), \tag{2.1}$$

where u is the wave amplitude and C is the wave speed. A typical example is electrodynamics described by the Maxwell equations. Written in terms of the scalar potential ϕ and the vector potential $\mathbf{A}$ and using the Lorentz gauge, the Maxwell equations without sources can be written as two linear wave equations in the form of equation (2.1) for ϕ and $\mathbf{A}$, plus the continuity equation

$$\frac{\partial \phi}{\partial t} + \nabla \cdot \mathbf{A} = 0.$$

Comparing equation (2.1) with equation (1.1), one can see that the wave equation has a second order time derivative, while the diffusion equation has a first order time derivative. This is the major difference between a wave propagation and diffusion.

Two physical quantities are conserved in a linear wave propagation process: (a) the total wave energy, H, defined as:

$$H = \int d\mathbf{x} \{(\frac{\partial u}{\partial t})^2 + C^2 (\nabla u)^2\}, \tag{2.2}$$

and (b) the total wave momentum $\mathbf{P}$:

$$\sum_a m_a \tilde{U}(\mathbf{x}, t) = U(\mathbf{x}, t).$$

The integer m_a is dependent not only on the velocity state $\hat{c}_a$ but also, in general, on $\mathbf{x}$ and t. Only one type of photon, each with the same value of σ, is emitted from a lattice site. All photons emitted have the same sign as their source $\tilde{U}(\mathbf{x}, t)$. Hence, $\sigma\tilde{U} \geq 0$. After emitting photons, the source decays obeying the following relationship

$$\tilde{U}(\mathbf{x}, t+1) - \tilde{U}(\mathbf{x}, t) = -G(\mathbf{x}, t+1),$$

where G is the decay rate of the source $\tilde{U}$. The change of the decay rate follows the relation

$$G(\mathbf{x}, t+1) - G(\mathbf{x}, t) = \sum_{a,\sigma} \sigma\{N_a^\sigma(\mathbf{x}, t) - N_a^\sigma(\mathbf{x} + \hat{c}_a, t)\}$$

which is equal to the difference between the net photon σ emitted and the net photon σ absorbed at time t. It is straightforward to see that the combination of the above two microscopic updating rules leads to a second order time derivative for the evolution of the averaged source $\tilde{U}$ in the continuous limit. As soon as the relationship between the photon number N_a^σ and the wave source $\tilde{U}$ is determined by fixing the factor m_a, the evolution of the wave process will be completely determined. With different choices of m_a we are able to formulate different wave processes. In general, m_a can be chosen to be a function of U. However, the given wave equation becomes the linear wave equation (2.1) when the integer m_a; $(a = 1, \ldots, B)$ is a constant everywhere.

The boundary conditions for the wave CA systems are easy to implement. For example, the "fixed" boundary condition $(U(\mathbf{x}_0) = 0)$ is realized by reversing the normal component (with respect to the boundary) of the velocity direction of the photons at the boundary while keeping the parallel component unchanged. The photon σ on the boundary is changed into $-\sigma$. Similarly, the "free" boundary condition $(\nabla_\perp U(\mathbf{x}_0) = 0)$ can be realized by leaving the photon spin value σ unchanged.

The choice of the microscopic rules for formulating a linear wave process is not unique. Thus a criterion for chosing the best process is desired. Using Huygen's principle, we wish to require the photon sources to be as isotropic as possible. We call this requirement the isotropic condition. This condition implies that we need to put $m_a B = m$, $(a = 1, \ldots, B)$. Furthermore, it should be emphasized that since we use photons as the only information carriers from one place to another, any physical information cannot have a transport speed greater than the photon speed. If we define the effective photon speed in a given direction to be the distance between two neighboring sites along this direction divided by the minimum number of time steps require for a photon to go from one site another, then it can be seen that this speed depends on direction. The effective photon speed depends on the discrete velocity directions on a discrete lattice. A photon cannot always travel in a straight line in an arbitrary direction, hence there is a minimum effective photon speed c/ν ($\leq c$) for each model. For example,

$$\mathbf{P} = 2 \int d\mathbf{x} \{ (\frac{\partial u}{\partial t}) \nabla u - u \nabla (\frac{\partial u}{\partial t}) \}. \tag{2.3}$$

It is important that both of these conservation laws are satisfied in any CA model for the wave equation.

The lattice gas model that we are presenting for the linear wave propagation process consists of a number of "photons" propagating on a lattice. The lattice is invariant under rotations with angles of $2\pi/B$. (For the two-dimensional square and triangular lattices, B equals 4 and 6, respectively. While for the one-dimensional lattice, B is equal to 2.) The distance between any two nearest neighbor sites of the lattice is c. A photon at a site of the lattice moves at each time step to one of its B nearest neighbors with speed c, if we set the time step equal to one time unit. Although we could in general let photons have many different speeds or even perform a random walk [4,5], only one speed is required for the linear wave.

We define two kinds of photons distinguished by a "spin" quantum number σ. One kind of photon has the values $\sigma = \xi$ and another has $-\xi$, respectively. They could be considered as particles and anti-particles. Moreover, we define a cancellation process: at a given position when there is a ξ photon and a $-\xi$ photon, a cancellation occurs so that both of these photons are destroyed. Therefore, after each cancellation only one kind photon is left, namely the kind originally having the larger number of photons. This macroscopic rule causes the total photon number to fluctuate. We let the total σ at a point in space represent the wave amplitude. The two different kinds of photons have the same magnitude but they differ in sign. The wave amplitude at a given location and time is defined to be the sum of the local σ. That is, if we define $N_a^\sigma(\mathbf{x}, t)$; $(\sigma = \xi, -\xi; a = 1, \dots, B)$ to be the number of photons with quantum σ at a particular site $\mathbf{x}$ and time t moving with velocity $\hat{c}_a$ ($|\hat{c}_a| = c$) in the direction a, then the microscopic wave amplitude is defined by

$$U(\mathbf{x}, t) = \sum_{a,\sigma} \sigma N_a^\sigma(\mathbf{x}, t).$$

For computational convenience, we further require that there are no more than N_0 photons with the same σ at any site of $\mathbf{x}$. This requirement causes the photons to behave somewhat like Fermi particles instead of Bosons. Only when N_0 is infinity do these CA photons become bosons. However, if we require $N_0 \gg$ (the total number of photons of either kind at any time), then these photons can be approximately considered as Bosons.

In order to formulate a wave model, the microscopic updating rules must be chosen. According to Huygen's principle, any spatial point on a wave front can be thought of as a new wave source. Hence, we consider at each site of the lattice there is a wave source which emits photons as isotropically as possible. The intensity of a source is defined as $\tilde{U}(\mathbf{x}, t)$. The number of photons emitted at time t from a source in direction c_a is equal to $m_a \times |\tilde{U}(\mathbf{x}, t)|$ ($m_a \geq 0; a = 1, \dots, B$). Using the wave amplitude definition above, we have under the isotropic condition, ν is equal to unity for one-dimensional lattice. $\nu = \sqrt{2}$ for two-dimensional square lattice associated with the diagonal direction of a two-dimensional square lattice. Likewise, we have $\nu = \sqrt{3}$ for three-dimensional square lattice. As a result, any physically meaningful wave propagation process should not have its wave speed C exceed c/ν. This sets an upper bound on allowed speeds for any information transfer. This condition can be violated if a wrong set of microscopic rules are selected. For instance, if the photon source intensity is too high and its emitted photons cannot attain enough speed to propagate away, cumulation will occur which in turn leads to instability. We refer to this requirement as the minimum speed condition, or the CA Courant condition. If both the isotropic condition and the minimum speed condition are satisfied, the evolution of the system obeys the following equation:

$$U(\mathbf{x}, t+1) - 2U(\mathbf{x}, t) + U(\mathbf{x}, t-1) = m\{\frac{1}{B}\sum_a U(\mathbf{x}+\hat{c}_a, t) - U(\mathbf{x}, t)\} \quad (2.4)$$

Equation (2.4) defines a deterministic microscopic CA wave process with discrete space and time. It happens to have the same form as a finite difference equation. However, unlike the usual finite difference approximation, it contains no roundoff error and no numerical instability. With a particular choice of m the CA system conserves the microscopic energy and momentum. Moreover, in the limit that the lattice cell size is small compared with characteristic lengths and the time step size is small compared with characteristic times, the continuous linear wave equation (2.1) is recovered after making an ensemble averaging $(\langle U(\mathbf{x}, t)\rangle = u(\mathbf{x}, t))$ and a Taylor expansion in time and space.

We have proved that the wave model conserves the following two global quantities, if the microscopic wave equation (2.4), expressed in terms of integers, can be exactly satisfied:

$$H = \sum_{\mathbf{x}}[U(\mathbf{x}, t)^2 - U(\mathbf{x}, t+1) * U(\mathbf{x}, t-1)]$$

and

$$\begin{aligned}
\mathbf{P} = \;& \frac{1}{c^2} \sum_{\mathbf{x}}\sum_a \hat{c}_a\{U(\mathbf{x}, t+1) * U(\mathbf{x}+c_a, t) + U(\mathbf{x}, t-1) * U(\mathbf{x}-\hat{c}_a, t) \\
& - U(\mathbf{x}, t) * [U(\mathbf{x}+\hat{c}_a, t+1) + U(\mathbf{x}-\hat{c}_a, t-1)]\}.
\end{aligned}$$

It can be shown that above H and $\mathbf{P}$ reduce to the usual H and $\mathbf{P}$ for the linear wave in the continuous space and time limit, as expressed in equations (2.2) and (2.3). Therefore these correspond to the microscopic energy and momentum for the CA wave system. The conservation of above two global quantities is equivalent to $H(t) = H(t+1)$ and $\mathbf{P}(t) = \mathbf{P}(t+1)$. That is, it can be shown that H and $\mathbf{P}$ expressed in terms of the quantities at time t are unchanged when we replace all the quantities by those at time $t+1$, using the microscopic wave equation (2.4).

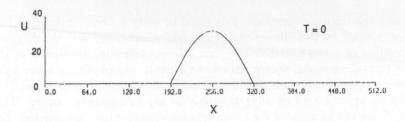

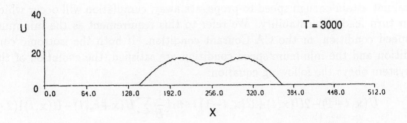

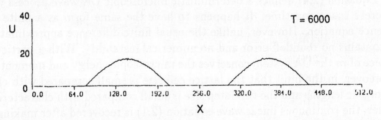

Figure 1: Evolution of a one-dimensional lattice gas wave at different time steps.

With the consideration of the CA Courant condition inequality, however, we have found that the microscopic energy H is a positive-definite quantity only when $B/m = \nu^2$ in equation (2.4). That is, the wave speed C must equal c/ν, the effective minimum photon speed. This leads to the conclusion that only the one-dimensional wave lattice gas model can satisfy the microscopic wave equation (2.4), since $\nu = 1$ in this case. Thus, it satisfies the microscopic conservation laws exactly. In higher dimensions the above global quantities H and $\mathbf{P}$ are conserved statistically. That is, we replace the microscopic wave amplitude U by its ensemble averaged value $u\ (= \langle U \rangle)$.

Results of a one-dimensional CA wave computer simulation are shown in figure 1 which describes the time evolution of a wave packet. This model is an exact microscopic wave process. The wave amplitude at $t < 0$ is set to be zero everywhere, while at $t = 0$ the amplitude is shown in the figure. As expected, the single wave packet evolves into a right-traveling wave packet and a left-traveling wave packet. Their shapes are preserved at all times.

For higher dimensional lattices the conservation laws are not exactly satisfied since $\nu > 1$, hence $m\tilde{U}(\mathbf{x}, t)$ may not always be evenly divided by B. It is in general impossible for the total number of photons emitted $(= m\tilde{U}(\mathbf{x}, t))$ to be distributed equally among all the B possible directions. In other words, $m_a; (a = 1, \ldots, B)$ can not be made exactly equal to m/B. Therefore equation (2.4) becomes only a statistical description of the wave CA for dimensions higher than 1.

Based on numerical experience, it is desirable, that both the isotropic condition and the CA Courant condition be satisfied as closely as possible in order to minimize microscopic fluctuations. One way to realize such a requirement in two-dimensional and three-dimensional, is first to distribute the total $m\tilde{U}(\mathbf{x}, t)$ $(m = 2)$ photons equally among all the B directions, then put the possible remaining two photons into two anti-parallel directions. These two directions are selected with equal probability among all $B/2$ pairs of directions. In this way, it can be shown that the microscopic equation of motion of the CA wave system is

$$U(\mathbf{x}, t+1) - 2U(\mathbf{x}, t) + U(\mathbf{x}, t-1) = \sum_a m_a U(\mathbf{x} + \hat{c}_a, t) - mU(\mathbf{x}, t)$$

where $\sum_a m_a = m$ and the ensemble averaged $\langle m_a \rangle = m/B$, $(a = 1, \ldots, B)$. Thus equation (2.4) becomes statistically valid and the two above conservation laws are satisfied if we replace U by its averaged value.

As a result, we can use the wave CA model to simulate light experiments. For example, figure 2 gives the results of the two-dimensional wave CA simulation for the double-slit experiment at a given instant. In this simulation, 256×256 lattice cells are used. At 64 cells away from the left boundary a wall with two holes each with width of 5 cells is inserted, so that photons can go through the holes in order to go from left region into the right region. Elsewhere on the wall, photons will be reflected back. Initially, we put a plane sine-wave with amplitude of 16 $|\sigma|$ and wave length of 32 cells in the left region. The right region is empty initially. A spatial averaging is used with the average super-cell size of 4×4. As expected the wave amplitude exhibits a spatial interference pattern.

It is easy to see that the microscopic fluctuations δU ($\equiv U - \langle U \rangle$) induced by the above method in two-dimensional and three-dimensional also approximately follow equation (2.4) except that an additional white noise source with maximum magnitude of $|\sigma|$ appears. Using standard mathematics it can be shown that the root mean square value of the fluctuation of a individual Fourier mode at large time t is

$$\sqrt{\langle (\delta U(\mathbf{k}, t))^2 \rangle} = |\sigma/\mathbf{k}|\sqrt{t},$$

where the wave number k satisfies $2\pi/c \geq |\mathbf{k}| \geq 2\pi/L$, and L is the system size. This indicates that the maximum spatial *rms* fluctuation $\sqrt{\langle (\delta U)^2 \rangle}$ is approximately $|\sigma|\sqrt{t}$, and the maximum spatial correlation of the fluctuation $\langle \delta U(\mathbf{x})\delta U(\mathbf{x} + \mathbf{r}) \rangle$ $(c \leq |\mathbf{r}| \leq L)$ is about $\sigma^2 ct/|\mathbf{r}|$ for three-dimensional

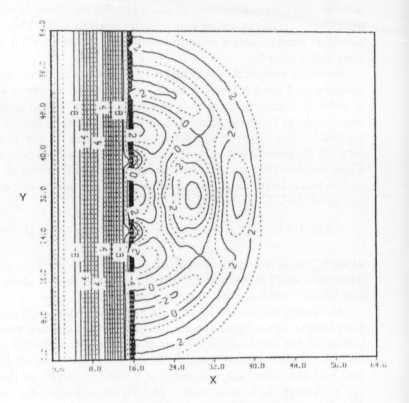

Figure 2: Spatial interference pattern of the wave amplitude in a two-dimensional wave lattice gas double-slit experiment at a given instant.

and $\sigma^2 ct \log(|\mathbf{r}|)$ for two-dimensional. Therefore, for a given number of time steps, a sufficiently large number of CA cells such that $L \gg c$ and the mean wave amplitude $|\bar{U}| \gg |\sigma|$, the noise is basically confined to small spatial scale for three-dimensional case and large scale wave structures remain undeformed. However, the fluctuation has stronger influence at large scales for two-dimensional.

3. Conclusions

In this report, we have presented a CA wave model for the linear wave equation. It is a many photon lattice gas system. Since the wave CA system conserves the macroscopic energy and momentum, its behavior follows a Hamiltonian dynamics. This is an essential requirement in formulating a CA wave process. It is foreseeable that more complicated wave processes may be constructed involving possibly many different kinds of nonlinear interactions. Since the CA provides a fast computational tool, many practical physical problems involving complicated geometries can be simulated.

Acknowledgements

We thank W. Mattheaus, D. Montgomery, H. Rose, and Y. T. Yan for their instructive discussions. The work is supported by the U. S. Department of Energy.

References

[1] U. Frisch, B. Hasslacher and Y. Pomeau, *Phys. Rev. Lett.*, **56** (1986) 1505.

[2] S. Wolfram, *J. Stat. Phys.*, **45** (1986) 471.

[3] D. Montgomery and G. Doolen, *Phys. Lett. A*, **120** (1987) 229.

[4] B. Boghosian and D. Levermore, *Complex Systems*, **1** (1987) 17.

[5] H. Chen and W. Matthaeus, *Phys. Rev. Lett.*, **58** (1987) 1845.

[6] T. Toffoli and N. Margolus, *Cellular Automata Machines*, (The MIT Press, Cambridge, MA, 1987).

[7] S. Wolfram, *Review of Modern Physics*, **55** (1983) 601.

[8] U. Frisch, D. d'Humières, B. Hasslacher, P. Lallemand, Y. Pomeau, and J. Rivet, *Complex Systems*, **1** (1987) 649.

Gary D. Doolen
Los Alamos National Laboratory, Los Alamos, NM 87545

Bibliography

This bibliography is a compilation of recent articles on lattice gas experimentation and theory, including those which appear in this volume. References are given in alphabetical order by primary author and some include the abstract of the paper. An attempt was made to include those articles through April 1989 which refer to the April 1986 paper by Frisch, Hasslacher, and Pomeau.

Abraham, Farid F. "Computational Statistical Mechanics Methodology, Applications and Supercomputing." *Adv. in Phys.* **35(1)** (1986):1–111.

Abstract. Computer simulation is adding a new dimension to scientific investigation, establishing a role of equal importance with the traditional approaches of experiment and theory. In this paper, we provide a text for understanding the computer simulation methodology of classical statistical mechanics. After developing the theoretical basis of the simulation techniques, the Monte Carlo and Langevin methods and various molecular dynamics methods are described. A very limited discussion is provided on interatomic potential functions, numerical integration schemes, and general simulation procedures for modeling different physical situations and for circumventing excessive computational burdens. The simulation methods are then illustrated using a variety of physical problems studied over the last several years at our laboratory. They include spinodal decomposition of a two-dimensional (2D) fluid, the melting of 2D and quasi-2D films, the structure and energetics of an incommensurate physisorbed film, and the roughening of a silicon solid-melt interface. Finally, we discuss the *supercomputer* issue, both in terms of super problems for supercomputers and parallel architectures leading to near-future computers that will be one thousand times faster than those currently available.

Aref, Hassan. "The Numerical Experiment in Fluid Dynamics." *J. Fluid Mech.* **173** (1986):15–41.

Abstract. Several aspects of the use of digital computers to generate solutions of equations of interest to fluid mechanics are discussed. The interdisciplinary nature of the field of computational fluid dynamics (CFD) is emphasized: the dependence on strides in computer technology, the impact of advances in algorithm development, the continuous interaction with laboratory experiment and analytical theory. The particular role of that mode of computer usage usually referred to as the numerical experiment is highlighted. 'Experiments' of this type have played a central role in establishing concepts such as the soliton and the strange attractor as paradigms within fluid mechanics. The ambitious goal of providing digital counterparts to laboratory equipment such as the wind tunnel is considered. The possibility of abandoning the Eulerian representation of flow fields in favor of following swarms of Lagrangian particles on a computer is stressed. Issues arising from and results of using this methodology are reviewed. Computer simulations are contrasted with computer-generated animation. The paper concludes with speculations on future developments.

Balasubramanian, K., F. Hayot, and W. F. Saam. "Darcy's Law from Lattice-Gas Hydrodynamics." *Phys. Rev. A.* **36(5)** (1987):2248–2253.

Abstract. Within the hexagonal lattice-gas model, we obtain Darcy's law for flow in the presence of scatterers. The associated momentum dissipation is described by an effective damping term in the Navier-Stokes equation, which we relate to the density of scatterers. The kinematic viscosity can be obtained from the Darcy velocity profile, once the permeability is determined. We also check that in the hexagonal lattice model, after coarse graining, velocity decay and plane-parallel Poiseuille flow occur as described by the macroscopic equations.

Baudet, C., and J. P. Hulin. "Lattice-Gas Automata: A Model for the Simulation of Dispersion Phenomena." *Phys. Fluids A* **1**(3) (1989):507.

> *Abstract.* The dispersion of a tracer in two-dimensional (2-D) parallel flow between two parallel plates has been studied numerically using a lattice-gas model with two different species of particles. After a stepwise change of concentration at one end of the model, the tracer distribution in the flow evolves theoretically, as predicted, toward a Gaussian profile in a time lapse in agreement with the predictions of the Taylor model. The mean square width of the front increases linearly with time after the stabilization period, as required for a diffusive spreading process. The longitudinal dispersion coefficient $D_\parallel$ has been determined in a range of Peclet number values Pe between 4.3 and 35.4. It varies as the square of the velocity, in agreement with the Taylor-Aris model; the molecular diffusion coefficient value ($D_m = 0.62$ in lattice and simulation step units) obtained from the proportionality coefficient is in good agreement with the values obtained by other independent methods.

Benzi, R., and S. Succi. "Bifurcations of a Lattice Gas Flow under External Forcing." IBM/ECSEC, 1987. Preprint.

Bernardin, D., and O. E. Sero-Guillaume. "Lattice Gas Mixtures Models for Mass Diffusion." LEMTA Nancy, 1989. Preprint.

Binder, P.-M. "Lattice Models of the Lorentz Gas: Physical and Dynamical Properties." *Complex Systems* **1** (1987):559–574.

Binder, P.-M. "Statistical Properties of Lorentz Lattice Gases." Los Alamos National Laboratory, 1987. Report LA-UR 87-3471.

Binder, P.-M., and D. d'Humières. "Self-Diffusion in a Tagged-Particle Lattice Gas." ENS, 1988. Preprint.

> *Abstract.* Collisions in lattice gas models do not conserve particle identity. Whenever one needs to follow individual particles, several choices are often consistent with a given model; these choices result in large variations of the diffusion coefficient. We illustrate this numerically and analytically.

Binder, P.-M. "The Properties of Tagged Lattice Fluids: I. Diffusion Coefficients." In *Discrete Kinematic Theory, Lattice Gas Dynamics, and Foundations of Hydrodynamics*, edited by R. Monaco. World Scientific, 1989, 28–37.

Binder, P.-M., D. d'Humières, and L. Poujol. "The Properties of Tagged Lattice Fluids: II. Velocity Correlation Functions." In *Discrete Kinematic Theory, Lattice Gas Dynamics, and Foundations of Hydrodynamics*, edited by R. Monaco. World Scientific, 1989, 38–43.

Binder, P.-M. "Numerical Experiments with Lattice Lorenz Gases." This volume, 471–480.

Boghosian, Bruce M., and C. David Levermore. "A Cellular Automaton for Burgers' Equation." *Complex Systems* 1 (1987):17–30. Reprinted in this volume, 481–496.

Abstract. We study the approximation of solution to the Burgers' equation,

$$\frac{\partial n}{\partial t} + c\frac{\partial}{\partial x}\left(n - \frac{n^2}{2}\right) = \nu\frac{\partial^2 n}{\partial x^2} \tag{1}$$

by spatially averaging a probabilistic cellular automaton motivated by random walks on a line. The automaton consists of moving "particles" on a one-dimensional periodic lattice with speed one and in a random direction subject to the exclusion principle that at most one particle may move in a given direction from a given lattice site, at a given time. The exclusion principle gives rise to the nonlinearity in Eq. (1) and introduces correlations between the particles which must be estimated to obtain statistical bounds on the error. These bounds are obtained in two steps. The first is showing that the ensemble average of the automaton is a stable explicit finite differencing scheme of Eq. (1) over the lattice with a second-order convergence in the lattice spacing. The numerical diffusion of this scheme plays an important role in relating the automaton rules to Eq. (1). The next step is showing that the spatial averaging of a single evolution of the automaton converges to the spatial averaging of the ensemble as $1/\sqrt{M}$ where M is the number of lattice sites averaged. Simulations are presented and discussed.

Boghosian, B., W. Taylor, and D. H. Rothman. "A Cellular Automata Simulation of Two-Phase Flow on the CM-2 Connection Machine Computer." *Proc. Supercomputing* 88 (1988).

Boghosian, B., and C. D. Levermore. "A Deterministic Cellular Automaton with Diffusive Behavior." In *Discrete Kinematic Theory, Lattice Gas Dynamics, and Foundations of Hydrodynamics*, edited by R. Monaco. World Scientific, 1989, 44–61.

Bonetti, M., A. Noullez, and J. P. Boon. "Viscous Fingering in a 2-D Porous Lattice." In *Discrete Kinematic Theory, Lattice Gas Dynamics, and Foundations of Hydrodynamics*, edited by R. Monaco. World Scientific, 1989, 394–398.

Bonetti, M., A. Noullez, and J. P. Boon. "Lattice Gas Simulations of Viscous Penetration." Preprint.

Boon, J.-P. "Lattice Gas Simulations: A New Approach to Fluid Dynamics." Université Libre de Bruxelles, 1989. Preprint.

Boon, Jean-Pierre, and Alain Noullez. "Lattice Gas Diffusion and Long Time Correlations." In *Discrete Kinematic Theory, Lattice Gas Dynamics, and Foundations of Hydrodynamics*, edited by R. Monaco. World Scientific, 1989, 309–407.

Abstract. Self-diffusion in lattice gases raises the problem of particle identification. Such an identification can be realized by introducing different types of particles and extending the cellular automata collision rules to include type conservation. Any particular set of collision rules induces specific diffusive behavior.

Simulations are performed to track a tagged particle using an extended version of the FHP model for colored automata. Mean-square displacements are measured for systems at low and moderate densities. Corresponding velocity autocorrelation functions are computed showing negative recorrelation at moderate density. Careful analysis of the long time behavior of the velocity autocorrelation functions has been conducted to investigate the long time tail effect. We conjecture that, within the limits of presently achievable accuracy, long time tails cannot be detected by direct measurement in lattice gases.

Bowler, K. C., A. D. Bruce, R. D. Kenway, G. S. Pawley, and D. J. Wallace. "Applications of Parallel Computing in Condensed Matter." *Physica Scripta* **T19** (1987).

Abstract. Computational methods permit the detailed study of microscopic properties and their macroscopic consequences in a host of problems in physics which may be inaccessible to direct experimental study and too complex for theoretical analysis. Reliable calculations from first principles, however, require enormous computing resources. In this talk we describe how parallel computing can provide a practical and cost-effective solution to this problem, illustrating the key ideas with examples from Monte Carlo and molecular dynamics simulations, electronic structure calculations and the analysis of experimental data.

Bowler, K. C., A. D. Bruce, R. D. Kenway, G. S. Pawley, and D. J. Wallace. "Exploiting Highly Concurrent Computers for Physics." *Physics Today* **Oct.** (1987):40.

Abstract. Architectures as varied as rigid arrays of many simple processors and reconfigurable networks of transputers are being used to solve problems as diverse as lattice gauge theories and neural networks.

Bowler, K. C., and R. D. Kenway. "Physics on Parallel Computers. Part 2: Applications." *Contemp. Phys.* **29(1)** (1988):33–55.

Abstract. Parallelism is an intrinsic feature of many physical systems. The design of high-performance computers is increasingly making use of the same concept by employing many processors working cooperatively to carry out a single computation. Consequently, the computer simulation of physical systems can rather naturally exploit the latest computer architectures to test theoretical models and to make measurements which are inaccessible to real experiments. In an earlier article we reviewed the design and use of parallel computers. Here we describe how they have been used to obtain insight into some fundamental problems in physics.

Burges, C., and S. Zaleski. "Buoyant Mixtures of Cellular Automata Gases." *Complex Systems* **1** (1987):31–50.

Burges, D., F. Hayot, and W. F. Saam. "Interface Fluctuations in a Lattice Gas." Ohio State University, 1988. Preprint.

Abstract. Within the framework of lattice gas hydrodynamics, we study fluctuations of an interface separating two immiscible fluids. The static fluctuations show the random walk behavior typical of systems with energies proportional to their lengths, whereas the time behavior of the interface can be described by a linear Langevin equation with noise. We discuss the thermodynamics of the interface fluctuating in the heat bath provided by the two fluids, measure and calculate the macroscopic and microscopic coefficients of surface tension, and check numerically the relevant fluctuation-dissipation relation.

Burges, D., F. Hayot, and W. F. Saam. "Model for Surface Tension in Lattice-Gas Hydrodynamics." *Phys. Rev. A* **38(7)** (1988):3589–3592.

Abstract. We introduce in lattice-gas hydrodynamics an interface between two fluids. The interface deforms locally under the impact of fluid particles. We define and measure a surface tension from Laplace's law applied to a circular bubble. Surface tension can be varied by changing the stiffness of the interface. A Boltzmann-type calculation gives an expression of surface tension in agreement with the numerical results.

Chang, Ching-Ray, and X. Y. Zhang. "The Dispersion Patterns of Acicular Particles in Viscous Medium." *IEEE Trans. on Magnetics* **23(5)** (1987):2886.

Abstract. A lattice model of ferrofluid with time-dependent viscosity has been developed to simulate the dispersion patterns of interacting fine ferromagnetic particles during the manufacturing process of magnetic recording medium. Closure and chaining of the magnetic particles in uniform magnetic field have been obtained. The length of the chain is dependent on the packing factor. For magnetic particles possessing electric charge, the agglomeration of particles will reduce. The hysteresis corresponding to different patterns are calculated using quasi-dynamic vectorial model.

Chen, Hudong, and William H. Matthaeus. "Cellular Automaton Formulation of Passive Scalar Dynamics." *Phys. Fluids* **30(5)** (1987):1235–1237.

Abstract. Cellular automata modeling of the advection of a passive scalar in a two-dimensional flow is examined in the context of discrete lattice kinetic theory. It is shown that if the passive scalar is represented by tagging or "coloring" automaton particles, a passive advection-diffusion equation emerges without use of perturbation expansions. For the specific case of the hydrodynamic lattice gas model of Frisch, Hasslacher, and Pomeau [*Phys. Rev. Lett.* **56**, 1505 (1986)] the diffusion coefficient is calculated by perturbation.

Chen, Hudong, and William H. Matthaeus. "New Cellular Automaton Model for Magnetohydrodynamics." *Phys. Rev. Lett.* **58(18)** (1987):1845–1848. Reprinted in this volume, 453–460.

Abstract. A new type of two-dimensional cellular automation method is introduced for computation of magnetohydrodynamic systems. Particle population is described by a 36-component tensor referred to a hexagonal lattice. By appropriate choice of the coefficients that control the modified streaming algorithm and the definition of the macroscopic fields, it is possible to compute both Lorentz-force and magnetic-induction effects. The method is local in the microscopic space and therefore suited to massively parallel computers.

Chen, H., S. Chen, Gary Doolen, and Y. C. Lee. "Simple Lattice Gas Models for Waves." *Complex Systems* **2** (1988):259–267. Reprinted in this volue, 497–508.

Abstract. A simple lattice gas model for solving the linear wave equation is presented. In this model a photon representation is used. Energy and momentum are shown to be conserved.

Chen, H., W. H. Matthaeus, and L. W. Klein. "An Analytic Theory and Formulation of a Local Magnetohydrodynamic Lattice Gas Model." *Phys. Fluids* **31(6)** (1988):1439.

Abstract. A theoretical description of the newly developed magnetohydrodynamic lattice gas model [*Phys. Rev. Lett.* **58**, 1845 (1987)] is presented. The model is a direct extension of the lattice gas model for incompressible Navier-Stokes fluids [*Phys. Rev. Lett.* **56**, 1506 (1986)]. In the present model the magnetic force and the magnetic induction effects are formulated with local microscopic dynamical rules only, using a bidirectional random walk process. The development of the theory strongly emphasizes the symmetries connecting the microscopic and macroscopic physics. A preliminary numerical test is described.

Chen, S., M. Lee, K. H. Zhao, and Gary D. Doolen. "A Lattice Gas Model with Temperature." *Physica D* (1989), to appear.

Abstract. A temperature-dependent lattice gas model is studied. The thermohydrodynamic equations are derived using the Chapman-Enskog expansion method. The model is applied to the Poiseuille flow and to heat conduction.

Chen, S., Zhen-Su She, L. C. Harrison, and Gary D. Doolen. "Optimal Initial Condition for Lattice-Gas Hydrodynamics." *Phys. Rev. A* **39(5)** (1989):2725.

Abstract. A method for minimizing the unphysical oscillations in simple lattice-gas hydrodynamic models is presented. Numerical simulations of two types of shear flows are reported that illustrate the usefulness of this method.

Chopard, Bastien, and Michel Droz. "Cellular Automaton Model for Heat Conduction in a Fluid." *Phys. Lett. A* **126(8,9)** (1988):476.

Abstract. A cellular automata model for a two-speed lattice gas is introduced to study heat propagation in a fluid at rest. Boltzmann and thermohydrodynamical equations are derived analytically. Explicit forms for the equation of state, the thermal conductivity and the Dufour coefficient are given. A good agreement is found between the theoretical predictions and numerical cellular automata simulations.

Chopard, B., and M. Droz. "Cellular Automata Model for Thermo-Hydrodynamics." In *Chaos and Complexity*, edited by R. Livi, S. Ruffo, S. Ciliberto, and M. Buiatti. World Scientific, 1988, 302–306.

Chopard, Bastien. "Strings: A Cellular Automata Model of Moving Objects." Proceedings of the workshop on Cellular Automata and Modeling of Complex Physical Systems, Les Houches, February 21–March 1, 1989. Berlin: Springer-Verlag, 1989.

Abstract. We propose a reversible and local rule for modeling large-scale moving objects called "strings." These are one-dimensional objects that can be thought of as a discrete version of a chain of masses and springs. Longitudinal and transverse motions can be defined, making the string move in a three-dimensional space with adjustable momentum and energy. Basic mechanisms for collisions between strings are discussed.

Clavin, P., D. d'Humières, P. Lallemand, and Y. Pomeau. "Cellular Automata for Hydrodynamics with Free Boundaries in Two and Three Dimensions." *Comptes Rendus de l'Académie des Sciences de Paris* **303** II (1986):1169. Reprinted in translation in this volume, 413–421.

Abstract. Cellular automata are used to simulate two-dimensional hydrodynamic flows with free boundaries as found in the Rayleigh-Taylor instability. We propose an extension of these rules for three-dimensional flows with free boundaries.

Clavin, P., P. Lallemand, Y. Pomeau, and G. Searby. "Simulation of Free Boundaries in Flow systems by Lattice-Gas Models." *J. Fluid Mech.* **188** (1988):437–464.

Abstract. It has been recently proved that lattice-gas models with Boolean particles can provide a very powerful method to study viscous flows at moderate Reynolds and small Mach numbers (d'Humières, Pomeau & Lallemand 1985; Frisch, Hasslacher & Pomeau 1986; d'Humières & Lallemand 1986). We present here algorithms for an extension of these models to provide a simple and efficient way to simulate a large variety of flow problems with free boundaries. This is done by introducing two different types of particles that can react following a specific kinetic scheme based on autocatalytic reactions. In order to check the powerful character and the reliability of the method we also present preliminary results of two-dimensional computer simulations concerning problems ranging from the competition between molecular diffusion and turbulent mixing in flows presenting a Kelvin-Helmholtz instability to the spontaneous generation of turbulence in premixed flame fronts subject to the Darrieus-Landau instability. The dynamics of an interface developing a Rayleigh-Taylor instability is also considered as well as some typical problems of phase transition such as spinodal decomposition and the nucleation process.

Clouqueur, Andre, and Dominique d'Humières. "R.A.P. 1, un Réseau d'Automates Programmables." ENS, 1986. Internal Report.

Clouqueur, Andre, and Dominique d'Humières. "RAP1, a Cellular Automaton Machine for Fluid Dynamics." *Complex Systems* 1 (1987):585–597. Reprinted in this volume, 251–266.

Abstract. RAP1 is a special purpose computer built to study lattice gas models. It allows the simulation of any model using less than 16 bits per node, and interactions restricted to first and second nearest neighbors on a 256×512 square lattice. The time evolution of the automaton is displayed in real time on a color monitor at a speed of 50 frames per second.

Clouqueur, A., and D. d'Humières. "R.A.P., A Family of Cellular Automaton Machines for Fluid Dynamics." Proceedings of the 12th Gwatt workshop on Complex Systems, October 13–15, 1988, Gwatt, Switzerland.

Clouqueur, A., and D. d'Humières. "Special Purpose Computers for Lattice Gases." Proceedings of the workshop on Cellular Automata and Modeling of Complex Physical Systems, les Houches, France, February 21–March 1. Berlin: Springer-Verlag, 1989.

Colvin, M. E., A. J. C. Ladd, and B. J. Alder. "Maximally Discretized Molecular Dynamics." *Phys. Rev. Lett.* 61 (1988):381.

Abstract. It is shown that the coarsest discretization of positions and velocities in molecular dynamics leads to qualitatively correct transport coefficients and quantitatively predictable, long time tails in the velocity autocorrelation function, but requires orders of magnitude less computer time than standard molecular dynamics methods.

Dahlburg, Jill P., David Montgomery, and G. D. Doolen. "Noise and Compressibility in Lattice-Gas Fluids." *Phys. Rev. A* 36(5) (1987):2471–2474.

Abstract. Computations are reported in which the hexagonal lattice gas is used to simulate two-dimensional Navier-Stokes shear flows. Limitations associated with noise in the initial loading and compressible effects associated with a velocity-dependent equation of state arise and interact with each other. A relatively narrow window in density and flow speed exhibits physical behavior.

Despain, A., C. E. Max, G. Doolen, and B. Hasslacher. "Prospects for a Lattice-Gas Computer." This volume, 211-218.

Abstract. A two-day workshop was held in June of 1988 to discuss the feasibility of designing and building a large computer dedicated to lattice-gas cellular automata. The primary emphasis was on applications for modeling Navier-Stokes hydrodynamics. The meeting had two goals: (1) to identify those theoretical issues which would have to be addressed before the hardware implementation of a lattice-gas machine would be possible; and (2) to begin to evaluate alternative architectures for a dedicated lattice-gas computer. This brief paper contains a summary of the main issues and conclusions discussed at the workshop.

D'Humières, D., P. Lallemand, and T. Shimomura. "Cellular Automata, a New Tool for Hydrodynamics." Los Alamos National Laboratory, 1985. Report LA-UR-85-4051.

D'Humières, D., Y. Pomeau and P. Lallemand. "Simulation of 2-D Von Karman Streets using a Lattice Gas." *Comptes Rendus de l'Académie des Sciences de Paris* 301 II (1985):1391. Reprinted in translation in this volume, 423–428.

Abstract. A lattice gas model is introduced in which particles travel along the links of a planar triangular lattice, with constant speed and synchronized to collide at the vertices of the lattice. The macroscopic hydrodynamic behavior of the gas is demonstrated for channel flow around a plate perpendicular to the mean flow velocity, leading to the creation of nonstationary eddies.

D'Humières, D., and P. Lallemand. "Flow of a Lattice Gas between Two Parallel Plates: Development of the Poiseuille Profile." *Comptes Rendus de l'Académie des Sciences de Paris* 302 II (1986):983. Reprinted in translation in this volume, 429–436.

Abstract. After recalling briefly the use of a lattice gas to simulate aerodynamical flows, we describe an experimental situation corresponding to a 2-D flow between two parallel plates. A series of velocity distributions shows the progressive deformation from a Blasius profile for a boundary layer on a flat plate to a parabolic Poiseuille distribution. The comparison of the present results to those computed by Slichting shows the excellent quality of the predictions that can be obtained with the new simulation method proposed by Frisch, Hasslacher, and Pomeau.

D'Humières, D., and P. Lallemand. "Lattice Gas Automata for Fluid Mechanics." *Physica* **140A** (1986):326–335.

> *Abstract.* A lattice gas is the representation of a gas by its restriction on the nodes of a regular lattice for discrete time steps. It was recently shown by Frisch, Hasslacher and Pomeau that such very simple models lead to the incompressible Navier-Stokes equation provided the lattice has enough symmetry and the local rules for collisions between particles obey the usual conservation laws of classical mechanics. We present here recent results of numerical simulations to illustrate the power of this new approach to fluid mechanics which may give new tools for numerical studies and build a bridge between cellular automata theory and complex physical problems.

D'Humières, D., and P. Lallemand. "2-D and 3-D Hydrodynamics on Lattice Gases." *Hel. Phys. Acta* **59** (1986):1231–1234.

D'Humières, D., P. Lallemand, and U. Frisch. "Lattice Gas Models for 3D Hydrodynamics." *Europhys. Lett.* **2(4)** (1986):291–297.

> *Abstract.* The 3D Navier-Stokes equations are obtained from two different lattice gas models. The first one has its sites on a cubic lattice and has particle speeds zero, one and $\sqrt{2}$. The second one is a 3D projection of a lattice gas implementation of the 4D Navier-Stokes equations, residing on a face-centered hypercube lattice.

D'Humières, D., P. Lallemand, and Y. Pomeau. "Simulation de l'Hydrodynamique Bidimensionnelle à l'Aide d'un Gaz sur Réseau." *Bull. Soc. Franç. Phys.* **60** (1986): 14–15.

D'Humières, D., Y. Pomeau, and P. Lallemand. "Écoulement d'un Gaz sur Réseau dans un Canal Bidimensionnel: Développement de Profil de Poiseuille." *C. R. Acad. Sci. Paris II* **301** (1986):983–988. Reprinted in translation in this volume, 429–435.

D'Humières, D., Y. Pomeau, and P. Lallemand. "Two-Dimensional Hydrodynamics Calculations with a Lattice Gas." In *Innovative Numerical Methods in Engineering, A Computational Mechanics Publication*. Berlin: Springer-Verlag, 1986, 241–248.

D'Humières, D., and P. Lallemand. "Hydrodynamical Simulations with Lattice Gas." In *Supercomputing*, edited by A. Lichnewsky and C. Saguez. North-Holland, 1987, 363–380.

D'Humières, Dominique, and Pierre Lallemand. "Numerical Simulations of Hydrodynamics with Lattice Gas Automata in Two Dimensions." *Complex Systems* 1 (1987):599–632. Reprinted in this volume, 297–332.

> *Abstract.* We present results of numerical simulations of the Frisch, Hasslacher, and Pomeau lattice gas model and of some of its variants. Equilibrium distributions and several linear and nonlinear hydrodynamics flows are presented. We show that interesting phenomena can be studied with this class of models, even for lattices of limited sizes.

D'Humières, Dominique, Pierre Lallemand, and Geoffrey Searby. "Numerical Experiments on Lattice Gases: Mixtures and Galilean Invariance." *Complex Systems* 1 (1987):633–647. Reprinted in this volume, 333-350.

> *Abstract.* In this paper, we first describe an extension of the standard Frisch, Hasslacher, Pomeau hexagonal lattice gas to study reaction-diffusion problems. Some numerical results are presented. We then consider the question of Galilean invariance from an "experimental" point of view, showing cases where the standard model is inadequate. Finally, we introduce a way to cure the Galilean disease and present some results of simulations for a few typical cases.

D'Humières, D., Y. Pomeau, and P. Lallemand. "Une Nouvelle Méthode de Simulation Numérique en Mécanique des Fluides: Les Gas sur Réseau." *Images de la Physique* 68 (1987):89–94.

D'Humières, D., P. Lallemand, J. P. Boon, A. Noullez, and D. Dab. "Fluid Dynamics with Lattice Gases." In *Chaos and Complexity*, edited by R. Livi, S. Ruffo, S. Ciliberto, and M. Buiatti. World Scientific, 1988, 278–301.

D'Humières, D., P. Lallemand, and Y. H. Qian. "Modèles Monodimensionels de Gaz sur Réseaux, Divergence de la Viscosité." *C. R. Acad. Sci. Paris II* **308** (1988):585–590.

D'Humières, D., P. Lallemand, and Y. H. Qian. "Review of Flow Simulations using Lattice Gases." Proceedings of the International Seminar on Hyperbolic Problems, Bordeaux, France, June 13–17. *Lecture Notes.* Springer, 1988, in press.

D'Humières, D., P. Lallemand, and G. Searby. "Dynamics of Two-Dimensional Bubbles by the Lattice Gas Method." In *Physicochemical Hydrodynamics: Intefacial Phenomena, NATA ASI Series* **B174**, edited by M. G. Velarde. New York: Plenum Press, 1988, 71–86.

D'Humières, D., Y. H. Qian, and P. Lallemand. "Invariants in Lattice Gas Models." In *Discrete Kinematic Theory, Lattice Gas Dynamics, and Foundations of Hydrodynamics*, edited by R. Monaco. World Scientific, 1989, 102–113.

Diemer, K., K. Hunt, S. Chen, T. Shimomura, and G. D. Doolen. "Density and Velocity Dependence of Reynolds Numbers for Several Lattice Gas Models." This volume, 137–178.

Doolen, G. D. "Lattice Gas Models for Fluid Dynamics." *Physics Today* 41 (1988): s.39–40.

Droz, M., and B. Chopard. "Cellular Automata Approach to Physical Problems." *Hel. Physica Acta* 61 (1988).

> *Abstract.* The main applications of cellular automata (C.A) in physical problems are briefly reviewed. After having defined what a C.A is, some considerations on the motivations to use C.A in physics are given. Several examples of their applications in the fields of hydrodynamics, equilibrium statistical mechanics of systems on a lattice, growth mechanisms, diffusion processes, pattern recognition, models of memory and deterministic dynamics are given. An explicit application to a model of nonequilibrium phase transition is considered in more details. It is shown how the C.A rules are constructed, how to define analytically a mean field approximation for the automaton and how the results of this mean field approximation compare with the exact numerical solution of the model. Finally some general remarks on the prospects and open problems in the field are made.

Droz, M., and B. Chopard. "Non-Equilibrium Phase Transitions and Cellular Automata." In *Chaos and Complexity*, edited by R. Livi, S. Ruffo, S. Ciliberto, and M. Buiatti. World Scientific, 1988, 307–317.

Dubrulle, B. "Gas sur Réseaux. Violation du Bilan Semi-Detaillé." Stage DEA, Observatoire de Nice, 1987.

Dubrulle, Bérengère. "Method of Computation of the Reynolds Number for Two Models of Lattice Gas Involving Violation of Semi-Detailed Balance." *Complex Systems* 2 (1988):577–609.

> *Abstract.* We show how the theory of lattice gases developed by Frisch, d'Humières, Hasslacher, Lallemand, Pomeau, and Rivet, can be extended to cases involving violation of semi-detailed balance. This allows further reduction of the viscosity. However, since the universality of the distribution

is lost, the function $g(\rho)$ becomes dependent on the collision laws and has to be evaluated by a suitable generalization of the work of Hénon on viscosities. Cases with and without rest particles are considered. The lattice Boltzmann approximation is used.

Dufty, J. W., and M. H. Ernst. "Hydrodynamics Modes and Green-Kubo Relations for Lattice Gas Cellular Automata." Utrecht University, 1989. Preprint.

Duong-Van, M., and M. D. Feit. "Comments on Power Spectra of Discrete Stochastic Time Series." *Phys. Lett. A* **119(8)** (1986):388.

Abstract. We show that the slope of slightly flatter than -2 seen in the power spectra of stochastic series is a consequence of finite discrete systems observed with limited temporal correlation.

Duong-Van, Minh, M. D. Feit, P. Keller and M. Pound. "The Nature of Turbulence in a Triangular Lattice Gas Automaton." *Physica* **23D** (1986):448–454.

Abstract. Power spectra calculated from the coarse-graining of a simple lattice gas automaton, and those of time-averaging other stochastic times series that we have investigated, have exponents in the range -1.6 to -2, consistent with observation of fully developed turbulence. This power spectrum is a natural consequence of coarse-graining; the exponent -2 represents the continuum limit.

Ernst, M. H., and P. M. Binder. "Lorentz Lattice Gases: Basic Theory." *J. Stat. Phys.* **51(5/6)** (1988):981.

Abstract. We present several ballistic models of the Lorentz gas in two-dimensional lattices with deterministic and stochastic deflection rules, and their corresponding Liouville equations. Boltzmann-level-equation results are obtained for the diffusion coefficient and velocity autocorrelation function for models with stochastic deflection rules. The long-time behavior of the mean square displacement is briefly discussed and the possibility of abnormal diffusion indicated. Even if the diffusion coefficient exists, its low density limit may not be given correctly by the Boltzmann equation.

Ernst, M. H., and J. W. Dufty. "Green-Kubo Relations for Lattice Gas Cellular Automata." Preprint.

Abstract. Green-Kubo relations are derived for linear transport coefficients (viscosities, diffusion) in lattice gases using the correlation function description of transport in fluids. The only ingredients are the local microscopic conservation laws without any further specifications of the microdynamical laws.

Eykholt, R., A. R. Bishop, P. S. Lomdahl, and E. Domany. "A Nonequilibrium-to-Equilibrium Mapping and Its Application to the Perturbed Sine-Gordon Equation." *Physica* **23D** (1986):102–111.

Abstract. Given a partial differential equation (pde) in one time and D spatial dimensions which is driven by noise $\zeta(\vec{x};t)$ with a known distribution $P[\zeta]$, one may find the distribution $P[\psi]$ of the solution $\psi(\vec{x};t)$. Furthermore, the distribution may be written as $P[\psi] \sim e^{-\beta H[\psi]}$ with $H[\psi]$ an effective Hamiltonian in $D+1$ dimensions (time having become an extra spatial dimension). Then, the most probable solution of the pde is that function which minimizes $H[\psi]$. Here, we describe this method and illustrate it for the damped, driven sine-Gordon equation in one spatial dimension ($D = 1$).

Falk, H. "Dynamical Spin System: Exact Solution and Mean Recurrence Time." *Physica D* **31** (1988):389–396.

Abstract. A model involving a chain of $N \geq 2$ spins $s_i = \pm 1$, $i = 1, \ldots, N$, evolving synchronously in discrete time t via a nonlinear, autonomous transformation $s_i(t+1) = s_i(t)s_{i+1}(t)$, $i = 1, \ldots, N-1$; $s_N(t+1) = s_N(t)$, is presented. The transformation equations are solved explicitly and the detailed decomposition of state space into ergodic sets is found. On the assumption of equally likely initial states, the mean recurrence time is calculated and its variance is discussed. The model displays a strikingly sensitive dependence on the number of spins, and this is reflected in the "staircase" behavior of the mean recurrence time. Remarks are made regarding the connection between the behavior of the model and the ground states of a related two-dimensional Ising model.

Frenkel, D. "Transfer Matrices and Time Correlation Functions in Lattice Gases." Proceedings of the workshop on Cellular Automata and Modeling of Complex Physical Systems, Les Houches, France, February 21–March 1. Berlin: Springer-Verlag, 1989.

Frenkel, D., and M. H. Ernst. "Simulation of Diffusion in a Two-Dimensional Lattice Gas Cellular Automaton: A Test of Mode-Coupling Theory." Unpublished.

Abstract. We compute the velocity autocorrelation function (VACF) of a tagged particle in a two-dimensional lattice-gas cellular automaton (LGCA) using a method that is about a million times more efficient than existing techniques. A t^{-1}-algebraic tail in the tagged-particle is found to agree quantitatively with the predictions of mode-coupling theory. However, the magnitude of logarithmic corrections to the t^{-1}-tail is much smaller than expected.

Frisch, U., B. Hasslacher, and Y. Pomeau "Lattice-Gas Automata for the Navier-Stokes Equation." *Physical Review Letters* **56(14)** (1986)1505–1508. Reprinted in this volume, 11–18.

Abstract. We show that a class of deterministic lattice gases with discrete Boolean elements simulates the Navier-Stokes equation, and can be used to design simple, massively parallel computing machines.

Frisch, U., and J. P. Rivet. "Lattice Gas Hydrodynamics, Green-Kubo Formula." *Comptes Rendus* **303** II (1986):1065. Reprinted in translation in this volume, 437–442.

Abstract. The Green-Kubo formalism, allowing the evaluation of transport coefficients of macroscopic systems in terms of equilibrium correlation functions, is extended to lattice gas hydrodynamics.

Frisch, Uriel, Dominique d'Humières, Brosl Hasslacher, Pierre Lallemand, Yves Pomeau, and Jean-Pierre Rivet. "Lattice Gas Hydrodynamics in Two and Three Dimensions." *Complex Systems* **1** (1987):649–707. Reprinted in this volume, 75–136.

Abstract. Hydrodynamical phenomena can be simulated by discrete lattice gas models obeying cellular automata rules [Frisch et al., *Phys. Rev. Lett.* **56** (1986):1505; D. d'Humières et al., *Europhys. Lett.* **2** (1986): 219]. It is here shown for a class of D-dimensional lattice gas models how the macrodynamical (large-scale) equations for the densities of microscopically conserved quantities can be systematically derived from the underlying exact "microdynamical" Boolean equations. With suitable restrictions on the crystallographic symmetries of the lattice and after proper limits are taken, various standard fluid dynamical equations are obtained, including the incompressible Navier-Stokes equations in two and three dimensions. The transport coefficients appearing in the macrodynamical equations are obtained using variants of the fluctuation-dissipation theorem and Boltzmann formalisms adapted to fully discrete situations.

Fritz, J. "On the Hydrodynamic Limit of a One-Dimensional Ginzburg-Landau Lattice Model. The *a priori* Bounds." *J. Stat. Phys.* **47** (1987):551–571.

Fritz, J. "Review of the Work of V. I. Osolodoc on the Stationary States of HPP-FHP-Type Cellular Automata." Unpublished.

Abstract. In a series of fairly long and nontrivial papers (CMP 1976–84) B. M. Gurevich and Yu. M. Suhov investigated the set of stationary states of infinite Hamiltonian systems. They have shown that every stationary state satisfying some regularity conditions is an equilibrium state. HPP-FHP automata mimic classical dynamics, and Oseledec managed to extend this method to some classes of cellular automata including FHP models with randomized collisions. In fact, he proves that every stationary Gibbs state with an additional mixing property is a Bernoulli measure. Here we sketch his nice proof and basic results for the simplest HPP-FHP models. Partly we follow a recent, still unpublished paper by Gurevich on asymptotically additive integrals. Oseledec's beautiful treatment of stochastic automata (FHP models) is also included.

Goles, Eric, and Gérard Y. Vichniac. "Invariants in Automata Networks." *J. Phys. A: Math. Gen.* **19** (1986):L961–L965.

> *Abstract.* We give two extensions of Pomeau's additive invariant for reversible cellular automata and networks.

Gunstensen, A. K. "A Fast Implementation of the FHP Lattice Gas." *MIT Porous Flow Project, Report n° 1* (1988).

Gunstensen, A. K. "A Galilean-Invariant Lattice-Gas Model for Immiscible Fluids." *MIT Porous Flow Project, Report n° 1* (1988).

Hardy, J., and Y. Pomeau. "Thermodynamics and Hydrodynamics for a Modeled Fluid." *J. Math. Phys.* **13** (1972):1042–1051.

Hardy, J., Y. Pomeau, and O. de Pazzis. "Time Evolution of Two-Dimensional Model System. I. Invariant States and Time Correlation Functions." *J. Math. Phys.* **14** (1973):1746–1759.

Hardy, J., O. de Pazzis, and Y. Pomeau. "Molecular Dynamics of a Classical Lattice Gas: Transport Properties and Time Correlation Functions." *Phys. Rev. A* **13** (1976):1949–1961.

Hasslacher, B. "Discrete Fluids." *Los Alamos Science* **15** (1988):175–200, 211–217.

Hatori, Tudatsugu, and David Montgomery. "Transport Coefficients for Magneto-hydrodynamic Cellular Automata." *Complex Systems* 1 (1987):735–752. Reprinted in this volume, 351–370.

Abstract. A Chapman-Enskog development has been used to infer theoretical expressions for coefficients of kinematic viscosity and magnetic diffusivity for a two-dimensional magnetohydrodynamic cellular automaton.

Hayot, Fernand. "The Effect of Galilean Non-Invariance in Lattice Gas Automaton One-Dimensional Flow." *Complex Systems* 1 (1987):753–761. Reprinted in this volume, 371–382.

Abstract. In the simple case of one-dimensional flow between plates, we show the effect of Galilean non-invariance of the usual hexagonal lattice gas mode. This effect leads to a distorted velocity profile when the velocity exceeds a value of 0.4. Higher-order corrections to the Navier-Stokes equations are considered in a discussion of the numerical importance of the distortion.

Hayot, F. "Unsteady, One-Dimensional Flow in Lattice-Gas Automata." *Phys. Rev. A* **35**(4) (1987):1774.

Abstract. We study numerically unsteady, one-dimensional flow between parallel plates in the hexagonal lattice-gas automaton. An initial tangential velocity instability is created at one plate and its propagation into the system is investigated. This propagation depends on the boundary conditions at the opposite plate. From the observed fluid behavior, viscosity is estimated.

Hayot, F. "Unsteady One-Dimensional Flow in Lattice Gas Automata." *Phys. Rev.* **A35** (1987):1774–1777.

Hayot, F. "Viscosity in Lattice Gas Automata." *Physica* **28D** (1987):210–214.

Abstract. Using the Green-Kubo formalism, we derive an expression, valid at any density, for the viscosity for hexagonal lattice gas automata. Estimates of an effective viscosity, valid for time scales relevant to numerical experiments, are given for the two-dimensional hexagonal lattice gas.

Hayot, F., and Raj Lakshmi. "Cylinder Wake in Lattice Gas Hydrodynamics." Ohio State University, 1989. Preprint.

Hayot, F., M. Mandal, and p. Sadayappan. "Implementation and Performance of a Binary Lattice Gas Algorithm on Parallel Processor Systems." *J. Compl. Phys.* **80** (1989):277–287.

Hénon, M. "Isometric Collision Rules for the Four-Dimensional FCHC Lattice Gas." *Complex Systems* **1** (1987):475–494.

Hénon, Michel. "Viscosity of a Lattice Gas." *Complex Systems* **1** (1987):763–789. Reprinted in this volume, 179–208.

Abstract. The shear viscosity of a lattice gas can be derived in the Boltzmann approximation from a straightforward analysis of the numerical algorithm. This computation is presented first in the case of the Frisch-Hasslacher-Pomeau two-dimensional triangular lattice. It is then generalized to a regular lattice of arbitrary dimension, shape, and collision rules with appropriate symmetries. The viscosity is shown to be positive. A practical recipe is given for choosing collision rules so as to minimize the viscosity.

Hénon, M. "On the Relation between Lattice Gases and Cellular Automata." In *Discrete Kinetic Theory, Lattice Gas Dynamics and Foundations of Hydrodynamics*, edited by R. Monaco. World Scientific, 1989, 160–161.

Hénon, M. "Optimization of Collision Rules in the FCHC Lattice Gas, and Addition of Rest Particles." In *Discrete Kinetic Theory, Lattice Gas Dynamics and Foundations of Hydrodynamics*, edited by R. Monaco. World Scientific, 1989, 146–159.

Abstract. Various collision rules for the FCHC 24-velocity lattice are discussed. Global rules give a Reynolds coefficient $R_*^{\max} \simeq 2$. Detailed rules, obtained by a fine-tuning optimization, give $R_*^{\max} = 7.57$.

Formulas are given for the FCHC lattice with the addition of rest particles, with arbitrary probabilities for each value of the number of particles present at a node. With a maximum of 1, 2, and 3 rest particles, the best values obtained for $R_*^{\max}$ are 8.46, 10.22 and 10.71.

Herrmann, Hans J. "Special Purpose Computers in Statistical Physics." *Physica* **140A** (1986).

Abstract. A new trend in physics, namely the building of special purpose computers (SPC), is reviewed. Special emphasis is given to the following questions: why does one build SPC's? When is it worthwhile to build an SPC? How does one proceed if one wants to build an SPC? Finally the most important results that have been obtained for statistical physics through SPC's up to now will be sketched.

Higuera, F. J., and J. Jimenez. "A Boltzmann Approach to Lattice Gas Simulations." 1988. Preprint.

Higuera, F. J., and S. Succi. "Simulating the Flow around a Circular Cylinder with a Lattice Boltzmann Equation." 1988. Preprint.

Higuera, F. J. "Lattice Gas Simulation Based on the Boltzmann Equation." In *Discrete Kinematic Theory, Lattice Gas Dynamics, and Foundations of Hydrodynamics*, edited by R. Monaco. World Scientific, 1989, 329–342.

Higuera, F. J., S. Succi, and R. Benzi. "Lattice Gas Dynamics with Enhanced Collisions." IBM ECSEC, 1989. Preprint.

Hogeweg, P. "Cellular Automata as a Paradigm for Ecological Modeling." *Appl. Math. & Comp.* **27** (1988):81–100.

Abstract. We review cellular automata as a modeling formalism and discuss how it can be used for modeling (spatial) ecological processes. The implications of this modeling paradigm for ecological observation are stressed. Finally we discuss some shortcomings of the cellular-automaton formalism and mention some extensions and generalizations which may remedy these shortcomings.

Huang, Jau-Inn, Yu-Hua Chu, and Chuan-Sheng Yin. "Lattice-Gas Automata for Modeling Acoustic Wave Propagation in Inhomogeneous Media." *Geophys. Res. Lett.* **15**(11) (1988):1239.

Abstract. We report recent developments in using lattice-gas automata to simulate acoustic wave propagation in inhomogeneous media, and give experimental results that include transmitted and reflected waves. Two sets of rules are adopted to govern the evolution of particles in the lattice; one applicable to lattice sites on the interface, the other to interior sites. Both sets of rules are simple and deterministic, thereby preserving the computational advantages of lattice-gas automata. Further investigation bears long-term promise of yielding a novel and efficient forward modeling tool for geologically realistic models.

D'Humieres, D. [listed under D instead of H]

Kadanoff, L. P., and J. Swift. "Transport Coefficients Near the Critical Point: A Master Equation Approach." *Phys. Rev.* **165** (1968):310–322.

Kadanoff, L. P. "On Two Levels." *Physics Today* **39** (1986):7–9.

Kadanoff, L., G. McNamara, and G. Zanetti. "From Automata to Fluid Flow: Comparisons of Simulation and Theory." University of Chicago, 1987. Preprint.

Kadanoff, Leo P., Guy R. McNarama, and Gianluigi Zanetti. "A Posieuille Viscometer for Lattice Gas Automata." *Complex Systems* **1** (1987):791–803. Reprinted in this volume, 383–398.

Abstract. Lattice gas automata have been recently proposed as a new technique for the numerical integration of the two-dimensional Navier-Stokes equation. We have accurately tested a straightforward invariant of the original model, due to Frisch, Hasslacher, and Pomeau, in a simple geometry equivalent to two-dimensional Poiseuille (Channel) flow driven by a uniform body force.

The momentum density profile produced by this simulation agrees well with the parabolic profile predicted by the macroscopic description of the gas given by Frisch et al. We have used the simulated flow to compute the shear viscosity of the lattice gas and have found agreement with the results obtained by d'Humières et al. [LANL reprint] using shear wave relaxation measurements, and, in the low density limit, with theoretical predictions obtained from the Boltzmann description of the gas [J. P. Rivet and U. Frisch, *C. R. Acad. Sci. Paris II* **302** (1986):732].

Kugelmass, S. D., R. Squier, and K. Steiglitz. "Performance of VLSI Engines for Lattice Gas Computations." *Complex Systems* **1** (1987):939–965.

Kugelmass, S. D., and K. Steiglitz. "Design and Construction of LGM-1: A Lattice Gas Machine with Linear Speedup." Proceedings of the 22nd Annual Conference on Information Sciences and Systems, Princeton University, March 16–18, 1988.

Ladd, Anthony J. C., Michael E. Colvin, and Daan Frenkel. "Application of Lattice-Gas Cellular Automata to the Brownian Motion of Solids in Suspension." *Phys. Rev. Lett.* **60**(11) (1988):975–978.

Abstract. An adaptation of lattice-gas cellular automata to the simulation of solid-fluid suspensions is described. The method incorporates both dissipative hydrodynamic forces and thermal fluctuations. At low solid densities, theoretical results for the drag force on a single disk and the viscosity of a suspension of disks are reproduced. The zero-shear-rate viscosity has been obtained over a range of packing fractions and results indicate that simulations of three-dimensional suspensions are feasible.

Ladd, A. J. C., and D. Frenkel. "Lattice-Gas Approach of Suspensions." Proceedings of the workshop on Cellular Automata and Modeling of Complex Physical Systems, Les Houches, France, February 21–March 1. Berlin: Springer-Verlag, 1989.

Lavallée, P., J. P. Boon, and A. Noullez. "Boundary Interactions in a Lattice Gas." In *Discrete Kinematic Theory, Lattice Gas Dynamics, and Foundations of Hydrodynamics*, edited by R. Monaco. World Scientific, 1989, 206–214.

Abstract. A method using the cellular automata approach is presented to evaluate the velocity profile in a fluid flow near solid boundaries. The usual method is to simulate the flow via the hexagonal lattice gas microdynamical equations from which the velocity profile is determined. The present approach is to solve the lattice Boltzmann equations using average, continuous values for the link populations. The velocity profile and the time evolution of the profile obtained in this way for a flow parallel to a flat plate are compared with actual simulations and with the boundary layer theory.

Lebowitz, J. L. "Microscopic Origin of Hydrodynamic Equations: Derivation and Consequences." *Physica* **140A** (1986):232–239.

Abstract. We describe some recent progress in deriving autonomous hydrodynamic-type equations for macroscopic variables from model stochastic microscopic dynamics of particles on a lattice. The derivations also yield the microscopic fluctuations about the deterministic macroscopic evolution. These grow, with time, to become infinite when the deterministic solution is unstable. A form of microscopic pattern selection is also found.

Lebowitz, Joel L., Enrico Presutti, and Herbert Spohn. "Microscopic Models of Hydrodynamic Behavior." *J. Stat. Phys.* **51(5/6)** (1988):841.

Abstract. We review recent developments in the rigorous derivation of hydrodynamic-type macroscopic equations from simple microscopic models: continuous-time stochastic cellular automata. The deterministic evolution of hydrodynamic variables emerges as the "law of large numbers," which holds with probability one in the limit in which the ratio of the microscopic to the macroscopic spatial and temporal scales go to zero. We also study fluctuations in the microscopic system about the solution of the macroscopic equations. These can lead, in cases where the latter exhibit instabilities, to complete divergence in behavior between the two at long macroscopic times. Examples include Burgers' equation with shocks and diffusion-reaction equations with traveling fronts.

Leko, T. D. "Comment on Lattice-Gas Automata for the Solution of Partial Differential Equations." *Math. Mech.* **68** (1988):T462–T463.

Levermore, D. "A Diffusive Automata Update." Proceedings of the workshop on Cellular Automata and Modeling of Complex Physical Systems, Les Houches, France, February 21–March 1.

Lim, H. A. "Cellular Automaton Simulations of Simple Boundary Layer Problems." Florida State University, 1988. Preprint.

Lim, Hwa A. "Lattice Gas Automata of Fluid Dynamics for Unsteady Flow." *Complex Systems* **2** (1988):45–58.

> *Abstract.* We study lattice gas automata of fluid dynamics in the incompressible flow limit. It is shown that the viscosity effect on the transition layer from the steady uniform velocity in one stream to the steady uniform velocity in another, adjacent stream produces the correct profile. We further study the intrinsic damping or smoothing action of viscous diffusion and show that the results agree with those obtained from the Navier-Stokes equation. In both cases, we obtain a kinetic viscosity $\nu \sim 0.65$, consistent with the prediction of the Boltzmann approximation.

Lim, H. A., G. Riccardi, and C. Bauer. "Applications of Cellular Models to Flow Patterns." Florida State University, 1988. Preprint.

Lindgren, K. "Entropy and Correlations in Lattice Systems." Proceedings of the workshop on Cellular Automata and Modeling of Complex Physical Systems, Les Houches, France, February 21–March 1. Berlin: Springer-Verlag, 1989.

Long, L. N., Robert M. Coopersmith, and B. G. McLachlan. "Cellular Automatons Applied to Gas Dynamic Problems." Proceedings of the AIAA 19th Fluid Dynamics, Plasma Dynamics and Lasers Conference, Honolulu, USA, June 1987.

> *Abstract.* This paper compares the results of a relatively new computational fluid dynamics method, cellular automatons, with experimental data and analytical results. This technique has been shown to qualitatively predict fluid-like behavior; however, there have been few published comparisons with experiment or other theories. Comparisons are made for a one-dimensional supersonic piston problem, Stokes First Problem, and the flow

past a normal flat plate. These comparisons are used to assess the ability of the method to accurately model fluid dynamic behavior and to point out its limitations. Reasonable results were obtained for all three test cases. While this is encouraging, the fundamental limitations of cellular automatons are numerous. In addition, it may be misleading, at this time, to say that cellular automatons are a computationally efficient technique. Other methods, based on continuum or kinetic theory, would also be very efficient if as little of the physics were included.

Maddox, J. "Mechanizing Cellular Automata." *Nature* **321** (1986):107.

Mareschal, M., and E. Kestermont. "Order and Fluctuations in Nonequilibrium Molecular Dynamics Simulations of Two-Dimensional Fluids." *J. Stat. Phys.* **48**(5/6) (1987):1187.

Abstract. Finite systems of hard disks placed in a temperature gradient and in an external constant field have been studied, simulating a fluid heated from below. We used the methods of nonequilibrium molecular dynamics. The goal was to observe the onset of convection in the fluid. Systems of more than 5000 particles have been considered and the choice of parameters has been made in order to have a Rayleigh number larger than the critical one calculated from the hydrodynamic equations. The appearance of rolls and the large fluctuations in the velocity field are the main features of these simulations.

Margolus, Norman, Tommaso Toffoli, and Gérard Vichniac. "Cellular-Automata Supercomputers for Fluid-Dynamics Modeling." *Phys. Rev. Lett.* **56**(16) (1986): 1694.

Abstract. We report recent developments in the modeling of fluid dynamics, and give experimental results (including dynamical exponents) obtained with cellular automata machines. Because of their locality and uniformity, cellular automata lend themselves to an extremely efficient physical realization; with a suitable architecture, an amount of hardware resources comparable to that of a home computer can achieve (in the simulation of cellular automata) the performance of a conventional supercomputer.

Margolus, Norman, and Tommaso Toffoli. "Cellular Automata Machines." This volume, 219–249.

Abstract. The advantages of an architecture optimized for cellular automata simulations are so great that, for large-scale CA experiments, it becomes absurd to use any other kind of computer.

In this article we discuss cellular automata machines in general, give some illustrative examples of the use of an existing machine, and then describe a much more ambitious architecture (which is under development) suitable for extensive three-dimensional simulations.

McCauley, Joseph L. "Chaotic Dynamical Systems as Automata." *Z. Naturforsch.* **42a** (1987):547–555.

Abstract. We discuss the replacement of discrete maps by automata, algorithms for the transformation of finite-length digit strings into other finite-length digit strings, and then discuss what is required in order to replace chaotic phase flows that are generated by ordinary differential equations by automata without introducing unknown and uncontrollable errors. That question arises naturally in the discretization of chaotic differential equations for the purpose of computation. We discuss as examples an autonomous and a periodically driven system, and a possible connection with cellular automata is also discussed. Qualitatively, our considerations are equivalent to asking when can the solution of a chaotic set of equations be regarded as a machine, or a model of a machine.

McNamara, Guy R., and Gianluigi Zanetti. "Use of the Boltzmann Equation to Simulate Lattice-Gas Automata." *Phys. Rev. Lett.* **61(20)** (1988):2332–2335. Reprinted in this volume, 289–296.

Abstract. We discuss an alternative technique to the lattice-gas automata for the study of hydrodynamic properties; namely, we propose to model the lattice gas with a Boltzmann equation. This approach completely eliminates the statistical noise that plagues the usual lattice-gas simulations that demand much less computer time. It is estimated to be more efficient than the lattice-gas automata for intermediate to low Reynolds number $R \lesssim 100$.

Molvig, K., P. Donis, J. Myczkowski, and G. Vichniac. "Continuum Fluid Dynamics from a Lattice Gas." MIT, 1988. Preprint.

Abstract. A lattice gas automata theory is presented which removes the discreteness artifacts that plague the current models. The automata are endowed with a genuine energy that is distinct from mass. The macroscopic equations possess Galilean invariance with allowance for compressibility and temperature variations, proper scalar pressure exhibiting equipartition (without the Mach number dependent anomaly), and an energy transport equation, all fully three dimensional.

Molvig, K., P. Donis, J. Myczkowski, and G. Vichniac. "Removing the Discreteness Artifacts in 3D Lattice-Gas Fluids." In *Discrete Kinematic Theory, Lattice Gas Dynamics, and Foundations of Hydrodynamics*, edited by R. Monaco. World Scientific, 1988, 408.

Molvig, K. "Realistic Fluid Behavior from a Lattice Gas: Theory and Simulation Results." Proceedings of the workshop on Cellular Automata and Modeling of Complex Physical Systems, Les Houches, France, February 21–March 1. Berlin: Springer-Verlag, 1989.

Molvig, K., P. Donis, J. Myczkowski, and G. Vichniac. "Thermodynamics of Multi-Species Lattice Gases." Proceedings of the workshop on Cellular Automata and Modeling of Complex Physical Systems, Les Houches, France, February 21–March 1. Springer-Verlag, 1989.

Molvig, K., P. Donis, J. Myczkowski, and G. Vichniac. "Multi-Species Lattice Gas Hydrodynamics." Unpublished.

Abstract. A new class of lattice gas automata are developed, based on multiple species of particles with energy exchanging interactions. These automata are endowed with a genuine energy that is distinct from mass and can be constructed so as to be free of all the artifacts of discretization that characterized previous lattice gas models. The model possesses true Galilean invariance with allowance for compressibility, proper scalar pressure exhibiting equipartition, and an energy transport equation. Its thermal behavior is explored and a practical method to achieve Galilean invariance is presented.

Montgomery, D., and G. D. Doolen. "Magnetohydrodynamic Cellular Automata." *Phys. Lett.* **A120** (1987):229–231

Montgomery, David, and Gary D. Doolen. "Two Cellular Automata for Plasma Computations." *Complex Systems* **1** (1987):831–838. Reprinted in this volume, 461–470.

Abstract. Plasma applications of computational techniques based on cellular automata are inhibited by the long-range nature of electromagnetic forces. One of the most promising features of cellular automata methods has been the parallelism that becomes possible because of the local nature of the interactions, leading (for example) to the absence of Poisson equations to be solved in fluid simulations. Because it is in the nature of a plasma that volume forces originate with distant charges and currents, finding plasma cellular automata becomes largely a search for tricks to circumvent this nonlocality of the forces. We describe automata for two situations where this appears possible: two-dimensional magnetohydrodynamics (2D MHD) and the one-dimensional electrostatic Vlasov-Poisson system. Insufficient computational experience has accumulated for either system to argue that it is a serious alternative to existing methods.

Nadiga, B. T., J. E. Broadwell, and B. Sturtevant. "Study of Multi-Speed Cellular Automaton." Caltech, 1988. Preprint.

Nickel, G. H. "Cellular Automaton Rules for Solving the Milne Problem." *Phys. Lett. A* **133(4-5)** (1988):219–224.

Abstract. The methodology for deriving cellular automaton rules to solve radiation transport problems is developed. Calculations using several of these rules to solve the Milne problem in cartesian geometry show good agreement with the exact solution.

Noullez, A., M. Bonetti, and J.-P. Boon. "Viscous Fingering in a 2-D Porous Lattice." Proceedings of the workshop on Cellular Automata and Modeling of Complex Physical Systems, Les Houches, France, February 21–March 1. Springer-Verlag, 1989.

Oono, Y., and S. Puri. "Computationally Efficient Modeling of Ordering of Quenched Phases." *Phys. Rev. Lett.* **58(8)** (1987):836.

Abstract. Computationally efficient discrete space-time models of phase-ordering dynamics of thermodynamically unstable systems (e.g., spinodal decomposition) are proposed. Two-dimensional lattice (100×100) simulations were preformed to obtain scaled form factors.

Oono, Y., and S. Puri. "Study of Phase-Separation Dynamics by Use of Cell Dynamical Systems. I. Modeling." *Phys. Rev. A* **38** (1987):434.

Abstract. We present a computationally efficient scheme of modeling the phase-ordering dynamics of thermodynamically unstable phases. The scheme utilizes space-time discrete dynamical systems, viz., cell dynamical systems (CDS). Our proposal is tantamount to proposing new *Ansätze* for the kinetic-level description of the dynamics. Our present exposition consists of two parts: part I (this paper) deals mainly with methodology and part II [S. Puri and Y. Oono, Phys. Rev. Lett. **38** 1542 (1988)] gives detailed demonstrations. In this paper we provide a detailed exposition of model construction, structural stability of constructed models (i.e., insensitivity to details), stability of the scheme, etc. We also consider the relationship between the CDS modeling and the conventional description in terms of partial differential equations. This leads to a new discretization scheme for semi-linear parabolic equations and suggests the necessity of a branch of applied mathematics which could be called "qualitative numerical analysis."

Oono, Y., and C. Yeung. "A Cell Dynamical System Model of Chemical Turbulence." *J. Stat. Phys.* **48(3/4)** (1987):593.

Abstract. A cellular-automaton-like caricature of chemical turbulence on an infinite one-dimensional lattice is studied. The model exhibits apparently "turbulent" space-time patterns. To make this statement precise, the following problems or point are discussed: (1) The infinite-system-size limit of such cell-dynamical systems and its observability is defined. (2) It is proved that the invariant state in the large-system-size limit of the "turbulent" phase exhibits spatial patterns governed by a Gibbs random field. (3) Potential characteristics of "turbulent" space-time patterns are critically surveyed and a working definition of (weak) turbulence is proposed. (4) It is proved that the invariant state of the "turbulent" phase is actually (weak) turbulent. Furthermore, we conjecture that the turbulent phase of our model is an example of a K system that is not Bernoulli.

Orszag, Steven A., and Victor Yakhot. "Reynolds Number Scaling of Cellular-Automaton Hydrodynamics." *Phys. Rev. Lett.* **56(16)** (1986):1691–1693. Reprinted in this volume, 269–274.

Abstract. We argue that the computational requirements for presently envisaged cellular-automaton simulations of continuum fluid dynamics are much more severe than for solution of the continuum equations.

Puri, S., and Y. Oono. "Study of Phase-Separation Dynamics by Use of Cell Dynamical Systems. II. Two-Dimensional Demonstrations." *Phys. Rev. A* **38** (1988):1542.

Abstract. We present detailed results on the form factors of two-dimensional systems undergoing phase-ordering processes, using both deterministic and stochastic cell dynamical systems. We show the robustness of the asymptotic form factors against quench depth, noise amplitude, etc. The effect of noise is essentially to delay the number of steps needed to reach the asymptotic behavior. In the case with a nonconserved order parameter, we demonstrate that the form factor obtained by T. Ohta, D. Jasnow, and K. Kawasaki [Phys. Rev. Lett. 49, 1223 (1982)] is asymptotically very accurate. We also present preliminary results for off-critical quenches.

Qian, Y. H., D. d'Humières, and P. Lallemand. "A Short Note on Green-Kubo Formula of Viscosity in One-Dimensional Lattice Gas Models of Single Particle Mass." ENS, 1989. Preprint.

Qian, Y. H., D. d'Humières, and P. Lallemand. "Simulations of Heat Conduction Flow Using a One-Dimensional Deterministic Lattice Gas Model and Its Analytical Steady Solution for Density." ENS, 1989. Preprint.

Rapaport, D. C., and E. Clementi. "Eddy Formation in Obstructed Fluid Flow: A Molecular-Dynamics Study." *Phys. Rev. Lett.* **57**(6) (1986):69.

> *Abstract.* Two-dimensional fluid flow past a circular obstacle has been simulated at the microscopic level by means of a molecular-dynamics approach. At sufficiently large Reynolds number the flow field is observed to exhibit characteristics common to real fluids, namely the appearance of eddies, periodic eddy separation, and an oscillatory wake. Very large systems—typically 160,000 particles—are required in order to provide adequate space for these flow patterns to develop.

Rapaport, D. C. "Microscale Hydrodynamics: Discrete-Particle Simulation of Evolving Flow Patterns." *Phys. Rev. A* **36**(7) (1987):3288.

> *Abstract.* The technique of molecular-dynamics simulation—in which the equations of motion of a system of interacting particles are solved numerically to yield the temporal evolution of the system—is used in a study of the flow of a two-dimensional fluid past a circular obstacle. The flow is observed to develop with time, passing through a series of well-defined patterns that bear a striking similarity with flow patterns observed experimentally in liquid and gas flow; the patterns include stationary eddies, periodic shedding of vortices, and a vortex street characterized by a Strouhal number close to the experimental value. Very large systems—by current molecular-dynamics standards—need to be used in order to accommodate the obstacle and the region occupied by the structured wake, and the present work includes the largest such simulations carried out to date. Though more extensive work is called for, the results suggest that continuum hydrodynamics is applicable down to much shorter length scales than hitherto believed, and that the molecular-dynamics approach can thus be used to study certain kinds of hydrodynamic instabilities.

Rem, P. C., and J. A. Somers. "Cellular Automata on a Transputer Network." In *Discrete Kinematic Theory, Lattice Gas Dynamics, and Foundations of Hydrodynamics*, edited by R. Monaco. World Scientific, 1988, 268–275.

Riccardi, G., H. A. Lim, and C. Bauer. "A Vectorized Cellular Automata Model of Fluid Flow." Florida State University, 1988. Preprint.

Rivet, J.-P. "Gaz sur Réseaux." Observatoire de Nice, 1986. Internal Report.

Rivet, J. P., and U. Frish. "Lattice Gas Automata in the Boltzmann Approximation." *Comptes Rendus* 302 II (1986):267–272. Reprinted in translation in this volume, 443–450.

Abstract. Shear and bulk viscosities are determined for two lattice gas automata simulating the two-dimensional Navier-Stokes equations.

Rivet, Jean-Pierre. "Green-Kubo Formalism for Lattice Gas Hydrodynamics and Monte-Carlo Evaluation of Shear Viscosities." *Complex Systems* 1 (1987):839–851. Reprinted in this volume, 399–414.

Abstract. A Green-Kubo formula, relating the shear viscosity to discrete time correlation functions, is derived via a Liouville equation formalism for a class of non-deterministic lattice gas models. This allows a Monte-Carlo calculation of the viscosity. Preliminary results are presented for the Frisch-Hasslacher-Pomeau two-dimensional lattice gas model.

Rivet, J.-P. "Simulation d'Écoulements Tridimensionnels par la Méthode des Gaz sur Réseau: Premiers Résultats." *C. R. Acad. Sci. Paris II* 305 (1987):751–756.

Rivet, J.-P. "Hydrodynamique par la Méthode des Gas sur Réseaux." Ph.D. thesis, Université de Nice, 1988.

Rivet, J.-P., M. Hénon, U. Frisch, and D. d'Humières. "Simulating Fully Three-Dimensional External Flow by Lattice Gas Methods." *Europhys. Lett.* 7(3) (1988): 231–236.

Abstract. We have built a three-dimensional 24-bit lattice gas algorithm with improved collision rules. Collisions are defined by a look-up table with 2^{24} entries, fine-tuned to maximize the Reynolds number. External flow past a circular plate at Reynolds number around 190 has been simulated. The flow is found to evolve from axi-symmetric to fully 3D. Such simulations take a few minutes of CRAY-2 per circulation time (based on plate diameter and upstream velocity).

Rivet, J.-P., M. Hénon, U. Frisch, and D. d'Humières. "Simulating Fully Three-Dimensional External Flow by Lattice Gas Methods." In *Discrete Kinematic Theory, Lattice Gas Dynamics, and Foundations of Hydrodynamics*, edited by R. Monaco. World Scientific, 1989, 276–285.

Rivet, J. P. "Three-Dimensional Lattice Gas Hydrodynamical Simulations: Initial Results." Unpublished.

Abstract. The "pseudo-4-D" lattice gas model of d'Humières, Lallemand, and Frisch with the collisions rules of Hénon is implemented. Viscosity measurements are presented and the Taylor-Green vortex is simulated at a Reynolds number close to one hundred.

Rothman, Daniel H. "Modeling Seismic *P*-Waves with Cellular Automata." *Geophys. Res. Lett.* **14(1)** (1987):17–20.

Abstract. Cellular automata are arrays of discrete variables that follow local interaction rules and are capable of modeling many physical systems. A successful recent application has been in fluid dynamics, in which the Navier-Stokes equations are solved by creating a model in which space, time, and the velocity of particles are all discrete. Acoustic waves can be obtained from these fluid models when perturbations of the idealized fluid are small. Because seismic *P*-waves can be approximated by the acoustic wave equation, cellular automata can be adapted for seismic wave computations. This study shows how to model *P*-waves in two dimensions by using a modified form of the cellular-automaton rules for fluids. Propagation, reflection, and the computation of synthetic seismograms are demonstrated. Because no arithmetic calculations are needed and each lattice site can be updated simultaneously, this method is well suited for implementation of massively parallel computers. Among the many potential advantages are unconditional stability, no round-off errors, and the possibility for devising novel approaches for modeling waves in inhomogeneous media.

Rothman, Daniel H. "Cellular-Automaton Fluids: A Model for Flow in Porous Media." *Geophysics* **53(4)** (1988):509–518.

Abstract. Numerical models of fluid flow through porous media can be developed from either microscopic or macroscopic properties. The large-scale viewpoint is perhaps the most prevalent. Darcy's law relates the chief macroscopic parameters of interest—flow rate, permeability, viscosity, and pressure gradient—and may be invoked to solve for any of these parameters when the others are known. In practical situations, however, this solution may not be possible. Attention is then typically focused on the estimation of permeability, and numerous numerical methods based on knowledge of the microscopic pore-space geometry have been proposed.

Because the intrinsic inhomogeneity of porous media makes the application of proper boundary conditions difficult, microscopic flow calculations have typically been achieved with idealized arrays of geometrically simple pores, throats, and cracks. I propose here an attractive alternative which can freely and accurately model fluid flow in grossly irregular geometries. This new method solves the Navier-Stokes equations numerically using the cellular-automaton fluid models introduced by Frisch, Hasslacher, and Pomeau. The cellular-automaton fluid is extraordinarily simple—particles of unit mass traveling with unit velocity reside on a triangular lattice and obey elementary collision rules—but is capable of modeling much of the rich complexity of real fluid flow. Cellular-automaton fluids are applicable to the study of porous media. In particular, numerical methods can be used to apply the appropriate boundary conditions, create a pressure gradient, and measure the permeability. Scale of the cellular-automaton lattice is an important issue: the linear dimension of a void region must be approximately twice the mean free path of a lattice gas particle. Finally, an example of flow in a 2-D porous medium demonstrates not only the numerical solution of the Navier-Stokes equations in a highly irregular geometry, but also numerical estimation of permeability and a verification of Darcy's law.

Rothman, Daniel H., and Jeffrey M. Keller. "Immiscible Cellular-Automaton Fluids." *J. Stat. Phys.* **52(3/4)** (1988):1119–1127. Reprinted in this volume, 275–282.

Abstract. We introduce a new deterministic collision rule for lattice-gas (cellular-automaton) hydrodynamics that yields immiscible two-phase flow. The rule is based on a minimization principle and the conservation of mass, momentum, and particle type. A numerical example demonstrates the spontaneous separation of two phases in two dimensions. Numerical studies show that the surface tension coefficient obeys Laplace's formula.

Rothman, D. H., and S. Zaleski. "Spinodal Decomposition in a Lattice-Gas Automaton." MIT Porous Flow Project, report n°1, 1988.

Rothman, D. H. "Immiscible Lattice Gases: New Results, New Models." Proceedings of the workshop on Cellular Automata and Modeling of Complex Physical Systems, Les Houches, France, February 21–March 1. Berlin: Springer-Verlag, 1989.

Rothman, D. H. "Lattice-Gas Automata for Immiscible Two-Phase Flow." In *Discrete Kinematic Theory, Lattice Gas Dynamics, and Foundations of Hydordynamics*, edited by R. Monaco. World Scientific, 1989, 286–299.

Rothman, Daniel H. "Negative-Viscosity Lattice Gases." MIT, 1989. Preprint.

Abstract. A new irreversible collision rule is introduced for lattice-gas automata. The rule maximizes the flux of momentum in the direction of the local momentum gradient, yielding a negative shear viscosity. Numerical results in 2-D show that the negative viscosity leads to the spontaneous ordering of the velocity field, with vorticity resolvable down to one lattice-link length. The new rule may be used in conjunction with previously proposed collision rules to yield a positive shear viscosity lower than the previous rules provide. In particular, Poiseuille flow tests demonstrate a decrease in viscosity by more than a factor of 2.

Rujàn, Pàl. "Cellular Automata and Statistical Mechanical Models." *J. Stat. Phys.* **49(1/2)** (1987):139.

Abstract. We elaborate on the analogy between the transfer matrix of usual lattice models and the master equation describing the time development of cellular automata. Transient and stationary properties of probabilistic automata are linked to surface and bulk properties, respectively, of restricted statistical mechanical systems. It is demonstrated that methods of statistical physics can be successfully used to describe the dynamic and the stationary behavior of such automata. Some exact results are derived, including duality transformations, exact mappings, "disorder," and "linear" solutions. Many examples are worked out in detail to demonstrate how to use statistical physics in order to construct cellular automata with desired properties. This approach is considered to be a first step toward the design of fully parallel, *probabilistic* systems whose computational abilities rely on the cooperative behavior of their components.

Sakaguchi, Hidetsugu. "Phase Transitions in Coupled Bernoulli Maps." *Prog. Theor. Phys.* Progress Letters **80(1)** (1988):7.

Abstract. A coupled map system is proposed which is deterministic and shows phase transitions. Our system is composed of a large number of the Bernoulli maps. Chaotic behaviors of the individual maps make the coupled system ergodic. By using two different models, equilibrium and nonequilibrium phase transitions are studied analytically and numerically.

Salem, J., and S. Wolfram. "Thermodynamics and Hydrodynamics with Cellular Automata." In *Theory and Applications of Cellular Automata*, edited by S. Wolfram. World Scientific, 1986, 362–366.

Searby, G., V. Zahnlé, and B. Denet. "Lattice-Gas Mixtures and Reactive Flows." In *Discrete Kinematic Theory, Lattice Gas Dynamics, and Foundations of Hydrodynamics*, edited by R. Monaco. World Scientific, 1989, 300–314.

Shimomira, Tsutomu, Gary D. Doolen, Brosl Hasslacher, and Castor Fu. "Calculations Using Lattice Gas Techniques." *Los Alamos Science* Special Issue (1987):201–210. Reprinted in this volume, 3–10.

Somers, J. A., and P. C. Rem. "The Construction of Efficient Collision Tables for Fluid Flow Computations with Cellular Automata." Proceedings of the workshop on Cellular Automata and Modeling of Complex Physical Systems, Les Houches, France, February 21–March 1. Berlin: Springer-Verlag, 1989.

Succi, S. "Cellular Automata Modeling on IBM 3090/VF." IBM/ECSEC, 1987. Preprint.

Succi, S. "Triangular Versus Square Lattice Gas Automata for the Analysis of Two-Dimensional Vortex Fields." *J. Phys. A: Math. Gen.* **21** (1988):L43–L49.

Abstract. The consequences of the lack of isotropy of the momentum flux tensor of the Hardy-Pomeau-De Pazzis (HPP) fluid are discussed. It is shown that this lack of isotropy is tantamount to introducing a force which is incompatible with a correct evolution of two-dimensional vortex configurations. In addition, a qualitative discussion is presented on the physical reasons why this problem can be cured by moving to the six-link lattice introduced by Frisch, Hasslacher and Pomeau (FHP).

Succi, S., R. Benzi and P. Santangelo. "An Investigation of Fractal Dimensions in Two-Dimensional Lattice Gas Turbulence.' *J. Phys. A Math. Gen.* **21** (1988):L771–776

Abstract. We investigate the dissipation mechanisms acting in a two-dimensional lattice gas automaton by inspecting the structure functions of the turbulent velocity field associated with the Boolean configuration of the automaton. In particular we investigate whether the Boolean noise produced by the automaton can promote fractal structures within the flow. We show that this is not the case and the presence of the noise only results in a non-analyticity of the flow field which can be progressively eliminated upon averaging the boolean field on coarser and coarser grids. As a result, we find that the non-fractal nature of homogeneous two-dimensional turbulence is not affected by the presence of the microscopic noise.

Succi, S., and D. d'Humières. "Lattice Gas Hydrodynamics on IBM 3090/VF." IBM/ECSEC, 1988. Preprint.

Succi, S., P. Santangelo, and R. Benzi. "High-Resolution Lattice-Gas Simulation of Two-Dimensional Turbulence." *Phys. Rev. Lett.* **60(26)** (1988):2738.

Abstract. The mechanisms of two-dimensional turbulence are investigated by means of a very high-resolution lattice-gas simulation. The results from this simulation are quantitatively compared with the direct integration of the Navier-Stokes equation. The dissipation of the flow simulated by the lattice gas is estimated by a simple scaling argument on the microscopic noisy field of the automaton.

Succi, S., R. Benzi, E. Foti, F. Higuera, and F. Szelényi. "Fluid-Dynamics Applications of the Lattice-Gas Equation on the IBM 3090/VF." Proceedings of the workshop on Cellular Automata and Modeling of Complex Physical Systems, Les Houches, France, February 21–March 1. Springer-Verlag, 1989.

Succi, S., R. Benzi, and F, Higuera. "Lattice Gas and Boltzmann Simulations of Homogeneous and Inhomogeneous Hydrodynamics." In *Discrete Kinematic Theory, Lattice Gas Dynamics, and Foundations of Hydrodynamics*, edited by R. Monaco. World Scientific, 1989, 329–342.

Succi, S., F. Higuera, and F. Szelénya. "Simulations of Three-Dimensional Flows with the Lattice Boltzmann Equation on the IBM 3090/VF." IBM ECSEC, 1989. Preprint.

Takesue, Shinji. "Reversible Cellular Automata and Statistical Mechanics." *Phys. Rev. Lett.* **59(22)** (1987):2499.

Abstract. Reversible cellular automata are used to investigate the thermodynamic behavior of large systems. Additive conserved quantities are regarded as the energy of these models. By the consideration of a large system as the sum of a subsystem and a heat bath, it is numerically shown that a canonical distribution is realized under certain conditions concerning the conserved quantities.

Tarnowski, D. "Les Supercalculateurs Bientôt Démodés?" *La Recherche* **174** (1986): 272–273.

Toffoli, Tommaso. "Information Transport Obeying the Continuity Equation." *IBM J. Res. Develop.* **32(1)** (1988):29.

> *Abstract.* We analyze nontrivial dynamical systems in which information flows as an additive conserved quantity—and thus takes on a strikingly tangible aspect. To arrive at this result, we first give an explicit characterization of equilibria for a family of lattice gases.

Toffoli, T. "Four Topics in Lattice Gases: Ergodicity; Relativity; Information Flow; and Rule Compression for Parallel Lattice-Gas Machines." In *Discrete Kinematic Theory, Lattice Gas Dynamics, and Foundations of Hydrodynamics*, edited by R. Monaco. World Scientific, 1989, 343–354.

Travis, Bryan J., Kenneth G. Eggert, Shi Yi Chen, and Gary Doolen. "Calculating Flow and Transport in Porous/Fractured Media using the Cellular Automata Approach." Preprint.

> *Abstract.* A unique feature of flow and transport in soils and rocks is the highly multiply connected typology and varied channel geometries through which fluids and solutes must pass. Phenomenological and averaged models have had some success in describing porous flow and transport, but a truly predictive capability will not exist until microphysical (pore scale) details can be modeled and well characterized and coupled to larger scale dynamics.
>
> To model flow and transport on a microphysics scale, two things are necessary: (1) an efficient means of generating complex 2-D and 3-D multiply connected structures; and (2) a highly efficient method for solving flow and transport equations in arbitrary geometries. The ability to perform these two tasks may now exist. The concept of fractal dimension has led to the development of fast algorithms for generating rough surfaces and realistic, sponge-like solids, while the emerging discipline of cellular automata appears capable of providing a true advance in our ability to model flow and transport in arbitrary geometries.
>
> This paper discusses the use of fractal generators and cellular automata methods for computing flow and transport in calculation of (1) permeability as a function of porosity and fractal dimension, (2) dispersion as a function of porosity and fractal dimension, (3) effective resistance in channel flow for channel width change, and (4) retardation of tracers due to diffusion across a fracture face.

van Velven, G. A., and M. H. Ernst. "Breakdown of the Approximation for a Lattice Lorentz Gas." In *Discrete Kinematic Theory, Lattice Gas Dynamics, and Foundations of Hydrodynamics*, edited by R. Monaco. World Scientific, 1989, 371–383.

Vichniac, G. "Cellular-Automata Fluids." In *Instabilities and Nonequilibrium Structures*, edited by E. Tirapegui and D. Villasael. Reidel, 1989.

Vives, E., and A. Planes. "Lattice-Gas Model of Orientable Molecules: Application to Liquid Crystals." *Phys. Rev.* **A38** (1988):5391–5400.

Wallace, D. J. "Scientific Computation on SIMD and MIMD machines." *Phil. Trans. R. Soc. Lond. A.* **326** (1988):481–498.

Abstract. The ICL Distributed Array Processor and Meiko Computing Surface have been successfully applied to a wide range of scientific problems. I give an overview of selected applications from experimental data analysis, molecular dynamics and Monte Carlo simulation, cellular automata for fluid flow, neural network models, protein sequencing and NMR imaging. I expose the problems and advantages of implementations on the two architectures, and discuss the general conclusions which one can draw from experience so far.

Wayner, Peter. "Modeling Chaos." *BYTE* **May '88** (1988):253.

Abstract. A parallel CPU architecture can take you where shorter clock ticks, smarter instructions, and more on-chip memory can't go.

Wolfram, Stephen. "Cellular Automaton Fluides 1: Basic Theory." *J. Stat. Phys.* **45(3/4)** (1986):471–526. Reprinted in this volume, 19–74.

Abstract. Continuum equations are derived for the large-scale behavior of a class of cellular automaton models for fluids. The cellular automata are discrete analogues of molecular dynamics, in which particles with discrete velocities populate the links of a fixed array of sites. Kinetic equations for microscopic particle distributions are constructed. Hydrodynamic equations are then derived using the Chapman-Enskog expansion. Slightly modified Navier-Stokes equations are obtained in two and three dimensions with certain lattices. Viscosities and other transport coefficients are calculated using the Boltzmann transport equation approximation. Some corrections to the equations of motion for cellular automaton fluids beyond the Navier-Stokes order are given.

Yakhot, Victor, Bruce J. Bayly, and Steven A. Orszag. "Analogy between Hyperscale Transport and Cellular Automaton Fluid Dynamics." *Phys. Fluids* **29** (1986):2025. Reprinted in this volume, 283–289.

Abstract. It is argued that the dynamics of a very large-scale (hyperscale) flow superposed on the stationary small-scale flow maintained by a force f(x) is analogous to the cellular automaton hydrodynamics on a lattice having the same spatial symmetry as the force f.

Zaleski, S. "Phase Transitions in Lattice Gases." Proceedings of the workshop on Cellular Automata and Modeling of Complex Physical Systems, Les Houches, France, February 21–March 1. Springer-Verlag, 1989.

Zaleski, Stéphane. "Weakly Compressible Fluid Simulations at High Reynolds Numbers." In *Discrete Kinematic Theory, Lattice Gas Dynamics, and Foundations of Hydrodynamics*, edited by R. Monaco. World Scientific, 1989, 384–393.

Abstract. Lattice gases have recently been introduced by Frisch, Hasslacher and Pomeau as an attractive method for the simulation of incompressible flow. These gases obey on the large scale a set of hydrodynamical equations with an artificial kind of compressibility. In this paper lattice gas simulations are compared with finite difference solutions of the same artificially compressible equations. In both methods the Mach number must be kept small to approximate incompressible flow, as in Chorin's classical scheme. It is found that the lattice gas has an excessively fine grid in certain classes of problems. In typical shear flow situations the lattice gas has a grid size of order $\lambda M c/\nu$ where λ is the boundary layer size, ν is the viscosity of the gas, M is the Mach number and c is the speed of sound. Thus it is reasonably efficient if the Mach number is not too small and the boundary layer needs to be finely resolved.

Zanetti, G. "The Hydrodynamics of Lattice Gas Automata," or "The Macroscopic Behavior of Lattice Gas Automata." University of Chicago, 1988. Preprint.

Zehnlé, V., and G. Searby. "Lattice Gas Experiments on a Non-Exothermic Diffusion Flame in a Vortex Field." *J. Physique*, May 1989.

Index

Printed and bound by CPI Group (UK) Ltd, Croydon, CR0 4YY

25/10/2024

01779256-0001